T0141275

Chalcogenadiazoles

Chemistry and Applications

Chalcogenadiazoles

Chemistry and Applications

Zory Vlad Todres

CRC Press
Taylor & Francis Group
Boca Raton   London   New York

CRC Press is an imprint of the
Taylor & Francis Group, an **informa** business

CRC Press
Taylor & Francis Group
6000 Broken Sound Parkway NW, Suite 300
Boca Raton, FL 33487-2742

First issued in paperback 2021

© 2012 by Taylor & Francis Group, LLC
CRC Press is an imprint of Taylor & Francis Group, an Informa business

No claim to original U.S. Government works

Version Date: 20110722

ISBN 13: 978-1-03-209925-5 (pbk)
ISBN 13: 978-1-4200-6607-4 (hbk)

This book contains information obtained from authentic and highly regarded sources. Reasonable efforts have been made to publish reliable data and information, but the author and publisher cannot assume responsibility for the validity of all materials or the consequences of their use. The authors and publishers have attempted to trace the copyright holders of all material reproduced in this publication and apologize to copyright holders if permission to publish in this form has not been obtained. If any copyright material has not been acknowledged please write and let us know so we may rectify in any future reprint.

Except as permitted under U.S. Copyright Law, no part of this book may be reprinted, reproduced, transmitted, or utilized in any form by any electronic, mechanical, or other means, now known or hereafter invented, including photocopying, microfilming, and recording, or in any information storage or retrieval system, without written permission from the publishers.

For permission to photocopy or use material electronically from this work, please access www.copyright.com (http://www.copyright.com/) or contact the Copyright Clearance Center, Inc. (CCC), 222 Rosewood Drive, Danvers, MA 01923, 978-750-8400. CCC is a not-for-profit organization that provides licenses and registration for a variety of users. For organizations that have been granted a photocopy license by the CCC, a separate system of payment has been arranged.

Trademark Notice: Product or corporate names may be trademarks or registered trademarks, and are used only for identification and explanation without intent to infringe.

Publisher's Note
The publisher has gone to great lengths to ensure the quality of this reprint but points out that some imperfections in the original copies may be apparent.

Visit the Taylor & Francis Web site at
http://www.taylorandfrancis.com

and the CRC Press Web site at
http://www.crcpress.com

Contents

About the Book

This book offers a timely and authoritative treatise on the chemistry and diverse applications of chalcogenadiazoles—five-membered rings containing two carbons, two nitrogens, and one chalcogen (an atom of oxygen, sulfur, selenium, or tellurium). Chalcogenadiazole oxides and species fused with other cyclic fragments are also considered. The book compares the effects of the chalcogen nature and the alternation manner of all atomic constituents on properties of these heterocyclic compounds. Chemical, physicochemical, and physical aspects as well as biomedical and technical applications are equally considered. Many of the already-available findings deserve to be widened in their usage and need to be advanced, but remain unknown to those working in boundary fields. As a whole, this book covers an important and rapidly developing branch of heterocyclic chemistry and can serve as an essential supplement to the corresponding textbooks for students, especially for those who are about to enter the job market.

Preface

This book considers the chemical, biomedical, and technical importance of chalcogenadiazoles. These compounds contain five-membered cycles composed of two carbons, two nitrogens, and one chalcogen—the latter as the key heteroatom. Chalcogens are chemical elements of VIA group of the periodic system, namely, oxygen (O), sulfur (S), selenium (Se), tellurium (Te), and polonium (Po). At present, all chalcogenadiazoles are known naturally except for the polonium species.

The external shells of the O, S, Se, and Te atoms have s^2p^4 configuration. These elements are isovalent. As the atomic number increases, covalent and ionic radii increase, and ionization energies decrease. At first glance, some similarities between analogously constructed chalcogenadiazoles might be expected. At a narrower glance, the fundamental properties of the key atoms in chalcogenadiazoles vary significantly, and these differences can lead to disparities between the molecules under consideration.

The number of different chalcogenadiazoles and their structural diversity make it difficult to gain a good understanding of the subject by studying an individual class of these hetrocycles. The book tries to identify and emphasize general features of this family of heterocycles. The combined presence of the mentioned atoms endows heterocyclic systems with some specific properties. The book concentrates on properties of each class of chalcogenadiazoles and their cycle-fused derivatives, considering chemical reactions of functional groups only in cases when these properties permit to characterize the heterocycles as substituents or in respect of their aromaticity. Nitrogen-oxides and chalcogen-oxides of these heterocycles are equally discussed.

The properties of chalcogenadiazoles are in many respects defined with electron affinity and electronegativity. For future consideration, it is useful to compare magnitudes of these characteristics. According to handbooks, the electron affinity (in eV) changes as follows: O—1.46, S—2.07, Se—2.02, and Te—1.96. The first ionization potential (eV) drops in the following order: O—13.62, S—10.36, Se—9.75, Te—9.01. Pauling's electronegativities are as follows: N—3.04, O—3.50, S—2.50, Se—2.40, and Te—2.10. From these values and keeping in mind the equal alternation of the atoms in heterocycles, it is possible to predict that the chemical behavior should be more or less similar for thia, selena, and tellura analogs and should differ for oxadiazoles. Namely, the N–O bond should be polarized with a partial negative charge on oxygen, whereas the N–S, N–Se, and N–Te bonds might be polarized in the opposite direction. For chalcogen atoms, oxidation states are variable. Thus, oxygen can bear from −2 to +2 charges, whereas sulfur can change its charge from −2 to +6. Consequently, in chalcogenadiazoles, oxygen must form two ordinary bonds with adjacent atoms, and sulfur can be bivalent or tetravalent. For selenium and tellurium, tetravalent states are possible but less stable.

The size of VIA group elements increases down in the series O < S < Se < Te. The covalent radii are 0.073, 0.103, 0.117, and 0.135 nm, respectively. The size difference can be crucial for crystal packing and for technical applications.

Technically, chalcogenadiazoles are used as dyestuffs, lubricating additives, corrosion inhibitors, and as components of microelectronic devices or solar cells. The heterocyclic compounds under consideration also find their applications in industrial monitoring (analytical) procedures. The material of the corresponding chapters will be analyzed from these points of view too.

Due to the increasing importance of chalcogenadiazoles as medication, their biomedical properties are described in reasonable detail. The focus is on trends in search of compounds with established bioactivity. The traditional drugs of this class can be found in the existing handbooks. Where relevant, new chalcogenadiazoles as agrochemicals or reagents for environmental and biochemical analysis are also considered.

In this book, different classes of chalcogenadiazoles are compared with respect to their degree of aromaticity. Aromaticity is unique in providing a comparison of oxygen, sulfur, selenium, and tellurium systems, with an approach intended to emphasize general concepts that will be helpful to the nonspecialist.

Each chapter is designed to be self contained, and there are extensive cross-references between chapters. The book is divided into four chapters devoted to compounds with oxygen, sulfur, selenium, and tellurium key heteroatoms. Chapters 1 through 3 contains four sections considering 1,2,3, 1,2,4, 1,2,5, and 1,3,4 isomers of the corresponding chalcogenadiazole. At the end of these chapters, a brief comparison of structure–property relationships is given. Telluradiazoles are an exception to this total correlation because the contemporary literature contains only data on 1,2,3 and 1,2,5 isomers. The concluding Chapter 5 compares chalcogenadiazoles from the standpoint of isosterism in chemical, technical, and biomedical aspects.

Each chapter can be read separately from the others. Nevertheless, the materials of different chapters are compared by means of a general outlook and by formulating problems for their future solution. The book is not loaded with mathematical considerations and contains as minimal quantum-chemical thoughtfulness as possible. Its main intent is to serve as a reference tool for future research.

The multidisciplinary approach (that combines data from organic, biological, medicinal, material science, and supramolecular practioners) presents the book as an important source of information not only for chemists in the field of organic, inorganic, organometallic chemistry but also for physicists, biochemists, pharmachemists, agrochemists, and other professionals, who in some way deal with chalcogenadiazoles. The treatise proposed can also serve as a supplement to a textbook on heterocyclic chemistry for senior and doctoral students. This may not be a book for faint-hearted undergraduates, but it can be a guide for students who are about to enter the job market.

We hope that this book will induce some curiosity and attract readers toward this branch of heterocyclic chemistry. On a personal note, I have been involved in some way or the other with various fields of organic, physical organic chemistry, and organic mechanochemistry. However, chalcogenadiazoles were what first attracted me to the field of chemistry, and this has now lasted a lifetime. I am deeply indebted to my student-years mentor, and whole-life esteemed friend, Professor Lev S. Efros, who made me interested in this elegant and practically important field of chemistry. This interest provided me the incentive for my PhD and ScD (habilitat) theses, although these works were not directly related to this field. This obviously testifies to the fruitful cross-pollination between heterocyclic chemistry and other branches of organic chemistry.

Author

Zory Vlad Todres holds an MSc and a PhD in heterocyclic chemistry as well as a doctor habilitatus in physical organic chemistry. He has authored several books for CRC Press, and his books have been well received by the chemical community. Through his books, he aspires to engraft new and practically important branches to the great tree of organic chemistry. Such titles are *Organic Ion-Radicals* (2003), *Organic Mechanochemistry* (2006), and *Ion-Radical Organic Chemistry* (2009). Presently, he is working on a book entitled *Organic Chemistry in Confinement*.

He is listed in the *Who's Who* of the United States and the United Kingdom sources.

E-mail: zorytodres@yahoo.com

1 Oxadiazoles

1.1 1,2,3-OXADIAZOLES

1,2,3-Oxadiazole systems hold a distinctive position in the oxadiazole series. There are few compounds really containing the 1,2,3-oxadiazole ring. Most of them exist in the open diazoketone tautomeric form. Sydnones and sydnone imines (that are at present considered as 1,2,3-oxadiazole derivatives) are stable and their chemistry and medical applications are very reach.

1.1.1 Formation

1,2,3-Oxadiazoles were first postulated as intermediates in the 1,3-dipolar cycloaddition of dinitrogen oxide (N_2O) to alkynes (Buckley and Levy 1951). Nevertheless, the unsubstituted 1,2,3-oxadiazole is still unknown and ab initio calculations predict that this compound cannot be isolated as a discrete species even in an inert matrix at low temperatures (Nguen et al. 1985). However, 1,2,3-oxadiazole-3-oxides and aryl-annulated oxadiazoles are stable products. Methods of their preparation are listed below.

1.1.1.1 From Alkynes, or Benzyl Cyanide, or Potassium Malonate

The substrates mentioned form 1,2,3-oxadiazole 3-oxides upon reaction with mononitrogen monoxide in THF at −78°C. Scheme 1.1 exemplifies the reaction with alkyne lithium (Sugihara et al. 2004). A computational study (Wu et al. 2005) explains features of the reaction. Mononitrogen monoxide is a free radical and it reacts with the triple bond as a dimer to form an adduct containing lithium. Water treatment of the adduct leads to the final product.

Scheme 1.2 illustrates formations of the corresponding oxides from benzyl cyanide (Bohle and Perepichka 2009) or from potassium malonate (Arulsamy and Bohle 2002). Both reactions were performed in methanol solution of potassium methylate.

The product of the reaction between potassium malonate and nitrogen oxide of Scheme 1.2 is remarkably stable in acidic or basic medium (Arulsamy et al. 2007).

1.1.1.2 From Nitrosated Mannich Bases

Condensation of aldehydes, hydroxylamine, and cyanide results in the formation of Mannich bases. Being treated with nitrous acid, the Mannich bases form 5-amino-1,2,3-oxadiazole-3-oxides (Scheme 1.3) (Gotz and Grozinger 1971).

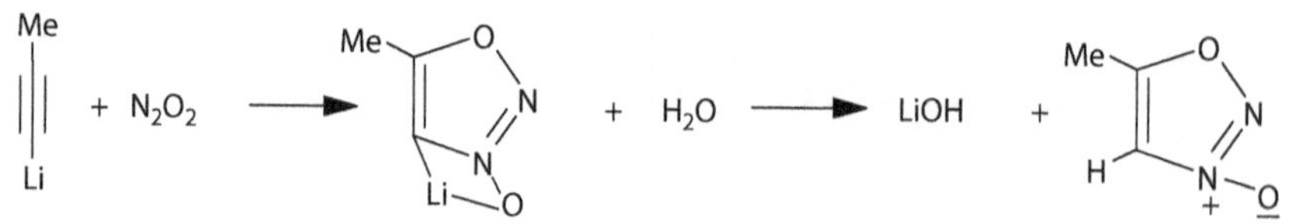

SCHEME 1.1

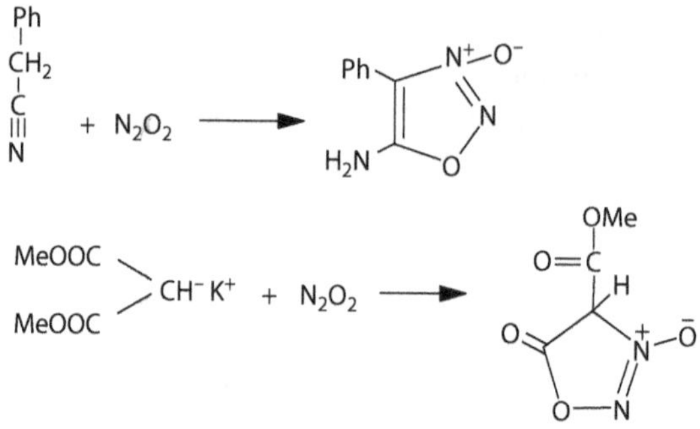

SCHEME 1.2

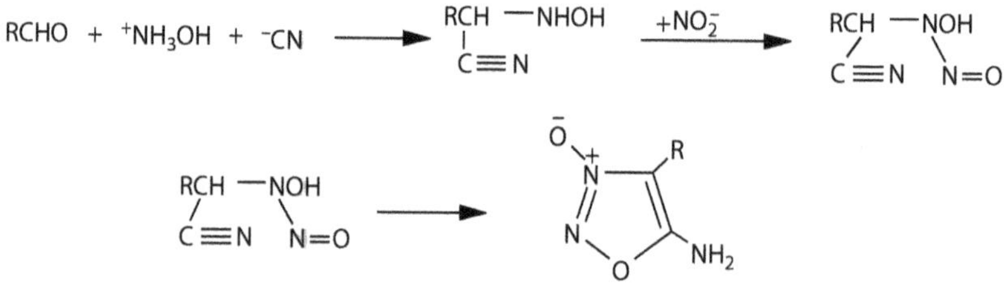

SCHEME 1.3

1.1.1.3 From *N*-Nitroso-α-Amino Acids

This reaction was proposed by Earl and Mackney (1935) and developed by Applegate and Turnbull (1988), then by Gao et al. (2008) and Browne et al. (2009). Cyclization is carried out with acetic or trifluoroacetic anhydride (see Scheme 1.4).

1.1.1.4 From Vicinal Hydroxyaryl Diazonium Salts

A diazonium salt prepared from 2-amino-3-naphthol forms naphtho-[2,3-*d*]-[1,2,3]-oxadiazole (Scheme 1.5). The reaction proceeds on basic alumina and the product can be sublimed at about 10°C under vacuum (Blocher and Zeller 1991).

The same synthetic approach has been used to generate benzene-fused 1,2,3-oxadiazole and its derivatives (Schulz and Schweig 1979, 1984).

1.1.2 Fine Structure

As has already been mentioned, 1,2,3-oxadiazole itself is not stable and exists entirely in the diazoketone form. According to calculations, total energy of the oxadiazole

SCHEME 1.4

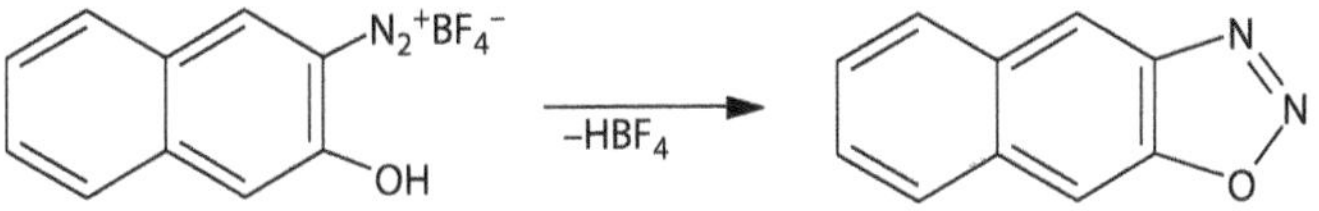

SCHEME 1.5

structure is higher than that of the diazoketone structure by 36 kJ/mol (Nguen et al. 1985, Semenov and Sigolaev 2004). The barrier to ring disclosure is 28 kJ/mol, and the cyclization barrier is 90 kJ/mol. As it was calculated for the unsubstituted 1,2,3-oxadiazole (Nguen et al. 1985), the nitrogen–oxygen bond is characterized by a particular weakness. It was experimentally established that the 1,2,3-oxadiazole heterocycle is somewhat stabilized when nitrogen at position 3 is bound with oxygen by the semi-polar bond.

Annulation of 1,2,3-oxadiazole to the benzene ring also results in an enhancement of the cyclic form stability: Total energy of the benzoxadiazole occurred to be by 4–5.6 kJ/mol lower than that of 6-diazacyclohexa-2,4-dienone. These values were obtained as experimentally, from UV photoelectron spectra (Schulz and Schweig 1979) as theoretically, from DFT calculations (Semenov and Sigolaev 2004). Annulation of 1,2,3-oxadiazole to naphthalene additionally stabilizes the fused heterocycle. Thus, naphtho-[2,3-*d*]-[1,2,3]-oxadiazole of Scheme 1.5 can be isolated and purified by sublimation. The product is stable for several days in the dark at −20°C. This compound keeps its structure in heptane but undergoes ring opening in methanol with the formation of 2,3-naphthoquinone diazide (Blocher and Zeller 1991). The same solvent effect on stability of the heterocycle was reported for benzo-fused oxadiazole (Schulz and Schweig 1984). Theoretical analysis of such a solvent effect showed that electrostatic polarization dominates over the aromatic stabilization in the benzo-fused oxadiazole (Semenov and Sigolaev 2004). The solvent nature affects the equilibrium between benzo-1,2,3-oxadiazole and benzo-1,2-quinone diazide.

Note, naphtho-[1,2-*d*]-[1,2,3]-oxadiazole-5-sulfonic acid is a commercially available product.

A separate mention should be made about 3-substituted 1,2,3-oxadiazolium-5-olates, the mesoionic compounds. Eade and Earl (1946) named them *sydnones* after the University of Sydney where the authors were working. 5-Imido-1,2,3-oxadiazoliums are named *sydnone imines* in the literature. These artificial names were adopted by Chemical Abstracts for few years only, but they are still used in chemical publications. Sydnones had been usually represented as structures with the ring-contour of π-electrons (Earl et al. 1947, see also p. 2791 in the 2004 review by Balaban et al.). Recently obtained experimental and theoretical results deny aromatic delocalization

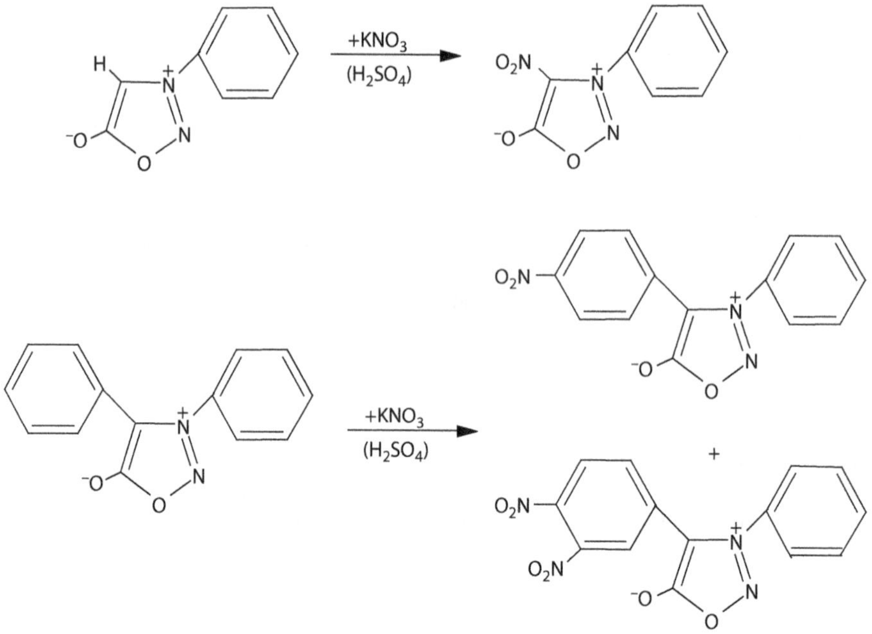

SCHEME 1.6

in sydnones. According to these results, sydnones are cyclic azomethine imines. (This is the reason for not picking sydnones out in a separate section, but to consider them as 1,2,3-oxadiazolium derivatives.) Naturally, the oxadiazolium structure of sydnones presupposes at least moderate C–H acidity at position 4. Indeed, Greco and O'Reilly (1970) estimated the pK_a of 3-phenylsydnone as 18–20.

In 3-phenylsydnone, the positive charge at position 3 voids the very possibility of the phenyl group nitration. Indeed, the nitronium ion preferentially reacts at position 4 of sydnone, rather at this phenyl ring (Scheme 1.6) (Baker et al. 1950). Note that position 4 is enriched with electron density due to donating effect of the C_5-olate group that is connected with position 4 by the double bond. For the same reason, nitration of 3,4-diphenylsydnone touches the phenyl group connected only with the electron-enriched position 4 of the heterocycle. The phenyl group connected with the electron-deficient position 3 remains intact (Scheme 1.6) (Schubert and Ellenrieder 1984).

1.1.3 REACTIVITY

1.1.3.1 Tautomerism

As has been already pointed out, the nitrogen–oxygen bond in the 1,2,3-oxadiazole ring is weak. Therefore, this heterocycle easily transforms into the diazoxide tautomer. The problem has been considered in Section 1.1.2. In Section 1.1.1, several examples have been given concerning preparation of 1,2,3-oxadiazole 3-oxides. Meanwhile, there was a point of view that 1,2,3-oxadiazole-3-oxide derivatives

SCHEME 1.7

exist as the corresponding 3-hydroxysydnones, see the delocalized structure on Scheme 1.7 (Earl et al. 1947, Applegate and Turnbull 1988), or exist in equilibrium with the oxadiazole oxide form. Recent studies established that the equilibrium has almost entirely shifted to the oxadiazole oxide form (Wu et al. 2005, Bohle and Perepichka 2009). In particular, 4-amino-1,2,3-oxadiazole 3-oxide of Scheme 1.2 behaves as a normal aromatic amine, forming azomethines with aldehydes or *N,N*-dialkylderivatives with alkyl halides.

Scheme 1.7 shows one typical example of the equilibrium. The oxadiazole-oxide structure of the compound depicted in Scheme 1.7 was established especially by x-ray study (Bohle and Perepichka 2009).

1.1.3.2 Salt Formation

Regioselective benzylation at nitrogen of ring position 2 takes place in the case of the tetrabutylammonium salt of 4-methylcarboxy-5-olato-1,2,3-oxadiazole 3-oxide (Scheme 1.8) (Bohle et al. 2008).

Methylation of the tetrabutylammonium salt of 4-methylcarboxy-5-olato-1,2,3-oxadiazole 3-oxide leads to different results depending on the nature of a methylating agent (see Scheme 1.9) (Bohle et al. 2007). Methyl iodide generates

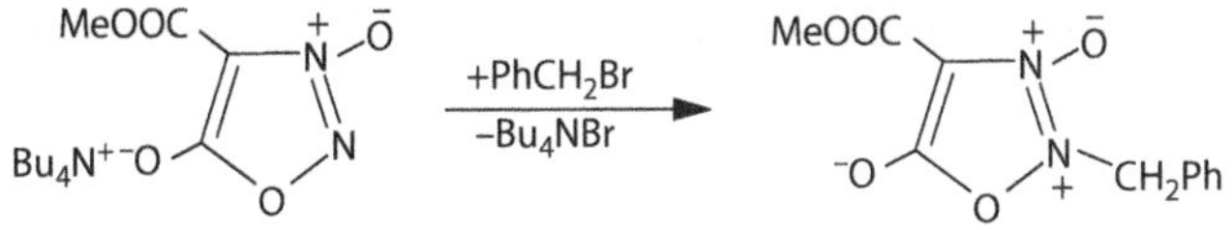

SCHEME 1.8

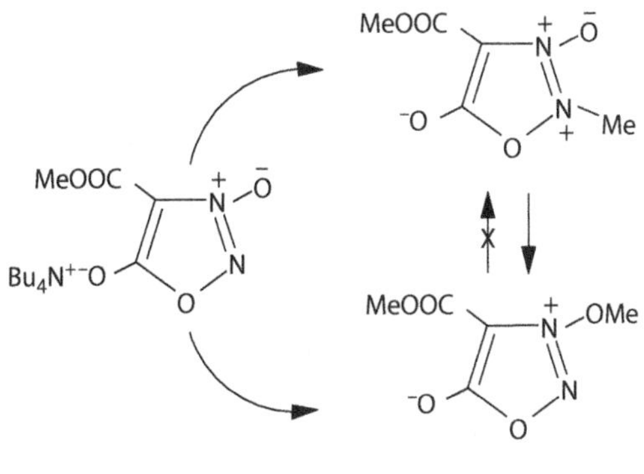

SCHEME 1.9

2-methyl-4-methylcarboxy-5-olato-1,2,3-oxadiazolium 3-oxide. Dimethylsulfate affords 3-methoxy-4-methylcarboxy-5-olato-1,2,3-oxadiazolium. (Trifluoromethane) methyl sulfonate gives the mixture of the two products mentioned. When a net melt of the latter product is heated to 140°C for 2 h, the former product is produced. In contrast, thermolysis of the former under these conditions does not result in the formation of the latter (see Scheme 1.9). In other words, the N-methoxy isomer corresponds to a kinetic product, and the N-methyl isomer corresponds to a thermodynamic product (Bohle et al. 2007).

Sydnone imine salts have two uncharged nitrogen atoms—the nitrogen at position 2 of the ring and the exocyclic nitrogen belonging to the amino group. These salts react, in the presence of bases, with a wide variety of electrophilic reactants. Nucleophiles attack only the amino group. The extensive body of such examples has been reviewed by Yashunskii and Kholodov (1980). Scheme 1.10 shows that the regioselectivity just mentioned is obvious.

1.1.3.3 Electrophilic Substitution

Halogenation, nitration, and sulfonation also proceed as substitution at position 4 of the sydnone molecule (Baker et al. 1950, Yashunskii et al. 1959, Ito and Turnbull 1996 respectively). Preston and Turnbull (1977) gave one interesting example of intramolecular electrophilic substitution: The diazocation of 3-(2-aminophenyl)sydnone in situ transforms it into a tricyclic compound (Preston and Turnbull 1977) (Scheme 1.11).

Hydrogen–metal electrophilic exchange in 1,2,3-oxadiazoles is well documented. Thus, sydnones with unsubstituted position 4 readily react with butyl lithium giving rise to 4-lithium derivatives (Fuchigami et al. 1986, Turnbull and Krein 1997). N-acylsydnone imines react with butyl lithium in the same way (Cherepanov and Kalinin 2000). With alkylmagnesium halides or with sodium hydride, 3-phenylsydnone undergoes metal–hydrogen exchange at position 4 (Greco and O'Reilly 1970).

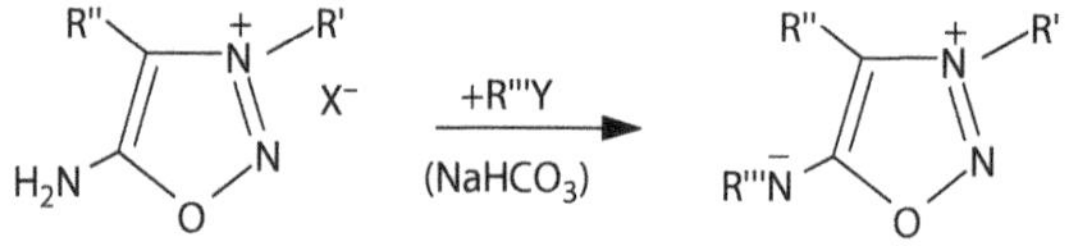

SCHEME 1.10

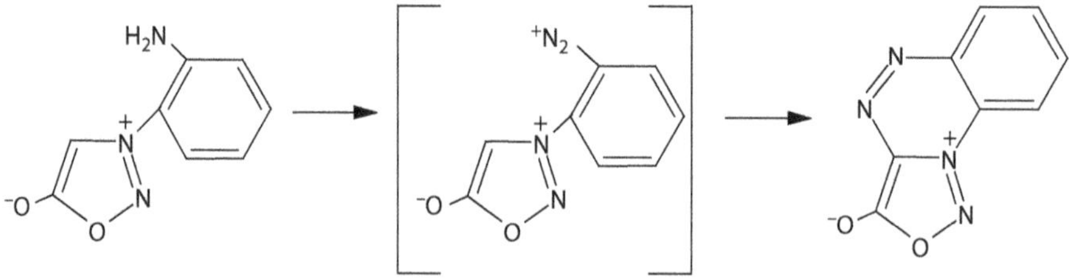

SCHEME 1.11

Upon action of mercuric acetate, this substrate forms 4-mercurio acetate with high yield (Scheme 1.12) (Yashunskii et al. 1959, Turnbull et al. 1994).

The acetylation of 3-arylsydnones with acetic anhydride upon acid catalysis had been described in 1972 by Tien and Ohta (Tien and Ohta 1972). Modern procedures consist of sonication of 3-arylsydnones with acetic anhydride and a catalytic amount of perchloric acid (Tien et al. 1992) and of convenient acetylation under catalysis of montmorillonite K-10 clay (Turnbull and George 1996). Carboxamidation of 3-arylsydnones was carried out with chlorosulfonyl isocyanate (Turnbull et al. 1998) (Scheme 1.13).

1.1.3.4 Nucleophilic Substitution

The bromo substituent in 4-bromo-3-arylsydnone is displaced with hydrazine (Ohta and Kato 1957) or aniline (Chadra et al. 2008). Scheme 1.14 illustrates transformation of 3-phenylsydnone into 3-phenyl-1,2,3-oxadiazolium-5-thiolate via 3-phenyl-5-ethoxy-1,2,3-oxadiazolium (Matsuda et al. 1979).

Scheme 1.15 represents another activated nucleophilic substitution with the formation of dicyano-α-(3-phenyl-1,2,3-oxadiazol-5-yl)methylide (Araki et al. 1985) upon action of trifluoromethanesulfonic anhydride (Tf$_2$O). With malodinitrile, the 5,5′-oxobis(3-phenyl-1,2,3-oxadiazolium) ditriflate obtained gives rise to the product of nucleophilic substitution. Araki et al. (1995) performed similar displacement reactions, using cyclopentadienide and indenide anions as nucleophiles.

1.1.3.5 Photolysis

UV irradiation of benzo[1,2-c]-[1,2,3]-oxadiazole in a noble gas matrix leads to the extrusion of dinitrogen and the formation of fulven-6-one (Scheme 1.16) (Schulz and

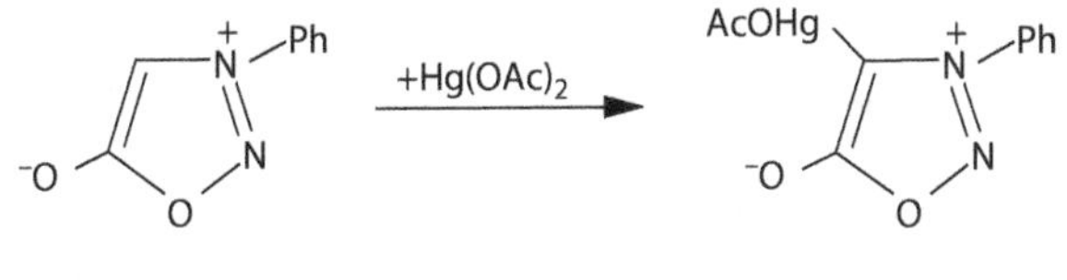

SCHEME 1.12

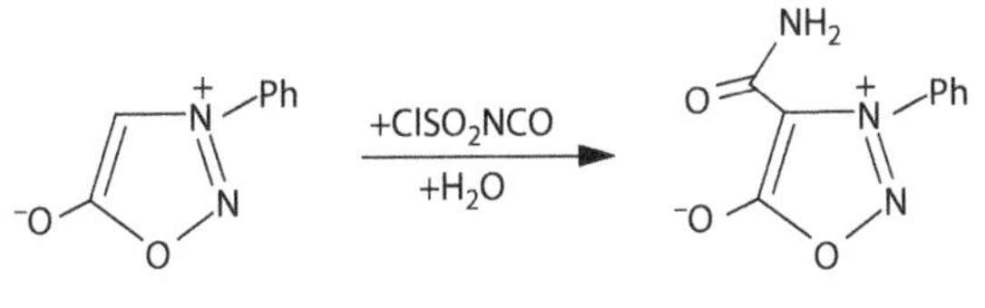

SCHEME 1.13

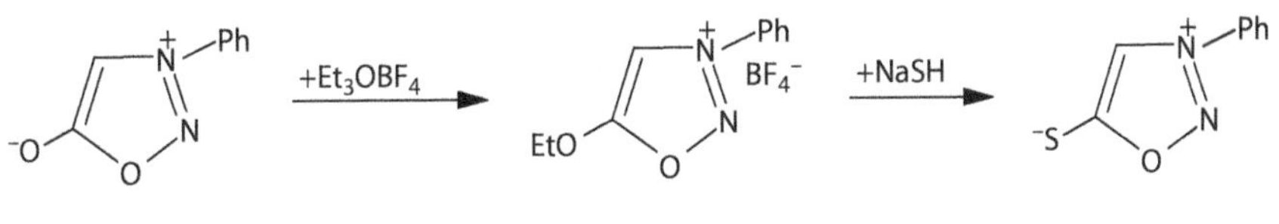

SCHEME 1.14

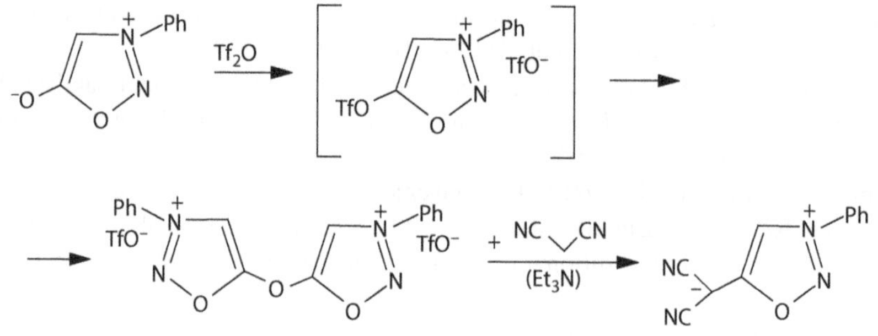

SCHEME 1.15

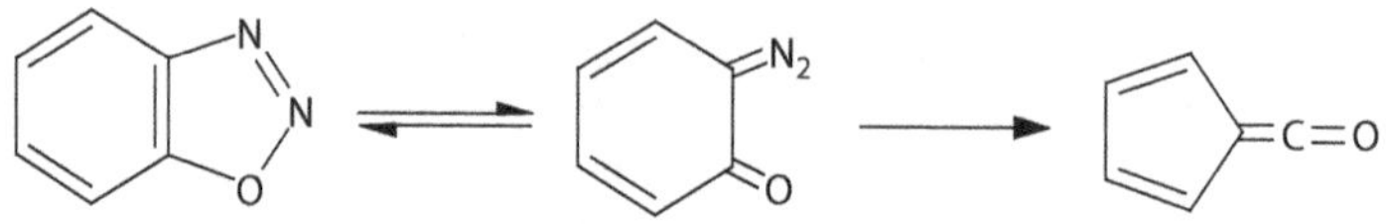

SCHEME 1.16

Schweig 1979). Similar results have been obtained with derivatives bearing various substituents on the benzenoid ring (Schweig et al. 1991).

Photolysis of naphtho-[2,3-*d*]-[1,2,3]-oxadiazole in methanol proceeds analogously. The main product 2-(methoxycarbonyl)indene is formed with 60% yield (Scheme 1.17) (Blocher and Zeller 1991, 1994).

1.1.3.6 Thermolysis

Gas-phase pyrolysis of benzo-[1,2-*c*]-1,2,3-oxadiazole at 320°C results in elimination of dinitrogen and exclusive formation of fulven-6-one (cf. Scheme 1.16) (Schulz and Schweig 1979).

1.1.3.7 Ring Transformation

Sydnones, as 1,3-dipoles, undergo cycloaddition with alkynes. Thus, 3,4-diphenyl-5-olato-1,2,3-oxadiazolium reacts with 4-ethynyl-*N*,*N*-dimethylaniline forming

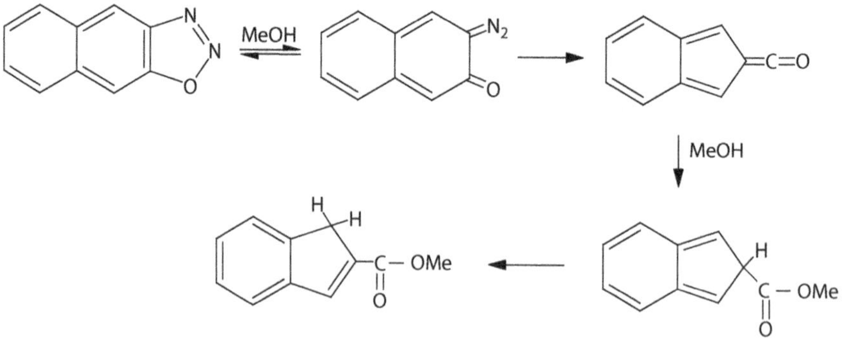

SCHEME 1.17

4-(1,5-diphenyl-1*H*-pyrazol-3-yl)-*N*,*N*-dimethylaniline (Scheme 1.18) (Browne et al. 2009).

Browne et al. (2010) give examples of reactions analogous to that depicted on Scheme 1.18 upon microwave irradiation. In xylene at 140°C, the transformations are completed in 30 min.

1,3-Dipolar cycloaddition of acetylenic sulfones to the sydnone of Scheme 1.19 was performed in organic solvents or on Merrifield resin (Gao et al. 2008). The reaction includes cycloaddition-cleavage and eventually results in the formation fused pyrrolo-pyrazoles.

Another principal example of ring transition consists in the formation of 3,4-dimethoxycarbonylfuroxan proceeding from the tetrabutyl ammonium salt of 4-methoxycarbonyl-5-olato-1,2,3-oxadiazole 3-oxide (Scheme 1.20) (Bohle et al. 2008). Scheme 1.20 also gives a possible mechanism of this recyclization: One-electron oxidation of the substrate leads to a free radical. The radical is unstable and eliminate nitrogen and carbon monoxides. The resulting nitriloxide dimerizes to form the final furoxan.

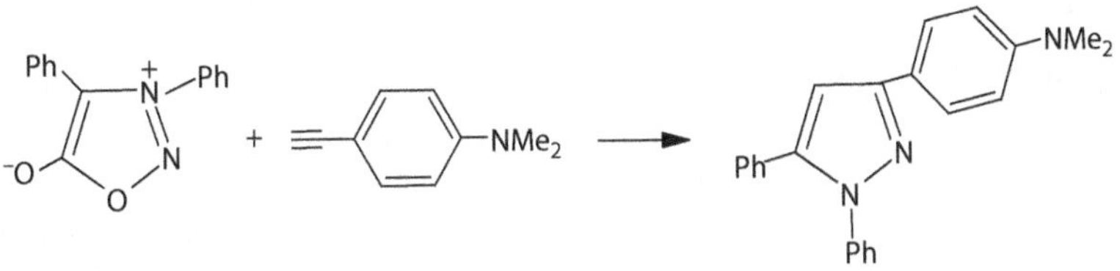

SCHEME 1.18

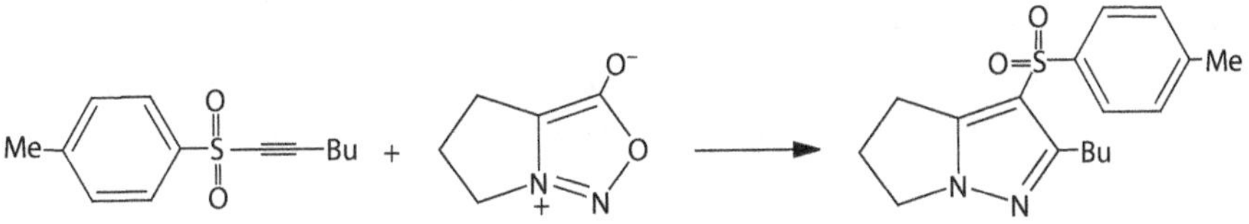

SCHEME 1.19

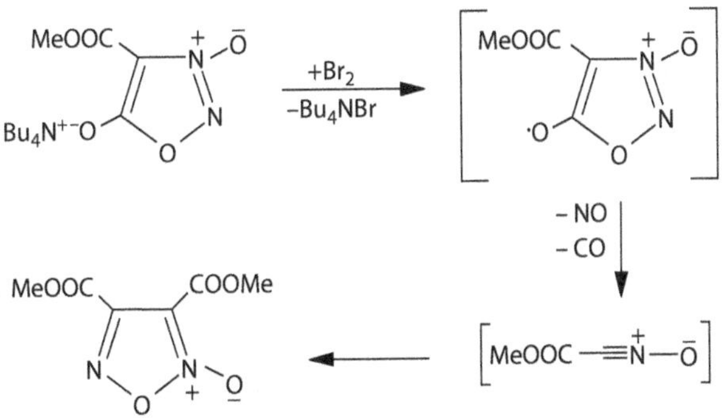

SCHEME 1.20

1.1.4 Biomedical Importance

Sydnone imines attract much attention due to their ability (in vivo) to cleave the sydnone ring and eventually release nitrogen monoxide (Levina et al. 2004).

One of the sydnone imines, 5-(phenylcarbamoylimino)-3-(1-phenylpropan-2-yl)-5H-1,2,3-oxadiazol-3-ium-2-ide, was recently proposed to enforce human resistance in extremely cold environments (Levina et al. 2006). New prospective sydnone-type drugs have also been proposed to develop antibacterials (Moustafa et al. 2004) and cancer inhibitors (Dunkley and Thoman 2003, Sapountzis et al. 2009).

1.1.5 Technical Applications

Technically applicable polysiloxane liquid crystals containing mesoionic sydnonyl moieties were reported by Chan et al. (2004). The polymers display a typical smectic schlieren (akin to slur) texture between 120°C and 200°C.

1.1.6 Conclusion

In majority, 1,2,3-oxadidiazoles exist in the open diazoketone tautomeric form. Although some representatives do exist, there are scarce data on chemistry of this class of heterocycles. Data on sydnones and sydnone imines are plenty. After years of discussion, it is understood that these compounds are not delocalized systems, but remain substituted oxadiazolium salts or zwitterions. Nevertheless, these names are still kept in the literature to underline the specificity of the corresponding compounds. Many sydnones and sydnone imines are pharmacologically active. Their bioactivity is frequently associated with ring cleavages in metabolic chains. The ring opening is an intermediary step and the final reaction leads to liberation off in vivo active nitrogen monoxide. Sydnonyl-containing polymers can seemingly form liquid-crystal domains, but the problem is hardly studied at present.

REFERENCES TO SECTION 1.1

Applegate, J.; Turnbull, K. (1988) *Synthesis*, 1011.
Araki, Sh.; Mizuya, J.; Batsugan, Y. (1985) *J. Chem. Soc., Perkin Trans. 1*, 2439.
Araki, Sh.; Mizuya, J.; Batsugan, Y. (1995) *J. Chem. Soc., Perkin Trans. 1*, 1989.
Arulsamy, N.; Bohle, D.S. (2002) *Angew. Chem. Intl. Ed.* **41**, 2089.
Arulsamy, N.; Bohle, D.S.; Perepichka, I. (2007) *Can. J. Chem.* **85**, 105.
Baker, W.; Ollis, W.D.; Poole, V.D. (1950) *J. Chem. Soc.* 1542.
Balaban, A.T.; Oniciu, D.C.; Katritzky, A.R. (2004) *Chem. Rev.* **104**, 2777.
Blocher, A.; Zeller, K.-P. (1991) *Angew. Chem. Intl. Ed.* **30**, 1476.
Blocher, A.; Zeller, K.-P. (1994) *Chem. Ber.* **127**, 551.
Bohle, D.S.; Ishihara, Y.; Perepichka, I.; Zhang, L. (2008) *Tetrahedron Lett.* **49**, 4550.
Bohle, D.S.; McQuade, L.E.; Perepichka, I.; Zhang, L. (2007) *J. Org. Chem.* **72**, 3625.
Bohle, D.S.; Perepichka, I. (2009) *J. Org. Chem.* **74**, 1621.
Browne, D.L.; Taylor, J.B.; Plant, A.; Harrity, J.P.A. (2009) *J. Org. Chem.* **74**, 396.
Browne, D.L.; Taylor, J.B.; Plant, A.; Harrity, J.P.A. (2010) *J. Org. Chem.* **75**, 984.
Buckley, G.D.; Levy, W.J. (1951) *J. Chem. Soc.* 3016.

Chadra, T.; Bhati, S.K.; Preetyadav; Agarwal, R.C.; Agarwal, T.P.; Kumar, A. (2008) *Orient. J. Chem.* **24**, 229.

Chan, W.L.; Yana, H.; Szeto, Y.Sh. (2004) *Mater. Lett.* **58**, 882.

Cherepanov, I.A.; Kalinin, V.N. (2000) *Mendeleev Commun.* 181.

Dunkley, Ch.S.; Thoman, Ch.J. (2003) *Bioorg. Med. Chem. Lett.* **13**, 2899.

Eade, R.A.; Earl, J.C. (1946) *J. Chem. Soc.* 591.

Earl, J.C.; Leake, E.W.; le Favre, R.J.W. (1947) *Nature* **160** (4063), 366.

Earl, J.C.; Mackney, A.W. (1935) *J. Chem. Soc.* 899.

Fuchigami, T.; Chen, C.-S.; Nonaka, T.; Yeh, M.Y.; Tien, H.-J. (1986) *Bull. Chem. Soc. Jpn.* **59**, 483.

Gao, D.; Zhai, H.; Parvez, M.; Back, Th.G. (2008) *J. Org. Chem.* **73**, 8057.

Gotz, M.; Grozinger, K. (1971) *Tetrahedron* **27**, 4449.

Greco, C.V.; O'Reilly, B.P. (1970) *J. Heterocycl. Chem.* **7**, 1433.

Ito, S.; Turnbull, K.J. (1996) *Synth. Commun.* **26**, 1441.

Levina, M.N.; Badyshtov, B.A.; Gan'shina, T.S. (2006) *Eksperim. Klinich. Farmakol.* **69**, 71.

Levina, V.I.; Grigor'ev, N.V.; Granik, V.G. (2004) *Khim. Geterotsikl. Soed.* 604.

Matsuda, K.; Adachi, J.; Nomura, K. (1979) *J. Chem. Soc. Perkin Trans. 1*, 956.

Moustafa, M.A.; Gineinah, M.M.; Nasr, M.N.; Bayoumi, W.A.H. (2004) *Arch. Pharm.* **337**, 427.

Nguen, M.Th.; Hegarty, A.F.; Elguero, J. (1985) *Angew. Chem. Intl. Ed.* **24**, 713.

Ohta, M.; Kato, H. (1957) *Nippon Kagaku Zasshi* **78**, 1653.

Preston, P.N.; Turnbull, K.J. (1977) *J. Chem. Soc. Perkin Trans. 1*, 1229.

Sapountzis, I.; Ettmayer, P.; Klein, Ch.; Mantoulidis, A.; Steegmaier, M.; Steurer, S.; Waizenegger, I. (2009) WO Patent 003,998.

Schubert, W.; Ellenrieder, W.J. (1984) *J. Chem. Res.* 256.

Schulz, R.; Schweig, A. (1979) *Angew. Chem. Intl. Ed.* **18**, 692.

Schulz, R.; Schweig, A. (1984) *Angew. Chem. Intl. Ed.* **23**, 509.

Schweig, A.; Baumgartl, H.; Schulz, R. (1991) *J. Mol. Struct.* **247**, 135.

Semenov, S.G.; Sigolaev, Yu.F. (2004) *Zh. Strukt. Khim.* **45**, 1128.

Sugihara, T.; Kuwahara, K.; Wakabayashi, A.; Takao, H.; Imagawa, H.; Nishizawa, M. (2004) *Chem. Commun.* 216.

Tien, H.-J.; Ohta, M. (1972) *Bull. Chem. Soc. Jpn.* **45**, 2944.

Tien, H.-J.; Yeh, W.-C.; Wu, S.-C. (1992) *J. Chin. Chem. Soc.* **39**, 443.

Turnbull, K.; Blackburn, T.L.; McClure, D.B. (1994) *J. Heterocycl. Chem.* **31**, 1631.

Turnbull, K.; George, J.C. (1996) *Synth. Commun.* **26**, 2757.

Turnbull, K.; Gross, K.C.; Hall, T.A. (1998) *Synth. Commun.* **28**, 931.

Turnbull, K.; Krein, D.M. (1997) *Tetrahedron Lett.* **38**, 1165.

Wu, Y.; Xue, Y.; Xie, D.; Yan, G. (2005) *J. Org. Chem.* **70**, 5045.

Yashunskii, V.G.; Kholodov, L.E. (1980) *Usp. Khim.* **49**, 54.

Yashunskii, V.G.; Vasil'eva, V.F.; Sheinker, Yu.N. (1959) *Zh. Obshch. Khim.* **29**, 2712.

1.2 1,2,4-OXADIAZOLES AND 1,2,4-OXADIAZOLE N-OXIDES

Unsubstituted 1,2,4-oxadiazole is an unstable compound, it was firstly prepared in 1962 by Moussebois et al. (Moussebois et al. 1962), although its derivatives have been known from 1884 (Tiemann and Krueger). Substituted 1,2,4-oxadiazoles are more stable and their chemistry and applications are well documented and discussed here using characteristic examples. Section 1.2.1 consists of these examples.

1,2,4-Oxadiazole N-oxides are considered separately, in Section 1.2.2, due to their distinctive chemical pattern that differs in fragility of the heterocyclic ring.

1.2.1 1,2,4-Oxadiazoles

1.2.1.1 Formation

1.2.1.1.1 From Amidoximes

The 1,2,4-oxadiazole moiety is often used to prepare peptidomimetic building blocks. In these cases, the oxadiazole ring is formed by dehydration of amidoxime precursors derived from an amino acid (see Scheme 1.21) (Jakopin et al. 2007). The step of cyclodehydration is qualifier. At this step, provision should be made for weakly basicity of the reaction solvent and the moderate reaction temperature. This allows avoiding racemization which usually occurs after the use of strong bases (Hamze et al. 2003). For instance, the chirality of the starting amino acids was retained, when succinic acid anhydride was used for cyclodehydration in dimethylformamide containing pyridine, at 110°C (Jakopin et al. 2007).

Propionic acid bound through a selenium linker with polystyrene-supported resin, produces 1,2,4-oxadiazoles as a result of treatment with amidoximes in the presence of *N,N'*-dicyclohexyl carbodiimide. Naturally, 1,2,4-oxadiazoles remain bound to resin-selenium through the ethan-1,2-diyl fragment. Using hydrogen peroxide, the heterocycles can be eliminated as the corresponding 5-vinyl derivatives (Wang et al. 2007). So, the polystyrene-supported resin not only facilitates separation of the products but also serves as provinyl safety-catch linker. The advantages of the method quoted include straightforward operation, lack of odor, stability, and high purity of the vinyl monomers.

Outirite et al. (2007) proposed the one-pot synthesis of 3.5-diaryl-1,2,4-oxadiazoles refluxing aromatic nitriles, hydroxylamine hydrochloride, and sodium carbonate in ethylene glycol. The products were obtained in good yields and in excellent state of purity. In this case, amidoxime is obtained from interaction of arylnitrile with hydroxylamine. As amidoxime forms, it reacts with the starting arylnitrile giving 3,5-diaryl-1,2,4-oxadiazole (Scheme 1.22).

A useful modification of the Scheme 1.22 consists of the preparation of diaryl-1,2,4-oxadiazoles through interaction of arylnitriles with hydroxylamine hydrochloride in the presence of magnesia-supported sodium carbonate at the first step. Then, the acylamidoxime formed is treated with aryl carboxychloride in the same vessel under solvent-free conditions using microwave irradiation. This modification allows one to prepare the desired products rapidly, in an environment-friendly method, with good yields (Kaboudin and Saadati 2007).

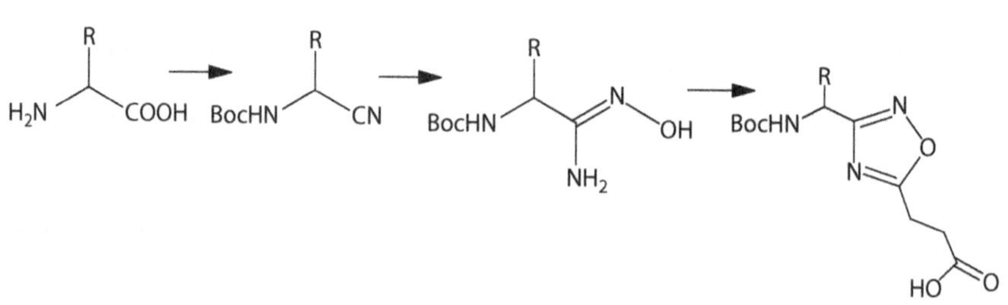

SCHEME 1.21

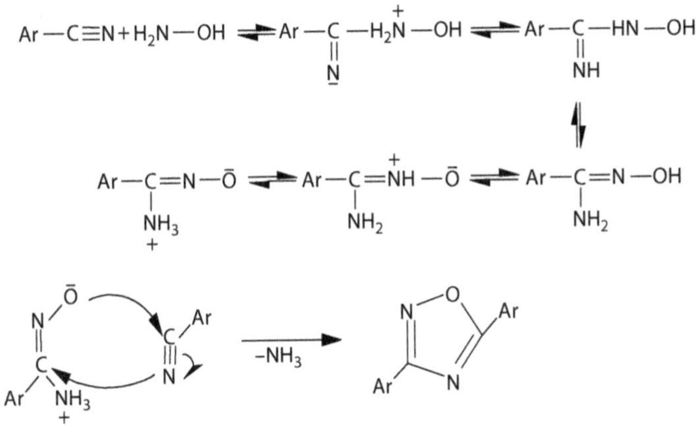

SCHEME 1.22

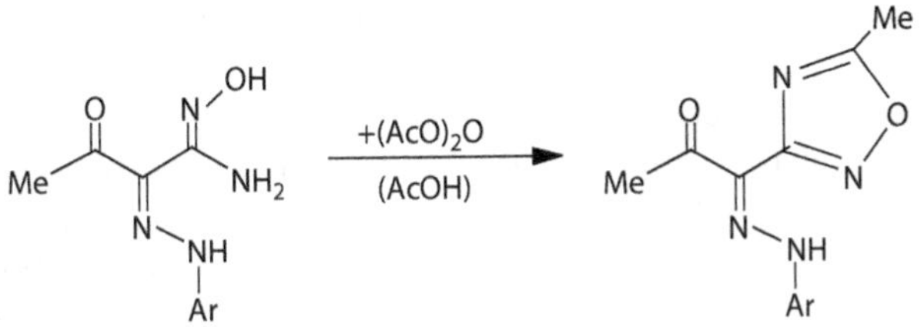

SCHEME 1.23

Scheme 1.23 describes the synthesis of 1,2,4-oxadiazole derivatives from amidoxime during a 2 h reflux in the mixture of acetic anhydride with acetic acid (Al-Mousawi et al. 2007).

Yan et al. (2007) used the method in Scheme 1.23 to prepare 1,2,4-oxadiazole-3-phenylpropionic acid derivatives as potent and selective immunosuppressants with enhanced pharmacokinetic properties. Takahashi (2007) claimed a method of acylamidioxime cyclization using trichloranhydide of 1,3,5-benzene tricarboxylic acid. The author recommended to perform this reaction in cyclohexane solution containing 2,6-lutidine.

Scheme 1.24 outlines the N-acylamidoxime cyclization into 1,2,4-oxadiazole derivatives. Cyclization proceeds spontaneously when N-acylamidoximes are formed (e.g., from oxazolones—see Kmetic and Stanovnik 1995).

Cyclization of O-acylamidoximes is often used to achieve library of 1,2,4-oxadiazole-containing biomolecules (Buscemi et al. 2006) or to modify commercial

SCHEME 1.24

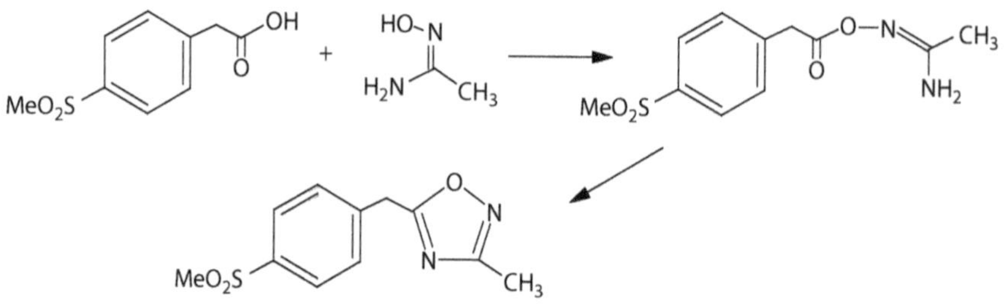

SCHEME 1.25

polymers such as polyacrylonitrile (Vega et al. 2006). Scheme 1.25 presents one example of this synthesis (Pipik et al. 2004).

In Scheme 1.25, the reaction of 4-(methylsulfonyl)phenyl acetic acid with methylamidoxime was activated with 1-hydroxybenzotriazole, 1-ethyl-3-(3-dimethylaminopropyl)carbodiimide, and hydrochloric acid. The O-acylated aminoamide formed was heated to give 3-methyl-5-[(methylsulfonyl)benzyl]-1,2,4-oxadiazole (90% yield, 30 kg amount). The last step (dehydration-cyclization) takes place without the addition of other reagents.

A similar solid-phase synthesis of 1,2,4-oxadiazole derivatives was published by Quan and Kurth (2004). Microwave-assisted methods of this cyclization have been also reported (Oussaid et al. 1995, Evans et al. 2003, Wang et al. 2005, Adib et al. 2006, de Freitas et al. 2007). Du et al. (2007) gave examples of one-pot synthesis of 3-substituted 5-carbonylmethyl-1,2,4-oxadiazoles from amidoximes and β-keto esters by simple heating of the components to 120°C–140°C for 2–4 h with no solvent and no additional reagents. Yields were perfect.

Ispikoudi et al. (2008) developed preparation of 5-amino-substituted 1,2,4-oxadiazole derivatives by one-step reactions of aryl, benzyl, cycloalkyl, and alkyl amidoximes with commercially available carbodiimides. The authors observed that all of the reactions required two equivalents of carbodiimide to achieve the complete consumption of stating amidoxime. This fact is easily explained with the mechanism of Scheme 1.26, proposed earlier by Kawashima and Tabei (1986).

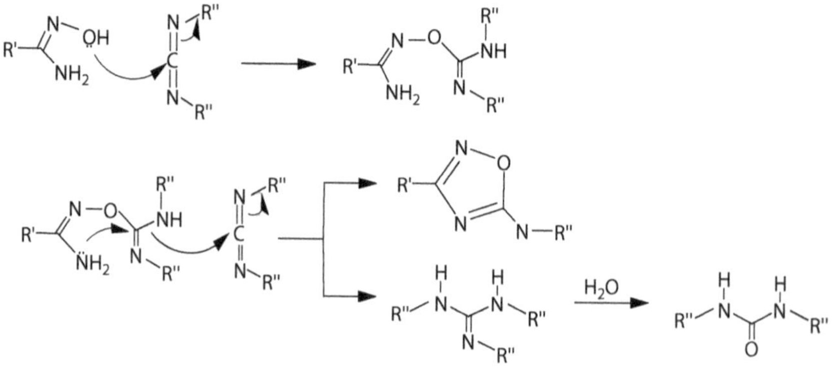

SCHEME 1.26

A nucleophilic attack of the hydroxyl group of amidoxime onto the electrophilic center of carbodiimide occurs initially, leading to the O-amidoxime adduct. The latter then undergoes an intramolecular attack by the amino group attached to the parent amidoxime. A second molecule of carbodiimide facilitates this transformation by abstracting the remaining alkyl- or aryl-amino moiety of carbodiimide. As a result, the 1,2,4-oxadiazole derivative is produced along with a guanidine-type intermediate, which is hydrolyzed, and delivers the corresponding urea derivative (Scheme 1.26).

Grant et al. (2008) have described the cyclization of O-acylamidoximes with aroyl chlorides in a continuous microreactor sequence to produce 1,2,4-oxadiazoles. The synthesis is completed in ca. 30 min, good yields of the products are achieved in quantities of 40–80 mg. Such amounts are sufficient for full characterization and for rapid library supply. Formerly, this kind of the oxadiazole synthesis had typically been carried out under sealed-tube conditions, in more than 70 h of the reaction time (Eloy and Lanaers 1962, Roppe et al. 2004).

Cyclization of N-hydroxy carboxyimidamide with acetyl chloride in the presence of pyridine needs 12 h to be completed (Shablykhin et al. 2007). Katritzky et al. (2005) proposed a rapid and easy method for the synthesis of 1,2,4-oxadiazoles, via O-acylation of amidoximes with (α-aminoacyl)benzotriazoles in the presence of triethylamine in ethanol. The reaction was completed immediately at room temperature. To obtain the aimed heterocycles, 3–5 min refluxing of O-acylamidoxime in EtOH + Et₃N was sufficient. Scheme 1.27 gives the summary of this synthesis route. As it can be seen, the oxadiazole ring forms with no participation of the triazole fragment. The role of benzotriazolyl moiety is only in the activation of the carboxylic acid functionality bound to benzotriazole (compare Schemes 1.25 and 1.27). Seemingly, reactions between amidoximes and dialkylureas are initiated by 1,1'-carbonyldiimidazole (Burton and Wadsworth 2007).

There is also an efficient method for the preparation of 1,2,4-oxadiazole from carboxylic acids and amidoximes using propylphosphonic cyclic anhydride (T3P) as a coupling agent and water scavenger (Augustine et al. 2009). Application of this anhydride in a pair with triethylamine affords 90%–95% yields of the oxadiazoles and produces only water soluble by-products, in most cases. An aqueous work up is sufficient to get the aimed oxadiazoles. T3P is characterized with low toxicity and low allergenic potential.

Bora and Farooqui (2007) proposed to perform O-acylamidoxime cyclodehydration in the presence of freshly dried molecular sieves in a fully neutral environment. Under such (mild and neutral) conditions, acid and base labile substrates can cyclize

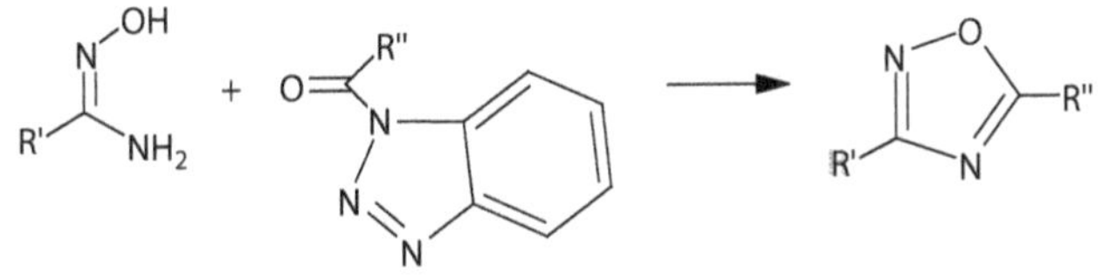

SCHEME 1.27

in quantitative yields. The method is simple, does not produce side products, and the workup is easy.

1.2.1.1.2 From Hydroxamoyl Chlorides

This reaction is good for the preparation of spiro- or fused 1,2,4-oxadiazoles (Awadallah 2006, Cortes et al. 2007) and is carried out in the presence of triethyl-amine. One simple example is shown in Scheme 1.28 (Awadallah 2006).

1.2.1.1.3 From O-Substituted Imidates

O-substituted imidates (generated from amidoximes and nitriles) undergo cycliza-tion into 1,2,4-oxadiazoles. Scheme 1.29 presents this method. It was proven that ring closure occurs through the fragment N–C–O–N–C and not through the alterna-tive fragment N–C–N–O–C (Yarovenko et al. 1990). The presence of freshly dried molecular sieves facilitates cyclization (Bora and Farooqui 2008).

1.2.1.1.4 From Nitriles and Nitrile Oxides or Oximes

These methods are also used to access 1,2,4-oxadiazoles. Aryl-activated nitriles are most commonly employed (see, e.g., Lenaers and Eloy 1963), but the relatively dull aliphatic nitriles can also be involved in the reaction if Lewis acids (Morrocchi et al. 1967) or metals (Lasri et al. 2007) are used as promotors.

Dondoni and Barbaro (1975) proposed the reaction between mesitylene nitrile oxide (MesCNO) and aromatic nitriles (ArCN) at elevated temperatures to obtain 3,5-disubstituted 1,2,4-oxadiazoles according to Scheme 1.30. Neidelin et al. (1998) carried out a reaction of aryl nitrile oxides with dicyanoketene ethylene acetal at room temperature under catalysis by boron trifluoride.

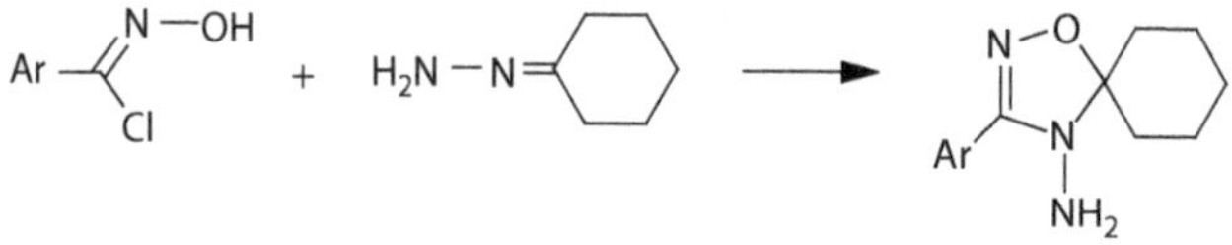

SCHEME 1.28

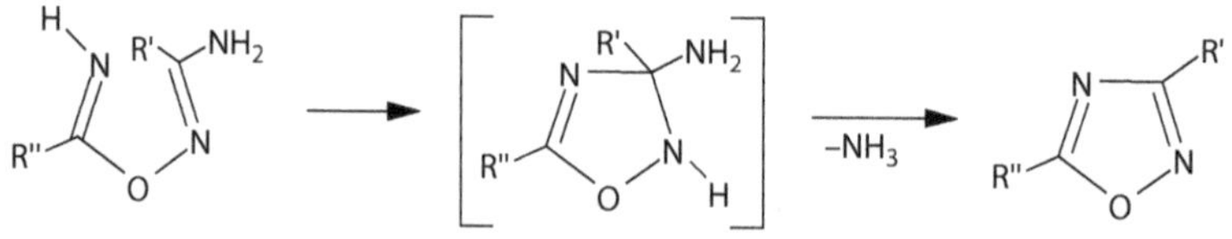

SCHEME 1.29

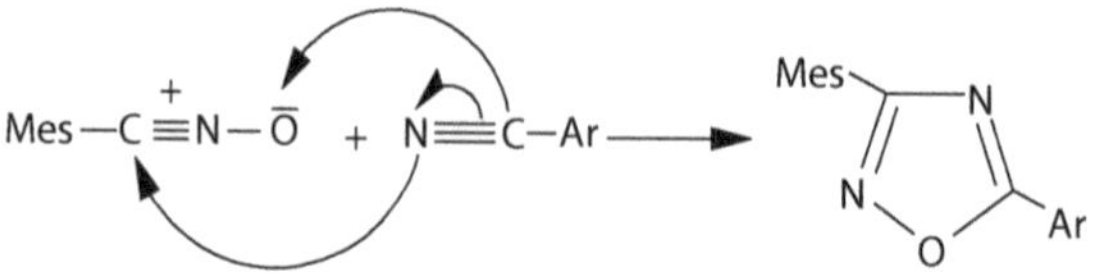

SCHEME 1.30

Regioselectivity of cycloaddition depicted by Scheme 1.30 can be, apparently, attributed to the fact that in the course of the reaction, the more positively charged fragment of nitrile, that is, the carbon atom of the nitrile group, points toward the negative center (the oxygen atom) of the dipole (nitrile oxide). At the same time, the more negative nitrogen atom of the nitrile group forms a bond with the partially positive carbon atom of nitrile oxide. (The carbon–nitrogen bond is more stable than the nitrogen–nitrogen bond and the five-membered cycle is more stable than the four-membered one.)

Schemes 1.31 and 1.32 concern the reactions between nitriles and ketones upon action of ferric nitrate (Itoh et al. 2005, Garfinkle et al. 2008) or yttrium trinitrate (Yu et al. 2007).

In the reaction outlined by Scheme 1.31, ketones initially experience enolyzation, and this process is accelerated by ferric nitrate. Then, nitration of the enols takes place. Dehydration of the nitroproducts formed leads to the production of nitrile oxides. The nitrile oxides add to the nitriles already present, forming oxadiazole derivatives. This method uses non-toxic and inexpensive ferric nitrate. Working up is simple; oxadiazoles are formed in a pure state and with good yields.

The same enolyzation followed with addition takes place in the case depicted in Scheme 1.32 (Itoh et al. 2005).

Scheme 1.33 presents a reaction that consists of highly regio- and diastereoselective nitrile oxide cycloaddition to 2,4-dimethyl-[3H]-1,5-benzodizepines (Nabih et al. 2004, Kumar and Perumal 2007). In this reaction, nitrile oxides are generated in situ from arylhydroximinoyl chloride and triethylamine.

Makarycheva-Mikhailova et al. (2007) proposed an original (and general, in principle) route to fused 1,2,4-oxadiazoles. The route consists of platinum(II)-mediated 1,3-dipolar cycloaddition of oxazoline N-oxides to nitriles (Scheme 1.34).

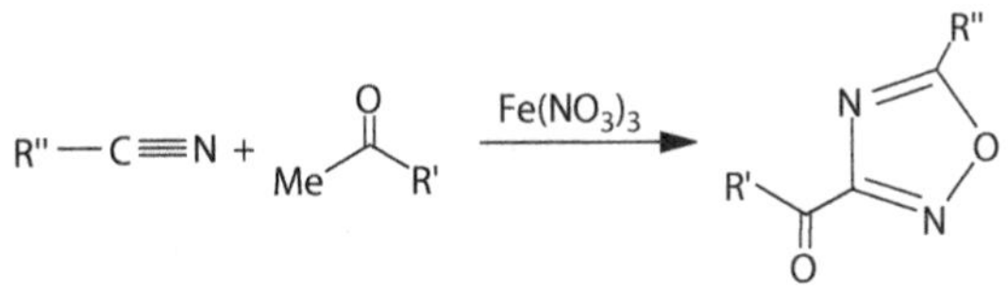

SCHEME 1.31

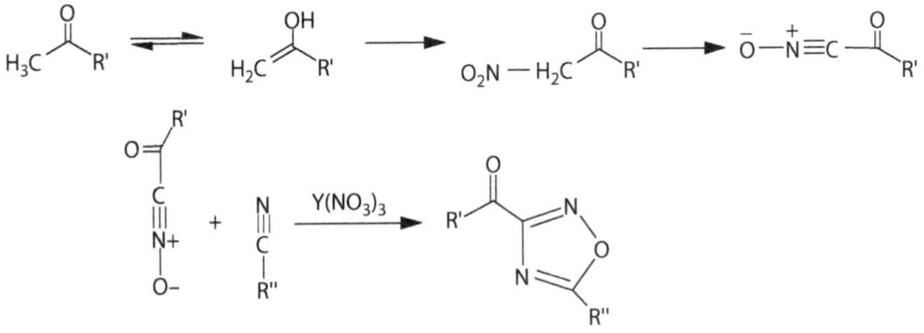

SCHEME 1.32

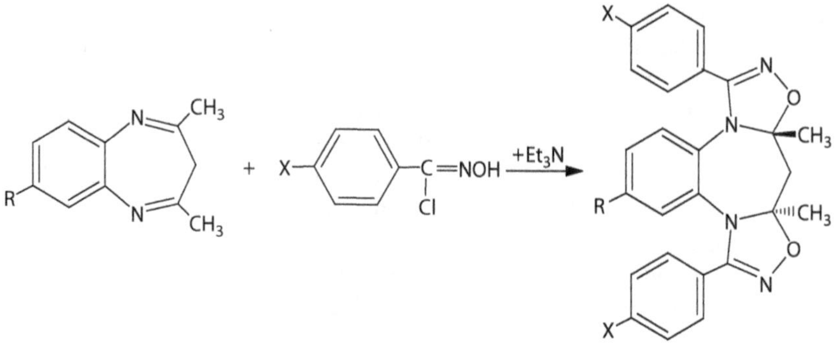

SCHEME 1.33

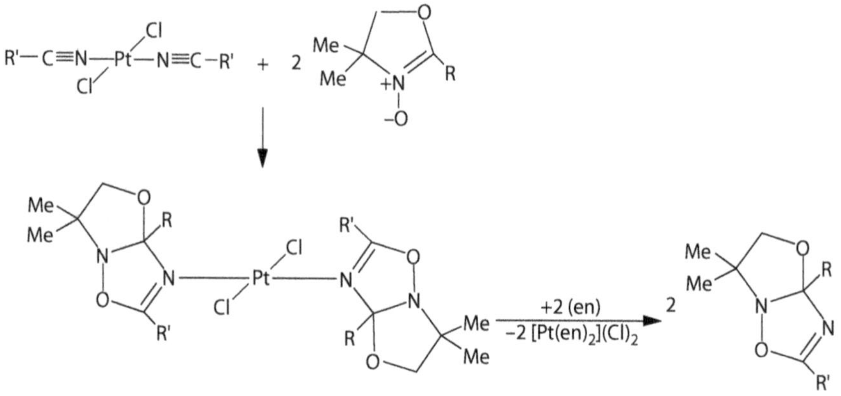

SCHEME 1.34

The reaction is unknown for free nitriles and oxazoline N-oxides, but under platinum(II)-mediated conditions, it proceeds smoothly. Treatment with sym-ethylenediamine (en) releases the oxadiazole derivatives from the platinum complexes. The expensiveness of the platinum starting material can be strongly reduced by platinum recycling, methods of which are given in the paper by Makarycheva-Mikhailova et al. (2007); reference 39 therein.

There is one more reaction of nitriles with oximes that is interesting in respect of its mechanism. The reaction proceeds in argon under initiation with the thianthrene cation-radical perchlorate and accelerated in the presence of 2,6-di(tert-butyl)-4-methylpyridine (Chiou et al. 1990). The first step of the process consists of single-electron transfer from the oxime to the thianthrene cation-radical. The resulting oxime cation-radical adds the nitrile. Cyclization of the adduct leads to the formation of an 1,2,4-oxadiazole derivative accompanied with deprotonation and regeneration of thianthrene (Scheme 1.35) (TH stands for thianthrene).

Naidu and Sorenson (2005) and Naidu (2008) prepared 1,2,4-oxadiazoline diester using an appropriate nitrile, N-alkylhydroxyl amine, and diethyl ethynedicarboxyl-ate (Scheme 1.36).

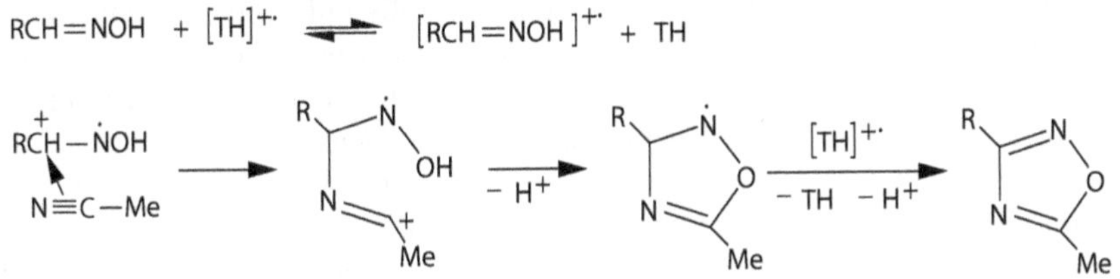

SCHEME 1.35

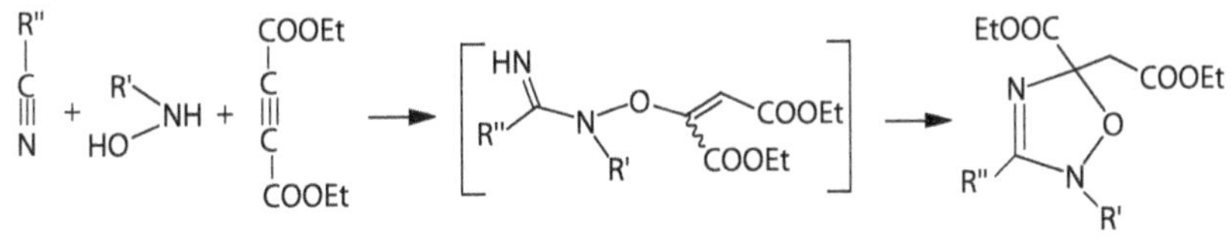

SCHEME 1.36

Rajanarendar et al. (2006) reported the formation of the 1,2,4-oxadiazole cycle as a result of interaction between nitrile oxides and the azomethine fragment of Schiff bases.

1.2.1.1.5 From Nitrones and Nitriles or Isocyanates

Reaction of nitriles with nitrones is another way to the 1,2,4-oxadiazole cycle. Nitrones are synthesized by the condensation of the corresponding aldehydes with N-methyl-hydroxylamine (Tyrrell et al. 2005). Scheme 1.37 describes the reaction of trichloroacetonitrile with a nitron bearing the systematic name of N-(2-methoxybenzylidene)methylamine N-oxide (Wagner and Garland 2008).

Plate et al. (1987) and Hermkens et al. (1988) noted that only nitriles bearing strong electron-withdrawing substituents (e.g., $N \equiv C–CCl_3$) are involved in cycloaddition with nitrones such as $Ar–CH=N(O)Me$. In contrast to $N \equiv C–CCl_3$, $N \equiv C–Me$ remains intact even under drastic reaction conditions. Hermkens et al. (1988) compared activity of nitrile and nitrones in this kind of cycloaddition.

Another method of nitrile activation involves complexation with metal via N atom (Sarju et al. 2008). The effect of the nitrile coordination to platinum(IV) is even stronger than the effect of the nitrile activation due to the introduction of a strong electron-withdrawing group (Kuznetsov and Kukushkin 2006). Platinum(IV) activates the nitrile ligand to an extent that the cycloaddition to nitrones becomes possible even with nitriles containing electron-releasing substituents (Wagner et al. 2001). Bokach et al. (2009) gave a theoretical explanation of the metal-coordination effect. Regrettably, the

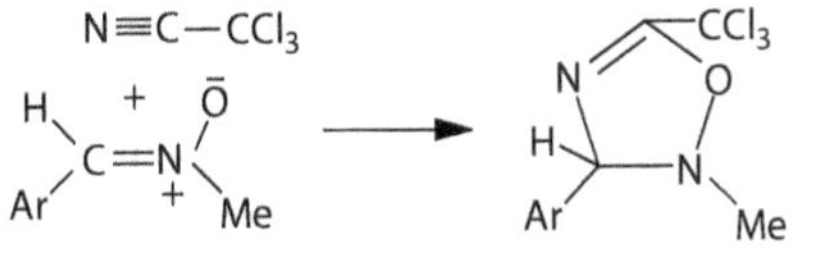

SCHEME 1.37

cycloaddition under consideration proceeds successfully only with the platinum(II), or platinum(IV), or palladium(II) nitrile complexes. Complexes of nitriles with titanium(IV), zirconium(IV), molybdenum(IV), and tungsten(IV) do not react with nitrones (Wagner 2004). Nevertheless, this is not a problem for the platinum-participate synthesis owing to the well-developed methods of precise-metal recovery.

Naturally, the 1,2,4-oxadiazole derivatives resulting from the metal-assisted condensation, remain bound with the metal. From complexes, heterocycles can be liberated via exchange of this ligand with amines; see, for example, Scheme 1.34 and the 2001 patent claimed by Kukushkin et al. (2001). Some of 1,2,4-oxidiazolic ligands cannot exist without metal center (Bokach et al. 2009). Meanwhile, there are platinum(II) 1,2,4-oxadiazoline complexes that are still active against those cancer cell lines that are sensitive or resistant in the presence of platinum-containing drugs (Coley et al. 2008).

This combination of nitrones with isocyanates is used to prepare precursors of pH-dependent spin probes. The latter are needed for noninvasive pH monitoring of pathophysiological conditions according to changes in electron spin resonance spectra. Scheme 1.38 introduces room-temperature synthesis of the spin-probe precursors, where the yields are 85%–99% (Polienko et al. 2008).

1.2.1.1.6 From N-Acylthioureas or O-Arylisoureas

The reaction of N-substituted N'-acylthioureas with hydroxylamine leads to the formation of the corresponding thioureas and 3-amino-1,2,4-oxadiazole derivatives (Scheme 1.39) (Boboshko et al. 2007). For N-aryl-N'-acylthioureas, the presence of an acceptor substituent in the aryl fragment improves the yields of the oxadiazoles formed. The authors assume that the oxadiazole formation is a result of the

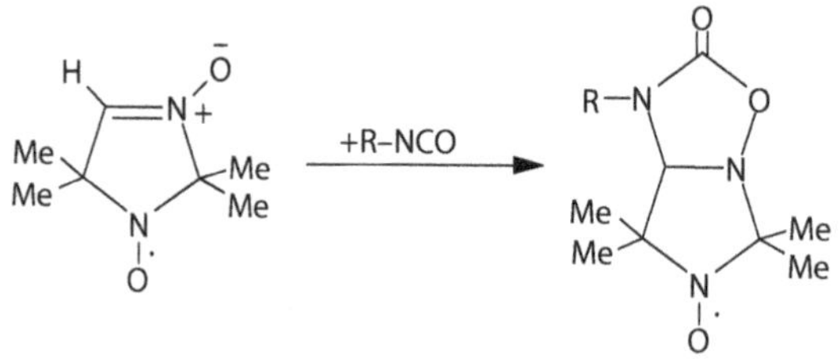

SCHEME 1.38

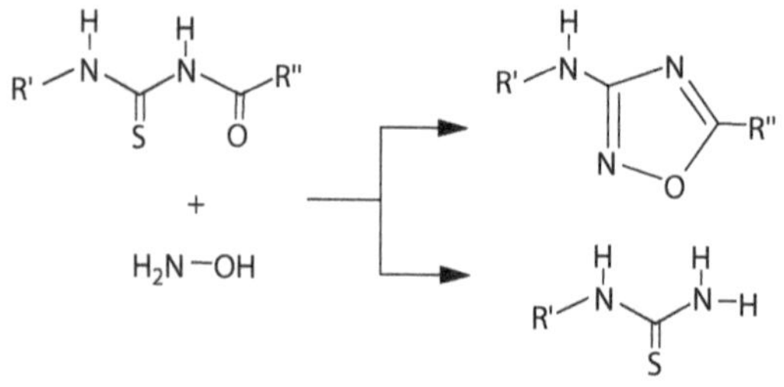

SCHEME 1.39

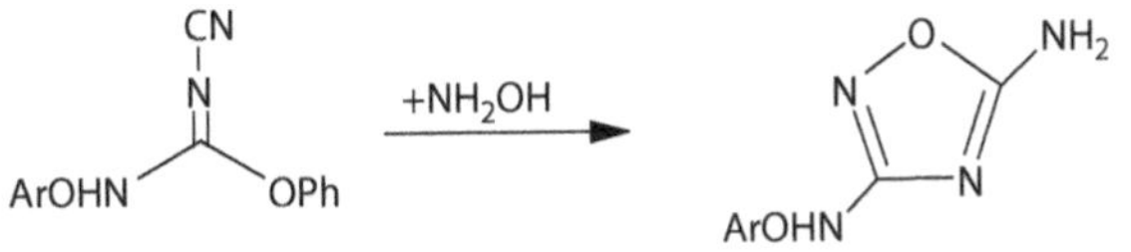

SCHEME 1.40

primary attack of hydroxylamine at the thiocarbonyl group of the acylthiourea with consequent ring closure. When hydroxylamine attacks the carbonyl group of the acylthiourea, the substituted thioureas are formed.

The reaction of O-arylisoureas with hydroxylamine under microwave irradiation is completed in 1 min and leads to aminooxadiazoles with excellent yields (Scheme 1.40) (Kurz et al. 2007).

1.2.1.1.7 From N-Acyl-α-Amino Acids and Their Derivatives

This method includes N-nitrosation of amino acids and their derivatives. It is applicable to the substrates bearing active hydrogen at a carbon atom. The resulting N-nitroso compound was isolated and irradiated by light. Upon irradiation, the nitroso group is shifted to the neighboring carbon atom. The C-nitroso isomer undergoes dehydration and decarbonylation to form 1,2,4-oxadiazole derivatives. Although yields achieved by this method are moderate, it is attractive because the starting materials are readily available from α-amino acids (Jochims 1996). Of course, N-substituted amides cannot be used for the oxadiazole formation. The need for light irradiation makes the method a little difficult from the preparation point of view. Moreover, many N-nitroso compounds are carcinogenic. Scheme 1.41 gives the sequence of the transformation mentioned (see Chow and Polo 1986). Scheme 1.41 should be compared to Scheme 1.24 concerning cyclization of N-acylamidoximes (Kmetic and Stanovnik 1995).

The amidoxime method was used to construct 1,2,4-oxadiazole-linked peptidomimetics. Peptidomimetics are being explored largely to circumvent some of the disadvantages typical for native peptides and to increase the bioavailability and potency of peptide-based drugs. The protocol (proposed by Sureshbabu et al. in 2008) employs a reaction between N-protected amino acid-derived amidoxime and N-protected amino acid fluoride in the presence of N-methylmorpholine as a promoter. O-Acylamidoxime is initially formed. Its dehydration is achieved by a 4 h reflux in ethanol containing sodium acetate. Scheme 1.42 makes the whole protocol chemically understandable.

1.2.1.1.8 From Ketonitrene Precursors

Ketonitrenes, generated in situ, cyclize to give 1,2,4-dioxadiazoles. Scheme 1.43 gives one of the examples: A mixture of trimethylsilyl azide and 4-chorobenzoyl

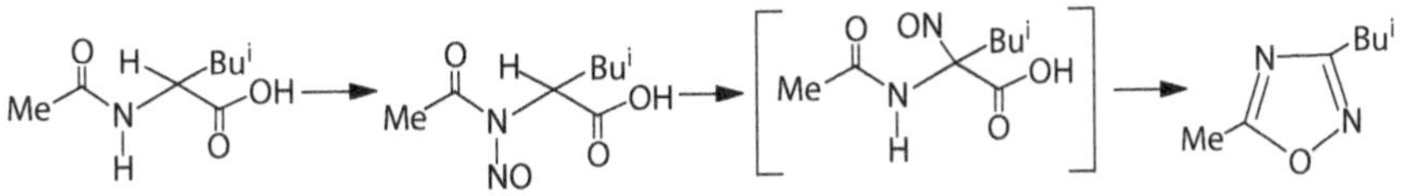

SCHEME 1.41

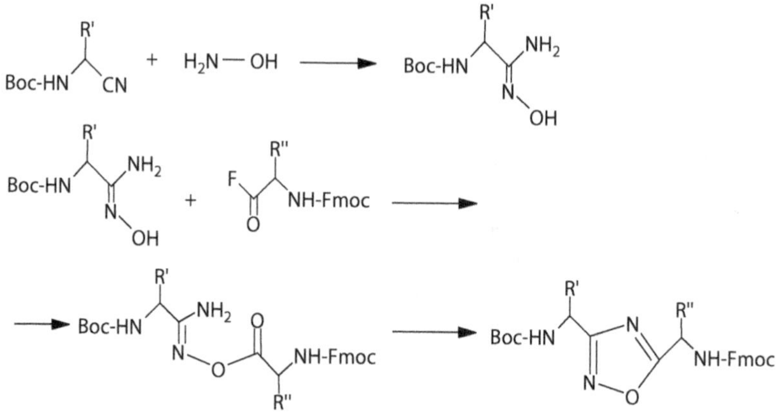

SCHEME 1.42

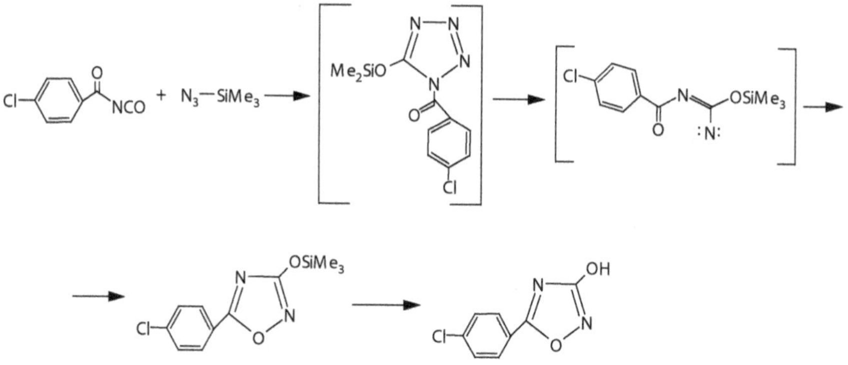

SCHEME 1.43

isocyanate was refluxed and the 3-siloxy-containing product is readily converted into 5-(4-chlorophenyl)-1,2,4-oxadiazole-3-ol, 85% yield (Tsuge et al. 1980).

Boulton–Katritzky rearrangement, leading to the formation of one 1,2,4-oxadiazole from another, also includes the ketonitrene step. Scheme 1.44 shows the conversion of 3-(acetylamino)-5-aryl-1,2,4-oxadiazoles into 3-(aroylamino)-5-methyl-1,2,4-oxadiazoles. The reaction needs light excitation (Buscemi et al. 1990).

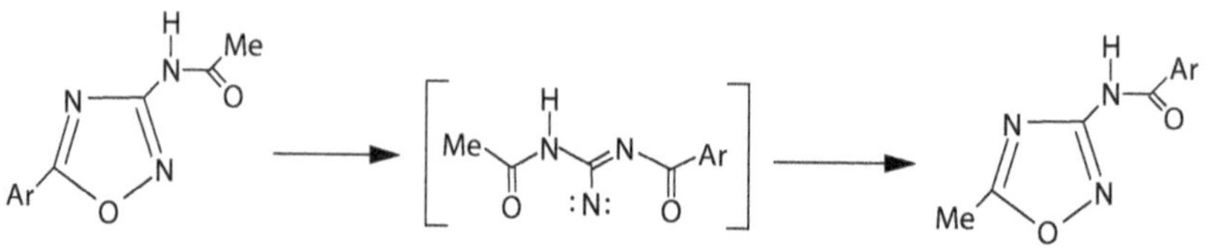

SCHEME 1.44

1.2.1.1.9 From Other Heterocycles

Scheme 1.45 introduces transformation of the pyrimidine ring into the 1,2,4-oxadiazole moiety. Notably, the formation of the new heterocycle proceeds regioselectively (Zhang et al. 1999).

1,2,4-Oxadiazoles can be prepared from tetrazoles (see Scheme 1.43). Plenkiewicz and Zdrojewski (1987) reported other relevant examples. On other side, treatment of 4(5)-phenyl(or alkyl)-2-aminoimidazoles with isoamyl nitrite in acetic acid afforded the corresponding 4(5)-substituted 2-amino-5(4)-hydroxyimino-5(4)H-imidazoles, which by heating in water were transformed into 3-benzoyl(or acyl)-5-amino-1,2,4-oxadiazoles (Cavalleri et al. 1973). The latter reference also directs to the preparation of (3-hydroxy-1,2,4-oxadiazol-5-yl)(phenyl)methanone from dibenzoyl furoxan upon mild hydrolysis.

Irradiation of (perfluoroalkanoylamino)furazans at 313 nm in a methanol solution of ammonia or some primary aliphatic amines afforded 1,2,4-oxadiazoles. This photoreaction is conjugated with thermal dehydration of the N-acylamino-amidoxime, which is supposedly formed (Scheme 1.46) (Buscemi et al. 2001).

1,2,4-Oxadiazoles were also successfully obtained from 4-hydroxyliminoimidazoles by 24 h refluxing in aqueous ethanol containing hydrochloric acid. The aimed products are formed through hydrolytic cleavage of the imidazole ring followed by recyclization, yields are 90%–95% (da Silva et al. 1999).

1.2.1.2 Fine Structure

The 1,2,4-oxadiazole ring is planar. Its aromaticity is relatively low and both carbon–nitrogen distances are close to those of the typical C=N bonds. Bond O–N is characterized by lability and is prone to be cleaved under thermal or light excitation. Both nitrogen atoms remain basic, but probably in different degrees (Jochims 1996).

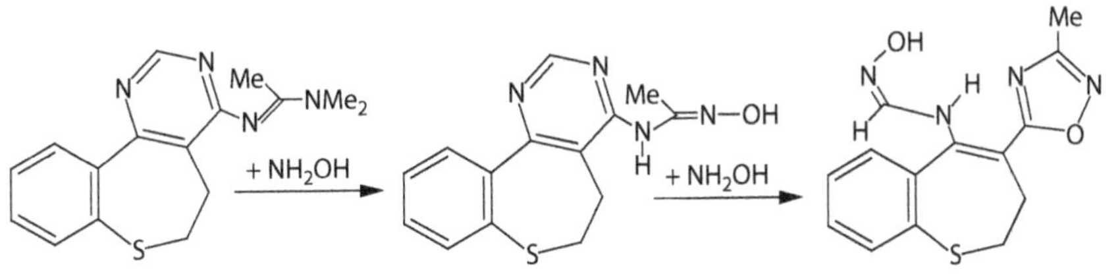

SCHEME 1.45

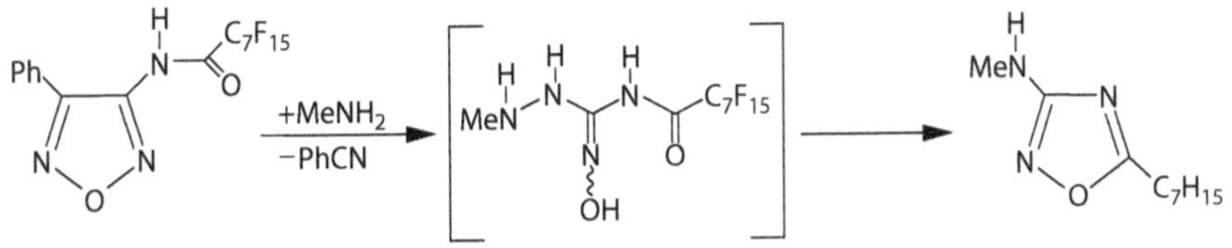

SCHEME 1.46

The aromaticity indices, I_A, calculated by Bird (1985) on the basis of bond lengths are 39 for 1,2,4-oxadiazole and 100 for benzene. Between oxadiazole isomers, 1,2,4-oxadiazole is the least aromatic and has a high tendency to rearrange into other, more stable, heterocycles.

1.2.1.3   Reactivity

1.2.1.3.1   Tautomerism

3-Hydroxy-5-phenyl-1,2,4-oxadiazole gives a striking example of solvent-dependent reversible proton transfer from the hydroxyl group to the ring nitrogens. In ethanol, acetone, or dimethylsulfoxide—the solvents capable of hydrogen bonding with this substituent—the compound under consideration exists predominantly as the hydroxy derivative. In chloroform solution and in the solid state the 3-keto form predominates. Scheme 1.47 presents the participants of the equilibrium mentioned (Katritzky et al. 1965).

1.2.1.3.2   Salt Formation

The reaction of methyl iodide with nitrogen heterocycles results in quaternization, this is trivial. However, the interaction between methyl iodide and the 1,2,4-oxadiazole derivative (Scheme 1.48) seems to be astonishing. The substrate has several amino groups of high basicity, its pyrazine fragment can also give its ring nitrogens for quaternization. Nevertheless, the reaction of Scheme 1.48 proceeds strictly regioselectively and the final product is obtained in the almost quantitative yield (Bock et al. 1986).

In the oxadiazole in Scheme 1.48, the nitrogen in the vicinal position to the oxygen turns to be more basic than the removal nitrogen and (that is more wonderful) than the nitrogen of the amino group. Probably, it is the quaternizing nitrogen that feels the pyrazine donor effect most efficiently. In this context, it should be noted that in 1,2,4-oxadiazole as it is, the N_4 remains the preferred site for protonation (Trifonov et al. 2005).

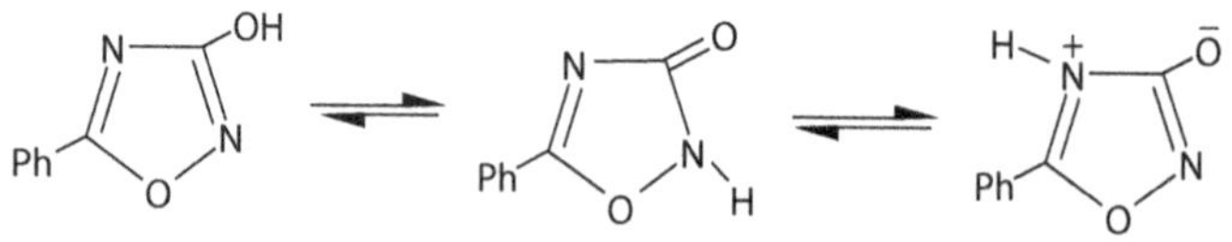

SCHEME 1.47

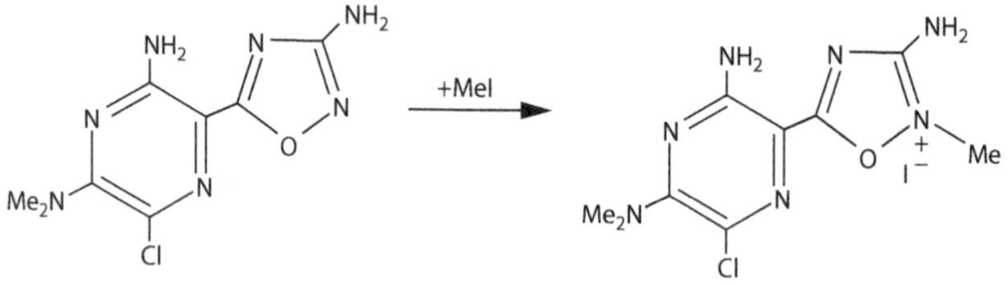

SCHEME 1.48

1.2.1.3.3 Reduction

Electrochemical reduction of 1,2,4-oxadiazoles on graphite electrode in aceto-nitrile is irreversible. Photo-induced single-electron transfer (in methanol) takes place from a sensitizer in its excited state to the oxadiazole in its ground state or from the triethylamine to the excited oxadiazole. The single-electron transfer leads to the oxadiazole anion-radical. This anion-radical is unstable and imme-diately splits off according to Scheme 1.49, which is based on the triethylamine case (Buscemi et al. 1999). The ring-cleaved intermediate then forms the quinaz-olin-4-one system or reduced products. The first way implies a back one-electron transfer from an intermediary quinazolin-4-one anion-radical, the second way implies hydrogen atom abstraction from the triethylamine cation-radical or from the methanol solvent.

The reaction of Scheme 1.49 confirms weakness of the bond O–N in 1,2,4-oxa-diazoles denoted in Section 1.2.1.

Whereas disubstituted 1,2,4-oxadiazoles do not change even in boiling aqueous sodium hydroxide solution, monosubstituted counterparts are readily hydrolyzed under the same conditions. Upon action of hydroxide, they undergo deprotonation from unsubstituted ring positions. This leads to ring cleavage. Reaction with methyl mercaptide proceeds similarly (Scheme 1.50) (Claisse et al. 1973).

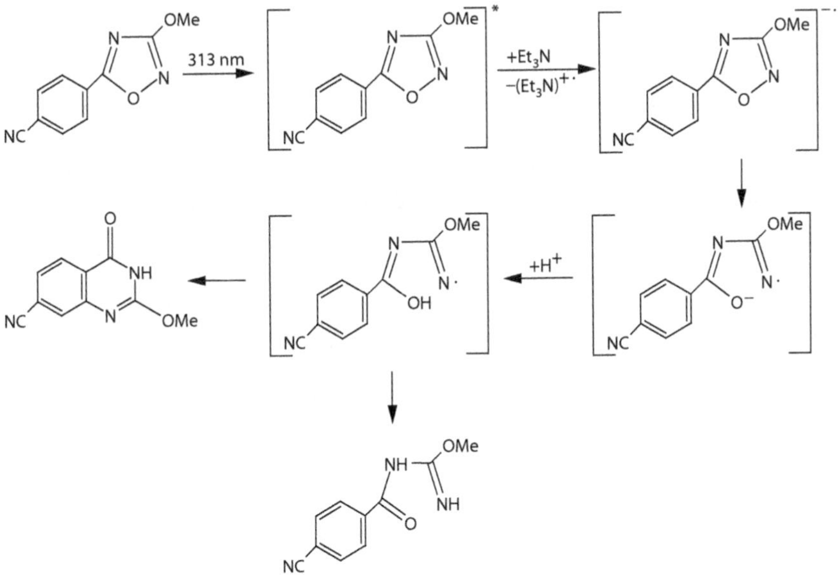

SCHEME 1.49

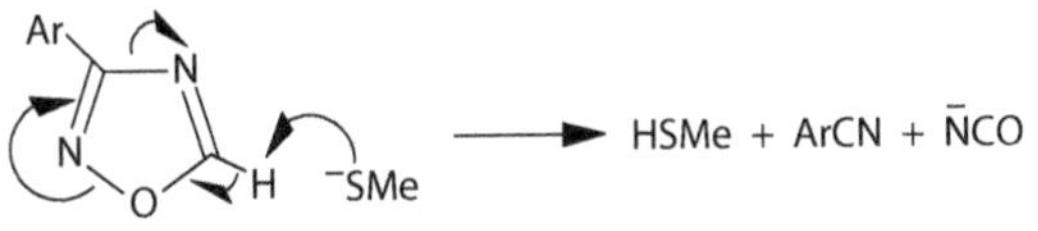

$$HSMe + ArCN + \bar{N}CO$$

SCHEME 1.50

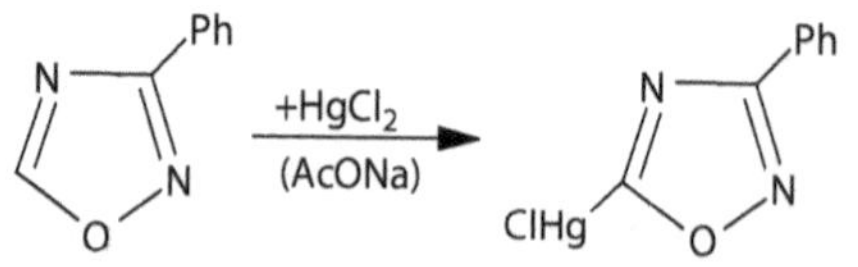

SCHEME 1.51

1.2.1.3.4 Electrophilic Substitution

1,2,4-Oxadiazoles and their derivatives hardly enter electrophilic substitution. Moussebois and Eloy (1964) described electrophilic mercuration of the 5-substituted system (Scheme 1.51).

1.2.1.3.5 Nucleophilic Substitution

Due to the electron withdrawing effect of both O_1 and N_4, the most electrophilic is C_5 position of 1,2,4-oxadiazole. Substitution of nucleophiles for the halo, alkoxy, aryl-sulfanyl, and trichloromethyl groups proceeds easily only when they are located at the carbon atom in position 5 of the ring. Nucleophilic substitution at position 3 does not proceed so readily: Reactions in this position may occur only in the presence of good leaving-groups such as chlorine (Jochims 1996).

1.2.1.3.6 Photolysis

In general, the photochemical reactivity of the 1,2,4-oxadiazole ring involves the cleavage of the nitrogen–oxygen bond. The resulting intermediates are the zwitterion of Scheme 1.52 and its valence tautomers—diradicals and nitrenes (Newman 1968, Piccionello et al. 2007). In particular, a nitrene from 3,5-diphenyl-1,2,4-oxadiazole was intercepted with 2,3-dimethylbutene that led to the final aziridine (Scheme 1.52). Piccionello et al. (2007) identified this product from spectroscopic and x-ray data.

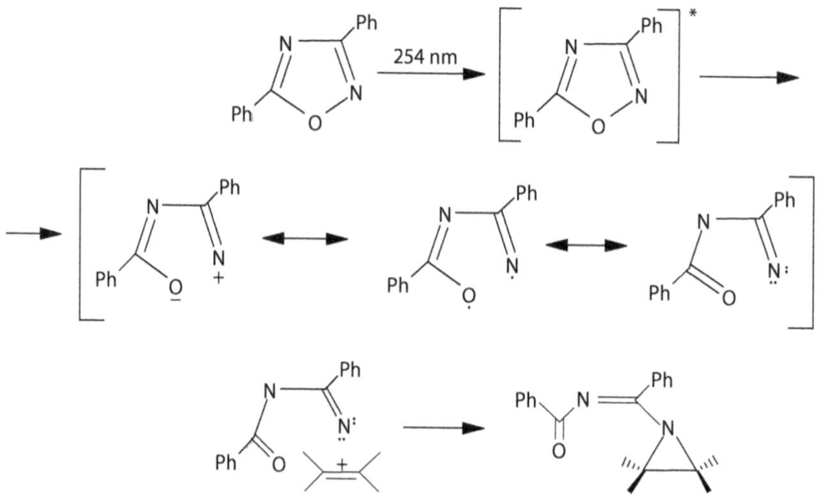

SCHEME 1.52

Such aziridines have biomedical applications as themselves (Baek and Lee 2000, Kalvins et al. 2001) and/or as synthons in preparation of heterocycles (see, for example, Johnson et al. 1996).

1.2.1.3.7 Thermolysis

1,2,4-Oxadiazoles with substituents at both the 3 and 5 positions are very resistant to thermolysis. When the ring bears only one substituent, thermal decomposition proceeds easily. Thus, 3-phenyl-1,2,4-oxadiazole decomposes rapidly at 200°C, whereas 3,5-diphenyl-1,2,4-oxadiazole keeps its integrity even at a high-temperature distillation of 296°C (Tiemann and Krueger 1884). However, 3,5-disubstituted 3H-1,2,4-oxadiazoles thermally open up, even at moderate temperatures. Thus, 3-(2-methoxyphenyl)-(3H)-5-(trichloromethyl)-1,2,4-oxadiazole transforms into the corresponding formamidine upon reflux in chloroform (60°C, 8 days). It is $N_2–O_1$ bond that is cleaved. The cleavage is accompanied by 1,2-aryl shift from carbon to the adjacent amino nitrogen. Scheme 1.53 illustrates the whole reaction (Wagner and Garland 2008).

The N–O bonds in compounds of the 1,2,4-oxadiazole family are the weakest ones and cleave upon thermolysis. The rearrangement of 3,4-disubstituted oxadiazolon-5-thiones into 3,4-disibstituted thiadiazol-5-ones is a prominent example of the thermal N–O breakage and the following recyclization. Scheme 1.54 presents the overall reaction (Pelter and Sumengen 1977, Durust et al. 2007).

Pyridooxadiazolone of Scheme 1.55 gives an example of the thermal breakage of not only N–O bond, but also N–C bond. During vacuum flash thermolysis, this compound loses carbon dioxide and transforms into 2-pyridylnitrene. The nitrene was

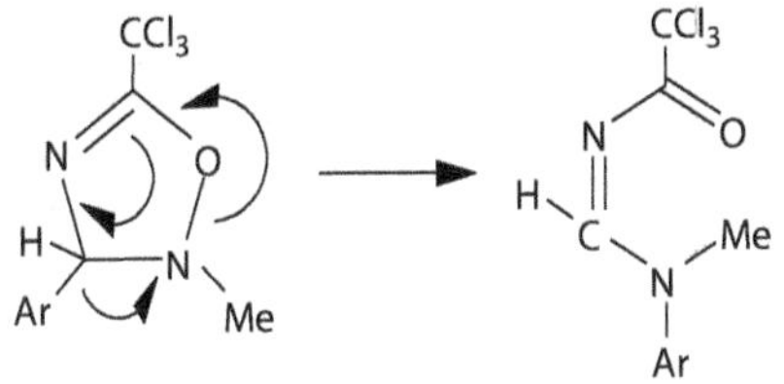

SCHEME 1.53

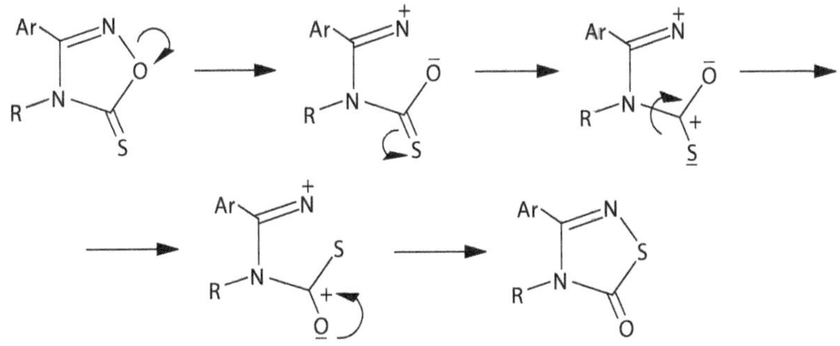

SCHEME 1.54

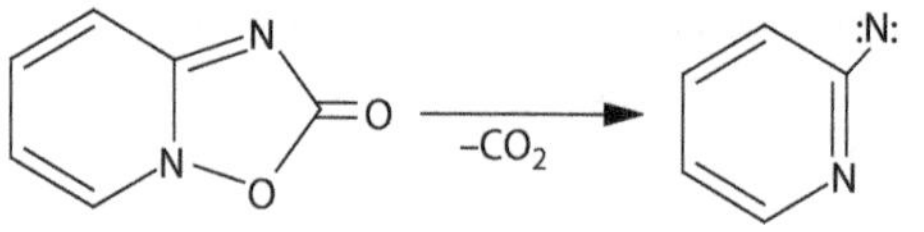

SCHEME 1.55

isolated in argon matrices and identified by electron spin resonance spectroscopy. Upon stabilization, the nitrene gives rise to a mixture of glutacononitrile, 2-amino-pyridine, pyrrole-2- and -3-carbonitriles (McCluskey and Wentrup 2008).

1.2.1.3.8 Coordination

Electrophilicity of 1,2,4-oxadiazole fits it with possibility to form charge-transfer complexes with donor counterparts, especially when oxadiazole bears electron acceptor substituents. With arylethenes, 3-phenyl-5-(trinitromethyl)-1,2,4-oxadiazole forms charge-transfer complexes. Within the complexes, the ethene pulls a nitro group out from the trinitromethyl substrate and, after that, the resulting components are covalently bound with each other. New compounds are unstable and break down giving nitroketones and 3-phenyl-1,2,4-oxadiazole-5-methylnitronitrolic acid according to Scheme 1.56 (Tyrkov 2004).

Scheme 1.56 gives an example of chemical reactions in confined environment. It would be interesting to study reactions of 1,2,4-oxadiazoles within inclusion complexes, too. For instance, molecular mechanics study by Al-Sou'od (2006) points out that the 3,3'-bis[4-alkyl-1,2,4-oxadiazol-5(4H)-one] bulks completely penetrate into the cavity of β-cyclodextrin. The driving forces for the formation of inclusion

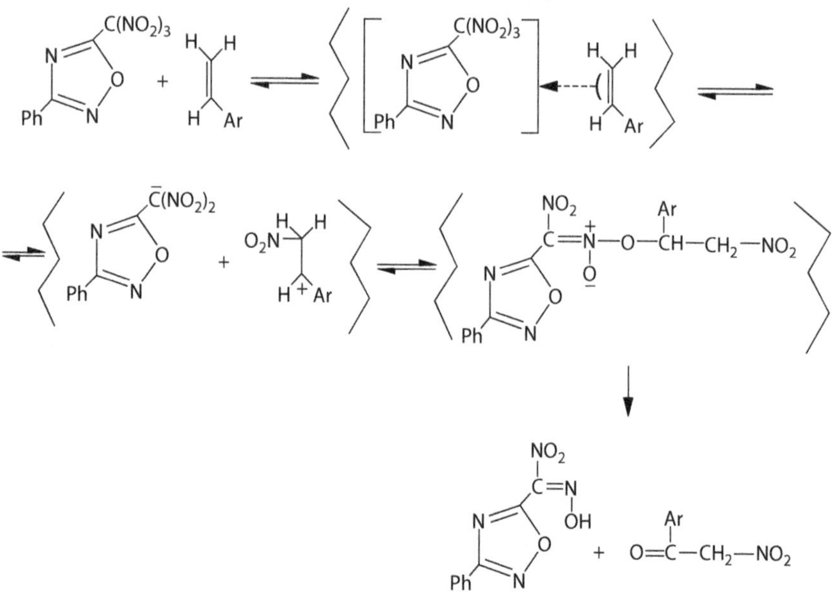

SCHEME 1.56

complexes are dominated by no bonded van der Waals host–guest interactions with some electrostatic contribution.

In the reaction with cupric acetate, 5-(2-hydroxyphenyl)-3-phenyl-1,2,4-oxadiazole forms bis[5-(2-oxyphenyl)-3-phenyl-1,2,4-oxadiazolyl]copper. With respect to the heterocycle, only nitrogen in position 4 of each ligand is involved in coordination (da Silva et al. 1999). The same manner of coordination to palladium dichloride or to silver mononitrate was observed for 3.3'-bis(1,2,4-oxadiazolyl). Palladium forms a mononuclear complex and binds with 4 and 4' nitrogens of 3.3'-bis(1,2,4-oxadiazolyl). Silver forms a one-dimensional coordination polymer, in which the bidentate ligand acts in a bridging mode, again with coordination through the each nitrogens at positions 4 and 4' (Richardson and Steel 2007).

1.2.1.3.9 Ring Transformation

1,2,4-Oxadiazole derivatives easily undergo the monocyclic Boulton–Katritzky rearrangement. In general, this rearrangement consists of interconversion between two five-membered heterocycles where a three-atom side chain and a pivotal annular nitrogen are involved (Boulton et al. 1967). Scheme 1.57 represents such a rearrangement of two 1,2,4-oxadiazole derivatives into each other. Because the initial and resulting isomers are both 1,2,4-oxadiazole derivatives, this kind of rearrangement proceeds reversibly. It should be particularly emphasized that this transformation includes a neutral form in their ground states. It should be also noted that the resulting 5-aryl-substituted component experiences aryloid stabilization. Therefore, the equilibrium is in reality shifted toward the right side (Pace et al. 2007).

Scheme 1.58 describes an irreversible Boulton–Katritzky rearrangement. In this case, the initial and resulting heterocycles differ from one another in their degrees

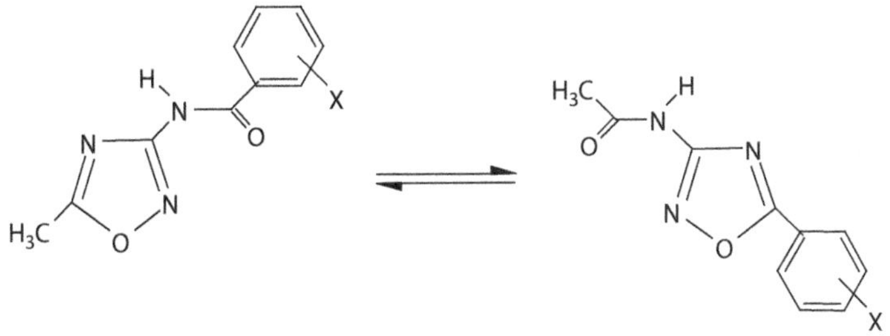

SCHEME 1.57

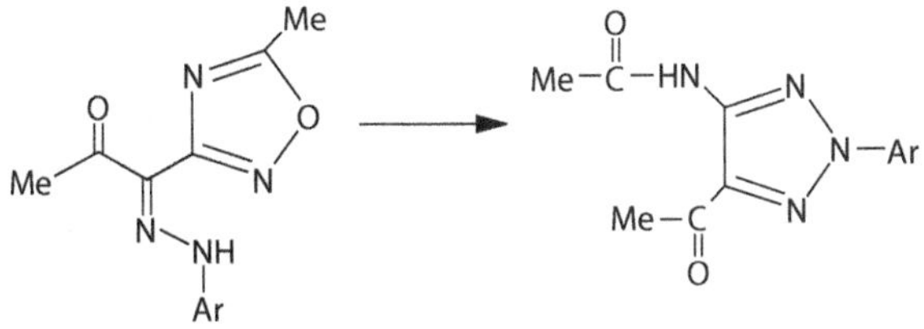

SCHEME 1.58

of aromaticity. Namely, a 1,2,4-oxadiazole derivative transforms into a derivative of 1,2,3-triazole under reflux in dimethylformamide (Al-Mousawi et al. 2007). In terms of the united aromaticity indexes (I_U based on the corresponding experimental heats of formation), 1,2,3-triazole is characterized with $I_U = 90$, whereas 1,2,4-oxadiazole has $I_U = 48$ only (Bird 1992).

Arylhydrazones derived from 3-benzoyl-5-phenyl-1,2,4-oxadiazole rearrange to 1,2,3-triazoles when heated to their melting point (Scheme 1.59) (Gilchrist 1985, p. 228, cf. Fontana et al. 2008).

Transformation of the 1,2,4-oxadiazole ring ($I_U = 48$) into the imidazole one ($I_U = 79$) was observed by Ruccia et al. (1974). Scheme 1.60 illustrates the reaction observed and gives a chance to consider its mechanism. The thermal reaction of 3-amino-5-phenyl-1,2,4-oxadiazole with pentan-2,4-dione leads to enaminoketone. Deprotonation of the compound formed results in the formation of the corresponding anion. The latter gives rise to the final imidazole derivative (Scheme 1.60). A recent example involves the reaction of 3-amino-5-phenyl-1,2,4-oxadiazoles with fluorinated β-diketones in the presence of montmorillonite K-10. Here, the formation of final imidazoles is the result of an acid-catalyzed Boulton–Katritzky rearrangement involving N–C–C side-chain. Importantly, the K-10 catalyst forms a complex with the oxadiazole ring of the starting material (Piccionello et al. 2008).

Computational studies on Boulton–Katritzky rearrangement gave potential energy surfaces that were in complete accordance with the relative aromaticity of

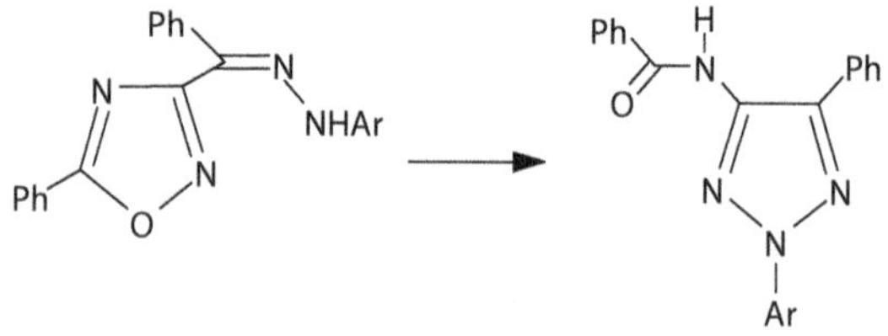

SCHEME 1.59

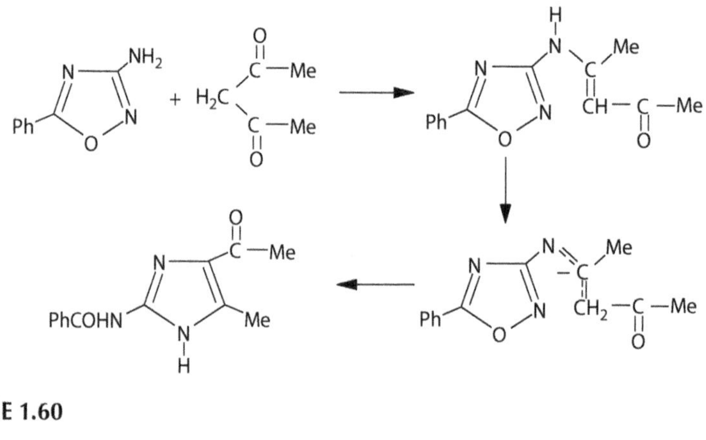

SCHEME 1.60

the initial 1,2,4-oxadiazole derivatives and that of the final compounds containing other heterocyclic moieties (Pace et al. 2009).

Scheme 1.49 gives one of examples for transformation of 5-aryl-1,2,4-oxadia-zoles to quinazolin-4-one derivatives. As seen, quinazolin-4-ones are formed by heterocyclization together with ring opening products, resulting from a formal reduction of the ring N–O bond. Depending on the substituents' nature and on the reaction conditions, yields of the products vary from moderate to good (Buscemi et al. 1999).

While the reaction of Scheme 1.49 needs photoinitiation, reactions of 3-ethoxycar-bonyl-5-perfluoroalkyl-1,2,4-oxadiazoles with hydroxylamine or N-methylhydroxy-lamine in dimethylformamide take place without any excitation. A perfluoroalkyl substituent sufficiently activates the whole substrate and facilitates the transforma-tion. Nucleophilicity and steric factors define the dichotomic behavior of the substrate. Namely, fluorinated oxadiazinone corresponds to the reaction with hydroxylamine whereas N-methylhydroxylamine generates 1,2,5-oxadiazole derivatives according to the mechanisms in Scheme 1.61 (Piccionello et al. 2009a). In fact, while the less hindered NH_2 nitrogen of hydroxylamine attacks exclusively position 5 of the oxa-diazole ring, the NHMe nitrogen reacts exclusively with the ester moiety.

The reaction between methylhydrazine and 3-(2-fluorophenyl)-5-phenyl-1,2,4-oxadiazole includes substitution of 2-methylhydrazin-1-yl for fluorine followed by Boulton–Katritzky rearrangement. Scheme 1.62 outlines the overall reaction that proceeds in dimethylformamide at room temperature. The yield of the resulting indazole comes to 95% (Piccionello et al. 2009b).

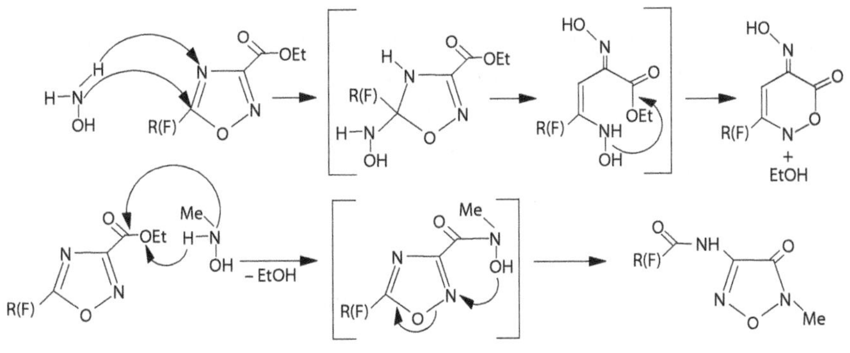

SCHEME 1.61

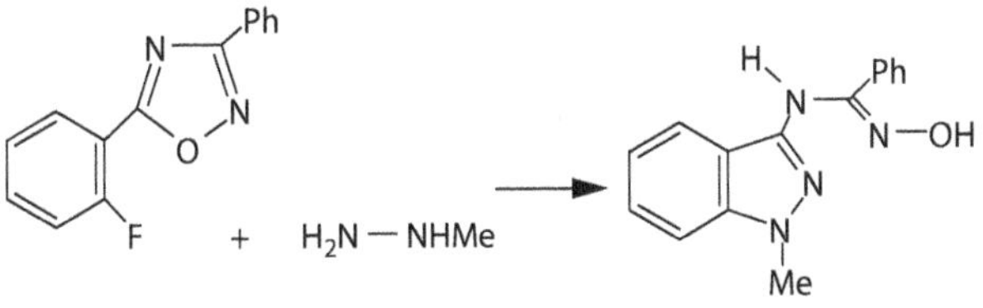

SCHEME 1.62

1.2.1.4 Biomedical Importance

1.2.1.4.1 Medical Significance

1,2,4-Oxadiazoles are compounds of marked medical significance. 1,2,4-Oxadiazoles are bioisosteric replacements for a hydrolytically unstable amide or ester functionality. Due to its electronic properties, the 1,2,4-oxadiazole nucleus is stable to hydrolysis and shows good balance between metabolic steadiness and bioavailability. This heterocycle supplies with several possibilities to form a hydrogen bond that is essential for permeability, extent of drug penetration into cells (see Toulmin et al. 2008 and references therein). Sometimes 1,2,4-oxadiazole derivatives occur to be highly susceptible to oxidation after penetration into microsomes (Roberts et al. 2008). Improved pharmacokinetic and in vivo performance make 1,2,4-oxadiazoles an important motif of new medications and put them forward as parts of numerous drug discovery programs.

For instance, the 3-methyl-1,2,4-oxadiazole moiety is metabolically more stable, biologically more active in comparison to the methyl ester group in alkenyldiaryl methanes as a drug to combat the progression of acquired immunodeficiency syndrome (Sakamoto et al. 2007). The 1,2,4-oxadiazol-5-one moiety was stated as a bioisostere of the carboxylic acid function in retinoid structures (Charton et al. 2009). The bioisosteric replacement of the carboxamide substituent in position 3 of pyrazole with 5-alkyl-1,2,4-oxadiazol-3-yl opens a way to new anti-obesity drugs with less toxicity for central nervous system (Chu et al. 2008). 3-Aryl-5-alkyl-1,2,4-oxadiazoles are recommended for use in the prevention or treatment of obesity (Ahn et al. 2008). The bioisosteric replacement of the amide functionality in lipoamides by the 3-(3,4-dimethoxyphenyl)-1,2,4-oxadiazole fragment results in strong neuroprotective activity with an improved pharmokinetical profile (Koufaki et al. 2007).

The search directions included analgesics (Merla et al. 2007, Farooqui et al. 2009), immunomodulators (Vu and Corpuz 1999, Albert et al. 2008, Boli et al. 2008, Nishi et al. 2008a,b) as well as anti-osteoarthritis (Wustrow et al. 2009) or anti-platelet and anti-thrombosis agents (Bethge and Petz 2005). The 1,2,4-oxadiazole-containing compounds were prepared for evaluation as anti-rhinoviral medicines (Fan et al. 2009).

Design, synthesis, and evaluation of 1,2,4-oxadiazole derivatives were carried out in the search of anti-scarring medicines (Fish et al. 2007, Bailey et al. 2008). It is 1,2,4-oxadiazole bearing the hydroxamate group in position 5 and the formamide group in position 3 that was first of all patented (Bailey et al. 2001). (3R)-6-Cyclohexyl-N-hydroxy-3-(3-{[(methylsulfonyl)amino]-methyl}-1,2,4-oxadiazol-5-yl) hexanamide was selected as a candidate for further preclinical evaluation as a topically applied, dermal anti-scarring agent (Fish et al. 2007).

Human sirtuin-type proteins have pathogenic roles in cancer, diabetes, heart failure, neurodegeneration, and aging. 1,2,4-Oxadiazole-carbonylaminoureas were proposed as these proteins inhibitors. 3-[3-(1-Naphthyl)]-1,2,4-oxadiazole-5-carboxamino]-1-[3-(trifluoromethyl)phenyl]thiourea occurred to be the most potent drug for these target proteins (Huhtiniemi et al. 2008). 3-Substituted 5-[(2-trifluoroacetyl)thien-4-yl]-1,2,4-oxadiazoles were discussed concerning cancer therapy (Muraglia et al. 2008).

As it was revealed, hexenopyranoside (dos Anjos et al. 2008) and carboxamide (Koryakova et al. 2008) derivatives of 1,2,4-oxadiazole presented impressive inhibition of cell growth. In these molecules, the presence of 1,2,4-oxadiazole plays a crucial role: Bioisosteric transformation of the 1,2,4-oxadiazole ring into, 1,3,4-oxadiazole nucleus in the carboxamides mentioned results in loss of the anticancer activity (Koryakova et al. 2008). Alkyldiazolyl aryl ethers also exhibit antitumor activity. This activity is superior to that of ethanecrynic acid (Zhao et al. 2008).

The development can be seen from a brief review of drug-related patents and papers concerning 1,2,4-oxadiazole derivatives. Oxadiazolyl piperidines were proposed for the treatment of pain and diarrhea (Tafesse 2007) as well as metabolic disorders (Barba et al. 2009). 5-Alkyl-3-(1-alkylaminoindol-3-yl)-1,2,4-oxadiazoles were recommended for pain control and prevention of cerebral ischemia (Moloney and Robertson 2002, Moloney et al. 2008). Diaryl and arylhetaryl oxadiazoles are claimed as useful for the treatment of autoimmune disease and multiple sclerosis (Harada et al. 2007, 2009, Ahmed et al. 2008). Alkyl hydroxyaryl oxadiazoles were found to be abnormally active against Alzheimer disease (Fukunaga et al. 2009, Sakai and Watanabe 2009), central and peripheral nervous system disorders (Silva et al. 2007, Learmonth et al. 2008). Newly prepared oxadiazolyl benzoic acids are useful for treating or preventing nonsense RNA suppression (Almstead et al. 2008a,b). Dosewise administration of these acids to cystic fibrosis patients results in a decrease in cough, in an increase mucus clearing, and general improvement in well-being (Hirawat and Miller 2008). Oxadiazolyl carboxamides were claimed as drugs for the treatment or prevention of thrombosis disorders (Zbinden et al. 2008) whereas the corresponding sulfonamides were patented as cytomegalovirus inhibitors (Svenstrup et al. 2007). Piperidyl thienyl oxadiazoles (Deprez et al. 2008) and oxadiazole Mannich bases (Ali and Shaharyar 2007) were prepared and successfully tested in treating mycobacterial infections such as tuberculosis. Pyridyl aryl oxadiazoles are recommended as neuronal acetylcholine receptors (Gopalakrishnan et al. 2008, Ji et al. 2008) or drugs against rheumatoid arthritis and active chronic hepatitis (Hobson et al. 2008). Oxadiazolones (Keil et al. 2005, 2007a,b,c, Kobayashi and Nakahira 2008) and triazolyl-bicycloctyl-oxadiazoles (Waddell et al. 2007) were found to be antidiabetic agents. Oxadiazolone derivatives were also claimed as compounds suitable for the treatment and/or prevention of cardiovascular illnesses (Bartel et al. 2007), disorders of fatty acid metabolism and glucose utilization as well as disorders of the central and peripheral nervous systems (Bernardelli et al. 2007, Keil et al. 2007a,b,c, Touaibia et al. 2007). A wide group of mental disorders such as schizophrenia, anxiety, depression, panic, bipolar illness, etc. can be treated by aryl heteryl 1,2,4-oxadiazoles (Gal et al. 2007). As found, 1H-[1,2,4]-oxadiazolo[4,3-a] quinoxalin-1-one inhibits neuritis outgrowth and causes neuritis retraction in nerves cells (Lee et al. 2009). This prevents inflammation of a nerve or nerves, often associated with a degenerative process that is accompanied by changes in sensory and motor activity in the region of the affected nerve.

A series of 5-substituted 3-(5-bromo-2,3-dimethoxyphenyl)-1,2,4-oxadiazoles was prepared and their anti-Parkinson's activity was evaluated. Some representatives were found to be active and free from neurotoxicity. According to results of partition study, they should easily cross the blood–brain barrier (Tiwari and Kohli 2008).

A wide range of alkylamine derivatives of oxadiazole was patented as calcium channel antagonist for treatment of many human disorders (Cuiman et al. 2008).

This patent gives data on modulate these channels upon action of the claimed oxadiazole derivatives. Ionotropic activity was also observed by Budriesi et al. (2009) during testing of 1,2,4-oxadiazolones.

Cheng et al. (2008) found 6-chloro-*N*-{3-[3-(2-chloro-4-fluorophenyl)-1,2,4-oxadiazol-5-yl propyl]quinonoline}-3-amine as an effective, highly selective and orally bioavailable analgesic/anti-inflammatory agent with excellent pharmacokinetic properties. In search for pain-relievers, without harmful side effects for the liver and central nervous system, DiMauro et al. (2008) identified 3-(2-chloro-4-fluorophenyl)-5-[4-piperidyl-1-(6-trifluoromethoxyquinol-3-yl)]-1,2,4-oxadiazole as a highly potent and selective drug that is metabolically stable in liver microsomes and displays low clearance and a long half-life period. In rodents, the drug exhibited promising pharmacokinetics. *N*,*N*-Dipropyloxantine bound, via 4-phenyloxymethylene, with 3-position of 5-(3-methoxyphenyl)-1,2,4-oxadiazole was discovered as a selective, high affinity medication for the potential treatment of asthma (Zablocki et al. 2005).

1.2.1.4.2   Imaging Agents and Oxygen Carriers

Among novel probes to imagine (in vivo) β-amyloid plagues in brain affected by Alzheimer's disease, 3,5-diphenyl-1,2,4-oxadiazole derivatives displayed sufficient uptake for imaging. However, the probe wash-out from the brain occurred to be impermissibly slow (Ono et al. 2008). This family of oxadiazoles should be improved to provide good balance between the binding affinity and the clearance rapidity. 3-(2-Pyridyl)-5-(2-napthyl)-1,2,4-oxadiazole received some acceptance in luciferase-based assays of bioluminescence (Auld et al. 2008).

Fluoropolymer based on a polyaspartamide containing 1,2,4-oxadiazole units is a potential artificial oxygen carrier. This polymer is water soluble. It dissolves enough amount of oxygen and fills the requirements of intravenous administration in order to enrich patients with oxygen, assisting in this sense to native blood (Mandracchia et al. 2007).

1.2.1.4.3   Agricultural Protectors

Oxadiazolyl-arylamides were claimed as bactericides for agricultural and gardening uses (Hara and Saiga 2007). 3-Phenyl-5-(2-thienyl)-1,2,4-oxadiazole was patented as a means for controlling nematodes that infest plants or animals (Williams et al. 2009). Pyrazolyl oxadiazolyl amides were claimed as high-effective pesticides or acaricides. Some of them exhibit 100% pesticidal activity against armyworms at 500 ppm (Xu et al. 2007).

1.2.1.5   Technical Applications

1,2,4-Oxadiazoles are applicable to the field of thermotropic liquid crystal. Asymmetry and the strong lateral O–N dipole widen the temperature interval of mesophase existence, decrease melting points and decomposition temperatures relatively to liquid crystals containing 1,3,4-oxadiazoles (Parra et al. 2006). Oxadiazolylpyridinium iodides with perfluoroalkyl tails at C-carbons of the oxadiazole rings form liquid crystal domains. Besides, these salts conduct electric current and exhibit thermochromism that opens the way to future applications in optoelectronics (Lo Celso et al. 2007).

Hockey-stick liquid crystals consisting of two terminal 10-carbon alkyl chains and the 1,2,4-oxadiazole fragment bearing a line aromatic–C≡C–aromatic constituent, form nonsymmetrical liquid crystal phases. In particular, their structures contain smectic and nematic phases typical of calamatic structures. In solution, these compounds exhibit strong blue fluorescence solution (Gallardo et al. 2008). Technical applicability of such properties is obvious.

Chiral liquid crystals, where the asymmetric carbon is not directly attached to the 1,2,4-oxadiazole, were obtained from enantiopure secondary alcohols linked, through an aromatic ester spacer, to the C(3) of the 1,2,4-oxadiazole (Parra et al. 2008). Pyridyl-1,2,4-oxadiazole derivatives and carboxylic acids form van der Waals (H-bonded) complexes, which are stable enough to organize the liquid crystalline phase. In the case of diacids, such liquid crystals manifest fluorescence in the visible region (Parra et al. 2005).

1.2.1.6   Conclusion

Being resistant to hydrolysis and electrophilic attack, 1,2,4-oxadiazoles are able to ring disclosure and transform into other heterocycles. This kind of reactivity depends on structural and ambient conditions such as temperature, solvent polarity, acidity or basicity of the media, exposure to light, etc.

It should be admitted that the actual bioactive 1,2,4-oxadiazoles can be conditioned by their reactivity in metabolic chains. As to technical applications, the literature gives a few examples of liquid crystals and luminescent materials. This aspect seems to be insufficiently investigated.

REFERENCES TO SECTION 1.2.1

Adib, M.; Jahromi, A.H.; Tavoosi, N.; Mahdavi, M.; Bijanzadeh, H.R. (2006) *Tetrahedron Lett.* **47**, 2965.

Ahmed, M.; Giblin, G.M.P.; Myatt, J.; Norton, D.; Rivers, D.A. (2008) WO Patent 128,951.

Ahn, M.J.; Jung, M.H.; Cho, N.O.; Lee, B.H. (2008) Kor. Patent 647,448.

Al-Mousawi, S.M.; Moustafa, M.Sh.; Elnagdi, M.H. (2007) *J. Chem. Res.* 515.

Al-Sou'od, Kh.A. (2006) *J. Inclus. Phenom. Macrocycl. Chem.* **54**, 123.

Albert, R.; Cooke, N.G.; Lewis, I.; Weiler, S.; Zecri, F. (2008) WO Patent 037,476.

Ali, M.A.; Shaharyar, M. (2007) *Bioorg. Med. Chem. Lett.* **17**, 3314.

Almstead, N.G.; Hwang, P.S.; Moon, Y.-Ch.; Welch, E.M. (2008a) US Patent 139,632.

Almstead, N.G.; Hwang, P.S.; Pines, S.; Moon, Y.-Ch.; Takasugi, J.J. (2008b) WO Patent 030,570.

Augustine, J.K.; Vairaperumal, V.; Narasimhan, Sh.; Alagarsamy, P.; Radhakrishnan, A. (2009) *Tetrahedron* **65**, 9989.

Auld, D.S.; Southall, N.T.; Jadav, A.; Johnson, R.L.; Diller, D.L.; Simeonov, A.; Austin, Ch.P.; Inglese, J. (2008) *J. Med. Chem.* **51**, 2372.

Awadallah, A.M. (2006) *Asian J. Chem.* **18**, 2151.

Baek, G.U.; Lee, B. (2000) Kor. Patent 067,156.

Bailey, S.; Billotte, S.; Derric, A.M.; Fish, P.V.; James, K.; Thomson, N.M. (2001) WO Patent 047,901.

Bailey, S.; Fish, P.V.; Billotte, S.; Bordener, J.; Grelling, D.; James, K.; McElroy, A.; Mills, J.E.; Reed, Ch.; Webster, R. (2008) *Bioorg. Med. Chem. Lett.* **18**, 6562.

Barba, O.; Bradley, S.F.; Fyfe, M.C.Th.; Hanrahan, P.E.; Krulle, Th.M.; Procter, M.J.; Reynet Maccormak, Ch. Schofield, K.L.; Smyth, D.; Stewart, A.J.W.; Swain, S.A. (2009) WO Patent O34,388.

Bartel, S.; Hahn, M.; Moradi, W.A.; Muenter, K.; Roelle, Th.; Stasch, J.-P.; Schlemmer, K.-H.; Wunder, F. (2007) WO Patent 045,366.

Bernardelli, P.; Keil, S.; Urmann, M.; Matter, H.; Wendler, W.; Glien, M.; Chandross, K.; Lee, J.; Terrier, C.; Minoux, H. (2007) WO Patent 039,172.

Bethge, K.; Petz, H.H. (2005) *Arch. Pharm.* **78**, 338.

Bird, C.W. (1985) *Tetrahedron* **41**, 1409.

Bird, C.W. (1992) *Tetrahedron* **48**, 335.

Boboshko, L.G.; Zubritskii, M.Yu.; Kovalenko, V.V.; Mikhailov, V.A.; Popov, A.F.; Rybakov, V.B.; Savyolova, V.A.; Taran, N.A. (2007) *Zh. Org. Farm. Khim.* **5**, 61.

Bock, M.G.; Smith, R.L.; Blaine, E.H.; Cragoe, E.J. (1986) *J. Med. Chem.* **29**, 1540.

Bokach, N.A.; Kuznetsov, M.A.; Haukka, M.; Ovcharenko, V.I.; Tretyakov, E.V.; Kukushkin, V.Yu. (2009) *Organometallics* **28**, 1406.

Boli, M.; Lehmann, D.; Mathys, B.; Mueller, C.; Nayler, O.; Vteiner, B.; Velker, J. (2008) WO Patent 035,239.

Bora, R.O.; Farooqui, M. (2007) *J. Heterocycl. Chem.* **44**, 645.

Bora, R.O.; Farooqui, M. (2008) *J. Indian Chem. Soc.* **85**, 569.

Boulton, A.J.; Katritzky, A.R.; Hamid, A.M. (1967) *J. Chem. Soc. C*, 2005.

Budriesi, R.; Cosimelli, B.; Ioan, P.; Ugenti, M.P.; Carosati, E.; Frosini, M.; Fusi, F.; Spisani, R.; Sapanara, S.; Cruciani, G.; Novellino, E.; Spinelli, D.; Chiarini, A. (2009) *J. Med. Chem.* **52**, 2352.

Burton, A.J.; Wadsworth, A.H. (2007) *J. Label. Comp. Radiopharm.* **50**, 273.

Buscemi, S.; Cusmano, G.; Gruttadauria, M. (1990) *J. Heterocycl. Chem.* **27**, 861.

Buscemi, S.; Pace, A.; Piccionello, A.P.; Vivona, N. (2006) *J. Fluorine Chem.* **127**, 1601.

Buscemi, S.; Pace, A.; Vivona, N. (2001) *Tetrahedron Lett.* **41**, 7977.

Buscemi, S.; Pace, A.; Vivona, N.; Caronna, T.; Galia, A. (1999) *J. Org. Chem.* **64**, 7028.

Cavalleri, B.; Bellami, P.; Lancini, G. (1973) *J. Heterocycl. Chem.* **10**, 357.

Charton, J.; Deprez-Poulan, R.; Hennuyer, N.; Tailleux, A.; Staels, B.; Deprez, B. (2009) *Bioorg. Med. Chem. Lett.* **19**, 489.

Cheng, Y.; Albrecht, B.K.; Brown, J.; Buchanan, J.L.; Buckner, W.H.; DiMauro, E.F.; Emkey, R.; Fremeau, R.T.; Harmange, J.-Ch.; Hoffman, B.J.; Huang, L.; Huang, M.; Lee, J.H.; Lin, F.-F.; Martin, M.W., Nguen, H.Q.; Patel, V.F.; Tomlinson, S.A.; White, R.D.; Xia, X.; Hitchcock, S.A. (2008) *J. Med. Chem.* **51**, 5019.

Chiou, Sh.; Hoque, A.K.M.M.; Shine, H.J. (1990) *J. Org. Chem.* **55**, 3227.

Chow, Y.L.; Polo, J.S. (1986) *J. Chem. Soc. Perkin Trans. 2*, 727.

Chu, Ch.-M.; Hung, M.-Sh.; Hsien, M.-Ts.; Kuo, Ch.-W.; Suja, T.D.; Song, J.-Sh.; Chiu, H.-H.; Chao, Yu.-Sh.; Shia, K.-Sh. (2008) *Org. Biomol. Chem.* **6**, 3399.

Claisse, F.A.; Foxton, M.W.; Gregory, G.I.; Sheppard, A.H.; Tiley, E.P.; Warburton, W.K.; Wilson, M.J. (1973) *J. Chem. Soc. Perkin Trans. 1*, 2241.

Coley, H.M.; Sarju, J.; Wagner, G. (2008) *J. Med. Chem.* **51**, 135.

Cortes, E.; Ramires, E.P.; Garcia-Mellado de Cortes, O. (2007) *J. Heterocycl. Chem.* **44**, 189.

Cuiman, C.; Duran, J.E.; Fors, K.S.; Hagen, T.J.; Holsworth, D.D.; Jalaie, M.; Leonard, D.M.; Poel, T.-J.; Quin, J.; Take, Y. (2008) WO Patent 117,148.

Da Silva, A.S.; de Silva, M.A.A., Carvalho, C.E.M.; Antunes, O.A.C.; Herrera, J.O.M.; Brinn, I.M.; Mangrich, A.S. (1999) *Inorg. Chim. Acta* **292**, 1.

de Freitas, J.J.R.; de Freitas, J.C.R.; da Silva, L.P.; de Freitas Filho, J.R.; Kimura, G.Y.V.; Srivastava, R.M. (2007) *Tetrahedron Lett.* **48**, 6195.

Deprez, B.; Willand, N.; Dirie, B.; Toto, P.; Villeret, V.; Locht, C.; Baulard, A. (2008) WO Patent 003,861.

DiMauro, E.F.; Buchanan, J.L.; Cheng, A.; Emkey, R.; Hitchcock, S.A.; Huang, L.; Huang, M.Y.; Janosky, B.; Lee, J.H.; Li, X.; Martin, M.W.; Tomlinson, S.A.; White, R.D.; Zheng, X.M.; Patel, V.F.; Fremeau, R.T. (2008) *Bioorg. Med. Chem. Lett.* **18**, 4267.

Dondoni, A.; Barbaro, G. (1975) *Gazz. Chim. Ital.* **105**, 701.

dos Anjos, J.V.; Sinou, D.; Srivastava, R.M.; do Nascimento, S.C.; de Melo, S.J. (2008) *J. Carbohydr. Chem.* **27**, 258.

Du, W.; Truong, Q.; Qi, H.; Guo, Ya.; Chobanian, H.R.; Hagmann, W.K.; Halle, J.J. (2007) *Tetrahedron Lett.* **48**, 2231.

Durust, Ya.; Altung, C.; Kilic, F. (2007) *Phosphorus Sulfur* **182**, 299.

Eloy, F.; Lanaers, R. (1962) *Chem. Rev.* **62**, 155.

Evans, M.D.; Ring, J.; Schoen, A.; Bell, A.; Edwards, P.; Bertelot, D.; Nicewonger, R.; Baldino, C.M. (2003) *Tetrahedron Lett.* **44**, 9337.

Fan, Sh.-Y.; Zeng, Zh.-B.; Mi, Ch.-L.; Zhou, X.-B.; Yan, H.; Gong, Z.-H.; Li, S. (2009) *Bioorg. Med. Chem.* **17**, 621.

Farooqui, M.; Bora, R.; Patil, C.R. (2009) *Eur. J. Med. Chem.* **44**, 794.

Fish, P.V.; Allan, G.A.; Bailey, S.; Blagg, J.; Butt, R.; Collis, M.G.; Greilling, D.; James, K.; Kendall, J.; McElroy, A.; McCleverty, D.; Reed, Ch.; Webster, R.; Whitlock, G.A. (2007) *J. Med. Chem.* **50**, 3442.

Fontana, A.; Guernelli, S.; Lo Meo, P.; Mezzina, E.; Morganti, S.; Noto, R.; Rizzato, E.; Spinelli, D.; Zappacosta, R. (2008) *Tetrahedron* **64**, 733.

Fukunaga, K.; Watanabe, K.; Barneoud, P.; Benavides, J.; Pratt, J. (2009) WO Patent 035,162.

Gal, K.; Weber, Cs.; Wagner, G.A.; Horvath, A.; Nyeki, G.; Vastag, M.; Keserue, G. (2007) WO Patent 039,781.

Gallardo, H.; Cristiano, R.; Vieira, A.A.; Filho, R.A.W.N.; Srivastava, R.M. (2008) *Synthesis*, 605.

Garfinkle, J.; Ezzili, C.; Rayl, T.J.; Hochstatter, D.G.; Hwang, I.; Boger, D.L. (2008) *J. Med. Chem.* **51**, 4392.

Gilchrist, T.L. (1985) *Heterocyclic Chemistry*. Pitman, London.

Gopalakrishnan, M.; Honore, M.P.; Lee, Ch.-H.; Malysz, J.; Ji, J.; Li, T.; Schrimpf, M.R.; Sippy, K.B.; Anderson, D.J. (2008) US Patent 167,286.

Grant, D.; Dahl, R.; Cosford, N.D.P. (2008) *J. Org. Chem.* **73**, 7219.

Hamze, A.; Hernandes, J.-F.; Fulcrand, P.; Martinez, J. (2003) *J. Org. Chem.* **68**, 7316.

Hara, Y.; Saiga, T. (2007) Jpn. Patent 029,793.

Harada, H.; Hattori, K.; Fujita, K.; Morita, M.; Imada, S.; Abe, Y.; Itani, H.; Morokata, T.; Tsutsumi, H. (2007) WO Patent 116,866.

Harada, H.; Hattori, K.; Fujita, K.; Morita, M.; Imada, S.; Abe, Y.; Itani, H.; Morokata, T.; Tsutsumi, H. (2009) US Patent 076,070.

Hermkens, P.H.H.; van Maarseveen, J.H.; Kruse, C.G.; Scheeren, H.W. (1988) *Tetrahedron* **44**, 6491.

Hirawat, S.; Miller, L. (2008) WO Patent 045,566.

Hobson, A.D.; Fix-Stenzel, Sh.; Cusak, K.P.; Bereinlinger, E.C.; Ansell, G.K.; Stoffel, R.H. (2008) WO Patent 076,356.

Huhtiniemi, T.; Suuronen, T.; Rinne, V.M.; Wittekindt, C.; Lahtela-Kakkonen, M.; Jarho, E.; Wallen, E.A.A.; Salminen, A.; Poso, A.; Leppaenen, J. (2008) *J. Med. Chem.* **51**, 4377.

Ispikoudi, M.; Litinos, K.E.; Fylaktakidou, K.C. (2008) *Heterocycles* **75**, 1321.

Itoh, K.-i; Sakamaki, H.; Horiuchi, C.A. (2005) *Synthesis*, 1935.

Jakopin, Z.; Roskar, R.; Dolenc, M.S. (2007) *Tetrahedron Lett.* **48**, 1465.

Ji, J.; Lee, Ch.-L.; Sippy, K.B.; Li, T.; Gopalakrishnan, M. (2008) US Patent 269,236.

Jochims, J.S. (1996) In: *Comprehensive Heterocyclic Chemistry II*, Vol. 4, p. 179. Katritzky, A.R.; Rees, C.W.; Scriven, E.F.V. (Eds.). Pergamon, Oxford.

Johnson, J.J.; Nwoko, D.; Hotema, M.; Sanchez, N.; Alderman, R.; Lynch, V. (1996) *J. Heterocycl. Chem.* **33**, 1583.

Kaboudin, B.; Saadati, F. (2007) *Tetrahedron Lett.* **48**, 2829.

Kalvins, I.; Adrianov, V.; Shestakova, I.; Kanepe, I.; Domracheva, I. (2001) WO Patent 021,585.

Katritzky, A.R.; Shestopalov, A.A.; Suzuki, K.S. (2005) *ARKIVOC* (vii), 36.

Katritzky, A.R.; Wallis, B.; Brownlee, R.T.C.; Topsom, R.D. (1965) *Tetrahedron* **21**, 1681.

Kawashima, E.; Tabei, K. (1986) *J. Heterocycl. Chem.* **23**, 1657.

Keil, S.; Bernardelli, P.; Urmann, M.; Matter, H.; Wendler, W.; Glien, M.; Chandross, K.; Lee, L. (2007a) WO Patent 039,178.

Keil, S.; Urmann, M.; Bernardelli, P.; Glien, M.; Wendler, W.; Chandross, K.; Lee, L. (2007b) WO Patent 039,177.

Keil, S.; Urmann, M.; Glien, M.; Wendler, W.; Chandross, K.; Lee, L. (2007c) WO Patent 039,176.

Keil, S.; Wendler, W.; Glien, M.; Goerlitzer, J.; Chandross, K.; McGarrgy, D.G.; Merrill, J.; Bernardelli, P.; Ronan, B.; Terrier, C. (2005) WO Patent 097,786.

Kmetic, M.; Stanovnik, B. (1995) *J. Heterocycl. Chem.* **32**, 1563.

Kobayashi, T.; Nakahira, H. (2008) Jpn. Patent 156,318.

Koryakova, A.G.; Ivanchenkov, Ya.A.; Ryzhova, E.A.; Bulanova, E.A.; Karapetian, R.N.; Mikitas, O.V.; Katrukha, E.A.; Karey, V.I.; Okun, I.; Kravchenko, D.V.; Lavrovsky, Ya.V.; Korzinov, O.M.; Ivachtchenko, A.V. (2008) *Bioorg. Med. Chem. Lett.* **18**, 3661.

Koufaki, M.; Kiziridi, Ch.; Nikoloudaki, F.; Alexis, M.N. (2007) *Bioorg. Med. Chem. Lett.* **17**, 4223.

Kukushkin, V.Yu.; Pombeiro, A.J.L.; Frausto da Silva, J.J.R.; Wagner, G. (2001) WO Patent 098,283.

Kumar, R.R.; Perumal, S. (2007) *Tetrahedron* **63**, 7850.

Kurz, Th.; Lolak, N.; Geffken, D. (2007) *Tetrahedron Lett.* **48**, 2733

Kuznetsov, M.L.; Kukushkin, V.Yu. (2006) *J. Org. Chem.* **71**, 582.

Lasri, J.; Charmier, M.A.J.; da Silva, M.F.C.G.; Pombeiro, A.J.L. (2007) *Dalton Trans.*, 3259.

Learmonth, D.A.; Kiss, L.E.; Palma, P.N.L.; Ferreira, H.D.S.; Silva, P.M.V. (2008) WO Patent 094,053.

Lee, H.G.; Kim, S.Y.; Kim, D.S.; Seo, S.R.; Lee, S.-I.; Shin, D.M.; de Smet, P.; Seo, J.T. (2009) *J. Neurosci. Res.* **87**, 269.

Lenaers, R.; Eloy, R. (1963) *Helv. Chim. Acta* **43**, 1067.

Lo Celso, F.; Pibiri, I.; Triolo, A.; Triolo, R.; Pace, A.; Buscemi, S.; Vivona, N. (2007) *J. Mater. Chem.* **17**, 1201.

Makarycheva-Mikhailova, A.V.; Golenetskaya, J.A.; Bokach, N.A.; Balova, I.A.; Haukka, M.; Kukushkin, V.Yu. (2007) *Inorg. Chem.* **46**, 8328.

Mandracchia, D.; Piccionello, A.P.; Pitaressi, G.; Pace, A.; Buscemi, S.; Giammona, G. (2007) *Macromol. Biosci.* **7**, 836.

McCluskey, A.; Wentrup, C. (2008) *J. Org. Chem.* **73**, 6265.

Merla, B.; Oberboersch, S.; Sundermann, B.; Englberger, W.; Hennies, H.-H.; Graubaum, H. (2007) Germ. Patent 102,005,061,427.

Moloney, G.P.; Angus, J.A.; Robertson, A.D.; Stoermer, M.J.; Robinson, M.; Wright, Ch.E.; McRae, K.; Christopoulus, A. (2008) *Eur. J. Med. Chem.* **43**, 513.

Moloney, G.P.; Robertson, A.D. (2002) WO Patent 036,590.

Morrocchi, S.; Ricca, A.; Velo, L. (1967) *Tetrahedron Lett.* **8**, 331.

Moussebois, C.; Eloy, F. (1964) *Helv. Chim. Acta* **47**, 838.

Moussebois, C.; Lenaers, R.; Eloy, F. (1962) *Helv. Chim. Acta* **45**, 446.

Muraglia, E.; Altamura, S.; Branca, D.; Cecchetti, O.; Feringo, F.; Orsale, M.V.; Palumbi, M.C.; Rowley, M.; Scarpelli, R.; Steinkuehler, Ch.; Jones, Ph. (2008) *Bioorg. Med. Chem. Lett.* **18**, 6083.

Nabih, K.; Baouid, A.; Hasnaoui, A.; Kenz, A. (2004) *Synth. Commun.* **34**, 3565.

Naidu, B.N. (2008) *SYNLETT*, 547.

Naidu, B.N.; Sorenson, M.E. (2005) *Org. Lett.* **7**, 1391.

Neidelin, R.; Kramer, W.; Li, S. (1998) *J. Heterocycl. Chem.* **35**, 161.

Newman, H. (1968) *Tetrahedron Lett.* **9**, 2417.

Nishi, T.; Nakamura, T.; Sekiguchi, Y.; Mizuno, Y.; Shimozato, T.; Nara, F. (2008a) US Patent 113,961.

Nishi, T.; Nakamura, T.; Sekiguchi, Y.; Mizuno, Y.; Shimozato, R.; Nara, F. (2008b) Jpn. Patent 120,794.

Ono, M.; Haratake, M.; Saji, H.; Nakayama, M. (2008) *Bioorg. Med. Chem.* **16**, 6867.

Oussaid, B.; Moeini, L.; Martin, B.; Villemin, D.; Garrigues, B. (1995) *Synth. Commun.* **25**, 1415.

Outirite, M.; Lebrini, M.; Lagrenee, M.; Bentiss, F. (2007) *J. Heterocycl. Chem.* **44**, 1529.

Pace, A.; Pibiri, I.; Piccionello, A.P.; Buscemi, S.; Vivona, N.; Barone, G. (2007) *J. Org. Chem.* **72**, 7656.

Pace, A.; Pierro, P.; Buscemi, S.; Vivona, N.; Barone, G. (2009) *J. Org. Chem.* **74**, 351.

Parra, M.; Hidalgo, P.; Barbera, J.; Alderete, J. (2005) *Liq. Crystl.* **32**, 573.

Parra, M.; Hidalgo, P.; Carrasco, E.; Barbera, J.; Silvino, L. (2006) *Liq. Crystl.* **33**, 875.

Parra, M.; Hidalgo, P.; Elgueta, E.Y. (2008) *Liq. Crystl.* **35**, 823.

Pelter, A.; Sumengen, D. (1977) *Tetrahedron Lett.* **18**, 1945.

Piccionello, A.P.; Pace, A.; Buscemi, S.; Vivona, N.; Giorgi, G. (2009a) *Tetrahedron Lett.* **50**, 1472.

Piccionello, A.P.; Pace, A.; Buscemi, S.; Vivona, N.; Pani, M. (2008) *Tetrahedron* **64**, 4004.

Piccionello, A.P.; Pace, A.; Pierro, P.; Pibiri, I.; Buscemi, S.; Vivona, N. (2009b) *Tetrahedron* **65**, 119.

Piccionello, A.P.; Pibiri, I.; Pace, A.; Raccuglia, R.A.; Buscemi, S.; Vivona, N.; Giorgi, G. (2007) *Heterocycles* **71**, 1529.

Pipik, B.; Ho, G.-J.; Williams, J.M.; Conlon, D.A. (2004) *Synth. Commun.* **34**, 1863.

Plate, R.; Hermkens, P.H.H.; Smits, J.M.M.; Nivard, R.J.F.; Ottenheijm, H.C.J. (1987) *J. Org. Chem.* **52**, 1047.

Plenkiewicz, J.; Zdrojewski, T. (1987) *Bull. Soc. Chim. Belg.* **96**, 675.

Polienko, J.F.; Schanding, Th.; Gatiolov, Yu.V.; Grigor'ev, I.A.; Voinov, M.A. (2008) *J. Org. Chem.* **73**, 502.

Quan, Ch.; Kurth, M. (2004) *J. Org. Chem.* **69**, 1470.

Rajanarendar, E.; Mohan, G.; Ramesh, P.; Rao, E.K. (2006) *Heterocycl. Commun.* **12**, 431.

Richardson, Ch.; Steel, P.J. (2007) *Inorg. Chem. Commun.* **10**, 884.

Roberts, L.R.; Bryans, J.; Conlon, K.; McMurray, G.; Stobie, A.; Whitlock, G.A. (2008) *Bioorg. Med. Chem. Lett.* **18**, 6437.

Roppe, J.; Smith, N.D.; Huang, D.; Tehrani, L.; Wang, B.; Anderson, J.; Brodkin, J.; Chang, J.; Fiang, X.; Kong, Ch.; Munoz, B.; Varney, M.A.; Prasit, P.; Cosford, N.D.P. (2004) *J. Med. Chem.* **47**, 4645.

Ruccia, M.; Vivona, N.; Cusmano, G. (1974) *Tetrahedron* **30**, 3859.

Sakai, D.; Watanabe, K. (2009) WO Patent 035,159.

Sakamoto, T.; Cullen, M.D.; Hartman, T.L.; Watson, K.M.; Buckheit, R.W.; Pannecouque, Ch.; de Clerq, E.; Cushman, M. (2007) *J. Med. Chem.* **50**, 3314.

Sarju, J.; Arbour, J.; Sayer, J.; Rohrmoser, B.; Scherer, W.; Wagner, G. (2008) *Dalton Trans.* 5302.

Shablykhin, O.V.; Brovarets, V.S.; Drach, B.S. (2007) *Zh. Obshch. Khim.* **77**, 842.

Silva, P.M.V.; Learmonth, D.A.; Kiss, L.E. (2007) Eur. Patent 1,845,097.

Sureshbabu, V.V.; Hemantha, H.P.; Naik, Sh. A. (2008) *Tetrahedron Lett.* **49**, 5133.

Svenstrup, N.; Zimmermann, H.; Karthaus, D.; Goeller, A.; Henninger, K.; Lang, D.; Paulsen, D.; Riedl, B.; Schohe-Loop, R.; Schuhmacher, J.; Wunberg, T. (2007) Germ. Patent 102,006,009,928.

Tafesse, L. (2007) WO Patent 057,229.

Takahashi, M. (2007) Jpn. Patent 084,449.

Tiemann, F.; Krueger, P. (1884) *Chem. Ber.* **17**, 1685.

Tiwari, Sh.B.; Kohli, D.V. (2008) *Med. Chem. Res.* **17**, 386.

Touaibia, M.; Djimde, A.; Cao, F.; Boilard, E.; Bezzine, S.; Lambeau, G.; Redeuilh, C.; Lamouri, A.; Massicot, F.; Chau, F.; Dong, Ch.-Zh.; Heymans, F. (2007) *J. Med. Chem.* **50**, 1618.

Toulmin, A.; Wood, J.M.; Kenny, P.W. (2008) *J. Med. Chem.* **51**, 3720.

Trifonov, R.E.; Volovodenko, A.P.; Vergizov, S.N.; Shirinbekov, N.I.; Gindin, V.A.; Koren, A.O.; Ostrovskii, V.A. (2005) *Helv. Chim. Acta* **88**, 1790.

Tsuge, O.; Urano, S.; Oe, K. (1980) *J. Org. Chem.* **45**, 5130.

Tyrkov, A.G. (2004) *Izv. Vyssh. Uchebn. Zaved., Khim. Khim. Tekhnol.* **47**, 103.

Tyrrell, E.; Allen, J.; Jones, K.; Beauchet, R. (2005) *Synthesis* 2393.

Vega, I.; Morris, W.; D'Accorso, N. (2006) *React. Funct. Polym.* **66**, 1609.

Vu, C.B.; Corpuz, E.G. (1999) *J. Med. Chem.* **42**, 20.

Waddell, Sh.T.; Balcovec, J.M.; Kevin, N.J.; Gu, X. (2007) WO Patent 047,625.

Wagner, G. (2004) *Inorg. Chim. Acta* **357**, 1320.

Wagner, G.; Garland, T. (2008) *Tetrahedron Lett.* **49**, 3596.

Wagner, G.; Haukka, M.; Frausto da Silva, J.J.R.; Pombeiro, A.J.L.; Kukushkin, V.Yu. (2001) *Inorg. Chem.* **40**, 264.

Wang, Y.; Miller, R.L.; Sauer, D.R.; Djuric, S.W. (2005) *Org. Lett.* **7**, 925.

Wang, Y.-G.; Xu, W.-M.; Huang, X. (2007) *J. Combin. Chem.* **9**, 513.

Williams, D.J.; Dimmic, M.W.; Haakenson, W.P.; Wideman, A.; Shortt, B.J.; Cheeseright, T.; Crawford, M.J. (2009) WO Patent 023,721.

Wustrow, D.J.; Belliotti, Th.R.; Capiris, Th.; Kneen, C.O.; Bryans, J.S.; Field, M.J.; Williams, D.; El-Kattan, A.; Buchholz, L.; Kinsora, J.J.; Lotarski, S.M.; Vartanian, M.G.; Taylor, Ch.P.; Donevan, S.D.; Thorpe, A.J.; Schwarz, J.B. (2009) *Bioorg. Med. Chem.* **19**, 247.

Xu, Sh.; Ni, J.; Guo, L.; Ma, H.; Wang, F.; Diao, Y.; Wan, Q. (2007) Chin. Patent 1,927,860.

Yan, L.; Huo, P.; Hale, J.J.; Mills, S.G.; Hajdu, R.; Keohane, C.A.; Rosenbach, M.J.; Milligan, J.A.; Shei, G.-J.; Chrelet, G.; Bergstrom, J.; Card, D.; Mandala, S.M. (2007) *Bioorg. Med. Chem. Lett.* **17**, 828.

Yarovenko, V.N.; Taralashvili, V.K.; Zavarzin, I.V.; Krayushkin, M.M. (1990) *Tetrahedron* **46**, 3941.

Yu, Ch.; Lei, M.; Su, W.; Xie, Y. (2007) *Synth. Commun.* **37**, 4439.

Zablocki, J.; Kalla, R.; Perry, T.; Palle, V.; Varkhedkar, V.; Xiao, D.; Piscopio, A.; Maa, T.; Gimbell, A.; Hao, J.; Chu, N.; Leung, K.; Zeng, D. (2005) *Bioorg. Med. Chem. Lett.* **15**, 609.

Zbinden, K.G.; Haap, W.; Hilpert, H.; Kuhn, B.; Panday, N.; Riklin, F. (2008) US Patent 103,143.

Zhang, Y.-X.; Sasaki, K.; Hirota, T. (1999) *J. Heterocycl. Chem.* **36**, 787.

Zhao, G.; Jing, Y.; Li, Ch.; Wang, R.; Zhang, Y.; Song, D.; Lou, H. (2008) Chin. Patent 101,108,832.

1.2.2   1,2,4-OXADIAZOLE N-OXIDES

The 1,2,4-oxadiazole contains two nitrogen atoms, in positions 2 and 4. Accordingly, there are two different oxides, namely, 1,2,4-oxadiazole 2-oxide and 1,2,4-oxadiazole 4-oxide. As to the 2-oxides, there is only one mention in a 1923 unapproachable publication by Parisi (see Ref. 2 in the review by Quadrelli and Caramella 2007), with full silence in the following literature. In contrary, the preparation and some reactions of the 4-oxides are documented although their fine structure and chemistry wait for extended investigation.

All the N_4-oxides can be considered as 1,2,4-oxadiazoles with the added atomic oxygen to the lone pair of electrons at N_4 of the heterocycle. Formally, the N-oxide group is neutral. However, nitrogen and oxygen possess positive and negative charges. The correct representation of the N-oxide functional is N^+-O^-. This representation is used in the following text.

1.2.2.1 Formation

Although 1,2,4-oxadiazole oxides correspond to N-oxidation products, they cannot be obtained via oxidation of the "parent" oxadiazoles. There are many reactions leading to these N-oxides, but they tend to mixtures of products. The N-oxides occur to be minor ones. Therefore, the most interesting are methods providing 1,2,4-oxadiazole oxides as the predominant products with the yield not lower than 70%.

1.2.2.1.1 From Nitrosolates

One such method consists of the reaction between potassium ethyl nitrosolate and substituted benzyl halides (Wiemer and Leonard 1976) (Scheme 1.63).

1.2.2.1.2 From Nitrile Oxides

Another effective method consists of dimerization of stabilized nitrile oxides. (The unstable nitrile oxides dimerize to furoxans just after formation, in situ.) The nitrile oxide of Scheme 1.64 is stable. Being activated by boron trifluoride etherate, it dimerizes to give the corresponding disubstituted 1,2,4-oxadiazole 4-oxide. The reaction proceeds during heating in benzene or in dichloromethane (Krayushkin et al. 1990).

1.2.2.2 Fine Structure

Yu et al. (2003) compared 1,2,4-oxadiazole 4-oxide and 1,2,5-oxadiazole 5-oxide within the density functional theory approach. Starting from the same nitrile oxide, activation energy of the 1,2,4-isomer formation was turned out to be 102.2 kJ/mol, which is 54.4 kJ/mol higher than that of the 1,2,5-isomer formation. When the 1,2,5-oxadiazole 5-oxide formation is advantageous kinetically, the formation of 1,2,4-oxadiazole 4-oxide is profitable thermodynamically. This is because the formation of the latter should be more exothermic than the formation of the former by 66.6 kJ/mol (Yu et al. 2003).

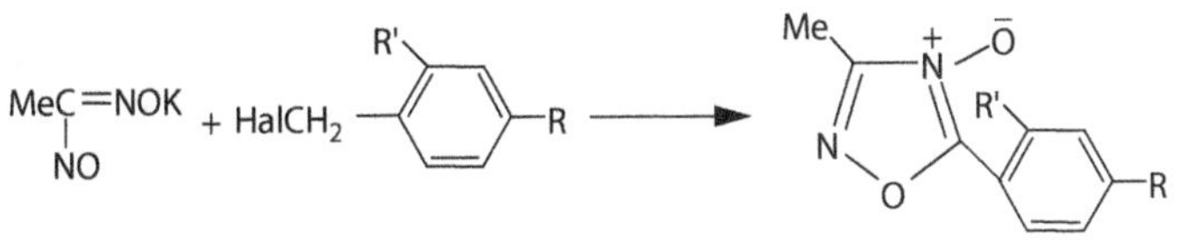

SCHEME 1.63

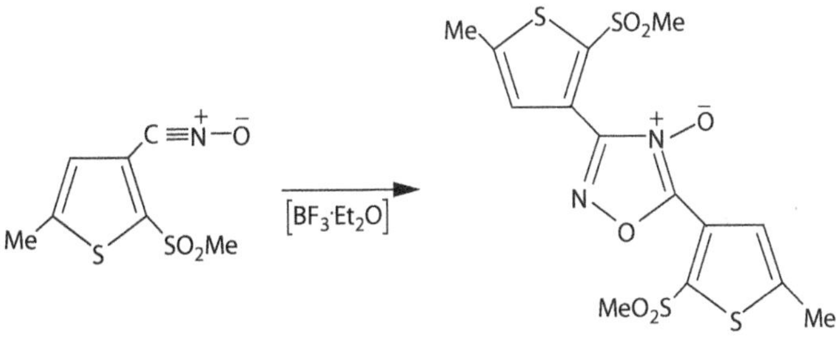

SCHEME 1.64

1.2.2.3 Reactivity

1.2.2.3.1 Salt Formation

The 1,2,4-oxadiazole ring is a basic moiety. As for N-oxides, their basic properties are understandably weakened. However, the oxides are capable of forming salts with hydrogen halides or Lewis acids. Upon action of slight bases such as alkali hydrogen carbonates, these salts convert into the parent N-oxides.

1.2.2.3.2 Deoxygenation

Deoxygenation can be considered as reduction of 1,2,4-oxadiazole oxides. As a matter of fact, zinc in acetic acid (Wieland and Bauer 1906) really reduced these oxides into the corresponding 1,2,4-oxadiazoles. As it has been shown by Quadrelli et al. (1999), 3,5-diaryl-1,2,4-oxadiazole 4-oxides undergo photolytic deoxygenation to 3,5-diaryl-1,2,4-oxadiazoles in a methanol degassed solution containing triethylamine. The reaction consists in electron-transfer reduction from the amine upon photoexcitation, as it is usual for photo-deoxygenation of heterocyclic N-oxides (see, e.g., Fasani et al. 1994).

Such chemicals as benzonitrile oxide or trialkylphosphites take the N-oxide oxygen away and are oxidized themselves. Scheme 1.65 outlines a sacrificial role of benzonitrile oxide: It adds to the N-oxide oxygen and then leaves with this oxygen thereby performing deoxygenation. The reaction proceeds at room temperature in methanolic solution and is completed in two-day standing (Quadrelli et al. 1997).

Upon action of trialkylphosphites, 1,2,4-oxadiazole 4-oxides transform into the corresponding oxadiazoles during 2 h refluxing of their solutions in benzene. As this takes place, the phosphites are converted into trialkyl phosphates (Boulton et al. 1969, Quadrelli et al. 1997) (Scheme 1.66).

3,5-Diaryl-1,2,4-oxadiazole 4-oxides react with trimethyl phosphite in the same manner. However, the 3,5-dimesityl derivative is an exception: Due to steric shielding from the dimesityl substituents, this compound does not undergo deoxygenation under the conditions mentioned for Scheme 1.66 (Quadrelli et al. 2000). Interestingly, 5-phenyl-1,2,4-oxadiazole 4-oxide docked at the Wang resin at the position 3 of the heterocyclic ring is easily deoxygenated by triethylphosphite to give the corresponding

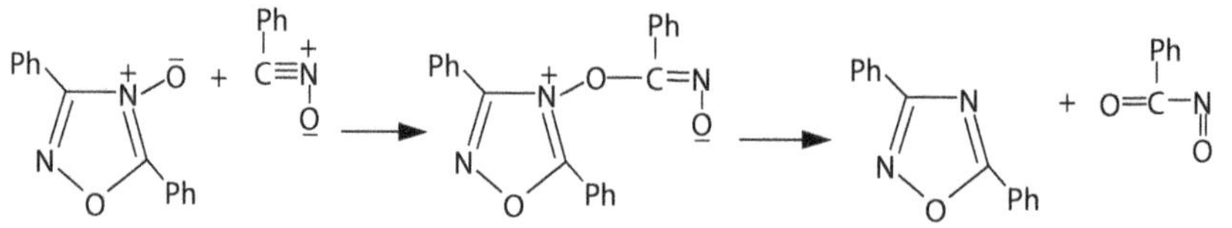

SCHEME 1.65

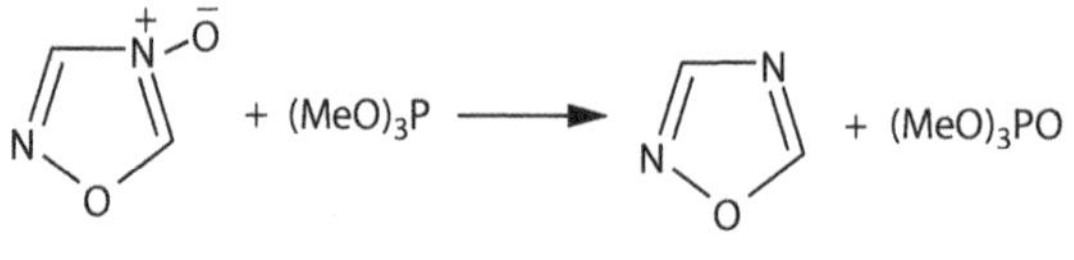

SCHEME 1.66

1,2,4-oxadiazole. Because 5-phenyl-1,2,4-oxadiazole 4-oxide is bound with the Wang resin through the phenyl-carboxyl (ester) group, the cleavage off the resin under trans-esterification conditions affords 3-(4-carbomethoxyphenyl)-5-phenyl-1,2,4-oxadiazole. Although the yield of this product was small (only 17%), the reaction remains to serve as a demonstrative one (Quadrelli et al. 2005).

1.2.2.3.3 Photolysis

Light sensitivity of 1,2,4-oxadiazole 4-oxides was noticed for both solid state and solution. Therefore, the corresponding materials should be kept in the dark to prevent degradation (de Sarlo and Guarna 1976). The 4-oxides absorb low-frequency ultraviolet light with $\lambda = 320$–350 nm and are prone to lose the N-oxide oxygen, forming 1,2,4-oxadiazoles (Wiemer and Leonard 1976).

According to Quadrelli et al. (1999), 3,5-diaryl-1,2,4-oxadiazoles are not single products forming during photolysis of the corresponding 4-oxides in methanol. Scheme 1.67 outlines the full product contents. The authors suggest the origin of each one; here, it is reasonable to direct this topic to the original paper.

Quadrelli et al. (2005) studied photolytic splitting of the 1,2,4-oxadiazole 4-oxide docked at the Wang resin through the carboxy phenyl substituent (Scheme 1.68). The reactions were conducted in methanolic suspension at room temperature. The top reaction depicted in Scheme 1.68 shows the results of 2 h irradiation by an ultraviolet lamp. The nitrosocarbonyl was generated on the resin while the benzonitrile was released in solution. The bottom reaction in Scheme 1.68 shows the results of 2 h

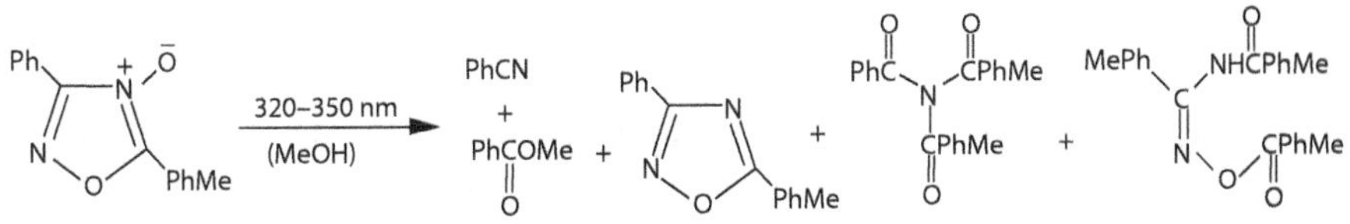

SCHEME 1.67

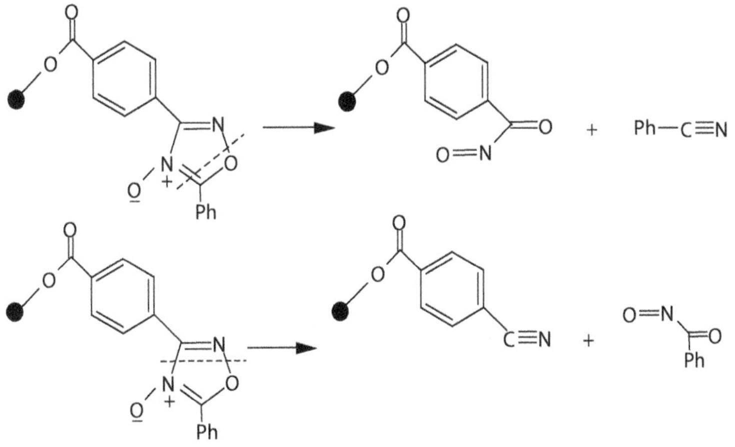

SCHEME 1.68

SCHEME 1.69

exposure to sunlight. In this case, the nitrosocarbonyl was released in solution while the benzonitrile was generated on the resin. Consequently, the splitting direction depends on an intensity of photoexcitation. It is significant that the resin-supported reactions did not lead to the deoxygenation of the N-oxide fragment as it does takes place upon light-irradiation of 1,2,4-oxadiazole 4-oxides undocked at the resin (cf. Scheme 1.67). Consequently, the resin-docking changes the reactivity of the oxides. This is some kind of confinement effect.

1.2.2.3.4 Thermolysis

Four-hour refluxing of 3,5-diaryl-1,2,4-oxadiazole 4-oxides in chlorobenzene (132°C) results in ring cleavage with the formation of arylnitrile and aroyl anhydride (Scheme 1.69). Thermolysis of 3,5-dimesityl-1,2,4-oxadiazole 4-oxide proceeds in the same way, but under softer conditions, at 80°C during 2 h refluxing in benzene (Quadrelli et al. 2000). The relatively easy cleavage of this compound is definitely caused by bulkiness of the aryl substituent: With bulky substituents, the thermal decomposition usually proceeds easier.

1.2.2.4 Conclusion

1,2,4-Oxadiazole oxides are a particular family of heterocycles closely related to the chemistry of nitrile oxides. The chemistry of these oxides differs with the fragility of their heterocyclic ring capable of transforming into the nitrolic acid and nitrosocarbonyl species upon photoexcitation or thermal initiation. Trapping such species can open routes to new medications.

REFERENCES TO SECTION 1.2.2

Boulton, A.J.; Hadjimihalakis, P.; Katritzky, A.R.; Majid-Hamid, A. (1969) *J. Chem. Soc. C,* 1901.
de Sarlo, F.; Guarna, A. (1976) *J. Chem. Soc., Perkin Trans. 1,* 1825.
Fasani, E.; Amer, A.M.; Albini, A. (1994) *Heterocycles* **37**, 985.
Krayushkin, M.M.; Kalik, M.A.; Loktionov, A.A. (1990) *Khim. Geterotsikl. Soed.* 909.
Parisi, E. (1923) *Atti Accad. dei Lincei, Sci. Fiz. Mat. Nat. Rendiconti.* **32**, 572.
Quadrelli, P.; Campari, G.; Mella, M.; Caramella, P. (2000) *Tetrahedron Lett.* **41**, 2019.
Quadrelli, P.; Caramella, P. (2007) *Curr. Org. Chem.* **11**, 959.
Quadrelli, P.; Invernizzi, A.G.; Falzoni, M.; Caramella, P. (1997) *Tetrahedron* **53**, 1787.
Quadrelli, P.; Mella, M.; Caramella, P. (1999) *Tetrahedron Lett.* **40**, 797.
Quadrelli, P.; Scrocchi, R.; Piccanello, A.; Caramella, P. (2005) *J. Comb. Chem.* **7**, 887.
Wieland, H.; Bauer, H. (1906) *Chem. Ber.* **39**, 1480.
Wiemer, D.F.; Leonard, N.J. (1976) *J. Org. Chem.* **41**, 2985.
Yu, Z.-X.; Caramella, P.; Houk, K.N. (2003) *J. Am. Chem. Soc.* **125**, 15420.

1.3   1,2,5-OXADIAZOLES AND 1,2,5-OXADIAZOLE N-OXIDES

This section operates with trivial names furazan and furoxan. Furazan means 1,2,5-oxadiazole. 1,2,5-Oxadiazole 2 (or 3)-oxide is often discussed under the name furoxan. Accordingly, benzofurazan means benzo[1,2-c]-[1,2,5]-oxadiazole and benzofuroxan means benzo [1,2-c]-[1,2,5]-oxadiazole 2 (or 3)-oxide. Furazans are considered in Section 1.3.1, while Section 1.3.2 is devoted to furoxans.

All the compounds of the furoxan class bear the N-oxide functional group as a result of the addition of atomic oxygen to the lone pair of electrons at one of the cyclic nitrogen atoms. Although N-oxide group is neutral, there is charge splitting between nitrogen and oxygen and the correct representation of the N-oxide functional is N^+–O^-. It is the representation that accepted in Sections 1.3.1 and 1.3.2.

Contrary to other heterocyclic N-oxides, furoxan or benzofuroxan cannot be formed by direct oxidation of the parent furazan or benzofurazan. These N-oxides are the products of special synthesis. Because of difference in their reactivity, furazans and furoxans are considered separately, in the framework of Sections 1.3.1 and 1.3.2, respectively.

1.3.1   1,2,5-OXADIAZOLES (FURAZANS)

Due to significant difference between monocyclic 1,2,5-oxadiazoles and 1,2,5-oxadiazoles fused with aromatic fragments, their formation, structure and chemistry will be considered separately.

1.3.1.1   Formation

1.3.1.1.1   Formation of Monocyclic Furazans

1.3.1.1.1.1   From vic-Dioximes   Dehydration of vicinal dioximes is the most frequently used method to prepare 1,2,5-oxadiazoles (see, e.g., patent by Yadav et al. 2006). Various reagents have been used to achieve the dehydration, including the anhydrides of acetic, phthalic, and succinic acids, or sulfuric acid, or aqueous ammonia, or alcoholic sodium hydroxide, or urea, or dicyclohexyl carbodiimide, or phenyliodine diacetate. The dehydration with succinic anhydride (co-melting) and with acetic anhydride (100°C, 1 h) can be applied to prepare unsubstituted and substituted furazans (Olofson and Michelman 1965). Only disubstituted furazans can be obtained by dehydration with the help of alkaline hydroxides. Furazan itself and its monoderivatives are unstable under alkaline conditions.

Prolong heating of vic-dioximes with silica gel was also reported to give furazans with good yields (Kamitori 1999). Dehydration of vic-dioximes with triphenylphosphine and diisopropyl azo diformate in refluxing toluene is another effective method (Tron et al. 2005). Sometimes, thionyl chloride can be used as a mild dehydrating agent. For instance, thermally labile acenaphtho[1,2-c]-[1,2,5]-oxadiazole had been prepared by that method (Boulton and Mathur 1973).

A somewhat unusual route to furazans consists of reaction of vic-dioximes with acetylene. The work of Zaitsev et al. (2006) gives an illustration (see Scheme 1.70). The initially formed O-vinyloxime may be involved in the cyclization according to Scheme 1.71. The reaction proceeds at 100°C, 1 h, 14 atm of HC≡CH, the furazan yield is 22%.

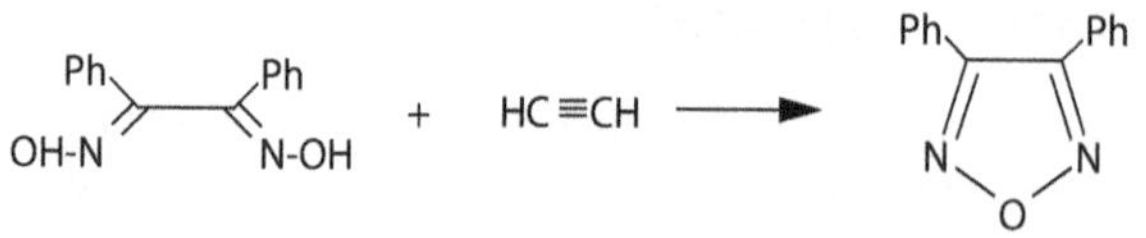

SCHEME 1.70

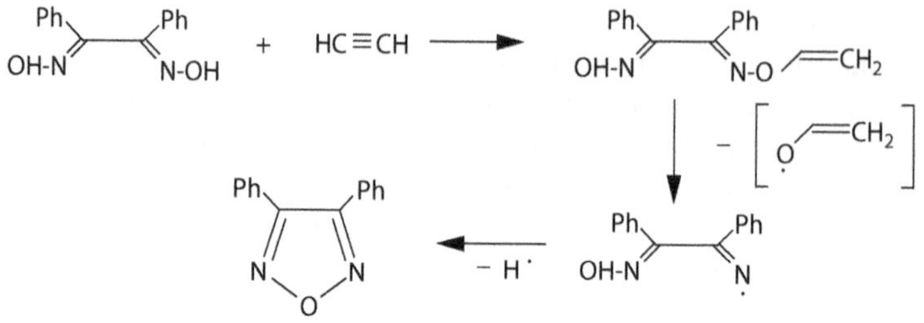

SCHEME 1.71

1.3.1.1.1.2 From Hydroxyamidines By refluxing with EtONa in EtOH or by reaction with POCl₃, hydroxyamidine derivative of Scheme 1.72 gives rise to the 4-substituted 3-aminofurazan. Upon refluxing in Ac₂O, the 3-(*N*-acetylamino)-furazan derivative occurs to be a product, whereas with ClCOOMe in Et₃N/THF *N*-acylated sulfonamide was obtained (Winterweber et al. 2006).

1.3.1.1.1.3 From Other Heterocycles The rearrangement of Z-1-(1,2,4-oxadiazol-3-yl)alkanone oximes is an effective route to 1,2,5-oxadiazole-3-amines. The Z-oximes usually rearrange spontaneously. The E-isomers are stable, but isomerize

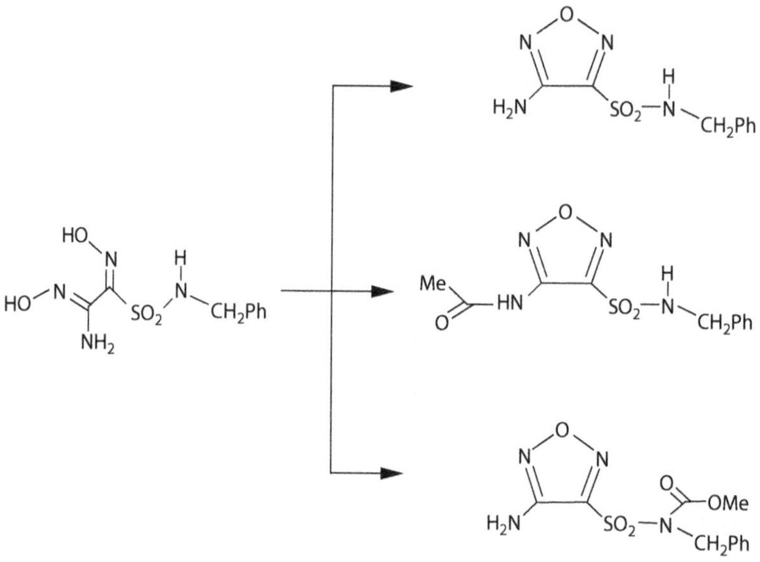

SCHEME 1.72

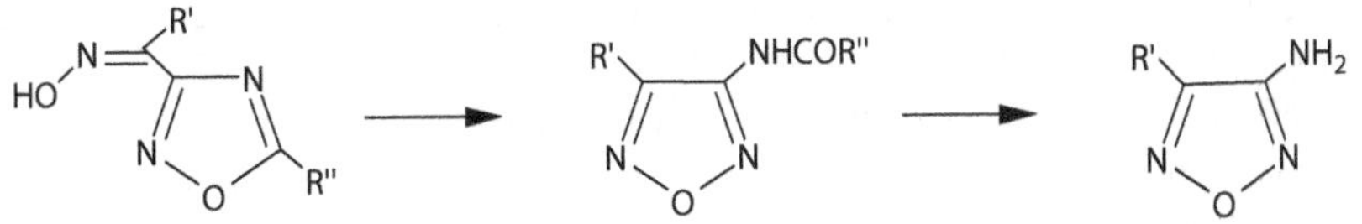

SCHEME 1.73

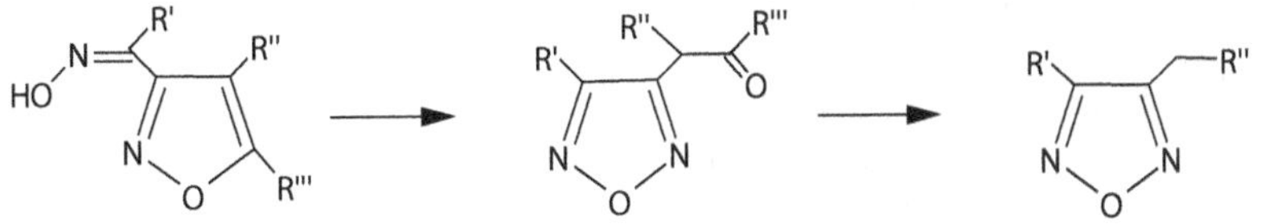

SCHEME 1.74

to Z-isomer by the addition of acid. And the acid also hydrolyzes the intermediary amides to produce the aimed amino compounds (Scheme 1.73) (Vivona et al. 1985).

The oximes of 1-(isoxazol-3-yl)alkanones rearrange into (2-oxoalkyl)-1,2,5-oxadiazoles in the presence of bases. Upon these conditions, deacylation of the product initially formed takes place, as it is depicted by Scheme 1.74 (Ruccia et al. 1981).

1.3.1.1.2 Formation of Fused Furazans

1.3.1.1.2.1 From vic-Quinone Dioximes The method consists in dehydration of arene-1,2-quinone dioximes upon action aqueous alkali, thionyl chloride, acetic anhydride, phenyl isocyanate, acyl chlorides, and sulfuric acid. Cyclization can also be achieved by thermolysis of the oxime diacetates or dibenzoate (Sliwa 1984).

1.3.1.1.2.2 From Aromatic Nitro, or Nitroso, or Amino Azides Vicinal compounds of these classes have been used to prepare substituted arenefurazans (di Nunno et al. 1973, Takakis and Hadjimihalakis 1990).

1.3.1.1.2.3 From vic-Dinitro Arenes This method required double molar amounts of sodium azide as regards the substrate. One mol of the reactant is consumed in substitution of the azido group for the nitro group that results in the formation of the corresponding furoxan. The second mole of sodium azide permits reduction of the furoxan to furazan (Scheme 1.75) (di Nunno et al. 1973).

1.3.1.2 Fine Structure

1.3.1.2.1 Monocyclic Furazans

The structure of the parent unsubstituted 1,2,5-oxadiazole has been determined by microwave spectroscopy (Stiefvater 1988), and its derivatives have been extensively

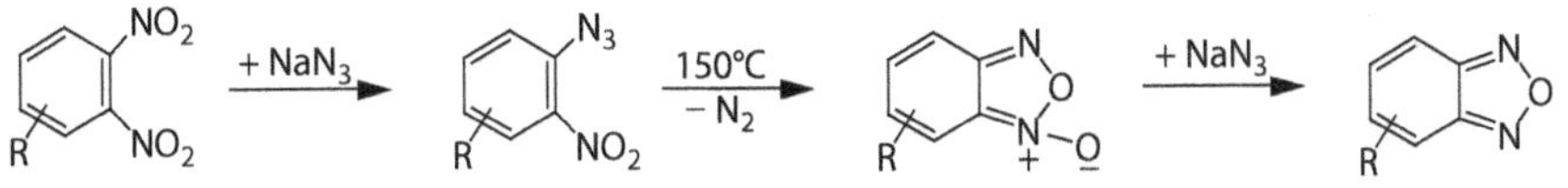

SCHEME 1.75

examined by the x-ray method. The ring has C_{2v} symmetry, and is essentially planar. The π-bond ("double" bond) orders are typically 70%–80% for C_3–N_2 and C_4–N_5, and 40%–50% for C_3–C_4, magnitudes that are consistent with significant delocalization of π electrons. In contrast, the N_2–O_1 and N_5–O_1 bonds are really ordinary bonds. Based on the bond lengths in 1,2,5-oxadiazole, Bird (1985) calculated its aromaticity index as $I_A = 43$. Based on experimental heat of formation, the united aromaticity index (I_U) was estimated to be 53 (Bird 1992). (Benzene with its $I_A = I_U = 100$ is a reference compound.) This "partial" aromaticity appears in aminofurazan ability of forming the corresponding diazonium salts upon the action of sodium nitrite and sulfuric and acetic acids (Rakitin et al. 1993, Tselinskii et al. 2001). As known, only aromatic amines are capable of forming more or less stable diazonium salts.

As a substituent, 1,2,5-oxadiazole ring possesses weak mesomeric and marked electron-withdrawing properties. The electron-withdrawing properties are reflected via facilitation of lithiation at the 3-methyl group (Micetich 1970) and via a marked decrease in basicity of the 3-amino group: This amino group forms salts with anhydrous mineral acids only (Sheremetev and Makhova 2001).

According to results of electrophilic reactions with participation of 3-substituted 4-phenylfurazan, the weak mesomeric effect of the furazan ring, however, can surpass its electron-withdrawing effect. Nitration of this compound gives a mixture of three possible mononitrophenyl regioisomers, the para-isomer being dominant (Calvino et al. 1983, Sheremetev et al. 2007). The same substrate acquires the corresponding para-phenyl substituent in results of bromination and chlorosulfonation (Cremlyn et al. 1985, Sheremetev et al. 2007). It should be, however, taken into account that this "sovereignty" of the phenyl substituent can be caused by its strong tearing out from the furazan plane and, consequently, by the loss of conjugation between the both molecular constituents.

1.3.1.2.2 Fused Furazans

Bond-length derived aromaticity index (I_A) for benzofurazan is 106.3 (Bird 1993). For comparison, the relative bicyclic molecules, quinoxaline and naphthalene, have I_A values 132 and 142 (Bird 1992). Meanwhile, benzofurazan is characterized with significant double-bond fixation between two pairs of carbons and nitrogens in the heterocyclic fragment and α,β-carbons in benzenoid ring (Brown et al. 1970, Gul'maliev et al. 1973). It is commonly accepted that the key oxygen heteroatom does not participate in over-molecular π-conjugation. However, one-electron transfer generates the benzofurazan anion-radical, in which the oxygen is drawn in delocalization of an unpaired electron (Strom and Russell 1965, Solodovnikov and Todres 1967). However, this introduced electron can occupy σ rather than π orbital (Gul'maliev et al. 1975). The same results were obtained for naphtofurazan (Solodovnikov and Todres 1967, Gul'maliev et al. 1973, 1975).

1.3.1.3 Reactivity

1.3.1.3.1 Monocyclic Furazans

1.3.1.3.1.1 *Salt Formation* Owing to the electrophilic character of the 1,2,5-oxadiazole ring, furazans hardly react with positively charged counterparts. Thus,

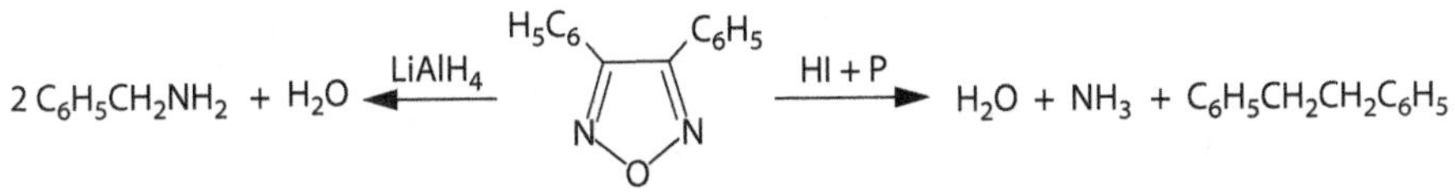

SCHEME 1.76

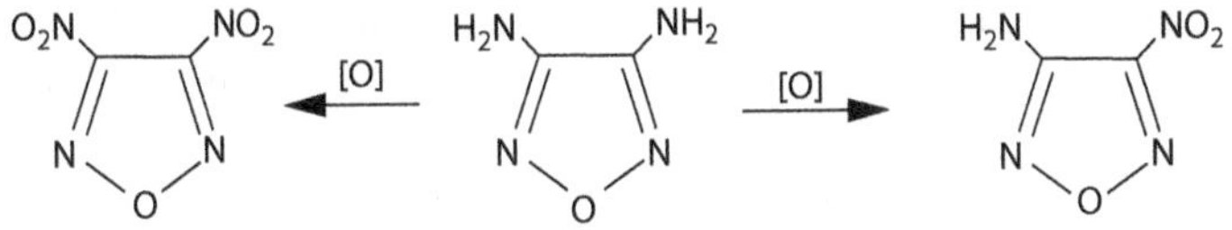

SCHEME 1.77

furazan itself react with dimethyl sulfate only on heating and undergoes only mono-methylation. The product is 2-methyl-1,2,5-oxadiazolium (Butler et al. 1995).

1.3.1.3.1.2 Reduction Reduction of furazans leads to ring opening, but the character of the product forming depends on the nature of a reducing agent. For instance, 3,4-diphenylfurazan gives benzylamine upon action of lithium aluminum hydride. The use of hydroiodic acid together with phosphorus results in the forma-tion 1,2-diphenylethane and ammonia (Scheme 1.76) (Gilchrist 1985).

1.3.1.3.1.3 Oxidation Furazans are resistant to oxidation. Thus, 3,4-diamino-furazan gives 3-amino-4-nitrofurazan, the heterocycle remains untouched (Scheme 1.77) (Li et al. 2008).

The oxidative reaction of Scheme 1.77 was performed using a mixture of hydro-gen peroxide, sodium tungstate, and methylsulfonic acid. To oxidize only one amino group, 30% solution of hydrogen peroxide was used, and 90% solution of hydro-gen peroxide was employed to oxidize the both amino groups. The product of the "binary" oxidation, 3,4-dinitrofurazan, is a very power explosive with a positive oxy-gen balance, but it is too reactive and shock-sensitive to be considered for practical use (Agrawal and Hodgson 2007).

1.3.1.3.1.4 Electrophilic Substitution Sheremetev et al. (2007) cited the known data on electrophilic substitution in 3-phenyl-4-R-furazans. Mononitration leads to ortho, meta, and para substitution in the phenyl moiety, although the para-nitration proceeds preferentially. Bromination proceeds at the expense of para position of the phenyl ring.

1.3.1.3.1.5 Nucleophilic Substitution These reactions proceed in cases when one of the substituents has good fugacity. Halides, nitrite, and sulfonyl are ready-leaving groups. 3-Chloro-1,2,5-oxadiazoles bearing at C_4 phenyl or 3-pyridyl react with alk-oxides giving rise to 3-alkoxy-4-phenyl-1,2,5-oxadiazole (Nash et al. 1969) or 3-alk-oxy-4-(3-pyridyl)-1,2,5-oxadiazole (Sauerberg et al. 1992). The yields are good.

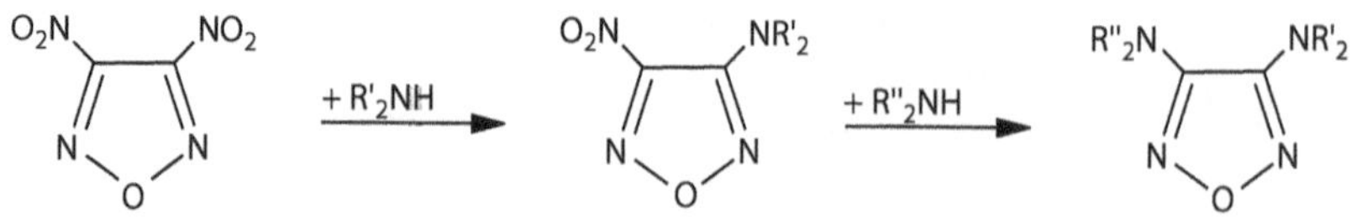

SCHEME 1.78

The two nitro groups in 3,4-dinitro-1,2,5-oxadiazole can be replaced stepwise. Scheme 1.78 shows that two differently substituted amino derivatives at the 3- and 4-positions can be prepared (Sheremetev and Makhova 2001).

Substitution of fluorine for one nitro group in 3,4-dinitrofurazan is another example of nucleophilic reaction. This reaction is interesting from the standpoint of specific conditions that make the substitution possible and that deserve to find wider application. Namely, to reach 3-fluoro-4-nitrofurazan, an anhydrous mixture of the substrate, $N(Et)_3 \cdot 3HF$ as a reactant, and 1,3-dibutylimidazolium tetrafluoroborate or hexafluorophosphate ionic liquid, as a solvent, was heated at 50°C with stirring 4–24 h under argon atmosphere. The product was separated with 50% yield from the residue by vacuum distillation or sublimation. Attempts to synthesize fluorofurazans from 3-nitro-4-R-furazans bearing electron-donating groups R, such as methyl and amino, were unsuccessful (Sheremetev et al. 2006).

Scheme 1.79 illustrates replacement of the nitro group by acetylenic alcohols in 3-amino-4-nitrofurazan. The reaction proceeded in aqueous dimethylformamide in the presence of potassium hydroxide (Berezina et al. 2001).

1.3.1.3.1.6   Ring Cleavage   Furazans having no substitution in the ring are sensitive to bases and, as a rule, split up. Scheme 1.80 gives one example of such a splitting (Olofson and Michelman 1964).

1.3.1.3.1.7   Photolysis   Upon photolysis in ether solution, furazan undergoes cleavage of two ring bonds (C–C and N–O) followed by subsequent rearrangement of the opened chain with formation of hydrogen cyanide (HCN) and hydrogen

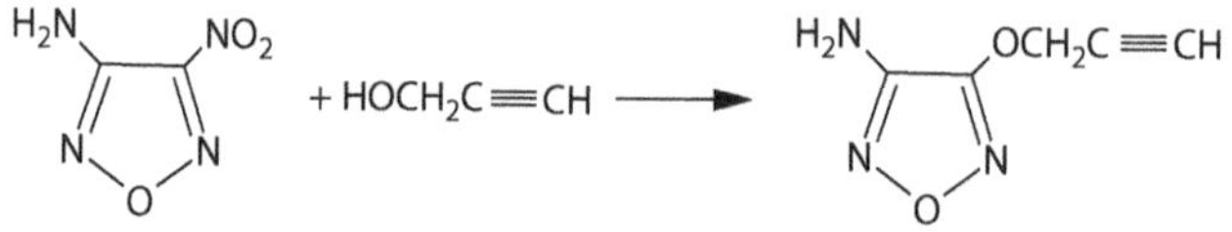

SCHEME 1.79

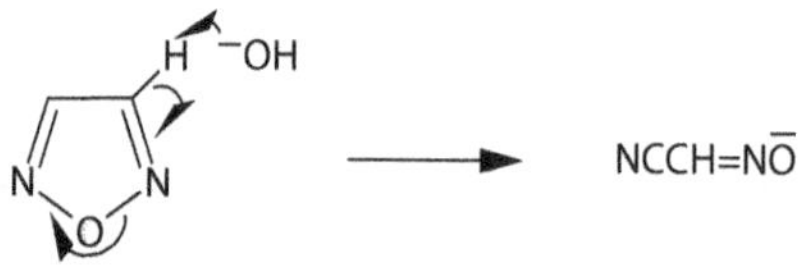

SCHEME 1.80

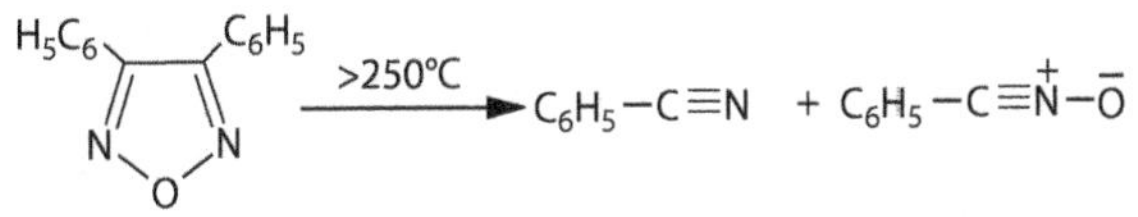

SCHEME 1.81

isocyanate (HNCO) as final products (Prokudin and Nazin 1987). Under gas-phase laser-flash photolysis, furazans undergo rupture of CN and NO bonds so that NO occurs to be the product of photolytic decomposition (Guo et al. 2008).

1.3.1.3.1.8 Thermolysis Furazans are more or less stable upon thermal action. In most cases, the thermal decomposition requires temperatures higher than 200°C. 3,4-Diphenylfurazan does not get destroyed up to 250°C. At higher temperature, it cleaves along C_3–C_4 and N–O bonds according to Scheme 1.81 (Boulton and Mathur 1973).

1.3.1.3.1.9 Ring Transformation Various 3-heteroallyl-substituted 1,2,5-oxa-diazoles undergo Boulton–Katritzky rearrangement, in which the oxadiazole is converted into a new five-membered heterocycle bearing a hydroxyiminoalkyl sub-stituent (Ruccia et al. 1981).

1.3.1.3.2 Fused Furazans

1.3.1.3.2.1 Salt Formation Similarly to monocyclic furazan, benzofurazan reacts with dimethyl sulfate only on heating and undergoes only monomethylation. From the kinetic measurements, the conclusion was inferred that benzofusion has no more than small rate-diminishing effect on quaternization (Davis et al. 1974).

At the same time, annulated furazans do not show any inclination to form hydrogen-bonded complexes: In aprotic and protic solvents, arene-fused furazans exhibit the same optical spectra. This means that their basicity is not sufficient for the formation of such complexes (Todres and Zajtsev 1971).

1.3.1.3.2.2 Reduction Treatment of benzofurazan with tin and hydrochloric acid affords 1,2-phenylenediamine, whereas catalytic hydrogenation takes place exclusively in the benzenoid ring with saturation of both its carbon–carbon double bonds (Gilchrist 1985).

1.3.1.3.2.3 Electrophilic Substitution Halogenation, nitration, and sulfonation of arene-fused furazans are directed to an aromatic α-carbon (α is the position in respect of the fused oxadiazole cycle—Bogdanov and Karavaev 1951, Bogdanov and Petrov 1954).

1.3.1.3.2.4 Nucleophilic Substitution Halide substituents in the homocyclic ring can be displaced by a variety of nucleophiles. In particular, nucleophilic substitution of amino, thioalkyl, phenoxy functions for halides in halonitrofurazans forms the basis of a powerful analytical technique applied to the biological samples (Imai et al. 1989, 1991). α-Chloro-β,α′-dinitrobenzofurazan is superelectrophilic and undergo facile substitution of weak carbonic nucleophile for chlorine. The weak nucleophiles successfully employed were indoles, 1,2,5-trimethylpyrrole, and azulene. Despite the fact that steric effects preclude a coplanarity of the donor and acceptor moieties, the resulting substitution products are subject to an intense intramolecular charge transfer (Rodriguez-Dafonte et al. 2009).

1.3.1.3.2.5 Photolysis Irradiation of benzofurazan in dimethyl ethynedicarboxylate results in formation of an isoxazole derivative. Scheme 1.82 shows the main event of the photolytic reaction, that is, interception of the ring-opened product with the ethyne (Yavari et al. 1975).

1.3.1.3.2.6 Thermolysis Annulation of furazans with aromatics enhances the ring strain that results in some decline in thermal stability. For instance, acenaphtho[1,2-c]-[1,2,5]-oxadiazole (acenaphthofurazan) is indefinitely stable at room temperature, but slowly decomposes to nitrile oxide, as shown in Scheme 1.83, on warming at 72°C in toluene. Brief (2–3 min) heating of acenaphthofurazan to 125°C provoked thermal decomposition according to Scheme 1.83 (Boulton and Mathur 1973). As seen from Schemes 1.81 and 1.83, the latter thermal reaction proceeds under significantly less-forcing conditions than that of the former one.

1.3.1.3.2.7 Coordination Benzofurazan forms mono- and binuclear complexes with chromium, molybdenum, and tungsten hexacarbonyls. As established, the metal pentacarbonyl fragment is bound with one or both nitrogen atoms of the oxadiazole ring. The oxygen key heteroatom remains free. The complexes are easily reduced to yield persistent anion-radicals with localization of the spin density in the benzofurazan framework (Kaim et al. 1989).

Being a component of a cryptand system, benzofurazan derivatives do not participate in coordination with metals, but show themselves as fluorophores. Such an action is caused by the electron-withdrawing character of benzofurazan

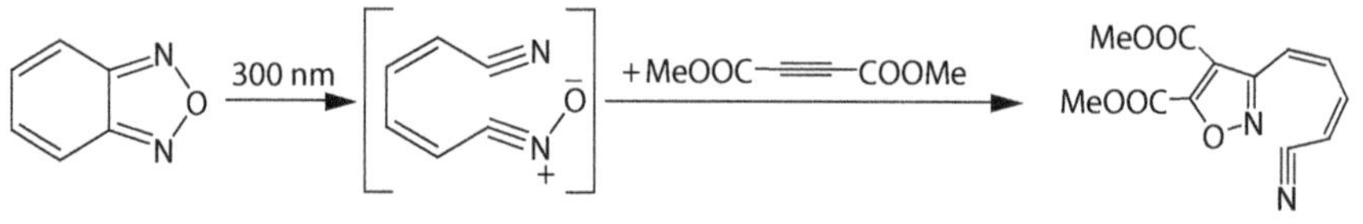

SCHEME 1.82

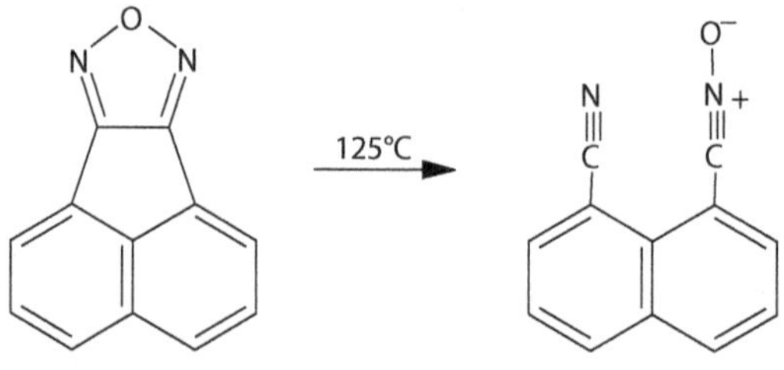

SCHEME 1.83

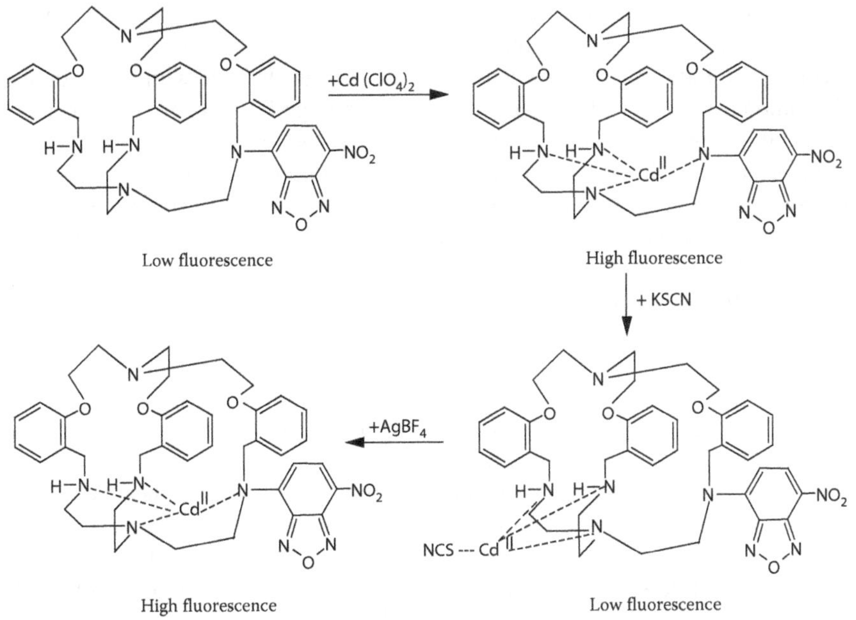

SCHEME 1.84 Adapted from Bag, B. and Bharadwaj, P.K., *Inorg. Chem.*, 43(5), 4627, 2004. With permission.

derivatives, especially of nitrobenzofurazan. Scheme 1.84 gives one striking example of this behavior (Bag and Bharadwaj 2004). The cryptand of Scheme 1.84 contains the α-nitrofurazan-α′-yl fragment that is covalently linked to a nitrogen atom of azacryptand. The authors monitored fluorescence spectra of the metal-free compound and its Cd^{II} complexes. Changes in fluorescence are explained by Scheme 1.84.

In the metal-free state, the nitrogen lone pair is blocked due to photo-induced electron transfer to the nitrobenzofurazan fragment, and the latter comes to the ground state via a nonradiative pathway. This causes quenching of fluorescence.

Photoelectron transfer becomes impossible when a metal ion uses the lone pair for coordination. This leads to recovery of fluorescence. With perchlorate or tetrafluoroborate salts of Cd^{II}, the metal ion enters the cavity. This metallocomplex exhibits high fluorescence. When a metal ion comes with coordinating anions such as thiocyanate, an out-cavity metallocomplex is formed. This reopen a possibility of photoelectron transfer from the cryptand nitrogen to the nitrofurazanyl neighbor: Fluorescence is lost.

Thus, translocation of Cd^{II} between the inside and outside of the cryptand cavity ensures a reversible fluorescence on/off-phenomenon. Systems similar to proposed by Bag and Bharadwaj (2004) can be useful for modulation of fluorescence signaling. They can be applicable as anion sensors. In the same laboratory, the cryptand of Scheme 1.84 was N-derivatized with the nitrobenzofurazan fragment and, additionally, with one or two anthracene fragments, bound with other N-atoms (Sadhu et al. 2007). Complexation with metals and protonation affect the fluorescence. In the case of the derivate with one anthracenyl additional fragment, Cd^{II}, Cu^{II}, Zn^{II}, and H^+ afford a large enhancement of fluorescence. Meanwhile, Fe^{II} and Ag^I exhibit one order of magnitude less enhancement. In the case of the derivate with two anthracenyl additional fragments, Cu^{II}, Ag^I, and H^+ give a large fluorescence enhancement. The metal nature effect is beyond of our scope, but the following peculiarity is, for us, important: The fluorescent enhancement is just observed in the nitrobenzofurazan moiety even when the anthracene fluorophore is excited. Substantial transfer of fluorescence resonance energy takes place from anthracene to the nitrobenzofurazan moiety (Sadhu et al. 2007).

The reaction of α-chloro-α'-nitrobenzofurazan with *N,N,N'*-tris(pyridine-2-yl-methyl)ethane-1,2-diamine was used to obtain a new fluorescent sensor (Qian et al. 2009). The sensor obtained has five basic nitrogen atoms belonging to the three pyridine moieties and the one ethane-1,2-diamine fragment. This product provides intense and stable visible-light fluorescent emission at pH 7 in aqueous solution. All the basic nitrogens coordinate to Zn^{2+} forming only mononuclear complex. During Zn^{2+} titration, fluorescent emission is linearly enhanced. This opens the way to determine zinc content in living cells. An enhancement factor of ~14 is observed when zinc-sensor ratio attains 1:1. Cation Cd^{2+} also induces an emission enhancement, but its effect is much less pronounced than that of Zn^{2+}. Due to the scarcity of a Cd^{2+} effect, this ion cannot interfere with Zn^{2+} in living cells. Such cations as K^+, Na^+, Ca^{2+}, and Mg^{2+}, which are abundant in cells, do not affect the Zn^{2+} response even when the $[M^{n+}]:[Zn^{2+}]$ ratio reaches 1000:1. These properties make the sensor constructed by Qian et al. (2009) an attractive candidate for intracellular Zn^{2+} imaging.

A similar effect was observed in a result of protonation that prevents intramolecular electron transfer from the *N*-(1-naphthyl)ethylenediamino fragment to the nitrobenzofurazanyl moiety in the compound depicted by Scheme 1.85 (Bem et al. 2007). Although this reaction does not belong to coordinative ones, it can serve as a relevant supplement to Scheme 1.84. Namely, in the neutral molecule of Scheme 1.85, photoelectron transfer prevents the fluorescence. The protonation avoids photoelectron transfer and the nitrobenzofurazanyl moiety restores its fluorescent possibility. The neutral compound of Scheme 1.85 contains two amino groups. The protonation touches only the amino group adjoining to naphthyl fragment. The strongly electron-withdrawing nitrofurazanyl group diminishes the basicity of the amino group adjacent to this fragment of the molecule. Besides, simulation by force-field molecular mechanics predicts the closed-sandwich geometry of the neutral molecule as a consequence of the intramolecular donor–acceptor interaction. This geometry transforms into an open one when the mentioned interaction disappears as a result of the protonation (Bem et al. 2007).

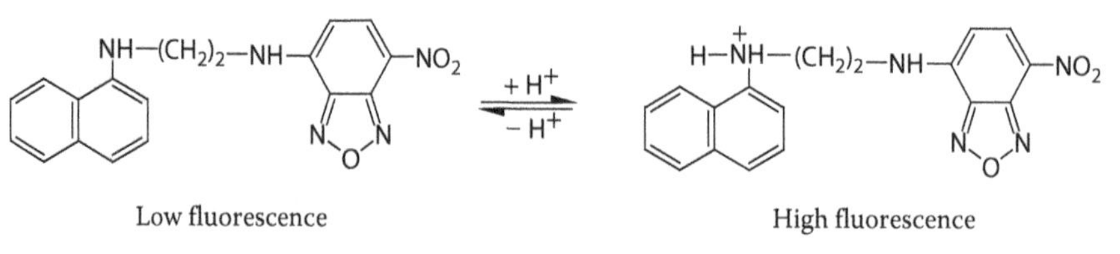

Low fluorescence   High fluorescence

SCHEME 1.85

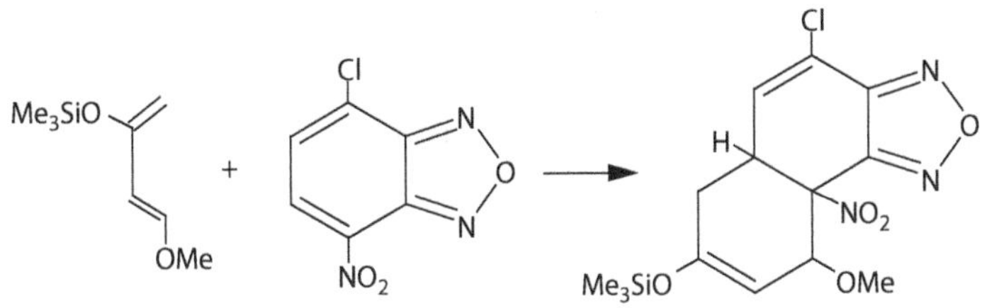

SCHEME 1.86

Interestingly, close location of aminonitrofurazans to silver-islanded Langmuir–Blodgett films enhances their fluorescence (Ray et al. 2006). In this case, the question is about an effect of the fluorescence probes on intensity of electromagnetic field of metal nanoparticles. These particles exhibit unique collective oscillations of electrons, which are, in physics, known as plasmon absorption of metal particles. These resonating plasmons create local electromagnetic fields close to particle surface. The electromagnetic field intensity decreases with increasing distance from the metal particles. Ray et al. (2006) used inert stearic acid layers to control the distance between the surface and the fluorophore. As it turned out, a 32-fold maximum enhancement of the probe fluorescent intensity was observed at direct contact with the silver-islanded film. (The possibility of metal–probe coordination was not considered.) The enhancement is reduced to fourfold when the probe is separated by 90 nm stearic acid layers.

1.3.1.3.2.8 Cycloaddition Vichard et al. (2001) described a cycloaddition reaction between nitrochlorobenzofurazan and butadiene derivatives. Scheme 1.86 outlines this reaction that proceeds quantitatively for 15 min in a room-temperature solution of methylene chloride.

1.3.1.3.2.9 Ring Transformation Heating of benzofurazans with ethanolamine at 150°C–170°C in the presence of a catalytic amount of *p*-toluenesulfonic acid leads to quinoxalines. This transformation supposedly includes dehydration of ethanolamine and addition of the enamine formed to benzofurazan with synchronous ring opening. Then, recyclization gives rise to quinoxaline N-oxide. An N-oxide oxygen atom, apparently, is eliminated upon treatment with ethanolamine. The reducing ability of the latter is known (Chung et al. 1984). Scheme 1.87 reflects the set of transformations (Samsonov 2007).

1.3.1.4 Biomedical Importance

1.3.1.4.1 Medical Significance

Cameron et al. (2004) described a series of monocyclic furazan carboxylic derivatives having antimalarial activity. The authors underlined that further development of these structures is prospective to find candidates suitable for consideration as new therapeutics for the treatment of malaria.

Several effective drugs of the furazan family were newly invented. Tested on a neuroblastoma cell line, the furazan bearing 3,4,5-trimethoxyphenyl at position 4 and 3-hydroxy-4-methoxyphenyl at position 5 showed high cytotoxicity and can be considered as an anticancer medication that is improved in comparison with the

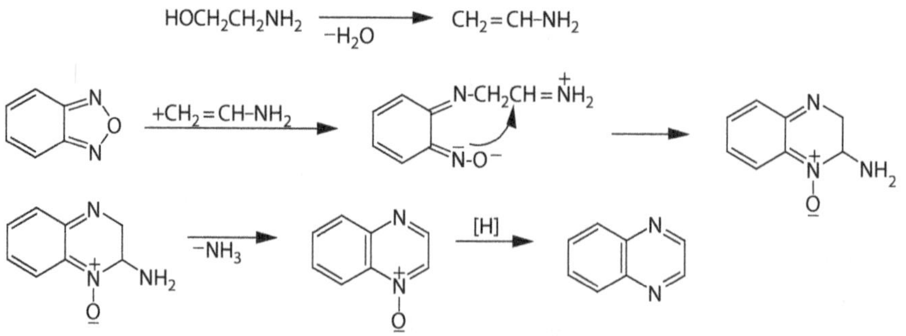

SCHEME 1.87

corresponding drugs currently in use (Tron et al. 2005). 4-Aminofurazan bearing at position 5 azabenzimidazole, substituted by ethoxymorpholine through oxyphenylaminoformyl spacer, dramatically lowers blood pressure in a rat model of hypertension (Stavenger et al. 2007). The analog having structure of 4-[2-(4-amino-1,2,5-oxadiazol-3-yl)-1-ethyl]-7{[(3S)-3-piperidinylmethyl]oxy}-1H-imidazole-[4,5-c]pyridine-4-yl)-2-methyl-3-butyn-2-ol inhibits human breast carcinoma xenografts (Heerding et al. 2008). A series of N-hydroxyamidinofurazans were claimed for the treatment of cancer and other diseases (Combs et al. 2007, 2008).

1.3.1.4.2 Applications to Environmental and Bioanalysis

Wanichachewa et al. (2009) proposed a sensor based on α-nitrobenzofurazan unit, for selective analysis of mercury. Mercury is extremely toxic even in very low concentration in water or edible objects because it can accumulate and damage human organisms. 2-[3-(2-Aminoethylthio)propylthio]ethanamine symmetrically bound with two α-nitrobenzofurazan-α'-yl units is able to detect Hg^{2+} up to 20 ppb, in an 80:20 acetonitrile/water solution. Other metallic ions do not prevent the determination. In the absence of Hg^{2+}, the fluorescence response of the sensor is at a maximum. In the presence of Hg^{2+}, this response decreases proportionally to the analyte concentration. Bound the mercuric cation, this sensor exhibits a redshift by 70 nm in the absorbance ultraviolet spectra with changes the solution color from yellow to pink. This can be observed by the naked eye. Sub-microconcentrations of Hg^{2+} can be observed in environmental and biological systems (Wanichachewa et al. 2009).

Amines in biological samples readily react with α-halo-α'-nitrobenzofurazans giving the corresponding products of nucleophilic substitution of halogen (Isobe and Mataga 2008). The fluoro analogs are the most reactive. The substituted products exhibit marked fluorescence and good separation by liquid chromatography. Therefore, the reactions are used for analysis of biological objects, for example, for determination of proline and hydroxyproline in blood plasma (Imai et al. 1989, 1994) or cysteine and its peptide in mammalian bodies (Asamoto et al. 2007).

Analogously, the adduct between 4,7-diphenyl-1,2,5-oxadiazolo[3,4-c]pyridine-6-carbonic acid and N-hydroxymaleimide exhibits fluorescence in the ultraviolet region. The maleimide fragment adds sulfhydryl groups of proteins in hen ovalbumin and milk casein to its double bonds forming bioconjugates. The conjugates exhibit fluorescent emission. The more the sulfhydryl content in the sample the stronger is the emission and the lower is detection limit. Thus, the fluorescent albumin conjugate was detected in the lane loaded with 10 mg protein, whereas the casein fluorescenyl conjugate was detected only in the lane loaded with 40 mg protein (Balasu and Popescu 2006).

Qualitative or quantitative determination of D-glucosamine from pharmaceutical samples was proposed on the basis of α-chloro-α′-nitrobenzofurazan as fluorophore. The amino compound displaces chlorine in the fluorophore, and the product is detected by fluorescence at 360 nm (Bem et al. 2008).

Fluorescence of ready-made aminobenzofurazans was patented as diagnosis tools to emphasize the polyamine transport system in cancerous cells and to select patients for the appropriate medical treatment (Annereau et al. 2009a,b).

Another important series of fluorescent probes is represented by benzofurazans bearing the diphenyl or diethyl phosphine group at α-position and the acetylamino or methylthio group at α′-position. Under photoirradiation, these probes exhibit ultraviolet fluorescence due to intramolecular electron transfer from donor to acceptor substituents. Reacting with hydroperoxides, the phosphine substituent transforms into the corresponding phosphine-oxide group. The oxidation enforces the electron transfer, which dramatically helps the fluorescence switching. The higher concentration of a hydroperoxide, the stronger is a fluorescence signal. This is the basis of a method to quantitatively detect hydroperoxides in biosamples (Onoda et al. 2005). The peroxide determination is important to control lipid peroxidation that is responsible for diseases including cancer, Alzheimer's, and atherogenesis.

Alkyl and alkylamino derivatives of nitrobenzofurazan bind with DNA duplex. In particular, $N′$-(7-nitrobenzo-[1,2-c]-[1,2,5]-oxadiazol-4-yl)propane-1,3-diamine was found to bind selectively to the thymine base of DNA. This fluorophore was recommended to control gene expression (Thiagarajan et al. 2010).

Benzofurazan bearing α-(N-substituted)-amine and α′-N,N-dimethylaminosulfonyl groups is capable of selectively binding to pyrimidine bases over purine bases. The binding event is accompanied by a significant enhancement of fluorescence emission due to the furazan component (Satake et al. 2007). The authors especially emphasize that the response of the probe is specific to pyrimidine bases, making it possible to detect pyrimidine/purine transreversion in DNA duplexes.

Amino nitro benzofurazans are well-known fluorescent probes that emit in a middle ultraviolet wavelength region, which avoids interference due to biomatrices. Intramolecular electron transfer from the donor to acceptor substituent is enforced upon photoirradiation and when solvent polarity is raised (Stevenson and Blanchard 2006). This type of probes was used, for example, to study interaction of bovine serum albumin with dipolar molecules (Bhattacharya et al. 2009), to elucidate the role of polycarboxylate dendrimers in formation of microcrystals (Bertorelle and Lavabre 2003), to rapidly determine the fate of biomolecules in vivo (Blois et al. 2008, Borisenko et al. 2008), to monitor accumulation of new active and low-toxic antimalarials in parasite infected erythrocytes (Ellis et al. 2008), and to visualize glycosphingolipid microdomains in biomembranes (Gege et al. 2008).

Derivatization of a peptide with a benzofurazan-containing fluorescent group made it possible to study the nature of interaction between the peptide and large unilamellar vesicles of various lipids (Perez-Lopez et al. 2009).

α-*N*-(2-acrloyloxyethyl)-α'-*N,N*-dimethylaminosulfonyl   benzofurazan   emits
ultraviolet fluorescence upon heating to 40°C. The fluorescence was disappeared
after cooling to room temperature (Nakada et al. 2008).

Reagents having a benzofurazan structure are often used to enhance detectability
of the target compounds and to improve the separation efficiency in high-performance
liquid chromatography. Among these are not only fluorogenic reagents, but also water-
soluble ones as well as reagents for the analysis of peptides and proteins, and reagent
for mass spectroscopic detection. A set of publications by Santa et al. (2007, 2008a,b)
describes recent progress in synthesis of benzofurazan derivatization reagents and
their application to biosamples. The Japanese patent by Isobe and Mataga (2008) also
introduces various furazan derivatives prepared applicable to biosamples.

1.3.1.4.3 Pesticidal Activity

Benzofurazan can be used as a pesticide preventing eclosion of housefly larvae
(Huang et al. 2008).

1.3.1.5   Technical Applications

1.3.1.5.1 Light Emitting

Introduction of nitraminobenzofurazan in back chain of polystyrenes allows fluores-
cently monitoring interfacial mobility of polymers on inorganic solids (Tanaka et al.
2009). The furazan-fused pyridine ring that contains $CONHCH_2CH=CH_2$ substituent was involved in homopolymerization. The homopolymer obtained shows yellow
fluorescence, which does not change after exposure to natural light under air at room
temperature for 3 weeks (Isobe et al. 2008).

Fluorescence   of   poly(*N*-isopropylacrylamide),   labeled   by   α-amino-α'-(*N,N*-
dimethylaminosulfonyl)benzourazan depends on temperature. Based on this dependence, a fluorescent polymeric thermometer was proposed (Gota et al. 2008). Change
of fluorescence from α-(1-methylaminoethanol)-α'-nitrofurazan thiobenzoate ester
was a basis to study radical termination in reversible addition-fragmentation chain
transfer-mediated polymerization of butyl acrylate. The method was used for identification of termination products (Geelen and Klumperman 2007).

The fluorescent dyes containing 4,7-bis(4-methoxyphenyl)-1,2,5-oxadiazolo[3,4-
c]pyridine-6-carboxylic acid were patented as components of cosmetic compositions. The compositions provide long-lasting brightness to nail and hair without
causing damage (Isobe 2008).

Polycarbazoles are utilized in various electronic and optical devices, such as polymeric light-emitting diodes, organic field electric transistors, and photovoltaic cells.
These applications are based on mobility of their holes. When fused 1,2,5-oxadia-
zole fragments are incorporated into backbones of polycarbazoles, working properties of these polymers are improved. As the fused parts, benzene and pyridine were
compared. The best results were obtained with α,α'-benzofurazan linker. This linker
provides the better structural organization in symmetric polymers of the polycarba-
zole class (Blouin et al. 2008). Introducing substituents into the benzenoid ring of
the benzofurazan linker offers potential for the fine-tuning of the polymer electronic
properties. In other words, it is possible to control the degree of energy migration to

a lower band-gap emission site without recourse to adjustment in feed ratios of comonomers (Bouffard and Swager 2008). Incorporation of benzofurazans into polymer chains leads to technically important materials with the electron-deficient properties that facilitate electron injection in light-emitting devices.

1.3.1.5.2 Explosiveness

Furazan derivatives that are rich in nitrogen and oxygen have attracted attention for a long time. As contemporaneous inventions, patents by Jacob et al. (2008a,b) can be exemplified. Some works are devoted to minimization of safety issues in scalable route to known explosives. For instance, such a solution was proposed for synthesis of β-(2-carboxypyridin-2-yloxy)benzofurazan. As compared to a previously practiced route, the way newly proposed one avoids the handling of β-hydroxybenzofurazan, which was found to decompose with a large energy release at relatively low temperatures. The formation of benzofurazan fragment was achieved by trivial perchlorate oxidation of the corresponding o-nitroaniline fragment in an intermediate in the five-step sequence designed by Ruggeri et al. (2003). This reference can direct to the original scheme, which is out of the book theme. It is only important to mention that pilot-plant scaling of the whole process showed its safety.

1.3.1.6 Conclusion

Compounds of 1,2,5-oxadiazole class are characterized with some extend of aromaticity and aromatic reactivity. Their participation in nucleophilic reaction is the most fruitful because it opens a way to derivatives with interesting and useful fluorescent ability. In this sense, optical results of coordination to metal should be especially underlined. Furazan technical applications, apart from explosives, are still not very wide. However, their biomedical importance is out of doubts. There are numerous valuable results of design and testing new medications. Fluorescent analytical methods have seriously been enriched by introduction of aminonitrobenzofurazan in various biosamples.

REFERENCES TO SECTION 1.3.1

Agrawal, J.P.; Hodgson, R. (2007) *Organic Chemistry of Explosives*. Wiley, Chichester, U.K.
Annereau, J.-Ph.; Barret, J.-M.; Guminskii, Y.; Imbert, Th. (2009a) WO Patent 013.360.
Annereau, J.-Ph.; Barret, J.-M.; Guminskii, Y.; Imbert, Th. (2009b) Fr. Patent 2,919,287.
Asamoto, H.; Ichibangase, T.; Saimaru, H.; Uchikura, K.; Imai, K. (2007) *Biomed. Chromatogr.* **21**, 999.
Bag, B.; Bharadwaj, P.K. (2004) *Inorg. Chem.* **43**, 4626.
Balasu, M.C.; Popescu, A. (2006) *Rev. Roum. Chem.* **51**, 847.
Bem, M.; Badea, F.; Draghichi, C.; Caproiu, M.T.; Vasilescu, M.; Voicescu, M.; Beteringhe, A. Caragheorgheopol, A.; Maganu, M.; Constantinescu, T.; Balaban, A.T. (2007) *ARKIVOC* (xiii), 87.
Bem, M.; Badea, F.; Draghichi, C.; Caproiu, M.T.; Vasilescu, M.; Voicescu, M.; Pencu, G.; Beteringhe, A.; Maganu, M.; Covaci, I.C.; Constantinescu, T.; Balaban, A.T. (2008) *ARKIVOC* (ii), 218.
Berezina, S.E.; Remizova, L.A.; Dielimoriba, D.; Domnin, I.N. (2001) *Zh. Org. Khim.* **37**, 1700.
Bertorelle, F.; Lavabre, D.; Fery-Forgues, S. (2003) *J. Am. Chem. Soc.* **125**, 6244.
Bhattacharya, B.; Nakka, S.; Guruprasad, L.; Samanta, A. (2009) *J. Phys. Chem. B* **113**, 2143.
Bird, C.W. (1985) *Tetrahedron* **41**, 1409.

Bird, C.W. (1992) *Tetrahedron* **48**, 335.

Blois, J.; Yuan, H.; Smith, A.; Pacold, M.E.; Weissleder, R.; Cantley, L.C.; Josephson, L. (2008) *J. Med. Chem.* **51**, 4699.

Blouin, N.; Michaud, A.; Gendron, D.; Wakim, S.; Blair, E.; Neagu-Plesu, R.; Belletete, M.; Durocher, G.; Tao, Y.; Leclerc, M. (2008) *J. Am. Chem. Soc.* **130**, 732.

Bogdanov, S.V.; Karavaev, B.I. (1951) *Zh. Obshch. Khim.* **21**, 1915.

Bogdanov, S.V.; Petrov, S.F. (1954) *Zh. Obshch. Khim.* **24**, 385.

Borisenko, G.G.; Kapralov, A.A.; Tyurin, V.A.; Maeda, A.; Stoyanovsky, D.A.; Kagan, V.E. (2008) *Biochemistry* **47**, 13699.

Bouffard, J.; Swager, T.M. (2008) *Macromolecules* **41**, 5559.

Boulton, A.J.; Mathur, S.S. (1973) *J. Org. Chem.* **38**, 1054.

Brown, N.M.D.; Lister, D.G.; Tyler, J.K. (1970) *Spectrochim. Acta A* **26**, 2113.

Butler, R.N.; Daly, K.M.; McMahon, J.M.; Burke, L.A. (1995) *J. Chem. Soc. Perkin Trans. 1*, 1083.

Calvino, R.; Ferrarotti, B.; Casco, A.; Serafino, A. (1983) *J. Heterocycl. Chem.* **20**, 1419.

Cameron, A.; Read, J.; Tranter, R.; Winter, V.J.; Sessions, R.B.; Brady, R.L.; Vivas, L.; Easton, A.; Kendrik, H.; Croft, S.L.; Barros, D.; Lavandera, J.L.; Martin, J.J.; Risco, F.; Garcia-Ochoa, S.; Gamo, F.J.; Sanz, L.; Leon, L.; Ruiz, J.R.; Gabarro, R.; Mallo, A.; Gomez de las Heras, F. (2004) *J. Biol. Chem.* **279**, 31429.

Chung, T.F.; Wu, Y.M.; Cheng, C.H. (1984) *J. Org. Chem.* **49**, 1215.

Combs, A.P.; Takvorian, A.; Zhu, W.; Sparks, R.B. (2007) WO Patent 075,598.

Combs, A.P.; Zhu, W.; Sparks, R.B. (2008) WO Patent 058,178.

Cremlyn, R.J.; Swinbourne, F.J.; Shode, O. (1985) *J. Heterocycl. Chem.* **22**, 1211.

Davis, M.; Deady, L.W.; Homfeld, E. (1974) *Aust. J. Chem.* **27**, 1917.

di Nunno, L.; Forio, S.; Todesco, P.G. (1973) *J. Chem. Soc. C*, 1954.

Ellis, G.L.; Amew, R.; Sabbani, S.; Stocks, P.A.; Shone, A.; Stanford, D.; Gibbons, P.; Davies, J.; Vivas, L.; Charnaud, S.; Bongard, E.; Hall, Ch.; Rimmer, K.I.; Lozanom, S.; Jesus, M.; Gargallo, D.; Ward, S.A.; O'Neill, P.M. (2008) *J. Med. Chem.* **51**, 2170.

Geelen, P.; Klumperman, B. (2007) *Macromolecules* **40**, 3914.

Gege, Ch.; Schumacher, G.; Rothe, U.; Schmidt, R.R.; Bendas, G. (2008) *Carbohydr. Res.* **343**, 2361.

Gilchrist, T.L. (1985) *Heterocyclic Chemistry*. Pitman, London.

Gota, Ch.; Uchiyama, S.; Yoshihara, T.; Tobita, S.; Ohwada, T. (2008) *J. Phys. Chem. B* **112**, 2829.

Gul'maliev, A.M.; Stankevich, I.V.; Todres, Z.V. (1973) *Khim. Geterotsikl. Soed.* 1473.

Gul'maliev, A.M.; Stankevich, I.V.; Todres, Z.V. (1975) *Khim. Geterotsikl. Soed.* 1055.

Guo, Y.Q.; Bhattacharya, A.; Bernstein, E.R. (2008) *J. Chem. Phys.* **128**, 034303.

Heerding, D.A.; Rhodes, N.; Leber, J.D.; Clark, T.J.; Keenan, R.M.; Lafrance, L.V.; Li, M.; Safonov, I.G.; Takata, D.T.; Venslawsky, J.W.; Yamashita, D.S.; Choudry, A.E.; Copeland, R.A.; Lai, Zh.; Schaber, M.D.; Tummino, P.J.; Strum, S.L.; Wood, E.R.; Duckett, D.R.; Eberwein, D.; Knick, V.B.; Lansing, T.J.; McConnel, R.T.; Zhang, Sh.Y.; Minthorn, E.A.; Concha, N.O.; Warren, G.L.; Kumar, R. (2008) *J. Med. Chem.* **51**, 5663.

Huang, Q.; Liu, M.; Feng, J.; Liu, Y. (2008) *Pesticide Biochem. Physiol.* **90**, 119.

Imai, K.; Uzu, S.; Kanda, S. (1994) *Anal. Chim. Acta* **29**, 3.

Imai, K.; Uzu, S.; Toyo'oka, T. (1989) *J. Pharm. Biomed. Anal.* **7**, 1395.

Isobe, Sh. (2008) Jpn. Patent 105,976.

Isobe, Sh.; Mataga, Sh. (2008) Jpn. Patent 156,556.

Isobe, Sh.; Mataga, Sh.; Mizuki, K.; Tanikaka, I.; Kawashima, Sh.; Tsukuda, T. (2008) Jpn. Patent 184,592.

Jacob, G.; Herve, G.; Cagnon, G.; Alvarez, F. (2008a) WO Patent 102,092.

Jacob, G.; Herve, G.; Cagnon, G.; Alvarez, F. (2008b) Fr. Patent 2,911,339.

Kaim, W.; Kohlmann, S.; Lees, A.J.; Zulu, M. (1989) *Zeitschr. anorg. allgem. Chem.* **575**, 97.

Kamitori, Y. (1999) *Heterocycles* **51**, 627.

Li, H.-Zh.; Zhou, X.-Q.; Li, J.-Sh.; Huang, M. (2008) *Youji Huaxue* **28**, 1646.

Micetich, R.G. (1970) *Can. J. Chem.* **48**, 2006.

Nakada, M.; Maeda, Sh.; Takahashi, Y. (2008) Jpn. Patent 063,701.

Nash, B.W.; Newberry, R.A.; Pickles, R.; Warburton, W.K. (1969) *J. Chem. Soc. C*, 2794.

Olofson, R.A.; Michelman, J.S. (1964) *J. Am. Chem. Soc.* **86**, 1863.

Olofson, R.A.; Michelman, J.S. (1965) *J. Org. Chem. Soc.* **30**, 1854.

Onoda, M.; Tokuyama, H.; Uchiyama, S.; Mawatari, K.-i.; Santa, T.; Kaneko, K.; Imai, K.; Nakagomi, K. (2005) *Chem. Commun.* 1848.

Perez-Lopez, S.; Vila-Romeu, N.; Esteller, M.A.A.; Espina, M.; Haro, I.; Mestres, C. (2009) *J. Phys. Chem. B* **113**, 319.

Prokudin, V.G.; Nazin, G.M. (1987) *Bull. AN SSSR, Div. Chem. Sci.* **36**, 1999.

Qian, F.; Zhang, Ch.; Zhang, Y.; He, W.; Gao, X.; Hu, P.; Guo, Z. (2009) *J. Am. Chem. Soc.* **131**, 1460.

Rakitin, O.A.; Zalesova, O.A.; Kulikov, A.S. (1993) *Izv. RAN, Ser. Khim.*, 1949.

Ray, K.; Badugu, R.; Lakowicz, J.R. (2006) *Langmuir* **22**, 8374.

Rodriguez-Dafonte, P.; Terrier, F.; Lakhdar, S.; Kurbatov, S.; Goumont, R. (2009) *J. Org. Chem.* **74**, 3305.

Ruccia, M.; Vivona, N.; Spinelli, D. (1981) *Adv. Heterocycl. Chem.* **29**, 141.

Ruggeri, S.G.; Bill, D.R.; Bourassa, D.E.; Castaldi, M.J.; Houk, T.L.; Ripin, D.H.B.; Wei, L.; Weston, N. (2003) *Org. Proc. Res. Develop.* **7**, 1043.

Sadhu, K.K.; Bag, B.; Bharadwaj, P.K. (2007) *Inorg. Chem.* **46**, 8051.

Samsonov, V.A. (2007) *Russ. Chem. Bull. Intl. Ed.* **56**, 2510.

Santa, T.; Al-Dirbashi, O.Y.; Ichibangase, T.; Fukushima, T.; Rashed, M.S.; Funatsu, T.; Imai, K. (2007) *Biomed. Chromatogr.* **21**, 1207.

Santa, T.; Al-Dirbashi, O.Y.; Ichibangase, T.; Rashed, M.B.; Fukushima, T.; Imai, K. (2008a) *Biomed. Chromatogr.* **22**, 115.

Santa, T.; Fukushima, T.; Ichibangase, T.; Imai, K. (2008b) *Biomed. Chromatogr.* **22**, 343.

Satake, H.; Nishizawa, S.; Teramae, N. (2007) *Nucleic Acid Symp. Ser.* No. 51, 297.

Sauerberg, P.; Olesen, P.H.; Nielsen, S.; Treppendahl, S.; Sheardown, M.J.; Honore, T.; Mitch, C.H.; Ward, J.C.; Pike, A.J.; Bymaster, F.P.; Sawyer, B.D.; Shannon, H.E. (1992) *J. Med. Chem.* **35**, 2274.

Sheremetev, A.B.; Aleksandrova, N.S.; Dmitriev, D.E. (2006) *Mendeleev Commun.* 163.

Sheremetev, A.B.; Konkina, S.M.; Dmitriev, D.E. (2007) *Izv. RAN, Ser.Khim.* 1516.

Sheremetev, A.B.; Makhova, N.N. (2001) *Adv. Heterocycl. Chem.* **78**, 65.

Sliwa, W. (1984) *Heterocycles* **22**, 1571.

Solodovnikov, S.P.; Todres, Z.V. (1967) *Khim. Geterotsikl. Soed.* 811.

Stavenger, R.A.; Cui, H.; Dowdell, S.E.; Franz, R.G.; Gaitanopoulos, D.E.; Goodman, K.B.; Hilfiker, M.A.; Ivy, R.L.; Leber, J.D.; Marino, J.P.; Oh, H.-J.; Viet, A.Q.; Xu, W.; Ye, G.; Zhang, D.; Zhao, Y.; Jolivette, L.J.; Head, M.S.; Semus, S.E.; Elkins, P.A.; Kirkpatrik, R.B.; Dul, E.; Khandekar, S.S.; Yi, T.; Jung, D.K.; Wright, L.L.; Smith, G.K.; Behm, D.J.; Doe, Ch.P.; Bentley, R.; Chen, Z.X.; Hu, E.; Lee, D. (2007) *J. Med. Chem.* **50**, 2.

Stevenson, S.A.; Blanchard, G.J. (2006) *J. Phys. Chem. A* **110**, 3426.

Stiefvater, O.L. (1988) *Zeitschr. Naturforsch. A* **43**, 597.

Strom, E.T.; Russell, G.A. (1965) *J. Am. Chem. Soc.* **87**, 3326.

Takakis, I.M.; Hadjimihalakis, P.M. (1990) *J. Heterocycl. Chem.* **27**, 177.

Tanaka, K.; Tateishi, Y.; Okada, Y.; Nagamura, T.; Doi, M.; Morita, H. (2009) *J. Phys. Chem. B* **113**, 4571.

Thiagarajan, V.; Rajendran, A.; Satake, H.; Nishizawa, S.; Teramae, N. (2010) *ChemBioChem* **11**, 94.

Todres, Z.V.; Zajtsev, B.E. (1971) *Khim. Geterotsikl. Soed.* 1036.

Tron, G.C.; Pagliai, F.; del Grosso, E.; Genazzani, A.A.; Sorba, G. (2005) *J. Med. Chem.* **48**, 3260.

Tselinskii, I.V.; Mel'nikova, S.F.; Romanova, T.V. (2001) *Zh. Org. Khim.* **37**, 1708.

Vichard, D.; Alvey, L.J; Terrier, F. (2001) *Tetrahedron Lett.* 42, 7571.

Vivona, N.; Buscemi, S.; Frenna, V.; Ruccia, M.; Condo, M. (1985) *J. Chem. Res.* 190.

Wanichachewa, N.; Siriprumpoothum, M.; Kamkaew, A.; Grudpan, K. (2009) *Tetrahedron Lett.* **50**, 1783.

Winterweber, M.; Geiger, R.; Otto, H.-H. (2006) *Monatshefte Chem.* **137**, 1321.

Yadav, M.R.; Giridhar, R.; Prajapati, H.B. (2006) Ind. Patent IN 2004 MU 109.

Yavari, I.; Esfandiari, S.; Mostashari, A.J.; Hunter, P.W.W. (1975) *J. Org. Chem.* **40**, 2880.

Zaitsev, A.B.; Schmidt, E.Yu.; Vasil'tsov, A.M.; Mikhaleva, A.I.; Petrova, O.V.; Afonin, A.V.; Zorina, N.V. (2006) *Khim. Geterotsikl. Soed.* 39.

1.3.2   1,2,5-Oxadiazole N-Monoxides (Furoxans)

In this section, monocyclic and fused furoxans are expeditious to consider together in order to follow effects of an annulated aromatic nucleus on the furoxan moiety. This is especially interesting in the sense of the ring open-ring closure tautomerism. Preparative approaches and reactivity remain general. The understandable exception concerns substitution into aromatic rings.

1.3.2.1   Formation

1.3.2.1.1   From Ethenes

Ponzio and Satta (1930), then Dinh et al. (2004) described transformation of isosafrole, that is, 5-(1-propenyl)-1,3-benzodioxole, into 4-methyl-5-(1,3-benzodioxol-5-yl) furoxan according to Scheme 1.88. The reaction was performed by addition sodium nitrite to isosafrole dissolved in acetic acid. This simplest reaction leads to the product with 90% yield.

1.3.2.1.2   From Alkyl Nitroacetates

Ethyl or methyl nitroacetate undergoes self-condensation during long-time (96 h) refluxing in chloroform with copper diacetate in the presence of 1-methylpiperidine. Diethyl or dimethyl 1,2,5-oxadiazole-3,4-dicarboxylate 2-oxide is formed, yields are 43% and 68%, respectively (Scheme 1.89) (Trogu et al. 2009).

1.3.2.1.3   From vic-Nitroazides

The method consists of nitrogen elimination and cyclization of the intermediary vic-nitronitrene (Scheme 1.90) (Al-Hiari et al. 2006). Heating conditions depend on structure of the starting vic-nitroazides and varies from refluxing acetic acid (118°C, 30 min) according to Tyurin et al. (2006) to refluxing bromobenzene (155°C, 30–45 min) according to Stadlbauer et al. (2000), Tinh and Stadlbauer (2008).

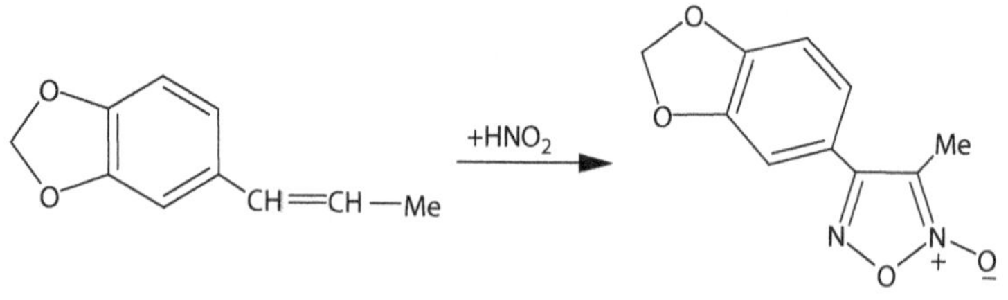

SCHEME 1.88

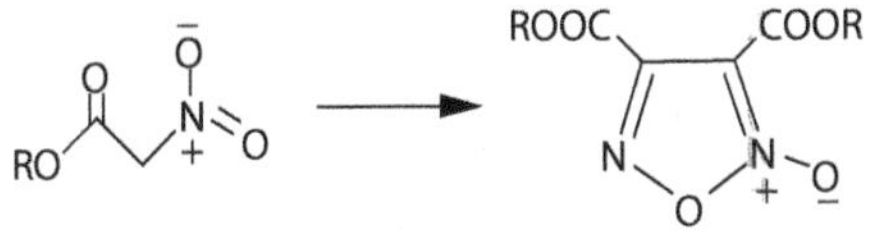

SCHEME 1.89

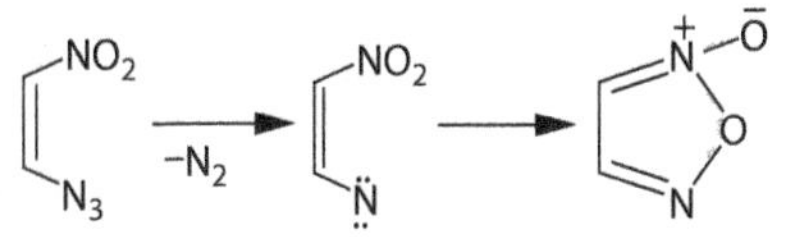

SCHEME 1.90

Avemaria et al. (2004) proposed a variant of solid-supported reaction. Namely, the triazene resin prepared from benzylaminomethyl polystyrene was treated with o-nitrobenzene diazonium salt and after that with trifluoroacetic acid–trimethylsilyl azide mixture. Evaporation of the solvent from the reaction filtrate led to o-nitrophenyl azide. Heating this intermediary product ca. 100°C gave rise to the final benzofuroxan in 100% yield and high purity.

1.3.2.1.4 From vic-Dioximes

The parent furoxan (1,2,5-oxadiazole 2-oxide) is obtained by oxidation of glyoxime with liquid dinitrogen tetroxide (Godovikova et al. 1994) or with gaseous nitrogen dioxide (Pasinszki et al. 2009); both reactions were performed in CH_2Cl_2 solutions. The oxidation can also be carried out upon action of chlorine or bromine (Vladykin and Trakhtenberg, 1980), tert-butoxychloride (Bohn et al. 1995) as well as by the treatment with alkaline solutions of alkali ferricyanide or hypochlorite (Armani et al. 1997). Ceric ions, nitric acid, manganese dioxide, lead tetraacetate, N-iodosuccinimide, and phenyliodo bis(trifluoroacetate) have also been used for vic-dioxime oxidative cyclization (Wang et al. 2002).

1.3.2.1.5 From Monoximes or vic-Nitroximes

Oxidation of 1,3-diarylpyrazole-4-carboxaldehyde oxime with phenyliodo bis(acetate) leads to dimerization of initially formed nitrile oxides according to generalized Scheme 1.91 (Prakash and Pannu 2007).

Dehydration/cyclization of α-nitro monoximes results in the formation of furoxans. This reaction is carried out by adding of the solution of a nitroxime

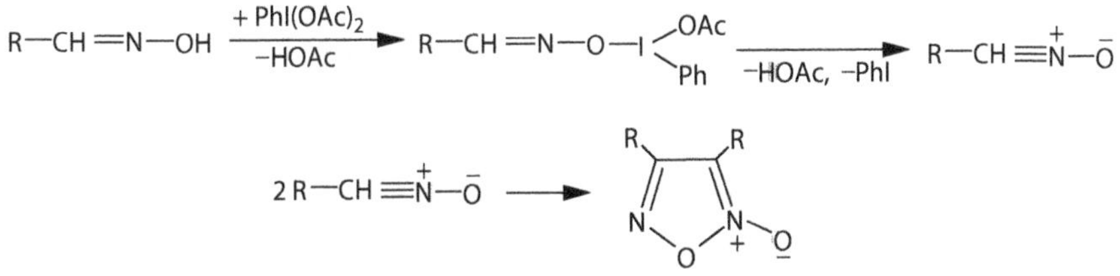

SCHEME 1.91

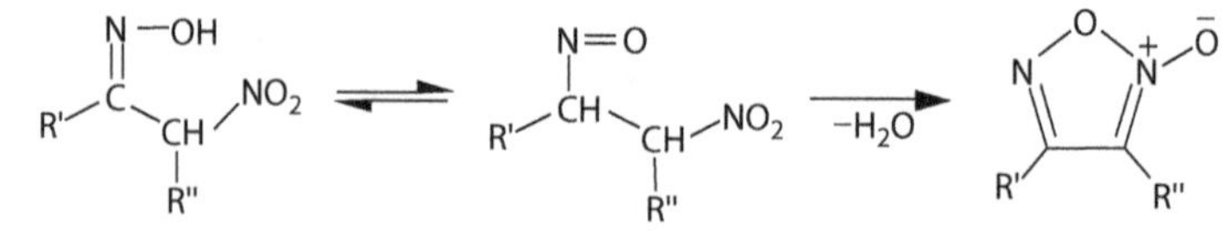

SCHEME 1.92

in acetonitrile to a suspension of acidic alumina. Yields of furoxans come up to 85%–95% (Curini et al. 2000). Alumina probably acts as a dehydration agent withdrawing α-nitro nitroso isomer from the tautomeric equilibrium with the formation of the furoxan ring system (Scheme 1.92).

1.3.2.1.6   From vic-Nitroarylamines

This is a standard and well-established method for the preparation of benzofuroxans from o-nitroarylamines by heating or UV irradiation as well as by treating with a mild oxidant like sodium hypochlorite in alkaline solution. Because of instability of some benzofuroxans under the alkaline conditions, oxidation of o-nitroarylamines with phenyliodo bis(acetate) was proposed (Dyall et al. 1992). Klyuchnikov et al. (2003) have discovered that the reaction of 2,3,4,6-tetranitroaniline with hydroxylamine hydrochloride in the presence of sodium acetate leads to 5-amino-4,6-dinitrobenzofuroxan and 3-amino-2,4,6-trinitrophenol. The formation of a furoxan ring proceeds through the replacement of the labile m-nitro groups in two 2,3,4,6-tetranitroanilines by hydroxylamine. The replacement results in coupling of the two trinitroaniline rests through the oxyiminobridge. The latter decomposes, forming aminodinitrobenzofuroxan and picrylamine according to Scheme 1.93.

1.3.2.1.7   From Nitrile Oxides

Dimerization of nitrile oxides leads to 3,4-disubstituted furoxans (Hawasi et al. 2005). Scheme 1.94 gives an example of one-pot synthesis of 3,4-bis(methyl carbonyl)-1,2,5-oxadiazole 2-oxide from acetone upon action of cerium ammonium nitrate (Itoh and Horiuchi 2004).

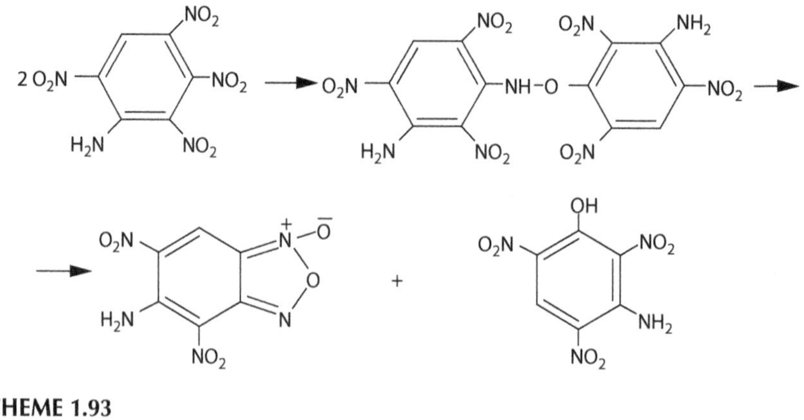

SCHEME 1.93

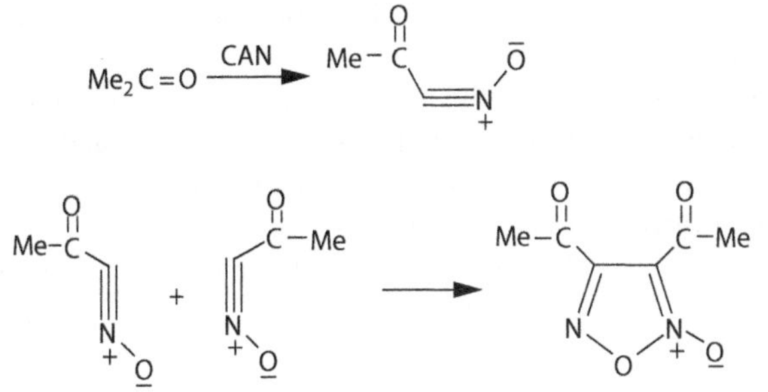

SCHEME 1.94

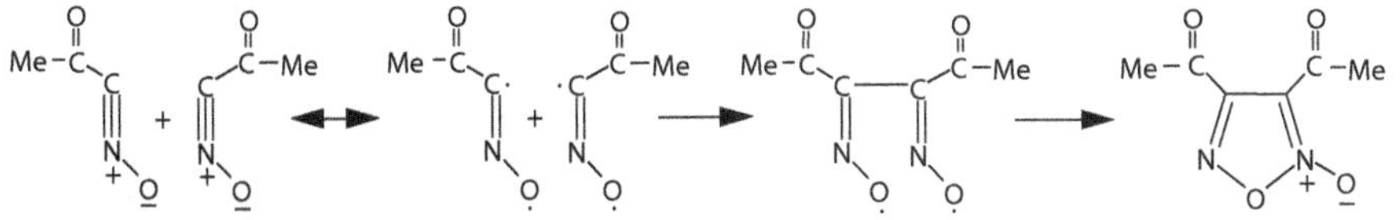

SCHEME 1.95

Similar mechanism is also considered in publications by Nishiwaki et al. (2003) and Krishnamurthy et al. (2006).

Dimerization of nitrile oxides, regardless of aliphatic or aromatic, can also proceed via diradical mechanism. Scheme 1.95 shows the sequence of the corresponding steps. This sequence was predicted by theoretical analysis at the B3LYP/6-31G level (Yu et al. 2003).

1.3.2.1.8 From Diazoketones

Cyclization of diazoketones with dinitrogen tetroxide was used in preparation of highly energetic 3,4-disubstituted furoxans according to Scheme 1.96 (Agrawal and Hodgson 2007).

1.3.2.2 Fine Structure

Comparative calculations show that the transformation of furazans into furoxans reduces the aromaticity indices (I_A, based on bond distances). Thus, I_A for 4,5-diphenylfurazan is 61.9, and for its N-oxide is 48.9 (reduction by 21%). For benzofurazan, $I_A = 106.3$, whereas for benzofuroxan $I_A = 81.0$ (reduction by 23%). For asymmetrically substituted furoxans and benzofuroxans, I_A's are essentially the

SCHEME 1.96

same for structures with localization of the exocyclic oxygen at the two different ring nitrogens (Bird 1985, 1992, 1993).

According to calculations by advanced quantum-chemical methods, the parent furoxan was predicted to be planar, with a strong exocyclic and a relatively weak endocyclic N–O bonds (Pasinszki et al. 2009). Magnitudes of dissociation energy for N^+–O^- bond in all furoxan derivatives are confined into the interval of 239–256 kJ/mol. These magnitudes are close to formation enthalpy of atomic oxygen, O, (250 kJ/mol), that is, practically lower by factor of 2 than the bond energy in molecular oxygen, O_2, (Matyushin et al. 2002 and references therein). From the data of natural population analysis, the furoxan moiety has a large negative charge (−0.46e) being electron rich. According to various aromaticity indices, furoxan itself can be considered as nearly as aromatic as furan and furazan (Pasinszki et al. 2009). We see that the aromaticity estimations for furoxan are ambiguous.

However, there is one cardinal peculiarity of furoxans. Formally belonging to the furazan family, furoxans are not true furazan N-oxides. As a matter of fact, the furoxan configuration is formed as a result of intramolecular coordinative interaction between the two vicinal nitroso groups. Oxygen of the one nitroso group interacts with nitrogen of the other nitroso group. This interaction takes place by an alternative manner so that the both nitroso groups change their kind of participation (see Scheme 1.97). Furoxans represent a case of ring-chain tautomerism. This tautomerism is energetically different for furoxans and benzofuroxans. It depends on the solvent and the temperature as well as on both the nature and position of substituents. The ring-chain tautomeric equilibration of benzofuroxans is observed at room temperature. In contrast, such a tautomerization of the non-fused furoxans takes place at elevated temperatures. There is a wide range of experimental works devoted to study this phenomenon. Some of them are here mentioned as examples. X-ray diffraction experiments established the vic-dinitroso nature of furoxans (Fruttero et al. 1997). One-electron reduction of benzofuroxan was failed to be fixed by the electron spin resonance method due to the low lifetime of this anion-radical (Todres 1970, Terrier et al. 2003). Multielectron reduction (in aprotic medium) takes place stepwise. Detachment of the N-oxide oxygen with the formation of the furazan anion-radical (Todres 1970) and cleavage of the furoxan ring with the formation of vicinal dioximes (Levin et al. 1966, Todres et al. 1974) were observed. Optical spectra testify for the aromatic character of the furoxan cycle

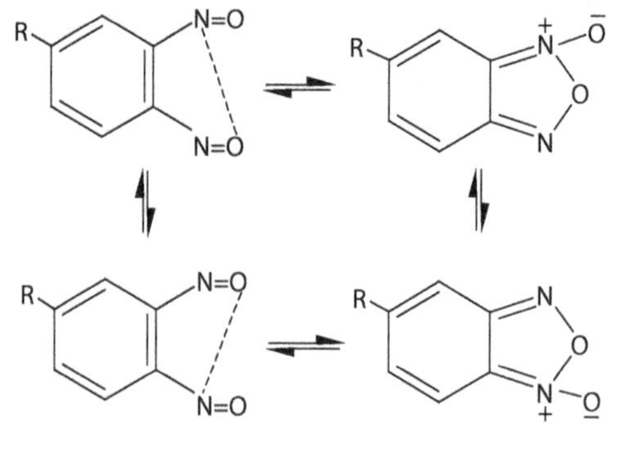

SCHEME 1.97

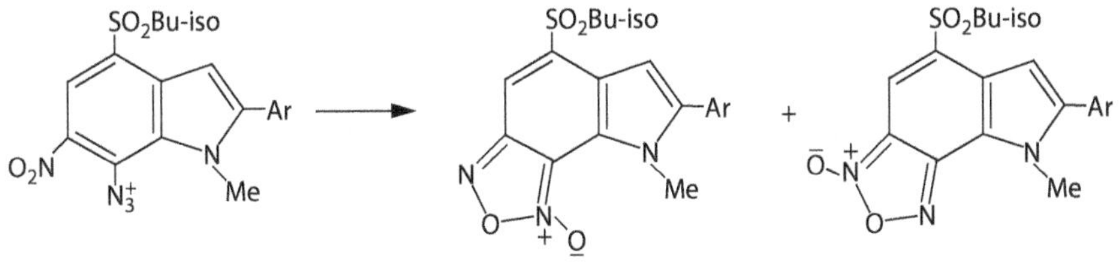

SCHEME 1.98 From Bastrakov, M.A. et al., *Mendeleev Commun.*, 19(1), 47, 2009. With permission.

temporarily formed upon the nitroso–nitroso coordination (Zaitsev et al. 1965). Fast tautomeric exchange in solution on NMR timescale was demonstrated (see, e.g., Diel et al. 1962, Godovikova et al. 1994, Aguirre et al. 2005).

In some fused furoxans, the equilibrium occurs to be shifted to the side of one tautomer. Thus, on refluxing in toluene, the indolazide of Scheme 1.98 undergoes intramolecular cyclization giving rise to the new tricyclic system—[1,2,5]-oxadiazolo[3,4-*g*] indole 3-oxide. This compound in dimethylsulfoxide exists as a mixture of Z and E isomers. (In a Z isomer, the furoxan N-oxide fragment is oriented to the indole N-methyl fragment. In an E isomer, these fragments are oriented to different flanks.) In the equilibrium of Scheme 1.98, the Z isomer predominates (Bastrakov et al. 2009). In crystalline form, only the Z isomer is present (Bastrakov et al. 2008). Smirnov et al. (2006) give a compendium of equilibrium states for other furoxans. Furoxan has outward-directed N-oxide when it is fused with quinoline or 7-bromobenzotetrazine 1,3-dioxide. Being fused with naphthalene in its 1,2-position, furoxan forms the dominant isomer with N-oxide oxygen oriented in an inward direction. There are also data concerning dependence of the equilibrium state on the solvent nature (Cillo and Lash 2004). It would be interesting to obtain an all-embracing explanation of the structure–equilibrium relationship for the whole furoxan family.

No appreciable transfer of the N-oxide oxygen from one of the ring nitrogen to another was also observed (even in dimethylsulfoxide solutions) for the σ-adduct obtained according to Scheme 1.99 (Terrier et al. 2005).

1.3.2.3 Reactivity

1.3.2.3.1 Salt Formation

Furoxan itself does not add protons to its ring nitrogen. Aminofuroxan is very weak as a base, but can react with ethoxycarbonyl isocyanate just at the expense of the furoxan amino substituent (Makhova et al. 2004).

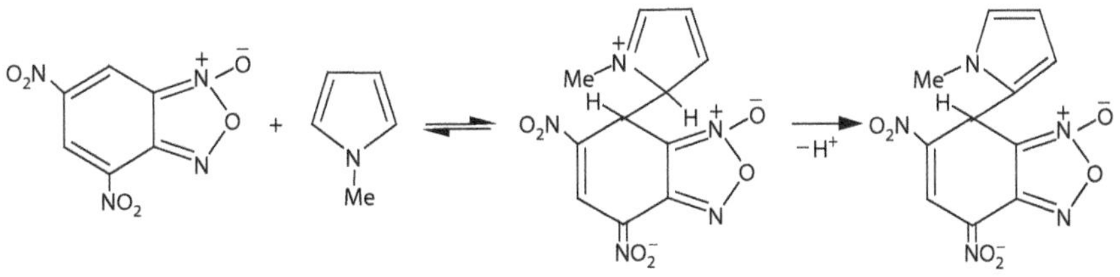

SCHEME 1.99

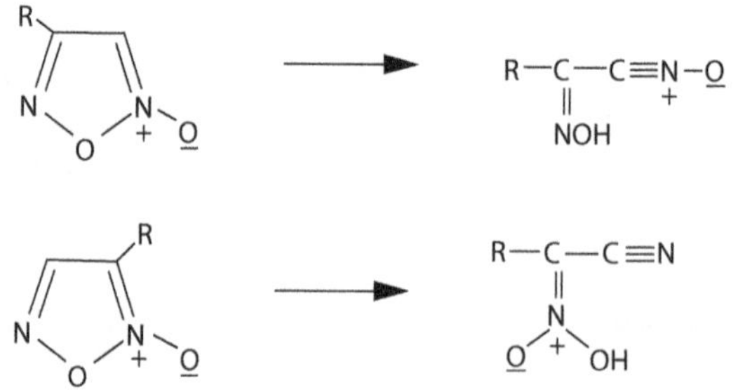

SCHEME 1.100

1.3.2.3.2 Reaction with Bases

Upon action of bases, 4-substituted furoxans transform into α-nitrile oxides of oximes, and 3-substituted furoxans into aciforms of α-nitroacylonitriles (Scheme 1.100) (Khmel'nitskii et al. 1983).

1.3.2.3.3 Reduction

The stepwise reduction of furoxans has already been considered in connection with their fine structure. It is reasonable to note that reducing furoxans does not always lead to ring cleavage. Thus, 4-substituted furoxans give 4-substituted furazans upon reduction with stannous chloride. 3-Substituted furoxans are not reduced under these conditions, but 3,4-derivatives do form the corresponding furazans. Triethyl phosphite reacts with 3,4-diphenylfuroxan forming triethylphosphate and 3,4-diphenylfurazan. Both products were obtained with yields of 93% and 94%, respectively (Mukayama et al. 1962).

Reduction of furoxans with lithium aluminum hydride or Grignard reactive results in the formation of amines or ketones so that furoxans with different substitutions in positions 3 and 4 give two different products according to Scheme 1.101 (Khmel'nitskii et al. 1983).

Reduction of furoxans with zinc in acetic acid, or sodium boron hydride, or hydroxyl amine in basic medium leads to the formation of glyoxime derivatives (Mukayama et al. 1962). Formation of 1,2-benzoquinone dioximes under electrochemical reduction of annulated furoxans has already discussed in Section 1.3.2.2.

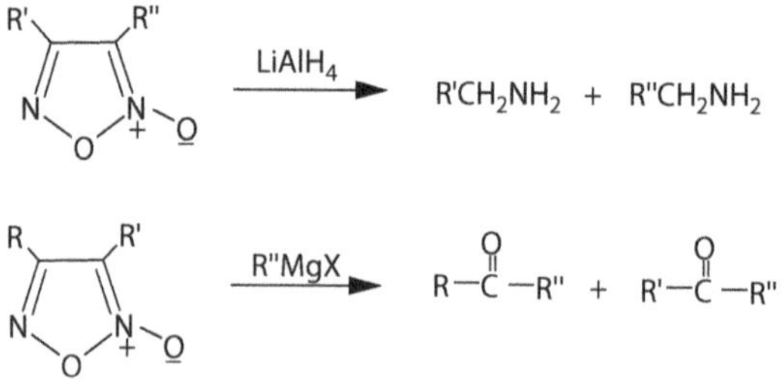

SCHEME 1.101

The reduction of the N-oxide moiety to produce furazans proceeds in mild condition by means of P(OAlk)$_3$ (Gut Ruggeri et al. 2003, Mandal and Maiti 1986) or PPh$_3$ (Das et al. 2006).

Hydrazine, hydroxylamine, and sodium azide are also used as reducing agents for furoxans and arenefuroxans (Khmel'nitskii et al. 1983). However, the above reducers are either difficult to access or highly toxic, or explosive and are not selective. Kondyukov et al. (2007) have found a new accessible, safe, and selective reducing agent for the transformation of benzofuroxans into benzofurazans. It is elemental sulfur. The reaction is performed in ethyleneglycol at 160°C for 2 h. Addition of morpholine to the reaction mixture enhances the reactivity of sulfur: Reduction is completed in 30 min and yields of benzofurazans increase doubly. Reduction of benzofuroxan nitroderivatives leads to the corresponding nitroderivatives of benzofurazan, the nitro groups remain intact.

Bioreduction of furoxans is also documented. Benzofuroxan forms 1,2-benzoquinone dioxime within mammal enzymatic system (Grosa et al. 2004). Benzofuroxans are able to be reduced by oxyhemoglobin to the corresponding 1,2-nitroaniline derivatives (Medana et al. 2001). This fact indicates that blood is a possible site for metabolism of benzofuroxans with the consequent methemoglobinemia.

Borah et al. (2009) have developed a simple and inexpensive approach to aryl-1,2-diamines by reduction of benzofuroxans with bakers' yeast under no fermenting condition in buffer solution of pH 7.0. Yields are excellent and halogen or nitro group (as substituents in homocyclic ring of benzofuroxan) remain intact during the reductive cleavage. The fact that the nitro group is not suffered occurs to be wonderful because there is an earlier report from Baik et al. (1994) on selective reduction of aromatic nitro compounds to aromatic amines by baker's yeast in basic solutions.

1.3.2.3.4 Oxidation

Although peracids such as performic or peracetic acid oxidize nitroso compounds into the nitro counterparts, furoxans do not change upon action of these oxidants (Kaufman and Picard 1959). Oxidation of benzofuroxan into o-dinitrobenzene was achieved only upon action of trifluoroperacetic acid. In this reaction, however, the product yield is small (all possible precautions had been taken). Furoxans fused with the naphthalene or phenathrene ring are not oxidized under these conditions (Boyer and Ellzey 1959). The inertness mentioned is a consequence of the definite aromaticity of the furoxan ring.

1.3.2.3.5 Electrophilic Substitution

Monocyclic furoxans are inert in respect of electrophiles. Aromatically annulated furoxans do react with electrophiles; see Schemes 1.102 (Chupakhin et al. 2004) and 1.103 (Agrawal and Hodgson 2007).

1.3.2.3.6 Nucleophilic Substitution

The furoxan ring is more susceptible to a nucleophilic attack than to an electrophilic one. As an example, substitution of the ethoxyl group for the phenylsulfonyl moiety is depicted in Scheme 1.104 (Fruttero et al. 1997).

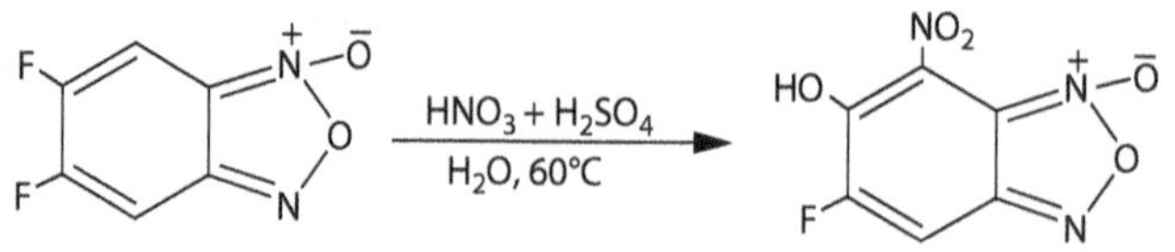

SCHEME 1.102

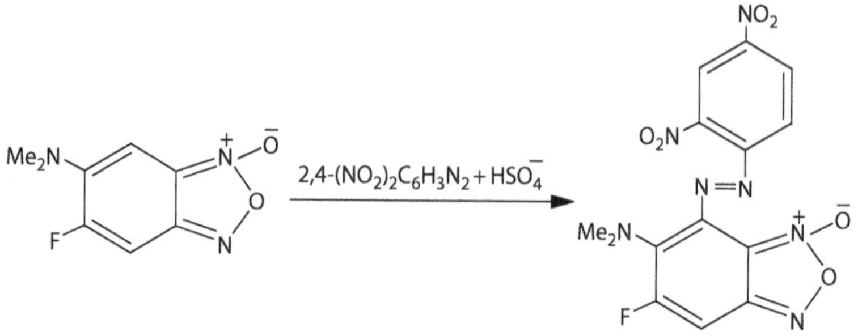

SCHEME 1.103

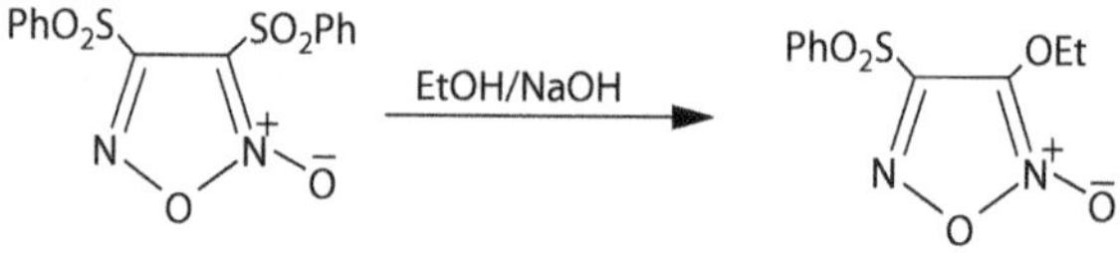

SCHEME 1.104

Being condensed with the furoxan moiety, the substituted benzenoid ring in benzofuroxans becomes to be active in nucleophilic reactions. Thus, β,β′-difluorobenzofuroxan enters in displacement of aromatically bound fluorine; (see Scheme 1.118) devoted to ring transformation (Vicente et al. 2008b). α,α′-Dinitrobenzofuroxan reacts as a superelectrophile with such very weak carbon nucleophiles as N-methylindole, azulene, and 1,2,5-trimethylpyrrole (Kurbatov et al. 2003, Rodrigues-Dafonte et al. 2009). The reactions result in the formation of stable Meisenheimer complexes. This superelectrophile is even able to displace the aryl-diazo fragments of the Meisenheimer complexes by 1,3,5-tris(N,N-dialkylamino)benzenes (Boga et al. 2007).

Scheme 1.105 shows that the superelectrophile forms the C–C Meisenheimer complex with 2-amino-4-methylthiazole. The new bond is formed with the thiazole cyclic carbon, but not at the expense of the amino group. The N-Meisenheimer complex is formed only in the case of 2-amino-4,5-dimethylthiazole that has no unsubstituted cyclic carbon (Forlani et al. 2006).

1.3.2.3.7 Photolysis

Upon photolysis, dimethylfuroxan forms two products, namely, acetonitrile N-oxide and 2,3-dinitrosobut-2-ene. The latter product can generate the former one (Himmel

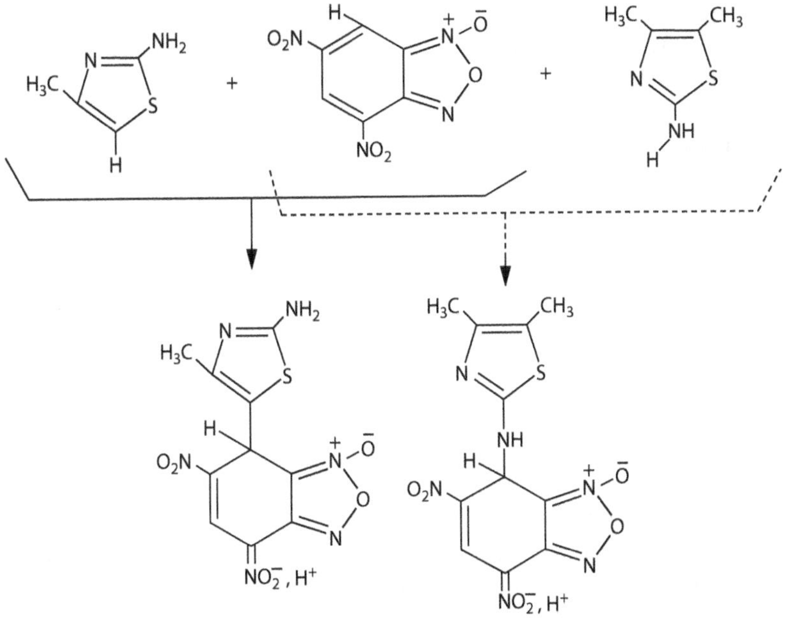

SCHEME 1.105

et al. 2003). The most interesting photolytic reaction of benzofuroxan leads to a lactame, namely, 1*H*-azepine-2,7-dione. This lactame results from irradiation using a high-pressure mercury lamp with a Pyrex filter in acetonitrile solution containing a small amount of water (Hasegawa and Takabatake 1991).

1.3.2.3.8 Thermolysis

The furoxan ring is known for its unique tendency toward thermal opening along bonds C_3–C_4 and N_2–O_1 to give two nitrile oxide fragments. Once generated, these fragments can dimerize into the corresponding furoxan (cf. Scheme 1.94). Thermolysis temperature varies with substituents at the ring C atoms. With bulky substituents the fragmentation is more facile (Chapman et al. 1976). On other hand, the substituent nature is also significant. Thus, the thermolysis temperature varies from 20°C for 3.4-dinitrofuroxan (Ovchinnikov et al. 1995) to 250°C for 3,4-diphenylfuroxan (Grundmann 1966). Dimethylfuroxan forms acetonitrile N-oxide with a high yield upon thermal decomposition (600°C) in the gas phase (Pasinszki and Westwood 2001). Norbornenofuroxan (IUPAC name 4,5,6,7-tetrahydro-4,7-methano-2,1,3-benzoxadiazole 1-oxide) at 100°C forms cyclopentane-1,3-di(nitrile oxide). This reaction was recommended for cross-linking in polymer synthesis (Kulikov et al. 2007).

Sometimes, the nitrile N-oxides formed isomerize into isocyanates at the pyrolysis temperature according to Scheme 1.106.

Unlike dimethylfuroxan, the parent furoxan, upon gas-phase thermolysis, does not split up to the monomer nitrile oxide, yielding only HNCO, HCN, CO_2, CO, NO, and H_2O decomposition products (Pasinszki et al. 2009). Heating 3- or 4-substituted

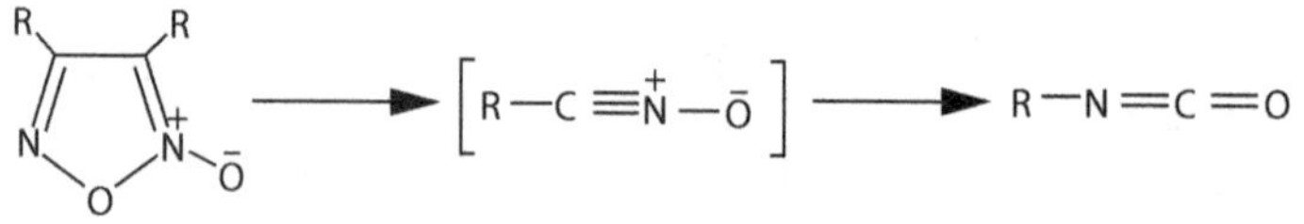

SCHEME 1.106

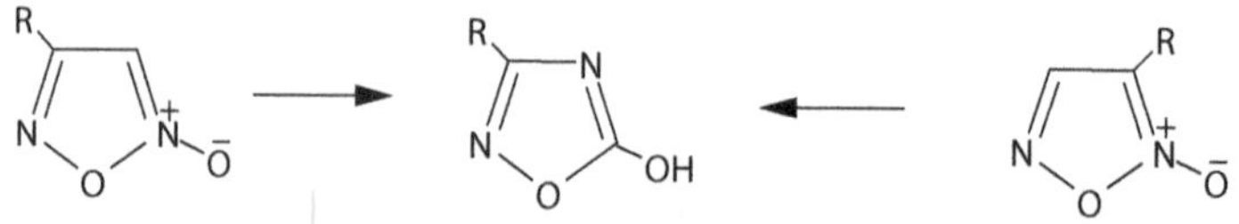

SCHEME 1.107

furoxan in protic acids at 140°C initiates their isomerization into 3-substituted 5-hydroxy-1,2,4-oxadiazole (Scheme 1.107) (Khmel'nitskii et al. 1983).

1.3.2.3.9 Cycloaddition

Furoxans participate in this reaction after ring opening to give the substructure C=N⁺–O⁻. Dipolarophiles add to these substructures yielding isoxazole derivatives (Shimizu et al. 1985, Calvino et al. 1993).

The diene-dienophile cycloaddition (the normal electron-demand Diels–Alder reaction) was performed by Vichard et al. (2001) (Scheme 1.108) and theoretically described by Ayadi et al. (2008) within B3LYP approach of density functional theory.

The multifaceted Diels–Alder reactivity of α,β'-dinitrobenzofuroxan toward isoprene and 2,3-dimethylbutadiene was studied by Kresze and Bathelt (1973), Goumont et al. (2004). Two consecutive normal electron-demand condensations involve the two nitro-activated double bonds of the furoxan benzoid moiety. The structures of the initially formed (short-lived) monoadducts were especially established. These monoadducts undergo subsequent addition of the second molecule of a diene, as illustrated in Scheme 1.109.

1.3.2.3.10 Ring Transformation

Condensation of benzofuroxan with nitrones in refluxing hydrocarbon solvents leads to benzimidazolium derivatives with high yields. These derivatives give nitroxyl radicals after one-electron oxidation with lead(IV) oxide or silver(I) triflate (Dooley

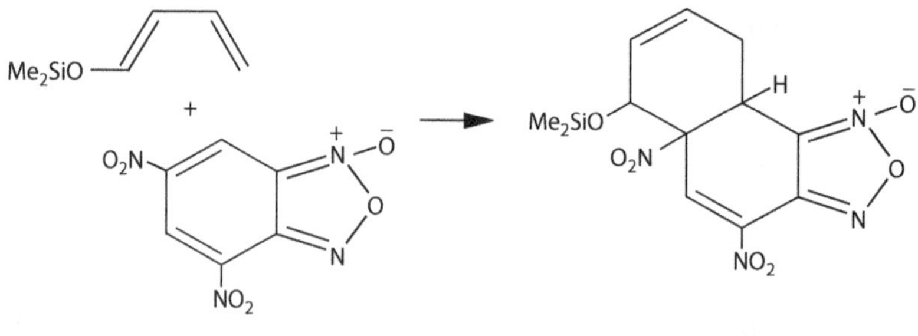

SCHEME 1.108

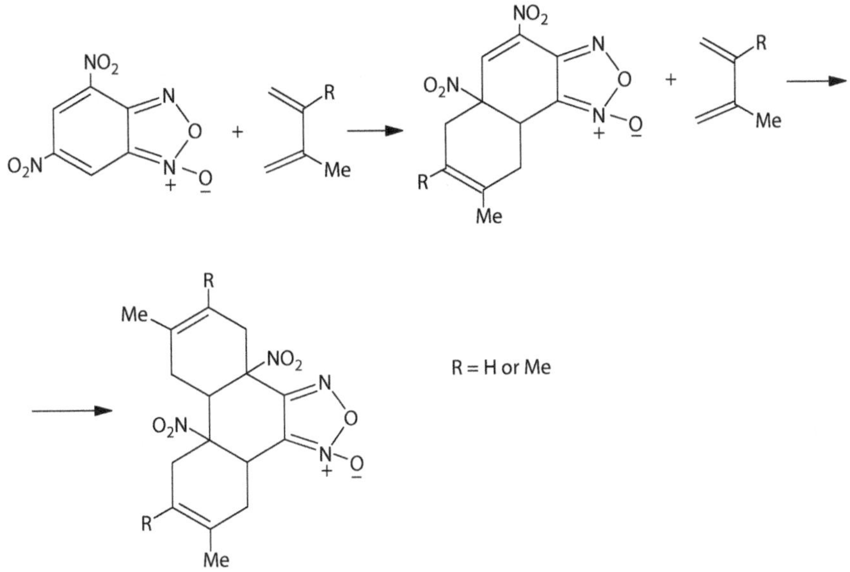

SCHEME 1.109

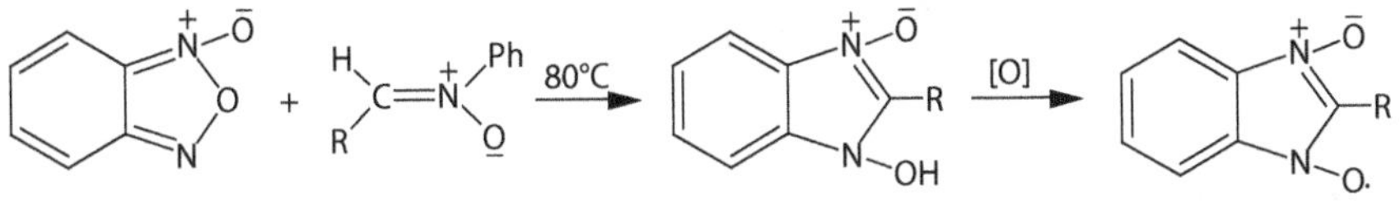

SCHEME 1.110

et al. 2007). Scheme 1.110 shows the reaction sequence, but no possible mechanism of this ring transformation was discussed. What is important, the radicals formed exhibit electronic transition in the near infrared region (850–900 nm) and very low (close to zero) reduction potentials. They are very easily reduced and act as excellent electron acceptors. Such properties are ideal for the development of multifunctional magnetic materials.

Boulton and Katritzky (1962) described ring-opening → ring-closure reactions of benzofuroxans with unsaturated groups at α position(s). Following studies discovered a wide body of such reactions. Schemes 1.111 (Karmakar et al. 1998), 1.112, and 1.113 (Makhova et al. 2004) give the corresponding examples.

The ring-opening → ring-closing tautomerism of Scheme 1.114 (Eckert et al. 1999) makes understandable the irreversible rearrangement of α-nitro-β-methylbenzofuroxan to α-nitro-α′-methylbenzofuroxan upon gently heating (see Scheme 1.115) (Katritzky and Gordeev 1993).

If the starting compound of Scheme 1.115 bears fluorine instead of methyl in β-position, no rearrangement occurs (Khmel'nitskii et al. 1983). Proximity of fluorine apparently prevents the nitro group from participating in rebuilding of the furoxan

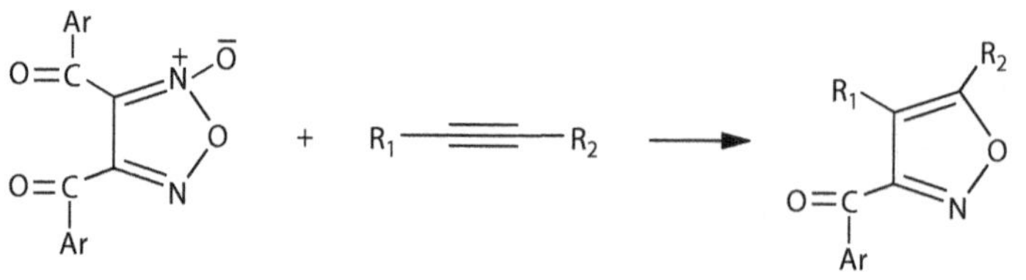

SCHEME 1.111

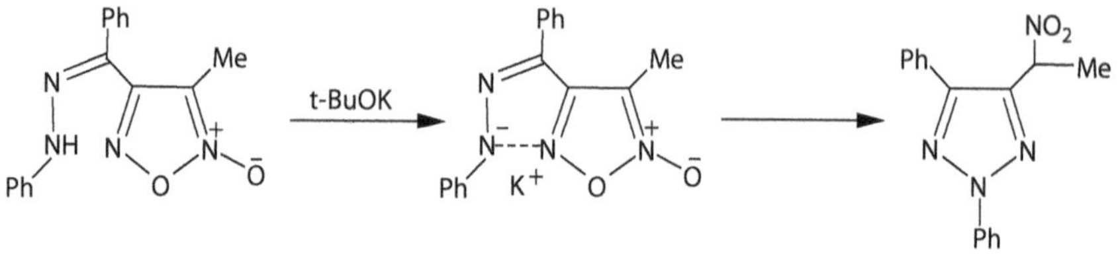

SCHEME 1.112

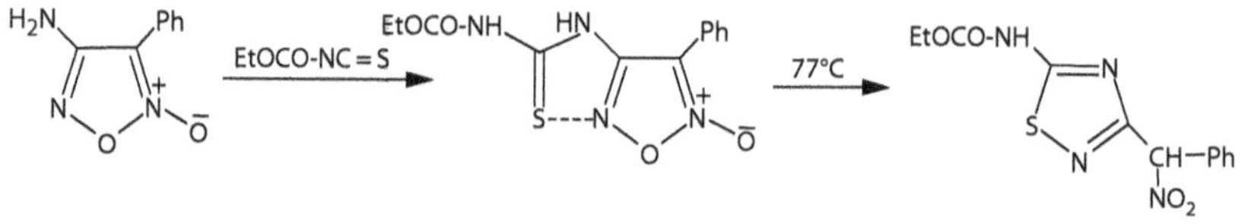

SCHEME 1.113

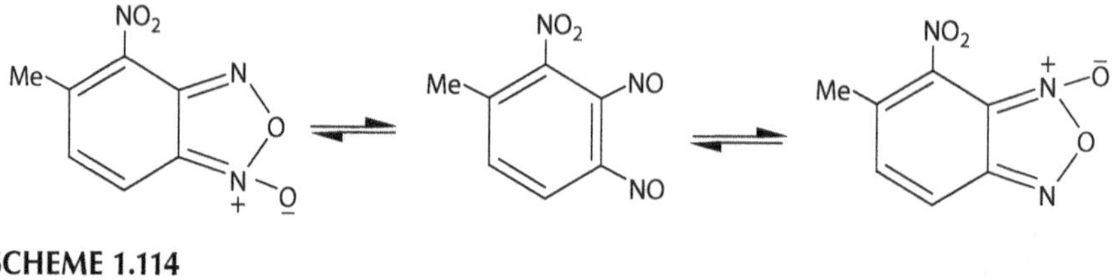

SCHEME 1.114

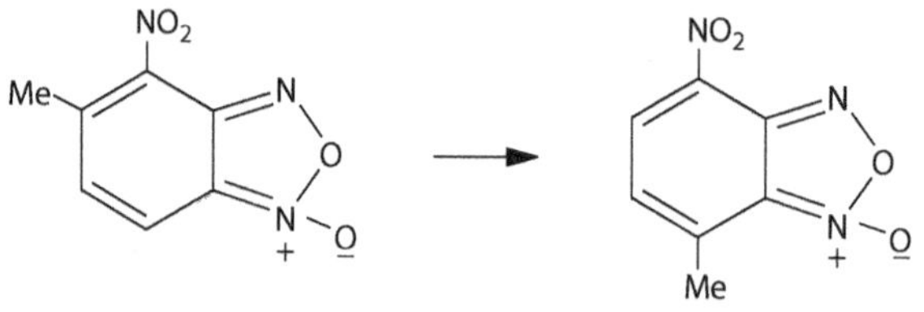

SCHEME 1.115

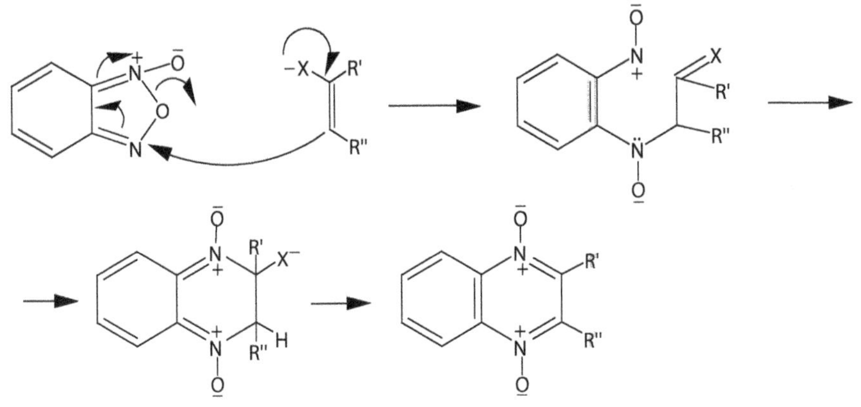

SCHEME 1.116

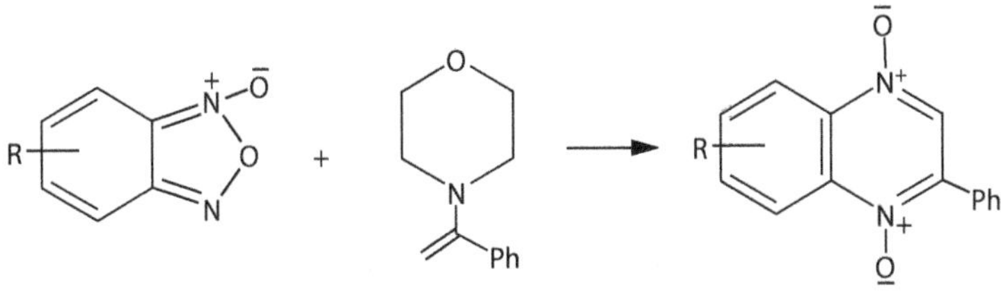

SCHEME 1.117

ring. At the same time, the vicinal dislocation of α-nitro and β-methylenesulfonyl-4-tolyl groups does not hinder the rearrangement, for some reason or other (Ostrowski and Wojciechowski 1990).

The Beirut rearrangement is a classical way to produce pharmaceutically required quinoxaline *N,N*-dioxides (see, e.g., Vicente et al. 2008a). The reaction had first been found by Haddadin and Issodorides (1965) in Beirut, the capital of Lebanon. The Beirut rearrangement involves condensation of benzofuroxan with enamines, dienes, aldehydes, α,β-unsaturated ketones, or enolates to form quinoxaline *N,N*-dioxides (see also review by Laursen and Nielsen 2004). Scheme 1.116 can depict the Beirut rearrangement, and the illustrative examples are represented by Schemes 1.117 (Rong et al. 2006) and 1.118 (Vicente et al. 2008b).

Dahbi et al. (2010) extended the Beirut rearrangement and prepared 2-phosphorylated quinoxaline 1,4-dioxides from benzofuroxan in the presence of molecular sieves (Scheme 1.119). In this case, good and reproducible yields up to 85% were obtained only when the initial suspensions of the sieves and reactants in tetrahydrofuran were transformed into pasty films by slow evaporation of ca. 90% of the initial solvent volume. The reactions run at room temperature and needs 3–5 days.

As has already been mentioned in Section 1.3.2.3.6 the reaction of Scheme 1.118 between difluorobenzofuroxan and benzyl acetone (in methanol upon ammonium bubbling) is accompanied by substitution of methoxide for fluorine. It should be

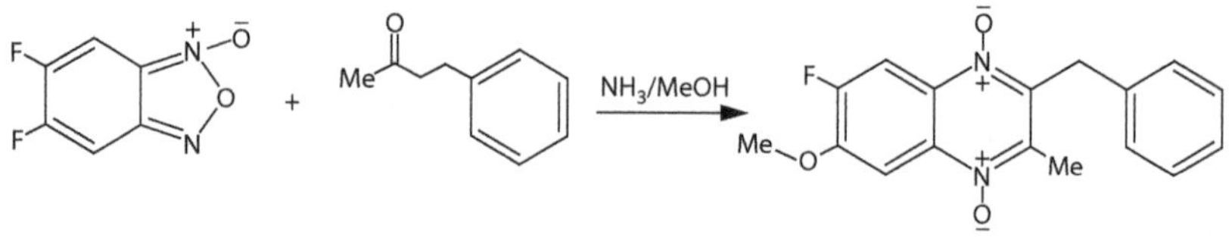

SCHEME 1.118

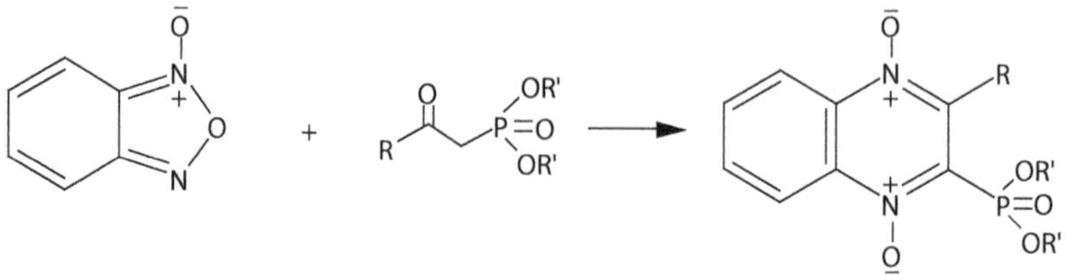

SCHEME 1.119

noted that the fluoro substituent is activated by the other (vicinal) fluorine together with the electron-withdrawing furoxan nucleus. The substitutions of nucleophiles for fluorine in aromatic compounds are rather rare. The fluorine must be activated by electron-withdrawing substituents. The displacing nucleophile must be strong enough. For instance, thioglycolate is able of fluorine displacing in 4-nitrofluoroben-zene (Dutov et al. 2007). According to Scheme 1.118, the fluorine displacement with the methoxide (another strong nucleophile) is a part of the combined reaction of the furoxan → quinoxaline dioxide transformation.

Transformation of benzofuroxans into substituted quinoxaline bases was described by Shi et al. (2008) (Scheme 1.120). Various biologically active quinoxa-line derivatives were efficiently synthesized in excellent yields by the reaction of 1,2-diketones with benzofuroxans promoted by tin dichloride. The latter is acting as a reducer and as a catalyst in this reaction.

Another rearrangement of furoxans deserves to be mentioned. This rearrange-ment is little known, but brings important opportunities in the field of intermedi-ates for synthetic dyes. The following materials will illustrate this statement. The rearrangement takes place in aromatic α,β-fused furoxans, the condensed ring of which is hydrogenated in positions 3 and 4 and contains the sodium sulfonate group in position 4. So-called bisulfite compounds of naphtho-1,2-furoxan (Bogdanov and Korolyova 1954, Bogdanov and Zil'berman 1956, Bogdanov and Todres 1961a,b,

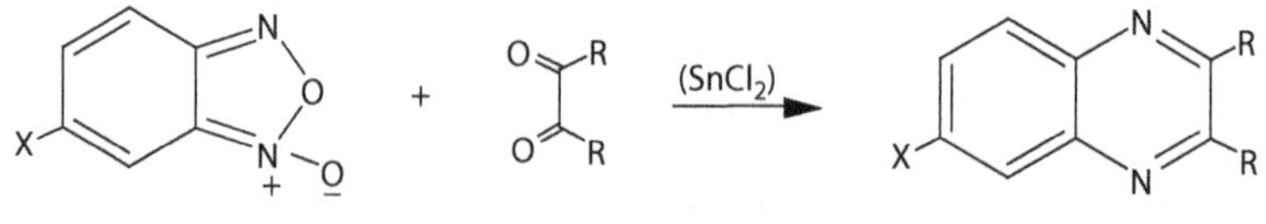

SCHEME 1.120

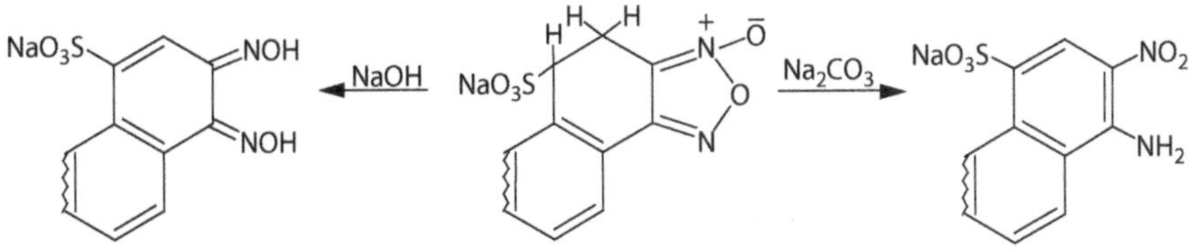

SCHEME 1.121

1962), anthra-1,2-furoxan (Bogdanov and Gorelik 1959), and phenanthro-1,2-fu-roxan (Bogdanov and Shibryaeva 1961) undergo this rearrangement according to Scheme 1.121. The Bogdanov rearrangement develops in two directions depending on basicity of an initiator. Upon the alkali hydroxide action, 1,2-quinone dioxime-4-sulfoacids are formed. Alkali carbonates initiate transformations into 2-nitro-1-amino-4-sulfoacids. Although mechanism of this rearrangement remains unclear, its practical importance is doubtless.

The Bogdanov rearrangement provides synthons for non-traditional aminon-aphthol sulfo acids. New aminonaphthol sulfo acids are attractive for synthesis of patent-pure azo dyes. The used repertoiry of intermediates from the aminonaphthol sulfo acid series contains γ-acid (2-amino-8-naphthol-7-sulfo acid), I-acid (2-amino-5-naphthol-7-sulfo acid), M-acid (1-amino-5-naphthol-7-sulfo acid), Chicago-S-acid (1-amino-8-naphthol-4-sulfo acid), Echt-acid (1-amino-2-napthol-4-sulfo acid), and B-acid (6-amino-2-napthol-4-sulfo acid). One unknown before Z-acid (7-amino-2-napthol-4-sulfo acid) was managed to obtain (Todres 1991). The foolproof structure of Z-acid was established by x-ray crystallographic analysis (Todres et al. 1982). This acid widens the repertoire mentioned and its applicability in preparation of patent-pure azo dyes was claimed (Todres et al. 1965).

1.3.2.3.11   Coordination

Among coordinative reactions to be considered, those are interesting that touch the furoxan ring. In this sense, coordination of 3,4-bis(2-pyridyl)furoxan to copper, palladium, or ruthenium deserves to be compared. The ligand forms palladium(II) and copper(II) complexes having seven-membered chelate rings with coordination via two pyridyl nitrogens. In contrast, the reaction with bis(2,2′-bipyridine)dichlororuthenium leads to an unusual ring-opening → ring-closing rearrangement to produce the 3-nitro-2-(2-pyridyl)pyrazolo[1,5-a]pyridine complex with bis(2,2′-bipyridine) dichlororuthenium (Scheme 1.122). The depicted structures of a copper complex and a ruthenium complex were established by x-ray crystallography (Richardson and Steel 2000).

The ruthenium part of Scheme 1.122 illustrates the very unusual rearrangement of 3,4-bis(2-pyridyl)furoxan into 3-nitro-2-(2-pyridyl)pyrazolo[1,5-a]pyridine. Complexation to ruthenium obviously facilitates this rearrangement, chemical sense of which can be understandable from Scheme 1.123.

Scheme 1.123 is a sketchy picture of the rearrangement. Participation of the ruthenium counterpart is necessary for the reaction to proceed. Absolutely no reaction takes place under the same conditions in the absence of the ruthenium reagent or in the presence of nickel(II) perchlorate (Richardson and Steel 2000).

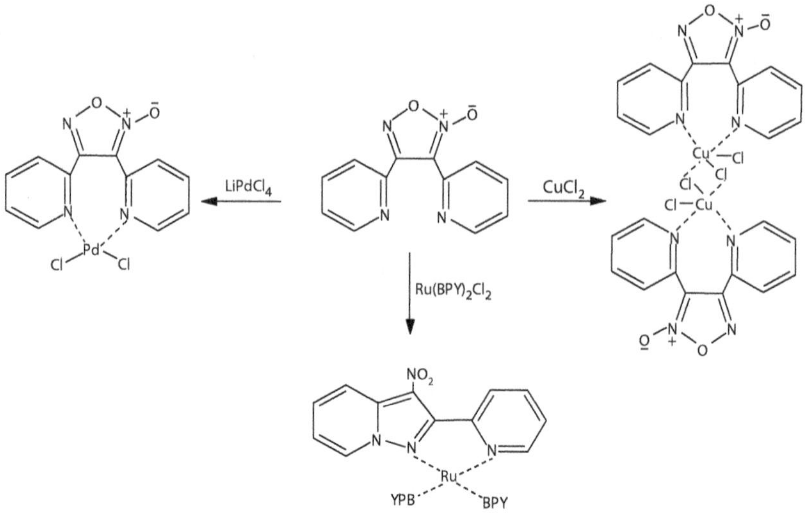

SCHEME 1.122

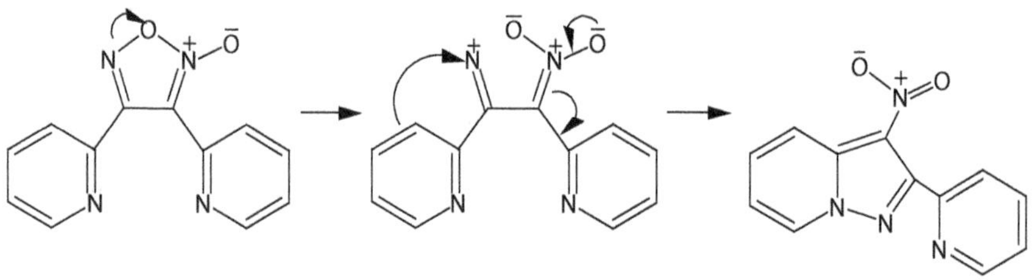

SCHEME 1.123

Likewise, oxidation of vic-dioximes is significantly facilitated if it is provided within copper complexes. Namely, the starting dioxime in suspension with triethylamine is introduced in acetonitrile solution containing cupric chloride and sodium perchlorate. Oxidation proceeds at room temperature and the furoxan formed as a part of its complex with copper. To isolate the furoxan, usual treatment with aqueous ammonia is sufficient. Yields of furoxans are about 90% (Das et al. 2008).

1.3.2.4 Biomedical Importance

1.3.2.4.1 Medical Significance

Furoxans and benzofuroxans are capable of transforming into the corresponding dinitroso compounds. Reacting with nucleophiles in biological conditions, these dinitroso compounds release nitrogen monoxide. In adequate tissues, nitrogen monoxide releasing can proceed in a controlled manner. This provides the furoxan family with properties of effective drugs. Pirogov et al. (2004) claimed 3,4-bis(furazan-3-yl) furoxans as inhibitors of platelet aggregation. The authors underline that this kind of activity is connected with generation of NO. (Furazans are also apt to generate NO, but less readily than furoxans are.) Coronary vasodilatation from furoxan N,N'-dialkylcarboxamides was proved to be caused by NO generation upon action

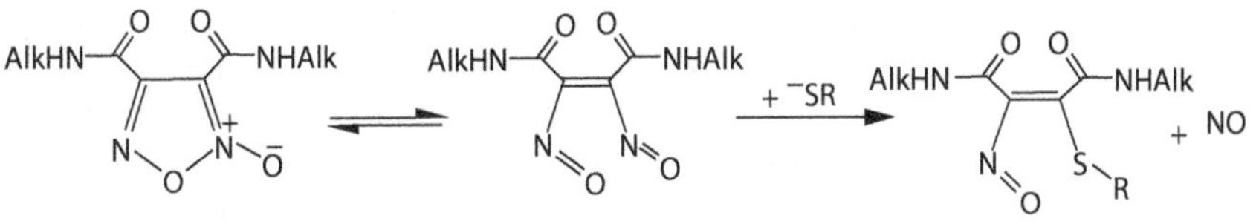

SCHEME 1.124

of endogenic thiols, RS⁻H⁺ such as L-cysteine (Feelisch et al. 1992). Scheme 1.124 explains the biochemical reactions mentioned.

Generally, nitrogen monoxide plays a crucial role in vascular homeostasis dilating arterial blood vessels. It inhibits platelet adherence and aggregation, attenuates leukocyte adherence and activation. Nitrogen monoxide facilitates the release of several neurotransmitters and hormones, stimulating the enzyme soluble guanylate cyclase. It places an important role in the immune response by cytotoxic action on macrophages and leukocytes. Wang et al. (2002) point out the nitrogen monoxide protective ability with respect to damage of cells associated with bio oxidation stress. Being a radical particle, nitrogen monoxide protects tissues such as lung epithelium from oxidants. Contemporarily, furoxans as NO donor agents are increasingly claimed: Hermann and List (1994), Narayanan (2002), Ellis (2005), Carvey and Ranatunge (2006), Ellis and Carvey (2007), and Buonsanti et al. (2007).

Sayed et al. (2008) identified various furoxans as new drugs for the control of schistosomiasis. This kind of furoxan activity was also associated with elimination of nitrogen monoxide. On the other hand, some furoxans are mutagenic and their toxic effect was ascribed to NO-releasing capacity (Aguirre et al. 2006).

Furoxans are active against very dangerous parasitic illnesses that infect many people in the world. Furoxans were tested against the parasites *Trypanosoma crusi* (*T. crusi*), which is responsible for American Trypanosomiasis, and *Plasmodium falciparum* (*P. falciparum*), which is responsible for malaria. Malaria is a universally known disease particularly widespread in underdeveloped countries. American Trypanosomiasis, in other words, Chagas' disease, is widely prevalent in the American hemisphere from southern United States to southern Argentina. It is endemic in 21 countries, with 18–20 million persons infected and 40 million people at risk of acquiring the disease. The disease is transmitted to humans by hematophagous *Reduviid* insects. From the insects, trypomastigotes are introduced through abraded skin at the site of biting. The trypomastigotes penetrate into human cells, rupture them, emerge from the ruptured ones, reach adjacent cells and bloodstream, and spread within the organism. The acute phase of the disease is usually seen in children and is characterized by fever, malaise, anorexia, adenopathy, or local sign of inflammation at the site of the insect bite. However, generally, there is no clear acute stage and persons infected may remain free of symptoms. In about one-third of acute cases, a chronic form develops approximately 10–20 years later, causing cardiomyopathy or gastrointestinal dysfunction. Patients with chronic disease die usually from heart failure. Recent studies indicate that there are ca. 200,000 new infected cases and 21,000 deaths by Chagas' disease every year (Urbina and Docampo 2003). At present, there is no effective treatment for such patients.

The drugs most frequently used in the treatment of Chagas' disease have very low antiparasitic activity in the chronic stage of the disease and bring serious side effects. The furoxan family representatives were tested in this respect. Mechanism of their biological action was studied, see Boiani et al. (2008a,b). Cerecetto et al. (1999) and Porcal et al. (2007) compared β-ethenyl derivatives of benzofuroxans and benzofurazans as in vitro anti–*T. crusi* agents. Only the furoxan samples manifest trypanocidal activity. For the furazan compounds (the N-oxide-free counterparts of furoxans) the lack of activity was observed. The N-oxide moiety occurred to be relevant in this sense. The furoxan compounds perturb the mitochondrial electron chain, inhibiting parasite respiration. In this connection, other paper by Porcal et al. (2008) should be mentioned. The authors developed robust and multigram-batch processes for the preparation of new antipyranosomal benzofuroxans, namely E and Z isomers of 5-arylethenylbenzo[1,2-*c*]-[1,2,5]-oxadiazole N_2-oxides. The processes developed differ in minimization of the generation of the benzofurazans as secondary (and unneeded) by-products.

Further studies identified 3-phenylsulfono-4-thio[2-(*N,N*-dimethylamino)ethyl] furoxan as effective anti-trypanosomal drug. Its inactivation ability was explained by the nitrogen monoxide release during metabolism (Ascenzi et al. 2004).

In respect of the malaria possible treatment, some furoxan 3-phenyl-4-sulfolano derivatives were found to be anti–*P. falciparum* agents. The corresponding furazan analogs turned to be ineffective (Galli et al. 2005). According to the authors, the furoxan activity is the result, at least in part, of the furoxan's NO-releasing ability.

One prospective way to construct new effective medicines consists in the combination of a furoxanyl or a benzofuroxanyl moiety (capable of NO-releasing) with another pharmacologically active substructure in one and the same molecule. Scheme 1.125 represents such a molecule. In preclinical tests, this compound behaves as a well-balanced hybrid with mixed calcium-channel blocking and NO-dependent vasodilatating activity (Rolando et al. 2002). The vasodilatating action is caused by the furoxan part whereas the calcium-channel blocking activity is introduced by the nitrophenyl dimethyl dihydropyridine biscarboxylate constituent.

In addition, a series of furoxan sulfonyl derivatives were obtained and characterized as preclinical candidates for the treatment of the following diseases: inflammation (Fang et al. 2007), thrombus formation, platelet deposition (Carvey and

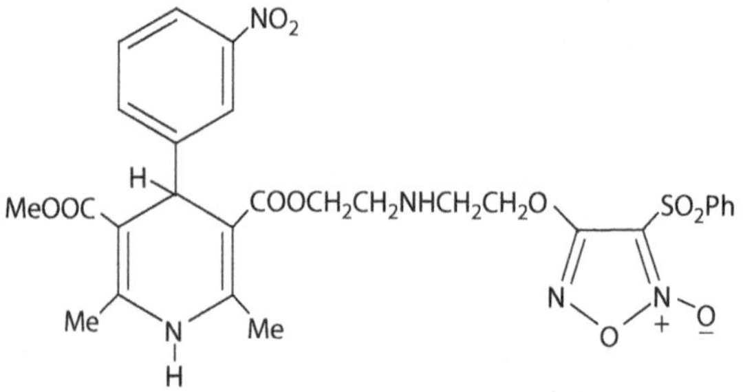

SCHEME 1.125

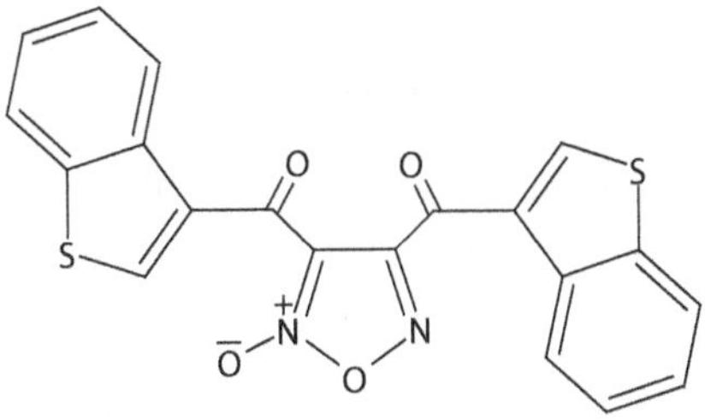

SCHEME 1.126

Ranatunge 2007) as well as atheroma-underpinned cardiovascular illness (Boschi et al. 2006). Furoxan containing arylsulfonyl and other therapeutically active constituents were designed and synthesized as potential anticancer agents (e.g., Moharram et al. 2004, Kong et al. 2008).

Chakravarti and Klopman (2008) worked out a program to quantitatively characterize furoxan differential cytotoxicity. This definition implies that a cancer chemotherapeutic agent does not only inhibit the growth or kill the cancer cells, but inhibits without eliciting unreasonable cytotoxic effects on the healthy cells. Furoxan of Scheme 1.126 was characterized as an anticancer drug of the high differential cytotoxic effect (Chakravarti and Klopman 2008). While the 1,4-diketo group provides a general cytotoxic effect, the furoxan fragment transforms this effect into a differential one.

In the preclinical treatment of cancer, tumor location is crucial and the corresponding radiopharmaceuticals are objects of intense search. Cerecetto et al. (2006) have synthesized an interesting complex, in which the monodentate ligand 3-mercaptomethyl-4-phenyl-furoxan and tridentate ligand N,N-bis(2-mercaptoethyl)-N',N'-diethylethylenediamine were coordinated to ^{99}Tc. In the complex, the tridentate ligand is bound via the mercapto sulfur and the amino nitrogens whereas the furoxan derivative is bound via its mercapto sulfur only. No furoxan nitrogens are involved. The complex obtained complies with many requirements but its tumor uptake and gastrointestinal activity are insufficient. The authors promise to design and synthesize technetium complexes with modified ligands, including new furoxanyl ones.

1.3.2.4.2 Agricultural Protectors

Benzofuroxan, α-nitrofuroxan and β-methoxyfuroxan were patented as insect feeding deterrents (Iwamoto et al. 1977). Sodium furoxan-β-carboxylate was claimed as rice protectant (Gibbs 1998). Monocyclic and benzene-fused substituted furoxans are herbicides. Fernandez et al. (2005) analyzed the relationship between physicochemical properties and herbicidal activity of these derivatives. The variance in their biological action was explained by changes in the lipophilicity and in the reduction potential.

1.3.2.5 Technical Applications

Furoxans and their condensed compounds with aromatic fragments found applications in a very wide spectrum. For instance, they were included in the formulation as rubber additives (Klyuchnikov et al. 2005), as inhibitors in the polymerization of aromatic vinyl monomers (Freedman et al. 1997), as components in the igniting composition for inflation of airbags (Zeuner et al. 2000), as solid propellants (Hong et al. 2001), as burn-rate modifiers (Shinde et al. 2003), and as liquid-crystalline materials (Bezborodov et al. 2004a).

1.3.2.5.1 Formation of Liquid-Crystalline State

As liquid crystals, 4,5-bis(4-alkoxyphenyl)furoxans form nematic phases over a very wide temperature range, greater than 100°C. Structural studies show that these furoxans are conical (not rod-like) (Bezborodov et al. 2004b). Therefore, the formation of enantiotropic nematic phase is surprising because the nematic phases are typical only for linear molecular structures characterized by a thread-like order with chaotic arrangement of molecular centers of gravity. Reasons of such exceptional behavior wait for their clarification.

1.3.2.5.2 Explosiveness

The furoxan ring is a highly energetic heterocycle. Furoxans and benzofuroxans are used as explosives (see, for instance, Pepekin and Smirnov 2001, Sikder and Sikder 2001, Sikder et al. 2002). Simple nitro furoxan compounds, such as 3,4-dinitro derivative, are unstable at room temperature and highly sensitive to impact. Dinitroazofuroxan, namely, 3,3'-diazene-1,2-diylbis(4-nitro-1,2,5-oxadiazole)-5,5'-dioxide has been described as a superpower explosive (Ovchinnikov et al. 1998). It has practically acceptable properties, that is, detonation velocity about 10 km/s and crystal density close to 2 g/cm^3. Such a high density is due to very efficient crystal packing. (High crystal density is an important demand for explosive.) This material was elegantly synthesized from the oxidative coupling of 4-amino-3-(azidocarbonyl)furoxan, followed by Curtius rearrangement and oxidation of the resulting amino groups to the nitro ones according to Scheme 1.127 (see review by Binnikov et al. 1999).

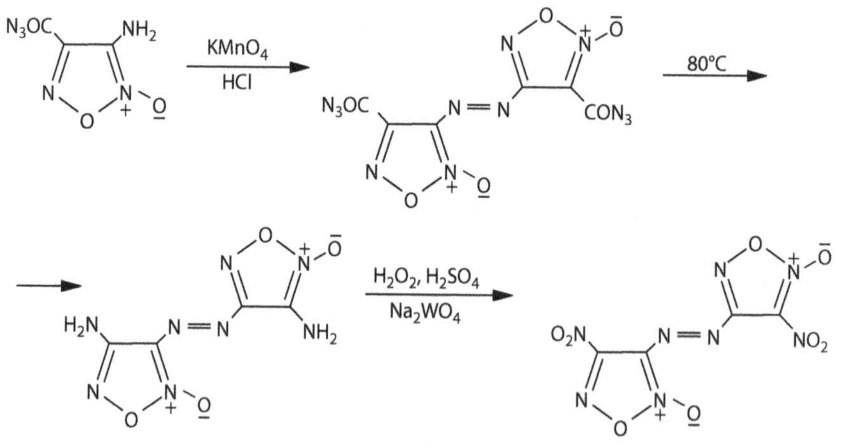

SCHEME 1.127

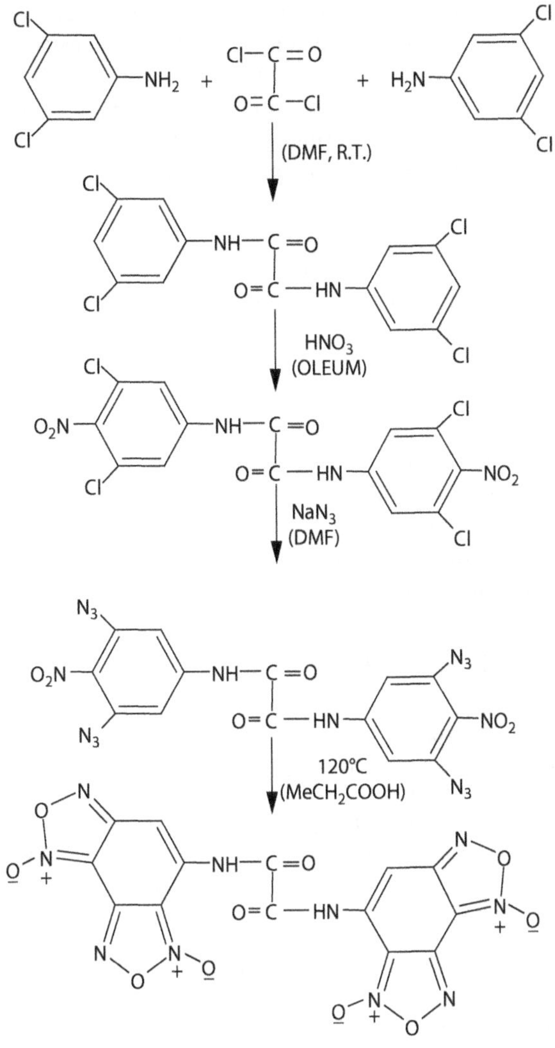

SCHEME 1.128

Benzofuroxans are more stable than non-fused furoxans and are more favorable for applications as explosives. Benzofuroxans containing amino functionality adjacent to a nitro group are of particular value. Such a group mutual disposition reduces impact sensibility of the corresponding explosives and increases their thermal stability (the properties important for transportation and storing of energetic materials). Analogously, α'-amino-α,β-dinitrobenzofuroxan is characterized with perfect explosive properties, namely, detonation velocity of ~7,9 km/s, crystal density of 1.90 g/cm³ and melting (decomposing) temperature of 270°C. This material is under advanced development in the United States and it was twice claimed (Norris 1988, Weber 1992).

Wang et al. (2008) synthesized *N,N'*-bis(difuroxano[3,4-*b*:3',4'-*d*]phenyl)oxalic amide (Scheme 1.128). The final compound was obtained with ca. 50% yield taking into account all four steps of the reaction. The product is high-energetic explosive with low sensitivity, its detonation velocity is 8.17 km/s, density is 1.92 g/cm^3 and melting (decomposing) temperature is 240°C.

The part of Section 1.3.5 connected with furoxans as high-energetic compounds must be closed with the following warning statement: It is both dangerous and illegal to participate in unauthorized experimentation with explosives.

1.3.2.6 Conclusion

Ring-to-open-chain transferability is the intrinsic and crucial property of furoxans. This transformation leads to nitroso compounds as intermediates. The later can cyclize once again, but isomeric furoxans are formed. This endows furoxan chemistry with definite specificity and with ability of NO-releasing in biosystems. Accordingly, furoxans remains very attractive for pharmaceutical evaluating and possible using. Uncommon and attractive properties of furoxans as liquid crystals also remain a prospective field of the future studies. Being rich with nitrogen and oxygen, many furoxans are explosives with acceptable exploitation characteristics. Engineering applications of furoxans are also promising. Having in mind the transformation of furoxans into other useful substances, it should be pointed out that their ability of giving dioximes or furazans depends on reduction conditions.

REFERENCES TO SECTION 1.3.2

Agrawal, J.P.; Hodgson, R.D. (2007) *Organic Chemistry of Explosives.* Wiley, Chichester, U.K.

Aguirre, G.; Boiani, L.; Boiani, M.; Cerecetto, H.; di Maio, R.; Ganzalez, M.; Porcal, W.; Denicola, A.; Piro, O.E.; Castellano, E.E.; Sant'Anna, C.M.R.; Barreiro, E.J. (2005) *Bioorg. Med. Chem.* **13**, 6336.

Aguirre, G.; Boiani, M.; Certecetto, H.; Fernandes, M.; Gonzalez, M.; Leon, E.; Pintos, C.; Raymondo, S.; Arredondo, C.; Pacheco, J.P.; Bosombrio, M.A. (2006) *Farmazie* **61**, 54.

Al-Hiari, Y.M.; Khanfar, M.A.; Qaisi, A.M.; Shuheil, M.Y.A.; El-Abadelah, M.M.; Boese, R. (2006) *Heterocycles* **68**, 1163.

Armani, V.; Dell'Erba, C.; Novi, M.; Petrillo, G.; Tavani, C. (1997) *Tetrahedron* **53**, 1751.

Ascenzi, P.; Bocedi, A.; Gentile, M.; Visca, P.; Gradoni, L. (2004) *Biochim. Biophys. Acta* **1703**, 69.

Avemaria, F.; Zimmermann, V.; Braese, S. (2004) *SYNLETT*, 1163.

Ayadi, S.; Essalah, Kh.; Krichane, S.; Abderraba, M. (2008) *Orient. J. Chem.* **24**, 121.

Baik, W.; Han, J.L.; Lee, K.; Lee, N.H.; Kim, B.H.; Hahn, J.-T. (1994) *Tetrahedron Lett.* **35**, 3965.

Bastrakov, M.A.; Starosotnikov, A.M.; Leontieva, M.A.; Shakhnes, A.Kh.; Shevelev, S.A. (2009) *Mendeleev Commun.* **19**, 47.

Bastrakov, M.A.; Starosotnikov, A.M.; Shakhnes, A.Kh.; Shevelev, S.A. (2008) *Izv. RAN, Ser. Khim.* 1508.

Bezborodov, V.S.; Kauhanka, M.M.; Lapanik, V.I. (2004a) *Vestsi Nats. AN Belarusi, Ser. Khim. Navuk* **3**, 61.

Bezborodov, V.S.; Kauhanka, M.M.; Lapanik, V.I. (2004b) *Liq. Crystl.* **31**, 295.

Binnikov, A.N.; Kulikov, A.S.; Makhova, N.N.; Ovchinnikov, I.V.; Pivina, T.S. (1999) *Proceedings of the 30th International Annual Conference of ICT*, Vol. 58, p. 1. (Karlsruhe, Germany 1999, June 29–July 2).

Bird, C.W. (1985) *Tetrahedron* **41**, 1409.

Bird, C.W. (1992) *Tetrahedron* **48**, 335.

Bird, C.W. (1993) *Tetrahedron* **49**, 8441.

Boga, C.; del Vecchio, E.; Forlani, L.; Tozzi, S. (2007) *J. Org. Chem.* **72**, 8741.

Bogdanov, S.V.; Gorelik, M.V. (1959) *Zh. Obshch. Khim.* **29**, 146.

Bogdanov, S.V.; Korolyova, I.N. (1954) *Zh. Obshch. Khim.* **24**, 1994.

Bogdanov, S.V.; Shibryaeva, L.S. (1961) *Zh. Obshch. Khim.* **31**, 522.

Bogdanov, S.V.; Todres, Z.V. (1961a) *Zh. VKhO* **6**, 584.

Bogdanov, S.V.; Todres, Z.V. (1961b) *Zh. VKhO* **6**, 585.

Bogdanov, S.V.; Todres, Z.V. (1962) *Zh. VKhO* **7**, 697.

Bogdanov, S.V.; Zil'berman, N.I. (1956) *Zh. Obshch. Khim.* **26**, 2071.

Bohn, H.; Brendel, J.; Martorana, P.A.; Schoenafinger, K. (1995) *J. Pharmacol.* **114**, 1605.

Boiani, L.; Aguirre, G.; Gonzalez, M.; Cerecetto, H.; Chidichimo, A.; Cazzulo, J.J.; Bertinaria, M.; Guglielmo, S. (2008a) *Bioorg. Med. Chem.* **16**, 7900.

Boiani, M.; Cerecetto, H.; Gonzalez, M.; Gasteiger, J. (2008b) *J. Chem. Inf. Model.* **48**, 213.

Borah, H.N.; Prajapati, D.; Boruah, R.C. (2009) *Synth. Commun.* **39**, 267.

Boschi, D.; Tron, G.C.; Lazzarato, L.; Chegaev, K.; Cena, C.; di Stilo, A.; Giorgis, M.; Bertinaria, M.; Fruttero, M.; Casco, A. (2006) *J. Med. Chem.* **49**, 2886.

Boulton, A.J.; Katritzky, A.R. (1962) *Rev. Chim. Acad. Rep. Pop. Roum.* **7**, 691.

Boyer, J.; Ellzey, S. (1959) *J. Org. Chem.* **24**, 2038.

Buonsanti, M.F.; Bertinaria, M.; di Stilo, A.; Gena, C.; Fruttero, R.; Casco, A. (2007) *J. Med. Chem.* **50**, 5003.

Calvino, R.; di Stilo, A.; Fruttero, R.; Gasco, A.M.; Sorba, G.; Gasco, A. (1993) *Farmaco* **48**, 321.

Carvey, D.S.; Ranatunge, R.R. (2006) WO Patent 055,542.

Carvey, D.S.; Ranatunge, R.R. (2007) WO Patent 059,311.

Cerecetto, H.; di Maio, R.; Gonzalez, M.; Risso, M.; Saenz, P.; Seoane, G.; Denicola, A.; Peluffo, G.; Quijano, C.; Olea-Azar, C. (1999) *J. Med. Chem.* **42**, 1941.

Cerecetto, H.; Gonzalez, M.; Onetto, S.; Risso, M.; Rey, A.; Giglio, J.; Leon, E.; Pilatti, P.; Fernandez, M. (2006) *Arch. Pharm.* **339**, 59.

Chakravarti, S.K.; Klopman, G. (2008) *Bioorg. Med. Chem.* **16**, 4052.

Chapman, J.A.; Crosby, J.; Cummings, C.A.; Rennie, R.C.A.; Paton, R.M. (1976) *J. Chem. Soc. Chem. Commun.*, 240.

Chupakhin, O.N.; Kotovskaya, S.K.; Romanova, S.A.; Charushin, V.N. (2004) *Russ. J. Org. Chem.* **40**, 1167.

Cillo, C.; Lash, T.D. (2004) *J. Heterocycl. Chem.* **41**, 955.

Curini, M.; Epifano, F.; Marcotullio, M.C.; Rosati, O.; Ballini, R.; Bosica, G. (2000) *Tetrahedron Lett.* **41**, 8817.

Dahbi, S.; Methnani, E.; Bisseret, Ph. (2010) *Tetrahedron Lett.* **51**, 5516.

Das, B.; Rudra, S.; Salman, M.; Rattan, A. (2006) WO Patent 035,283.

Das, O.; Paria, S.; Paine, T.K. (2008) *Tetrahedron Lett.* **49**, 5924.

Diel, P.; Christ, H.; Mallory, F. (1962) *Helv. Chim. Acta* **195**, 504.

Dinh, N.H.; Ly, N.Th.; Van, L.Th.Th. (2004) *J. Heterocycl. Chem.* **41**, 1015.

Dooley, B.M.; Bowles, S.E.; Storr, T.; Frank, N.L. (2007) *Org. Lett.* **9**, 4781.

Dutov, M.D.; Serushkina, O.V.; Shevelev, S.A.; Lyssenko, K.A. (2007) *Mendeleev Commun.* **17**, 347.

Dyall, L.K.; Harvey, J.J.; Jarman, T.B. (1992) *Aust. J. Chem.* **45**, 371.

Eckert, F.; Rauhut, G.; Katritzky, A.R.; Steel, P.J. (1999) *J. Am. Chem. Soc.* **121**, 6700.

Ellis, J.L. (2005) WO Patent 070,006.

Ellis, J.L.; Carvey, D.S. (2007) WO Patent 016,677.

Fang, L.; Zhang, Y.; Lehmann, J.; Wang, Y.; Ji, H.; Ding, D. (2007) *Bioorg. Med. Chem. Lett.* **17**, 1062.

Feelisch, M.; Schoenafinger, K.; Noak, E. (1992) *Biochem. Pharmacol.* **44**, 1149.

Fernandez, L.A.; Santo, M.R.; Reta, M.; Giacomelli, L.; Cattana, R.; Silber, J.J.; Risso, M.; Cerecetto, H.; Gonzalez, M.; Olea-Azar, C. (2005) *Molecules* **10**, 1197.

Forlani, L.; Tocke, A.L.; del Vecchio, E.; Lakhdar, S.; Goumont, R.; Terrier, F. (2006) *J. Org. Chem.* **71**, 5527.

Freedman, H.S.; Abruscato, G.J.; DeMassa, J.M.; Gentile, A.V.; Grossi, A.V. (1997) US Patent 5,659,095.

Fruttero, R.; Sorba, G.; Ermondi, G.; Lolli, M.; Gasco, A. (1997) *Farmaco* **52**, 405.

Galli, U.; Lazzarato, L.; Bertinaria, M.; Sorba, G.; Gasco, A.; Parapini, S.; Taramelli, D. (2005) *Eur. J. Med. Chem.* **40**, 1335.

Gibbs, D.E. (1998) US Patent 5,773,454.

Godovikova, T.I.; Golova, S.P.; Strelenko, Y.A.; Antipin, M.Yu.; Struchkov, Yu.T.; Khmel'nitskii, L.I. (1994) *Mendeleev Commun.* 7.

Goumont, R.; Sebban, M.; Marrot, J.; Terrier, F. (2004) *ARKIVOC* (iii), 85.

Grosa, G.; Galli, U.; Rolando, B.; Fruttero, R.; Gervasio, G.; Gasco, A. (2004) *Xenobiotica* **34**, 345.

Grundmann, C. (1966) *Fortschr. Chem. Forsch.* **7**, 62.

Gut Ruggeri, S.; Bill, D.R.; Bourassa, D.E.; Castaldi, M.J.; Houck, T.L.; Brown Ripin, D.H.; Wei, L.; Weston, N. (2003) *Org. Proc. Res. Develop.* **7**, 1043.

Haddadin, M.J.; Issodorides, C.H. (1965) *Tetrahedron Lett.* **6**, 3253.

Hasegawa, M.; Takabatake, T. (1991) *J. Heterocycl. Chem.* **28**, 1079.

Havasi, B.; Pasinszki, T.; Westwood, N.P.C. (2005) *J. Phys. Chem. A* **109**, 3864.

Hermann, F.; List, N. (1994) Germ. Patent 4,305,881.

Himmel, H.-J.; Konrad, S.; Friedrichsen, W.; Rauhut, G. (2003) *J. Phys. Chem. A* **107**, 6731.

Hong, W.; Tian, D.; Liu, J.; Wang, F. (2001) *Guti Huojian Jishu* **24**, 41.

Itoh, K.; Horiuchi, C.A. (2004) *Tetrahedron* **60**, 1671.

Iwamoto, R.; Sakata, H.; Okumura, K.; Hongo, A.; Sekiguchi, S. (1977) Jpn. Patent 52,007,055.

Karmakar, D.; Prajapati, D.; Sandhu, J.S. (1998) *Synth. Commun.* **28**, 2415.

Katritzky, A.R.; Gordeev, M.F. (1993) *Heterocycles* **35**, 483.

Kaufman, J.; Picard, J. (1959) *Chem. Rev.* **59**, 429.

Khmel'nitskii, L.I.; Novikov, S.S.; Godovikova, T.I. (1983) *Chemistry of Furoxans. Reactions and Applications (Khimiya Furoxanov. Reaktsii I Primenenie).* Nauka, Moscow, Russia.

Klyuchnikov, O.R.; Deberdeev, R.Ya.; Berlin, A.A. (2005) *Dokl. RAN* **400**, 491.

Klyuchnikov, O.R.; Starovoitov, V.I.; Khairutdinov, F.G.; Golovin, V.V. (2003) *Khim. Geterotsikl. Soed.* 142.

Kondyukov, I.Z.; Karpychev, Yu.V.; Belyaev, P.G.; Khisamutdinov, G.Kh.; Valeshnii, S.I.; Smirnov, S.P.; Il'in, V.P. (2007) *Zh. Org. Khim.* **43**, 636.

Kong, X.W.; Zhang, Y.H.; Dai, L.; Ji, H.; Lai, Y.Sh.; Peng, S.X. (2008) *Chinese Chem. Lett.* **19**, 149.

Kresze, G.; Bathelt, H. (1973) *Tetrahedron* **29**, 1043.

Krishnamurthy, V.N.; Talawar, M.B.; Vyas, S.M.; Kusurkar, R.S.; Asthana, S.N. (2006) *Defence Sci. J.* **56**, 551.

Kulikov, A.S.; Epishina, M.A.; Ovchinnikov, I.V.; Makhova, N.N. (2007) *Izv. RAN, Ser. Khim.* 1521.

Kurbatov, S.; Rodriguez-Dafonte, P.; Goumont, R.; Terrier, F. (2003) *Chem. Commun.* 2150.

Laursen, J.B.; Nielsen, J. (2004) *Chem. Rev.* **104**, 1663.

Levin, E.S.; Fodiman, Z.I.; Todres, Z.V. (1966) *Elektorokhimiya* **2**, 175.

Makhova, N.N.; Ovchinnikov, I.V.; Kulikov, A.S.; Molotov, S.I.; Baryshnikova, E.L. (2004) *Pure Appl. Chem.* **76**, 1691.

Mandal, B.K.; Maiti, S. (1986) *Eur. Polym. J.* **22**, 447.

Matyushin, Y.N.; Lebedev, V.P.; Chironov, V.V.; Pepekin, V.I. (2002) *Khim. Fiz.* **21**, 58.

Medana, C.; Visentin, S.; Grosa, G.; Fruttero, R.; Gasco, A. (2001) *Farmaco* **56**, 799.

Moharram, S.; Zhou, A.; Wiebe, L.I.; Knaus, E.E. (2004) *J. Med. Chem.* **47**, 1840.

Mukayama, T.; Nambu, H.; Okamoto, M. (1962) *J. Org. Chem.* **27**, 3651.

Narayanan, A.S. (2002) Ind. Patent 188,020.

Nishiwaki, N.; Okajima, Y.; Tamura, M.; Asaka, N.; Hori, K.; Tohda, Y.; Ariga, M. (2003) *Heterocycles* **60**, 303.

Norris, W.P. (1988) US Patent H 000,476.

Ostrowski, S.; Wojciechowski, K. (1990) *Can. J. Chem.* **68**, 2239.

Ovchinnikov, I.V.; Makhova, N.N.; Khmel'nitskii, L.I.; Kuzmin, V.S.; Akimova, L.N.; Pipekin, V.I. (1998) *Dokl. RAN* **359**, 499.

Ovchinnikov, I.V.; Popov, N.A.; Makhova, N.N. (1995) *Mendeleev Commun.* 231.

Pasinszki, T.; Havasi, B.; Hajgato, B.; Westwood, N.P.C. (2009) *J. Phys. Chem. A* **113**, 170.

Pasinszki, T.; Westwood, N.P.C. (2001) *J. Phys. Chem. A* **105**, 1244.

Pepekin, V.I.; Smirnov, A.S. (2001) *Khim. Fiz.* **20**, 78.

Pirogov, S.V.; Mel'nikova, S.F.; Tselinskii, I.V.; Romanova, T.V.; Spiridonova, N.P.; Betin, V.L.; Postnikov, A.B.; Kots, A.Y.; Khropov, Y.V.; Gavrilova, S.A.; Grafov, M.A.; Medvedeva, N.A.; Pytakova, N.V.; Severina, I.S.; Bulgarina, T.V. (2004) Russ. Patent 2,240,321.

Ponzio, G.; Satta, C. (1930) *Gazz. Chim. Ital.* **60**, 150.

Porcal, W.; Hernandez, P.; Aguirre, G.; Boiani, L.; Boiani, M.; Merlino, A.; Ferreira, A.; di Maio, R.; Castro, A.; Gonzalez, M.; Cerecetto, H. (2007) *Bioorg. Med. Chem.* **15**, 2768.

Porcal, W.; Merlino, A.; Boiani, M.; Gerpe, A.; Gonzalez, M.; Cerecetto, H. (2008) *Org. Proc. Res. Develop.* **12**, 156.

Prakash, O.; Pannu, K. (2007) *ARKIVOC* (viii), 28.

Richardson, Ch.; Steel, P.J. (2000) *Aust. J. Chem.* **53**, 93.

Rodrigues-Dafonte, P.; Terrier, F.; Lakhdar, S.; Kurbatov, S.; Goumont, R. (2009) *J. Org. Chem.* **74**, 3305.

Rolando, B.; Cena, C.; Caron, G.; Marini, E; Grosa, G.; Fruttero, R.; Gasco, A. (2002) *Med. Chem. Res.* **11**, 322.

Rong, L.C.; Li, X.Y.; Yao, C.S.; Wang, H.Y.; Shi, D.Q. (2006) *Acta Crystallogr.* **E62**, O1959.

Sayed, A.A.; Simeonov, A.; Thomas, C.J.; Inglese, J.; Austin, Ch.; Williams, D.L. (2008) *Nat. Med.* **14**, 407.

Shi, D.-Q.; Dou, G.-L.; Ni, S.-N.; Shi, J.-W.; Li, X.-Y. (2008) *J. Heterocycl. Chem.* **45**, 1797.

Shimizu, T.; Hayashi, Y.; Taniguchi, T.; Teramura, K. (1985) *Tetrahedron* **41**, 727.

Shinde, P.D.; Salunke, R.B.; Agrawal, J.P. (2003) *Propellants Explos. Pyrotech.* **28**, 77.

Sikder, A.K.; Salunke, R.B.; Sikder, N. (2002) *J. Energ. Mater.* **20**, 39.

Sikder, A.K.; Sikder, N. (2001) *Indian J. Heterocycl. Chem.* **11**, 149.

Smirnov, O.Yu.; Tyurin, A.Yu.; Churakov, A.M.; Strelenko, Yu.A.; Tartakovsky, V.A. (2006) *Izv. RAN, Ser. Khim.*, 133.

Stadlbauer, W.; Fiala, W.; Fisher, M.; Hojas, G. (2000) *J. Heterocycl. Chem.* **37**, 1253.

Terrier, F.; Lakhdar, S.; Boubaker, T.; Goumont, R. (2005) *J. Org. Chem.* **70**, 6242.

Terrier, F.; Mokhtari, M.; Goumont, R.; Halle, J.C.; Buncel, E. (2003) *Org. Biol. Chem.* **1**, 1757.

Tinh, D.V.; Stadlbauer, W. (2008) *J. Heterocycl. Chem.* **45**, 1695.

Todres, Z.V. (1970) *Izv. AN SSSR, Ser. Khim.*, 1749.

Todres, Z.V. (1991) *Phosphorus Sulfur* **61**, 351.

Todres, Z.V.; Espenbetov, A.A.; Yanovskii, A.I.; Struchkov, Yu.T. (1982) *Zh. Org. Khim.* **18**, 1512.

Todres, Z.V.; Fodiman, Z.I.; Levin, E.S. (1974) *Khim. Geterotsikl. Soed.* 604.

Todres, Z.V.; Moor, V.I.; Abramova, N.I.; Gryzlova, Z.A. (1965) Russ. Patent 167,915.

Trogu, E.; Cecchi, L.; de Sarlo, F.; Guideri, L.; Ponticelli, F.; Machetti, F. (2009) *Eur. J. Org. Chem.* 5971.

Tyurin, A.Yu.; Smirnov, O.Yu.; Chirakov, A.M.; Strelenko, Yu.A.; Tartakovsky, V.A. (2006) *Izv. RAN, Ser. Khim.* 341.

Urbina, J.A.; Docampo, R. (2003) *Trends Parasitol.* **19**, 495.

Vicente, E.; Lima, L.M.; Bongard, E.; Chamaud, S.; Villar, R.; Solano, B.; Burguete, A.;
    Perez-Silanes, S.; Aldana, I.; Vivas, L.; Monge, A. (2008a) *Eur. J. Med. Chem.* **43**, 1903.
Vicente, E.; Villar, R.; Burguete, A.; Solano, B.; Ancizu, S.; Perez-Silanes, S.; Aldana, I.;
    Monge, A. (2008b) *Molecules* **13**, 86.
Vichard, D.; Alvey, L.J.; Terrier, F. (2001) *Tetrahedron Lett.* **42**, 7571.
Vladykin, V.I.; Trakhtenberg, S.I. (1980) Russ. Patent 721,430.
Wang, J.-L.; Lu, L.-Y.; Ou, Y.-X. (2008) *Chin. J. Chem.* **26**, 190.
Wang, P.G.; Xian, M.; Tang, X.; Wu, X.; Wen, Z.; Cai, T.; Janczuk, A.J. (2002) *Chem. Rev.*
    102, 1091.
Weber, J.F. (1992) US Patent 5,136,041.
Yu, Zh.-X.; Caramella, P.; Houk, K.N. (2003) *J. Am. Chem. Soc.* **125**, 15420.
Zaitsev, B.E.; Pozdyshev, V.A.; Todres, Z.V. (1965) *Khim. Geterotsikl. Soed.* 825.
Zeuner, S.; Schropp, R.; Hofmann, A.; Roedig, K.H. (2000) Germ. Patent 19,840,993.

1.4   1,3,4-OXADIAZOLES

1.4.1   FORMATION

1.4.1.1   From Thiosemicarbazides

A facile and general protocol for the preparation of 5-substituted 2-(*N*-alkyl or aryl amino)-1,3,4-oxadiazoles was proposed by Dolman et al. (2006). The method relies on a tosylchloride/pyridine-mediated cyclization of thiosemicarbazides, which is readily prepared by the acylation of a given hydrazide with the appropriate isothiocyanate. Yields of the aimed oxadiazoles are 80%–99% (Scheme 1.129).

Another widely used method of thiosemicarbaside cyclization consists of oxidation with iodine in alkaline aqueous solution of potassium iodide at room temperature (see, as examples, works by Chaudhary et al. 2007, Gosselin et al. 2010). Reacting with carboxylic acids, thiosemicarbazides give rise to 1,3,4-oxadiazoles. One interesting example of this reaction is depicted in Scheme 1.130 (Ono and Saito 2008). The treatment of thiosemicarbazide with 5-tert-butylisophthalic acid led to the macrocycle containing four oxadiazole groups, although the formation of a polymer with the 2,5-diaryl-1,3,4-oxadiazole main chain can, at first glance, be anticipated. What is technically important is that this oxadiazole-based "crown" compound crystallizes as nanohoses, comprising stacks of the macrocycles, and, as nanowires, forming bundles of nanohoses (Ono and Saito 2008).

A protocol has also been described for a rapid assembly of 2-amino-1,3,4-oxadiazoles from thiosemicarbazides and carboxylic acid derivatives. The proposed reaction mechanism for this process involves the formation of acylthiosemicarbazide as an intermediate (cf. Scheme 1.131). The distinctive feature of Scheme 1.131, as compared to Scheme 1.129, is the use *N*-(3-dimethylaminopropyl)-*N'*-ethylcarbodiimide (the universally adopted abbreviation is EDCI) as a cyclization reagent (Piatnitski Chekler et al. 2008).

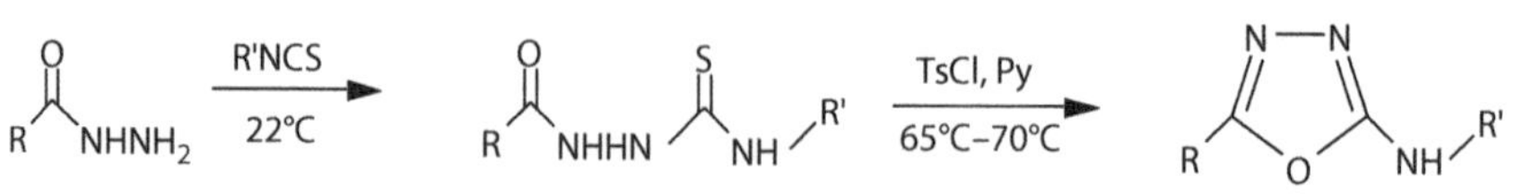

SCHEME 1.129

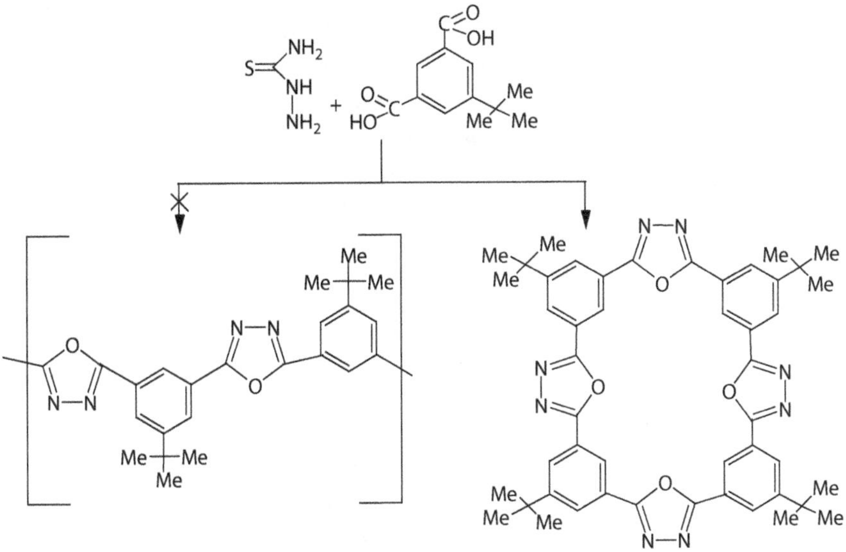

SCHEME 1.130

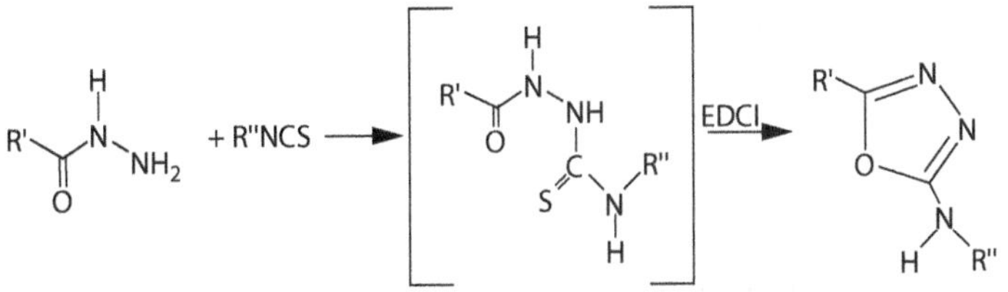

SCHEME 1.131

Franzini and Kool (2008) developed a DNA templated reaction that leads to the 1,3,4-oxadiazole derivative of Scheme 1.132 from Rhodamine B thiosemicarbazide under action of the phenylmercury. The driving force of the reaction consists in the thiophilicity of mercury. Detaching sulfur from thiosemicarbazide, the mercury cation initiates unlocking of the rhodamine spirolactam ring and induces a cyclization with generation of 1,3,4-oxadiazole ring. The oxadiazole-containing component provides an increase in the sample fluorescence that can be used in bioanalysis. For the sake of simplicity, Scheme 1.132 depicts only transformation of the spirolactam into the oxadiazole derivative.

It is the thiophilicity of mercury that makes understandable the formation of an oxadiazole derivative from an acylthiosemicarbazide upon action of mercury dichloride (Gavrilyuk et al. 2008) or mercury diacetate (Scheme 1.133) (Barbuceanu et al. 2008).

Because thiophilicity is also peculiar to silver cations, AgNO$_3$ is used for desulfurization cyclization of thiosemicarbazides, see, for example, Feng et al. (2007).

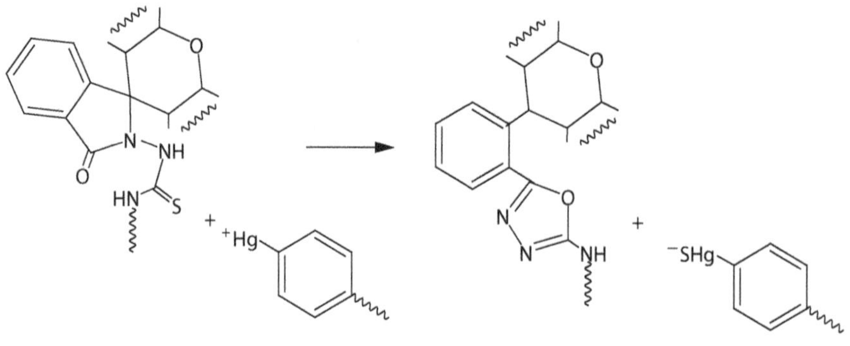

SCHEME 1.132

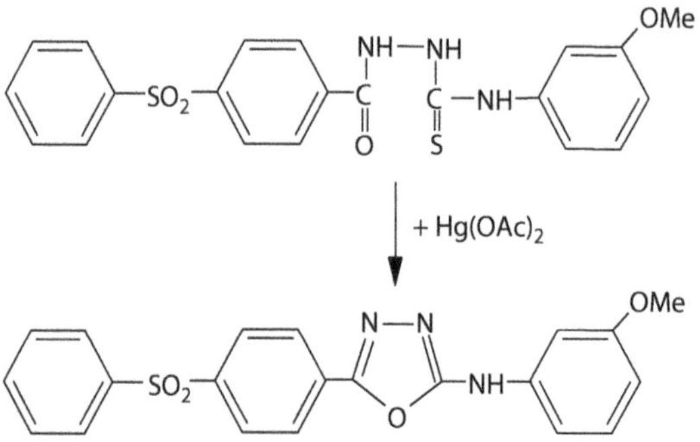

SCHEME 1.133

1.4.1.2 From Semicarbazones

Oxidative cyclization of semicarbazones give rise to 1,3,4-oxadiazoles. The oxidation can be performed by bromine in acetic acid containing sodium acetate (Werber et al. 1977). Chloramin T (*N*-chloro-4-methylbenzenesulfonamide, sodium salt) was also used as an oxidant for this purpose (Sridhar et al. 2006).

1.4.1.3 From Hydrazides or Hydrazones

One-pot synthesis of symmetrical 2,5-diaryl-1,3,4-oxadiazoles was developed by Bentiss and Lagrenee (1999). The synthesis involves heating aromatic carboxylic acids with hydrazine hydrochloride in a mixture of orthophosphoric acid, phosphorus pentoxide, and phosphorus oxychloride. Naturally, this reaction includes the formation of aroylhydrazide with the subsequent cyclization. Refluxing of 2-(naphthalene-2-yloxy)acetohydrazide in ethanol during 1.5 h with cyanogen bromide and sodium hydrogen carbonate results in the formation of 2-amino-5-(2-naphthyloxymetyl)-1,3,4-oxadiazole with 65% yield. This product was further used to prepare a series of anticonvulsants (Rajak et al. 2010).

Refluxing of acylhydrazides in triethyl orthoformate for 10–12 h also leads to 1,3,4-oxadiazoles (Joshi et al. 2007, 2008). Besides, 2,5-disubstituted 1,3,4-oxadiazoles can be obtained in a one-pot procedure from the reaction of acylhydrazides,

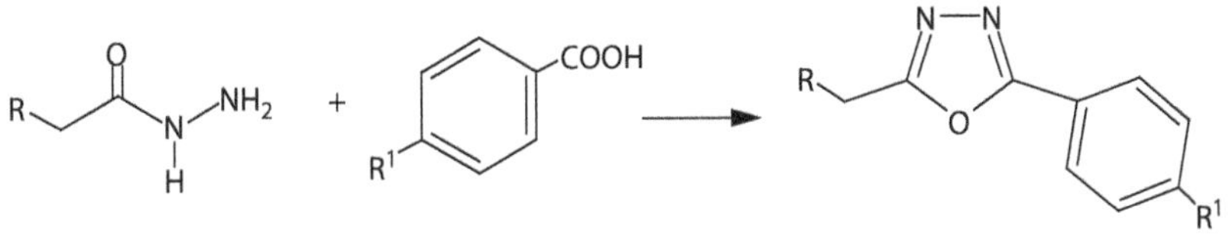

SCHEME 1.134

acylhalides, and phosphorus pentoxide in acetonitrile at room temperature. High yield, short reaction time (10–15 min), mild condition, and easy work-up are advantages of this methodology (Rostamizadeh and Ghamkhar 2008). Noguchi and Koyama (2008) claimed to perform this reaction with triethyl orthoformate in the presence of a Lewis acid in an aprotic polar solvent. In dimethyl acetamide and in the presence of boron trifluoride tetrahydrofuranate at 45°C–50°C, the process is completed after 2 h. The solvent-free variant of the process was also proposed: Microwave irradiation, 80°C, 10 min, Nafion NR 50 as a reusable catalyst (Polshettiwar and Varma 2008). Importantly, refluxing N-protected phenylglycine hydrazides with triethyl orthoesters in glacial acetic acid results in the formation N-protected 2-(1-amino-1-phenylmethyl)-1,3,4-oxadiazoles. This reaction is not accompanied with elimination of the N-protective group, despite its sensitivity to acids (Kudelko and Zielinski 2009).

5-Substituted-2-vinyl-1,3,4-oxadiazoles (as monomers) were prepared by refluxing diacylhydrazides bound to selenium resin in phosphorus oxychloride (12 h). The cyclodehydration proceeds without elimination of oxadiazole derivatives from the selenium resin. The oxadiazole derivatives occur bound with the resin throughout the CH_2–CH_2 fragment. When this resin is treated with hydrogen peroxide, 5-substituted-2-vinyl-1,3,4-oxadiazoles are released (Wang et al. 2007a).

The reaction of acylhydrazides with aromatic acids (Scheme 1.134) was performed in the presence of thionyl chloride during microwave heating (Kidwai et al. 1997, Saeed 2007, Saeed and Mumtaz 2008).

Several 5-substituted 2-amino-1,3,4-oxadiazoles were prepared from acylhydrazides and isothiocyanates through one-pot reaction (Coppo et al. 2004). The reaction proceeds in the presence of the following resin-bound reagents: cyclohexylcarbodiimide, propyl amine, and cyclic phosphazene. Scheme 1.131 shows the sequence of transformations, the intermediary 3-acylthiosemicarbazide enter the cyclization without its isolation.

Refluxing of aroylhydrazide with aryl carboxylic acid in phosphorus oxychloride leads to the corresponding 2,5-diaryl 1,3,4-oxadiazole (Wang et al. 2007b). Instead of reflux, microwave irradiation of an aroylhydrazide-aryl carboxylic acid mixture can be used. Addition of montmorillonite K-100 catalyzes the microwave reaction (Khatkar et al. 2007). Dehydration of N,N'-diaroylhydrazides to form 2,5-diaryl-1,3,4-oxadiazoles was attained using 2-chloro-1,3-dimethylimidazolium chloride (Isobe and Ishikawa 1999) or boron trifluoride-etherate (Tandon and Chhor 2001).

Cyclization of bis(acylhydrazide)s is easily performed upon action of carbon disulfide in ethanol solution containing potassium hydroxide (Tumosiene and Beresnevicius 2007, Joshi et al. 2008). Al-Talib and Tashtoush (1999) found one

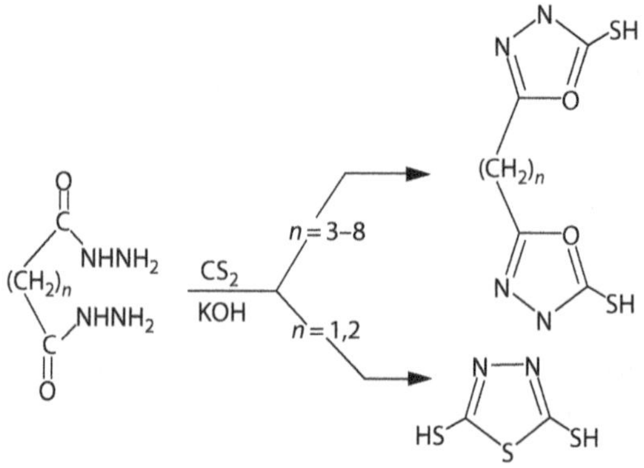

SCHEME 1.135

unprecedented feature of this reaction, namely dependence of its results on the number of methylene groups bridging both the acylhydrazine fragments. This feature can be seen from Scheme 1.135, additional experiments are needed for its explanation.

Oxidation of aroylhydrazide alone with potassium permanganate in acetone is used frequently (Reddy and Reddy 1987, Mashraqui et al. 2007). Oxidation the aroylhydrazide with cupric chloride in methanol is not a method widely used. The reaction leads to the formation of the oxadiazole fixed in the complex with copper (Scheme 1.136) (Singh et al. 2009a).

A series of *N*-acylhydrazones was transformed into 1,3,4-oxadiazoles during oxidation with cupric perchlorate in acetonitrile (Li et al. 2009). In this work, the reaction products were treated with aqueous solution of ethylenediamine. Oxadiazoles (got rid of copper) were isolated through extraction with ethyl acetate.

Another, curious, method of the 1,3,4-oxadiazole formation was described by Colquhoun et al. (2007). The method consists of the reaction between aroyl hydrazides and the tungsten-bound dichlorodiazomethane. The resulting products are (2-aryl-1,3,4-oxadiazol-5-yl)diazenido-tungsten complexes.

Hydrazine interacts with chloroacetyl chloride giving 2,5-bis(chloromethyl)-1,3,4-oxadiazole (Zhang et al. 2003, Tang et al. 2008). Hydrazine sulfate reacts with aryl carboxylic acids in the presence of phosphoric anhydride and phosphoric acid.

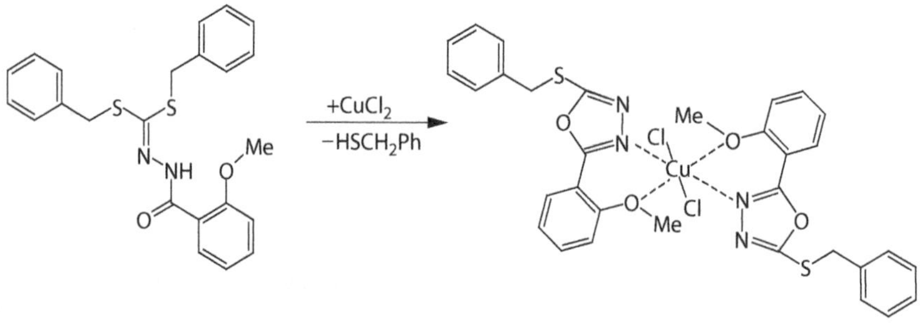

SCHEME 1.136

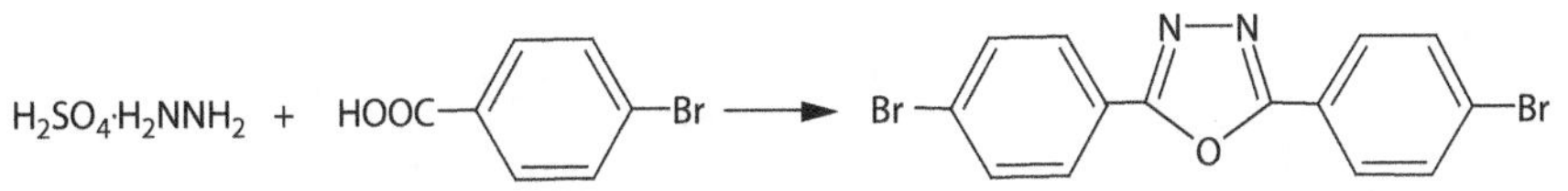

SCHEME 1.137

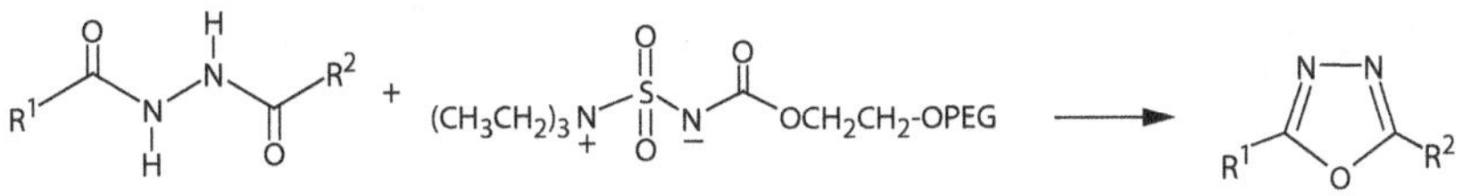

SCHEME 1.138

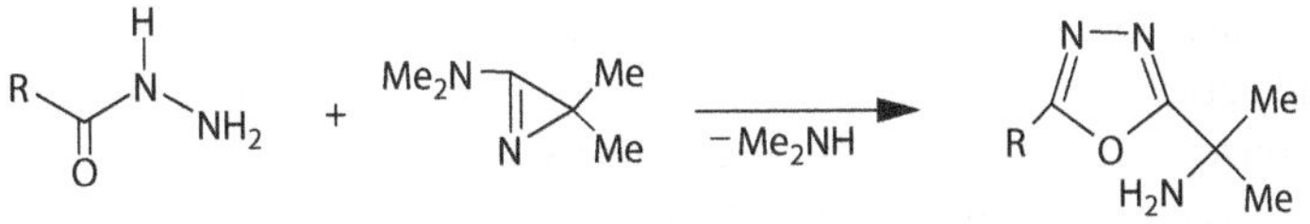

SCHEME 1.139

The final products are 2,5-diaryl-1,3,4-oxadiazoles. Scheme 1.137 gives one of the examples (Liu et al. 2007a). *N*-acylhydrazones are intermediates in these reactions.

Derivatives of 1,3,4-oxadiazole can also be prepared from acylhydrazines and aromatic aldehydes under oxidation by cerium ammonium nitrate. The reaction requires 11 h refluxing in methylene chloride. Oxadiazoles were obtained in moderate to good yields (Dabiri et al. 2006). High yields of oxadiazoles were obtained from 1,2-diacylhydrazines and the zwitterion supported on polyethyleneglycol during 2 min microwave heating in THF (Scheme 1.138) (Brain et al. 1999).

Reacting with acylhydrazines, substituted azirines form the oxadiazole derivatives. Scheme 1.139 gives one example of these reactions: The azirine provides onering carbon of the oxadiazole, and dimethylamine is eliminated. Yields are good (Link 1978).

Augustine et al. (2009) proposed a very efficient synthesis of 1,3,4-oxadiazole from diacylhydrazides, starting from carboxylic acids, alkyl hydrazone, and a couple of triethylamine with T3P. (T3P is propylphosphonic cyclic anhydride. It is frequently used as an active coupling agent and water scavenger.) Scheme 1.140 outlines the sequence of this one-pot synthesis that leads to 95% of 1,3,4-oxadiazoles.

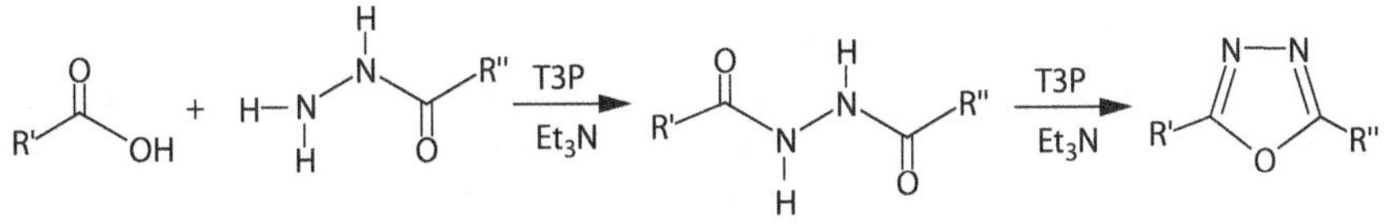

SCHEME 1.140

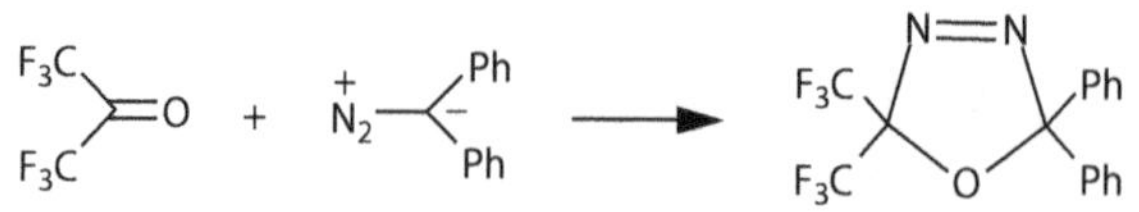

SCHEME 1.141

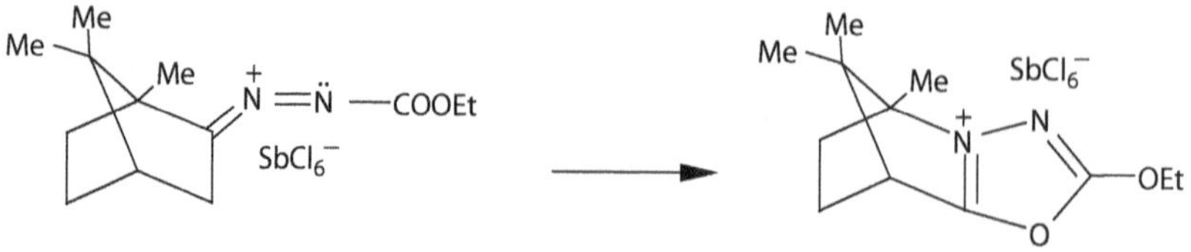

SCHEME 1.142

Hydrazones, readily available from hydrazine and aldehydes, is a source for 1,3,4-oxadiazoles, being oxidized with lead tetraacetate. The following conditions of this reaction are recommended: Two equivalents of the oxidant, benzene as a solvent, overnight stirring at room temperature. The yield of 2,5-diphenyl-1,3,4-oxadiazole from benzal azine was fixed in 70% (Gillis and LaMontagne 1967).

1.4.1.4   From Diazocompounds

Diphenyl diazomethane adds to perfluoroacetone so that the terminal nitrogen of the diazocompound bonds with the carbonyl carbon, and the oxygen atom of the carbonyl group bonds to the diazocompound carbon atom (Scheme 1.141) (Shimizu and Bartlett 1978).

The reaction of Scheme 1.141 is finished almost instantly at −78°C and the isolable product is formed in 95% yield. 2,5-Dihydro-1,3,4-oxadiazoles containing two heteroatom substituents at one and the same carbon atom of the heteroring, are thermally unstable and generate bis(heteroatom-substituted) carbenes (Warkentin 2009), see Section 1.4.3.

What is more, 1-aza-2-azoniaallene hexachloroantimonate of Scheme 1.142 undergoes cyclization to the corresponding fused 1,3,4-oxadiazolium salt (Hassan 2007). At least formally, this reaction should be placed in this category considering the oxadiazole formation from diazocompounds.

1.4.1.5   From (N-Isocyanimino) Triaryl Phosphoranes

In methylene chloride at room temperature, anthranilic acid reaction with (N-isocyanimino) triphenyl phosphorane proceeds smoothly to afford 2-(2-aminophenyl)-1,3,4-oxadiazole in high yield (Scheme 1.143) (Souldozi et al. 2007).

1.4.1.6   From Other Heterocycles

Triazole derivatives can be converted into derivatives of 1,3,4-oxadiazole (Reid and Heindel 1976). Scheme 1.144 shows one of these reactions that eventually results in synthesis of pharmacologically interesting thioethers (Bijev and Prodanova 2007).

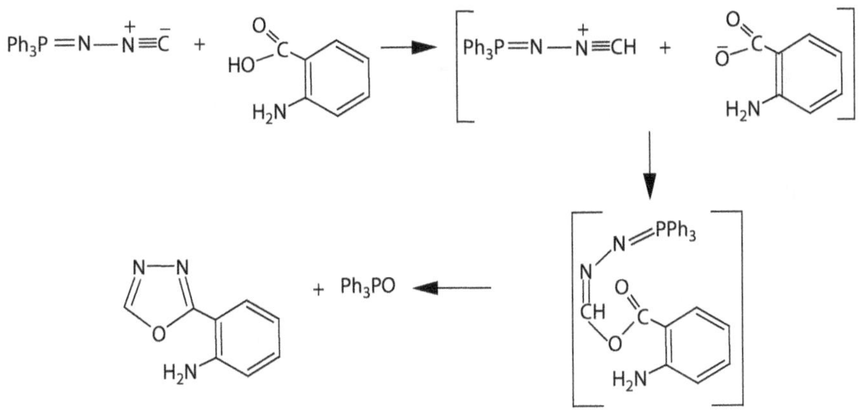

SCHEME 1.143

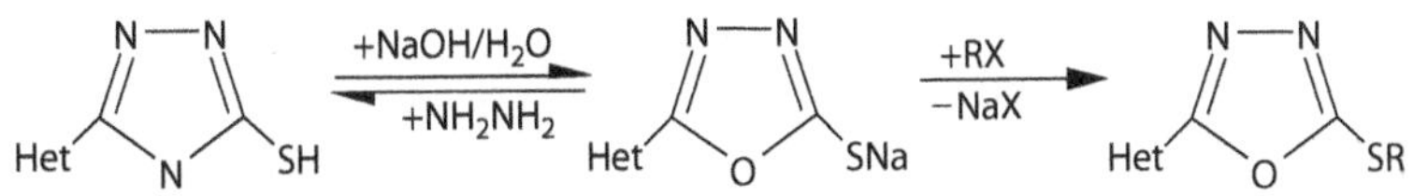

SCHEME 1.144

5-Substituted tetrazoles react with acetic acid anhydride at refluxing temperature to give 2-substituted -5-trifluoromethyl-1,3,4-oxadiazoles. Yields are 85%–95%. The method is suitable for 300 g scale preparation (Jursic and Zdravkovski 1994). Trifluoroacetic anhydride allows performing the reaction at 20°C–25°C (Scheme 1.145) (Vereshchagin et al. 2007).

With 5-vinyltetrazole as a starting material (cf. Scheme 1.145), 5-vinyl-2-(trifluoromethyl)-1,3,4-oxadiazole is formed (Kizhnyaev et al. 2008). The authors noted that the tetrazole cycle in reactions of this type undergoes acylation. This initiates nitrogen elimination and recyclization to form the oxadiazole ring. By varying acylating agents, a wide series of 5-substituted 2-vinyl-1,3,4-oxadiazoles can be prepared and further used in polymerization reactions.

Semenov and Smushkevich (2008) claimed a method for synthesis of 5-substituted 2-methyl-1,3,4-oxadiazoles from benzonitrile, or 1*H*-indole-3-carbonitrile, or 1*H*-indole-3-acetonitrile upon action of tributylstannyl chloride and sodium azide. Both azide and tributylstannyl ions take part in addition to a starting nitrile. The obtained 5-substituted 1-(tributyl)stannyl-1,2,3,4-tetrazoles are treated with acetic

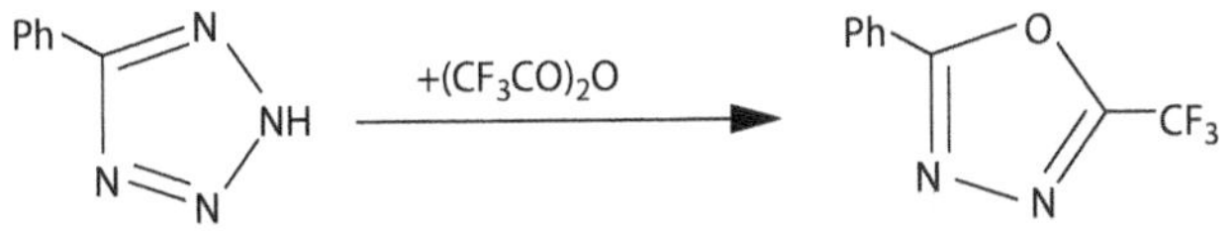

SCHEME 1.145

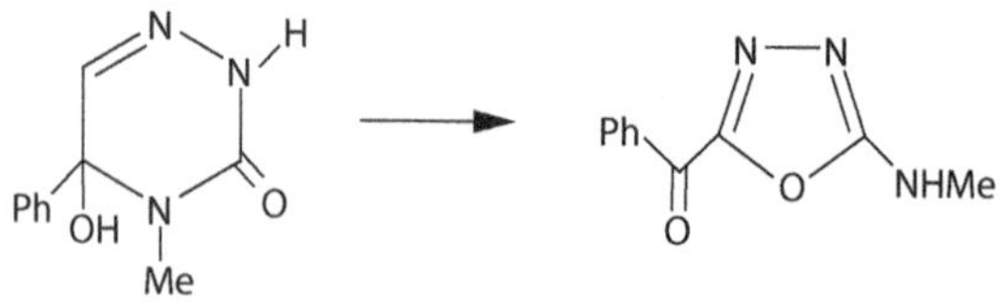

SCHEME 1.146

anhydride. The method provides preparing tin-free 2-methyl-1,3,4-oxadiazoles sub-stituted with bulky groups at position 5 by simplified technology, without prelimi-nary removal of the tributylstannyl fragment. The participation of tributylstannyl chloride thereby predetermines the success of the azide addition to the starting nitrile despite steric shielding of the nitrile fragment. (Permission is of 09.20.10, authorization of the summary given here is of 09.22.10.)

Upon reaction with bromine in acetic acid containing sodium acetate at 115°C, substituted 2,3,4,5-tetrahydro-1,2,3-triazine-3-ones undergo ring contraction to give 2(5)-acyl substituted 1,3,4-oxadiazoles. Thus, 5-hydroxy-5-phenyl-4,5-dihydro-1,2-triazin-3(2H)-one was converted into 2-(methylamino)-5-benzoyl-1,3,4-oxadiazole in 40% yield (Scheme 1.146) (Werber et al. 1979). Despite the yield is moderate, this method is useful because 2(5)-acyl substituted 1,3,4-oxadiazoles are difficult to synthesize by other methods.

1.4.2 FINE STRUCTURE

In comparison to benzene as a perfectly aromatic compound, 1,3,4-oxadiazole occurs as a half-aromatic species. According to calculations based on the corresponding bond lengths, the aromaticity index, I_A, is 100 for benzene and 50 for 1,3,4-oxadia-zole (Bird 1985). In accordance with this estimation, calculations of aromatic stabi-lization energy, differential heat of hydrogenization, and magnetic criteria indicate that the oxadiazole ring can be treated as non-aromatic as slightly aromatic (Erdem et al. 2005). Electron density in the 1,3,4-oxadiazole ring is concentrated mainly on the oxygen and nitrogen atoms (Malek et al. 2008).

Ono et al. (2007) experimentally studied physical, optical, and electrochemical properties of symmetrically substituted 2,5-diaryl-1,3,4-oxadiazoles. Between the two phenyl or naphthalenyl or anthracenyl moieties, π-conjugation through the oxa-diazole spacer does exist, but it is not strong.

Affinity of 1,3,4-oxadiazoles to dienes or dienophiles is another experimental characteristic of their aromaticity or nonaromaticity. 2,5-Bis(trifluoromethyl)-1,3,4-oxadiazole reacts with 2,3-dimethyl-1,3-butadiene at the expense of its one C=N bond (Scheme 1.147) (Vasil'ev et al. 2007). With norbornene, the same oxadiazole reacts as a diazabutadiene, giving bis(trifluoromethyl)diazinobornene (Scheme 1.148) (Warrener et al. 1977, 1995).

The intramolecular proton transfer converts 2-(2-hydroxyphenyl)-5-phenyl-1,3,4-oxadiazole and its 1,3,4-thiadiazole analog into the corresponding keto forms. As seen from Scheme 1.149, this tautomerization proceeds at the expense of the ring bond systems. Being very unlikely in the ground states, such kind of proton transfer is

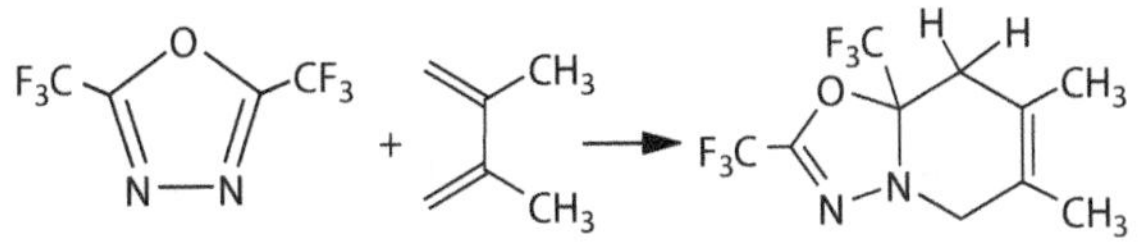

SCHEME 1.147

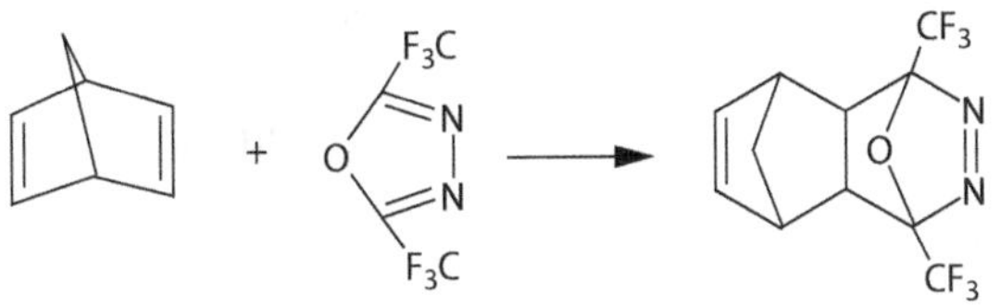

SCHEME 1.148

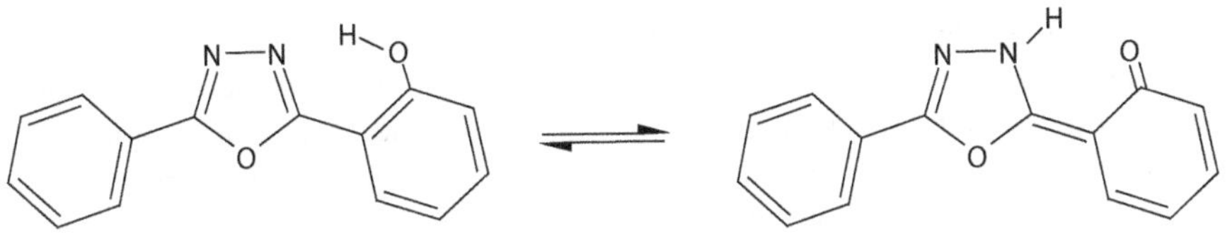

SCHEME 1.149

energetically favored in the case of photoexcitation. 1,3,4-Oxadiazole has less aromaticity index I_A than the 1,3,4-thiadiazole, which are equal to 50 and 63, respectively (Bird 1985). Accordingly, the direct energy barrier for the isomerization into the keto form is lower for the oxadiazole hydroxyphenyl derivative than for the hydroxyphenylthiadiazole (Yang et al. 2007b). It should be useful to mention that photoexcitation usually facilitates the proton transference and the formation of phototautomers.

Kizhnyaev et al. (2008) compared polymerization activity of 2-methyl-5-vinyl derivatives of 1,3,4-oxadiazole and 1,2,3,4-tetrazole. Electron acceptor effect of the oxadiazole cycle is lower than that of the tetrazole cycle. This reflects in activation energies of polymerization that are 95 and 81 kJ/mol for the oxadiazole and tetrazole monomers, respectively. Replacement of methyl substituent with trifluoromethyl in position 2 of 5-vinyl-1,3,4-oxadiazole, reduces the activation barrier from 95 to 88 kJ/mol. (Acceptor inductive effect of CF_3 is stronger than that of CH_3.)

1.4.3 REACTIVITY

1.4.3.1 Salt Formation

Structurally, the 1,3,4-oxadiazole ring system contains two adjacent nitrogen atoms and an isolated oxygen atom. These three electronegative heteroatoms endow the cycle with electron deficiency. Therefore, the basicity of 1,3,4-oxadiazoles is weak. Protonation can be performed only with very strong anhydrous acids and in appropriate mediums. For instance, 2,5-diphenyl-1,3,4-oxadiazolium perchlorate was

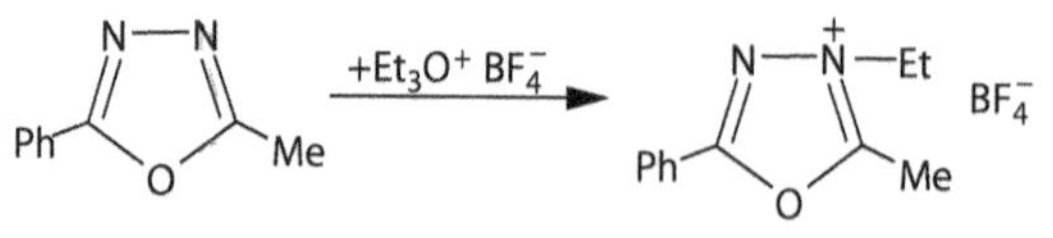

SCHEME 1.150

prepared by treating of 2,5-diphenyloxadiazole with perchloric acid in the solution of acetic acid. Naturally, one nitrogen of 1,3,4-oxadiazoles adds one incoming proton. The salt precipitate recrystallizes from acetonitrile without destruction. Protonation in water results in the oxadiazole ring opening to form the corresponding acylhydrazine (Boyd and Dando 1970).

Quaternization can be performed with triethyloxonium tetrafluoroborate in methylene chloride at room temperature. Under these conditions, 2-methyl-5-phenyl-1,3,4-oxadiazole gives 2-methyl-3-ethyl-5-phenyloxadiazolium tetrafluoroborate (Scheme 1.150) (Boyd and Dando 1970). It is steric shielding from the phenyl group that directs the quaternization to the more available N_3 atom of the heteroring.

5-(4,6-Dimethyl-2-pyrimidinyl)-1,3,4-oxadiazole-2-thiol represents one interesting case of salt formation. Transforming into oxadiazole-2(3H)-thione, this compound reacts with butylamine (as well as with morpholine or piperidine) like NH-acid, giving salts according to Scheme 1.151. Instead of the salt formation, aniline transforms the substrate into 1-(4,6-dimethyl-2-pyrimidinylcarbonyl)-4-phenylthiosemicarbazide. Reaction with phenylhydrazine also leads to the ring

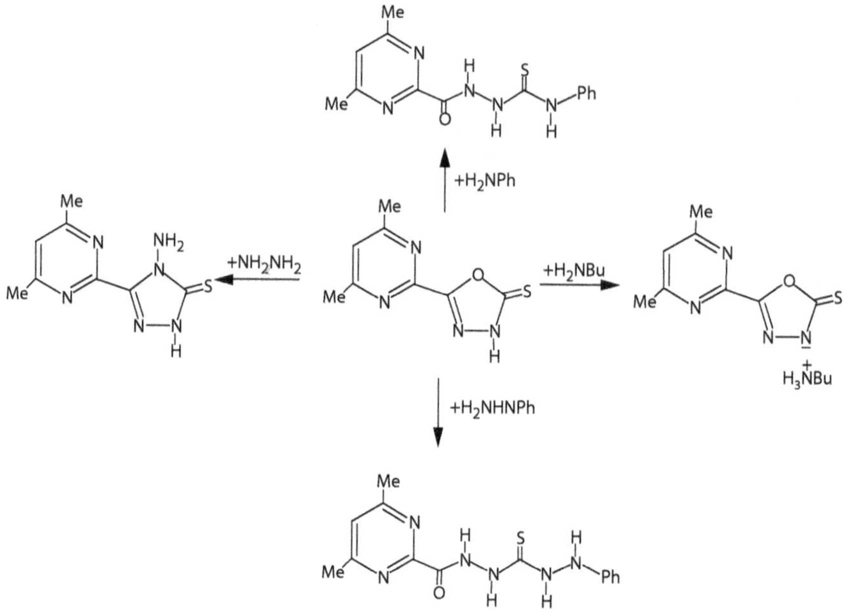

SCHEME 1.151

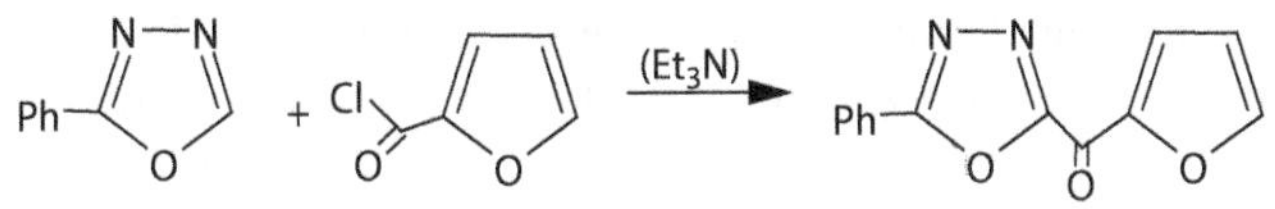

SCHEME 1.152

opening with the formation of 1-(4,6-dimethyl-2-pyrimidinylcarbonyl)-5-phenyl-thiocarbazide. However, with hydrazine hydrate, the substrate recyclization takes place giving rise to 4-amino-5-(4,6-dimethyl-2-pyrimidinyl)-2,4-dihydro-(3H)-1,2,4-triazole-3-thione. These transformations are also depicted in Scheme 1.151 (Mekuskiene et al. 2007).

1.4.3.2 Electrophilic Substitution

Reaction of 2-phenyl-1,3,4-oxadiazole with 2-furoyl chloride follows as C-acylation of the oxadiazole ring rather than the phenyl substituent (Scheme 1.152) (Regel 1977). Although the yield of the product is low, the direction mentioned on Scheme 1.152 is symptomatic.

Wide reactivity in respect of various electrophiles was observed for 2-aryl-5-(trimethylsilyl)-1,3,4-oxadiazoles. These compounds readily react with chlorine, bromine, aliphatic acyl chlorides, 2-nitrobenzensulfenyl chloride, and some reactive isocyanates with replacement of the trimethylsilyl group by the corresponding electrophile (Zarudnitski et al. 2008).

1.4.3.3 Nucleophilic Substitution

The chlorine substituent at a carbon atom of 1,3,4-oxadiazole is readily replaced by nucleophiles. Scheme 1.153 represents one of such reactions with hydrazine (Singh et al. 1984).

Anthranilic acid is an especially interesting nucleophile: In this case, displacement of the chlorine atom is accompanied by the formation of a tricyclic product. Scheme 1.154 shows the structure of the tricyclic compound (Singh et al. 1984).

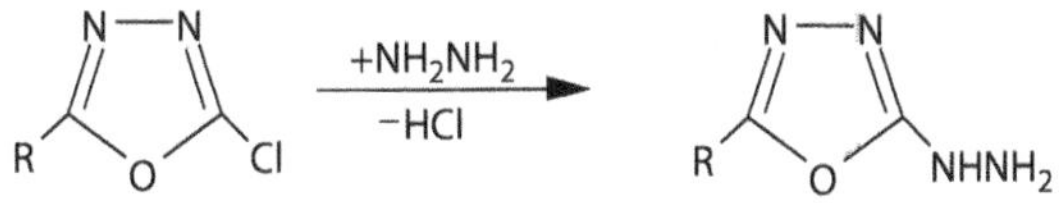

SCHEME 1.153

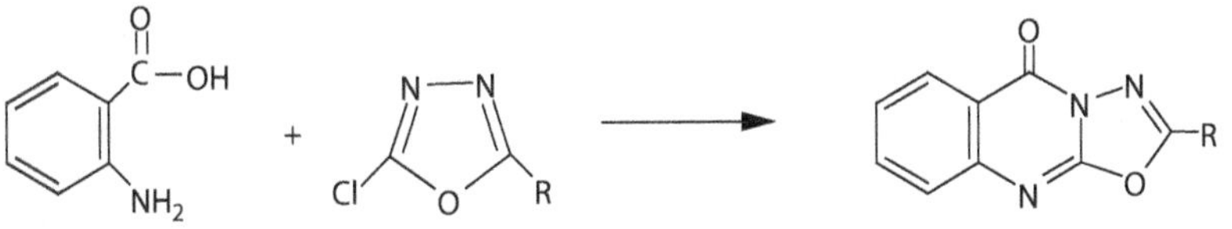

SCHEME 1.154

It is worthwhile mentioning one apparent nucleophilic reaction that consists of direct amination of 2-phenyl-1,3,4-oxadiazole with 4-chloromorpholine. The reaction leads to 2-phenyl-5-(4-morpholino)-1,3,4-oxadiazole with 80% yield. The conditions used in this process were following: Toluene as the solvent, 25°C as the reaction temperature, 2 h as time of duration, lithium tert-butylate and 2,2′-bipyridine as the bases, copper diacetate as the catalyst (Kawano et al. 2010). The authors noted that the most plausible pathway would involve base-assisted cupration of oxadiazole with the formation of a (oxadiazolyl)-copper intermediate. The latter supposedly adds chloromorpholine and forms the amination product that eventually comes out from the complex. The method proposed by Kawano et al. (2010) has several advantages compared to the known amination reactions: (1) the intermolecular direct C–H amination is possible even at room temperature, (2) the use of chloroamines enables the external-oxidant-free conditions, and (3) aminooxadiazoles, which have great biomedical importance, are now readily accessible.

1.4.3.4   Transformation into Ion-Radical States

The 1,3,4-oxadiazole systems are resistant against oxidation. Cation-radical forms are uncharacteristic for them. A typical example is anodic oxidation of two electron-donating triphenylamine moieties that are separated by two electron-withdrawing 1,3,4-oxadiazole groups, which are grafted in meta configuration on a phenylene ring then attached to the spirobifluorene core. Two-electron oxidation touches only the triphenylamine moieties. The presence of 1,3,4-oxadiazoles as electron-withdrawing moieties suppresses the delocalization of spin density generated in each triphenylamine fragment. This allows efficient electropolymerization through feasible triphenylamine cation-radical dimerization (Natera et al. 2007).

Another example concerns transformation of 1,3,4-oxadiazole moiety in the anion-radical state. Fujiwara et al. (2008) considered donor–acceptor dyad containing tetrathiafulvalene (TTF) attached to 2,5-diphenyl-1,3,4-oxadiazole (DPOD) through para position of one phenyl substituent. Molecular orbital calculations by B3LYP approach showed that atomic coefficients of the highest occupied and lowest unoccupied molecular orbitals are mainly localized on TTF and DPOD moieties, respectively. Photoinduced intramolecular electron transfer process quenches the electroluminescent emission so typical for 2,5-diaryl-1,3,4-oxadiazoles. The point is that photoinduced electron transfer takes place with generation of the cation-radical dyad with the anion-radical, namely TTF^+–$DPOD^-$. Although this dyad cannot be used in electroluminescent devices, the photoelectrochemical measurements indicate that cathodic photocurrents can be generated from thin films prepared from the material (Fujiwara et al. 2008). Analogously, 2,5-bis[4-diphenyl-(4′-*N*,*N*-diphenylamino)]-2,5-oxadiazole was claimed as bipolar carrier transporting material (Yang et al. 2008). When the electron donating triphenylamine units and the electron-accepting oxadiazole unit of 2,5-bis[2-diphenyl-(4′-*N*,*N*-diphenylamino)]-2,5-oxadiazole are linked in the ortho configuration, the twisted structure is formed. Within the twisted structure, intramolecular electron transfer is at loss and electroluminescent ability is restored (Tao et al. 2009). Interestingly, the twisted structure formed by linking the 9-position of carbazole with the ortho position in phenyl rings of 2,5-diphenyl-1,3,4-oxadiazole shows excellent electroluminescence (Tao et al. 2008a).

1.4.3.5 Cycloaddition

Participation of 1,3,4-oxadiazoles in the reaction of cycloaddition deserves consideration in some details. First of all, only oxadiazoles bearing acceptor substituents enter into cycloaddition with enes and dienes. Oxadiazoles with donor substituents do not give cycloaddition products. Accordingly, the electron-rich 2,5-dimethyl-1,3,4-oxadiazole does not react with electron-deficient dienophiles or dienes (Vasil'ev et al. 2006). Meanwhile, 2,5-bis(trifluoromethyl)-1,3,4-oxadiazole enters into cycloaddition and the first steps of these reactions with 2,3-dimethyl-1,3-butadiene (Vasil'ev et al. 2007) and norbornene (Warrener et al. 1995) are depicted in Schemes 1.147 and 1.148, respectively. Both the products depicted are unstable and undergo dinitrogen elimination, transforming into more stable compounds. Thus, diazinorbornen (Scheme 1.148) transforms according to Scheme 1.155 (Warrener et al. 1995).

Wolkenberg and Boger (2002) reported an elegant synthesis of anhydrocorinone according to Scheme 1.156. In this reaction, the methoxyethenyl fragment adds to the oxadiazole 2,3-diaza-1,3-butadienyl substructure. The primary adducts eliminates dinitrogen and transforms into the intermediary furan. Further heating promotes a second cycloaddition to produce anhydrocorinone, that is, the final arene.

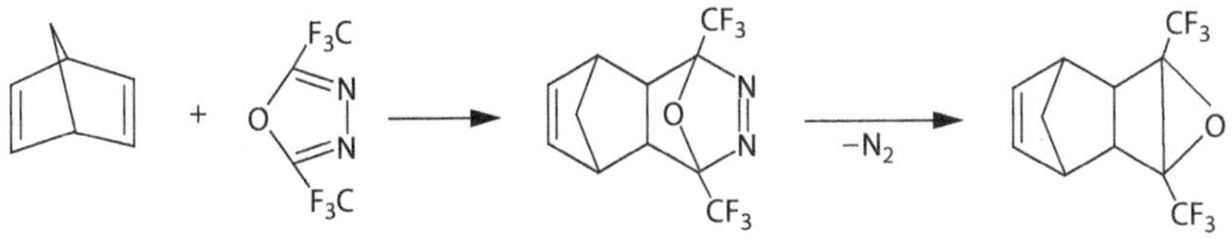

SCHEME 1.155

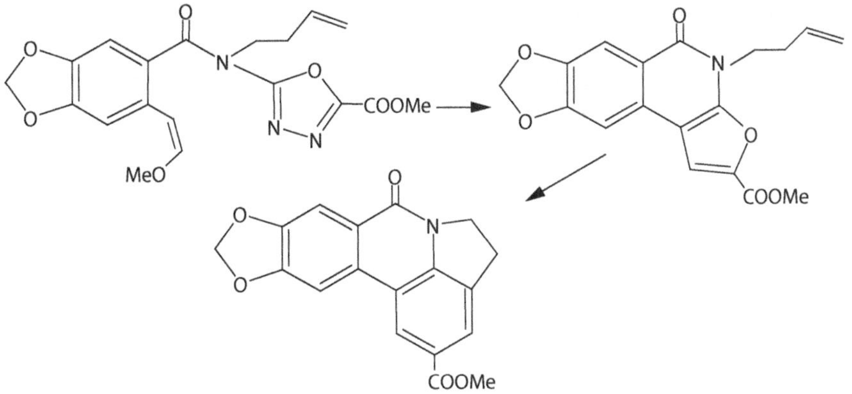

SCHEME 1.156

Intermolecular cycloaddition was described for 1,3,4-oxadiazoles with norbornenes under high pressure. The 1,4-diazabutadiene fragment of the heterocycle is involved in this reaction (Warrener et al. 1977).

1.4.3.6 Ring Cleavage

Some reactions of this type are depicted in Scheme 1.151 (Mekuskiene et al. 2007). 2,5-Dihydro-1,3,4-oxadiazoles with heteroatom substituents at C_2 were reviewed by Warkentin (2009) as very useful materials for the thermal generation of acetoxy(alkoxy)-, dialkoxy, alkoxy(arylalkoxy)-, diaryloxy-, alkoxy(alkylthio)-, bis(alkylthio)-, and alkoxy(amino)-carbenes. Such carbenes are relatively nucleophilic and react with a variety of electrophilic functional groups. This opens a way to many synthetically useful reactions, see Warkentin (2009).

1.4.3.7 Coordination

X-ray study of 2,5-diphenyl-1,3,4-oxadiazole established that this molecule arranges in planes, 0.3–0.4 nm away from each other. At this orientation, the phenyl ring (donor) in one plane interacts with the oxadiazole ring (acceptor) and forms π-complexes within the stacks (Franco et al. 2003). Hydrogen bonding and van der Waals interaction also play a role in the specific packing motif (Du et al. 2006a).

Stack structures were compared for 2-(4-dimethylamino)-1,3,4-oxadiazoles and for 5-methyl-1,3,4-oxadiazoles bearing at position 2 phenyl, or (4-dimethylaminophenyl) fragments (Emmerling et al. 2008). The flat molecular conformation occurs to be common to all of the three compounds. This satisfies to the condition for π ··· π interactions between the molecules. Nevertheless, no π ··· π interactions were found for the first of the structure mentioned (only C–H ··· π interactions are achieved). For the second structure mentioned, π ··· π interactions play a significant role, whereas for the third structure mentioned, π ··· π interactions are strong and solely substantial. To underline, π ··· π interactions take place between the oxadiazole ring (electron acceptor) and the phenyl ring (electron donor) thus indicating that the change of the substitution manner leads to different packing motifs.

Concerning complexation of 1,3,4-oxadiazole ligands to metals, it is interesting to identify coordinative places in polyfunctional ligands and their dependence on the metal nature. 5-(4-Pyridyl)-1,3,4-oxadiazole-2-thiol is specifically interesting because it is a rigid unsymmetrical ligand with three potential binding sites. The oxadiazole thiols tautomerize into the thione forms (Almajan et al. 2006), but both the forms are capable of coordinating to metals (Du et al. 2006b). Naturally, the pyridinium nitrogen can also be involved in coordination to metal as well as the oxadiazole ring. Coordination of this ligand to cobalt(II) and cadmium(II) was considered by Wang et al. (2007c). Scheme 1.157 represents coordinative modes found for the complexes with cobalt(II) and cadmium(II). The mentioned modes of coordination

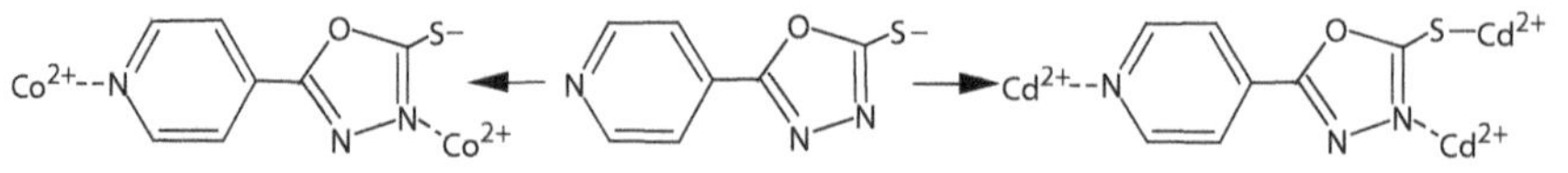

SCHEME 1.157

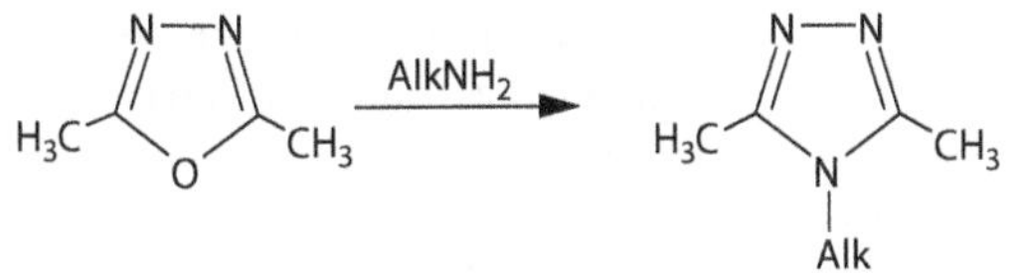

SCHEME 1.158

to cadmium(II) and cobalt(II) were corroborated by Wang et al. (2007d) and Zhang et al. (2007), respectively. Coordination of 2-mercapto-5-(4-pyridyl)-1,3,4-oxadiazole ligand to nickel(II) (Wang and Tang 2007, Zhang et al. 2007), to zinc(II) (Wang et al. 2007d) and to manganese(II) (Singh et al. 2009b) follows the same manner as to cobalt(II). In contrast, mercury(II) occurs to be bound to the sulfur and to the pyridyl nitrogen rather than to the nitrogen of oxadiazolyl (Wang et al. 2007d).

When pyridyls are connected with the 1,3,4-oxadiazole ring through symmetrically bound bridges, only the pyridyl nitrogen participates in complexation. For instance, a clip of the ester, obtained from 2,5-bis(2-hydroxyphenyl)-1,3,4-oxadiazole and 3-pyridinecarboxylic acid, forms silver(I) complex just through the two 3-pyridyl nitrogens (Dong et al. 2006). The same situation was established for transition metal complexes with a bent dipyridyl ligand, namely, 2,5-bis(3-pyridyl)-1,3,4-oxadiazole (Du et al. 2007). An incomprehensible exception is presented by mercury(II) one-dimensional coordination polymer, in which the 2,5-bis(3-pyridyl)-1,3,4-oxadiazole ligand binds to the metal through only one pyridinium nitrogen (Fard-Jahromi et al. 2008). For 2,5-bis(4-pyridyl)-1,3,4-oxadiazole, the same manner of coordination with mercury(II) was observed when the mercury salt used bore bromide, iodide, thiocyanate, or nitrite anion. In the case of azide anion, the mercury salt forms complex at the expense of the oxadiazole nitrogen atom (Mahmoudi and Morsali 2007).

2,5-Diphenyl-1,3,4-oxadiazole coordinates to rhenium pentacarbonylchloride forming binuclear complex, in which the two rhenium centers are bound with two bridging chlorides and with both nitrogen atoms of one and the same oxadiazole (Mauro et al. 2008).

1.4.3.8 Ring Transformation

Upon heating with alkylamines, 2,5-dimethyl-1,3,4-oxadiazole undergoes recyclization resulting in the formation of 4-alkyl-1,2,4-triazoles (Scheme 1.158) (Hetzheim and Moeckel 1966). The same reaction can be performed with 1,3,4-oxadiazoles incorporated in the main polymeric chain (Kizhnyaev et al. 2008).

1.4.4 BIOMEDICAL IMPORTANCE

1.4.4.1 Medical Significance

Biomedical applications of 1,3,4-oxadiazoles are diverse. Because this book is chemically oriented, the data provided in this section are grouped according to structural criteria rather than in the context of applicability.

Among drugs destined for intervention of cell-proliferate diseases, hydroxamate derivatives attract attention. However, these compounds suffer from less than ideal solubility and metabolic stability. These problems are due in a large part to the key hydroxamate moiety. The 1,3,4-oxadiazole fragment behaves as a bioisostere for the hydroxamate group, leading to a more metabolically stable and efficacious cell-growth inhibitor (Warmus et al. 2008).

Hughes et al. (2008) have synthesized 2-[(morpholin-4-yl)ethylamino]-5-[*N*-(4-benzyloxy-3'-chlorophenyl)-4,6-diaminopyrimidinyl]-1,3,4-oxadiazole, in which the oxadiazole ring was considered as an isosteric replacement of an ester functional group. Usually, esters are metabolically unstable to hydrolysis. The oxadiazole sample is hydrolytically stable, active as an anti-cancer medication and its extensive exploration is anticipated.

The 1,3,4-oxadiazole derivative bearing an amino acid substituent is a potent drug for the treatment of diabetes of type 2. A large-scale synthesis of this compound was recently described (Kim et al. 2008). 4-Methyl-(4-aminoaryl)-2-[4-(4'-chloro- or 2',4'-dichloro-benzyloxy)phenyl]-1,3,4-oxadiazolin-5-thiones inhibit leishmania up to 85% (Rastogi et al. 2006a,b). Amidine derivatives of 1,3,4-oxadiazole were claimed as useful for the treatment of cancer and viral infection (Combs and Yue 2008).

Carboxamide derivatives of 1,3,4-oxadiazole exhibit activity as enzymatic regulators (Dobson and Grundy 2007, Johnstone and Plowright 2007, Young et al. 2007, Back and Whaley 2008, Birch et al. 2009). One of these carboxamides, *N*-(4-fluorobenzyl)-5-hydroxy-1-methyl-2-(1-methyl-1-{[(5-methyl-1,3,4-oxadiazol-2-yl)carbonyl]amino}ethyl)-6-oxo-1,6-dihydropyridine-4-carboxamide, is a drug to treat infection by human immunodeficiency virus (Summa et al. 2008). Oxadiazole carboxamides can also be used in local anesthesia (Rajak et al. 2008), for the treatment of diabetes and obesity (Butlin and Plowright 2007), as medicines for endocrinal (Dahl et al. 2008) and gastrointestinal or neurodegenerative diseases (Poitout et al. 2008, Scopes et al. 2008). 1,3,4-Oxadiazoles bearing the carboxamide and carboxylic or keto groups were recommended for the treatment of diabetes and obesity (Butlin and Plowright 2007) or osteoporosis and osteosclerosis (Palmer et al. 2006).

2-(3-Pyridyl)-5-(3-cyanophenyl)-1,3,4-oxadiazole was patented as a neuroregulator (Gopalakrishnan et al. 2008). Derivatives of 1,3,4-oxadiazolyl-pyrrolidinyl ketone are able of crossing blood–brain barrier that is fundamental for the neuropathology treatment (Boss et al. 2009). These ketones also differ in their ability to inhibit pancreatic cell proliferation (Koo et al. 2007).

Isoxazole derivatives of 1,3,4-oxadiazole were evaluated as components of the human immunodeficiency viral infection (Cullen et al. 2007). The authors distinguished 2-methyl-1,3,4-oxadiazole substituted at position 5 with {5-(3-benzothioate)-5-(3-methoxy-7-methylbenzo[*d*]-isoxazol-5-yl)}pent-5-en-1-yl, as a candidate for future development. It was the compound that exhibited not only antiviral activity, but also superior cytoprotective and metabolic stability profiles. 2-Methyl-1,3,4-oxadiazole at position 5 bearing 4-amino-2-methyl-6-phenylpyridazin-5-yl-3(2*H*)-one is a potent analgesic agent (Giovannoni et al. 2007). 2-(1,2-*O*-isopropylidene-α-D-xylo-tetrafuranos-4-yl)-5-(4-bromophenyl)-1,3,4-oxadiazole is active against Dengue fever (Barradas et al. 2008).

A wide spectrum of activity was established for 1,4-diaza-bicyclo[3.2.2]nonyl-1,3,4-oxadiazolyl derivatives: They are appropriate for the treatment of diseases or disorders related to smooth muscle contraction, inflammation, pain, and withdrawal symptoms caused by the termination of use or abuse from chemical substances (Peters et al. 2007a,b, 2009).

Thioethers of the 1,3,4-oxadiazole series were nominated as candidates for preclinical evaluation. Antitumor, antiviral, antimalarial, and antibacterial activities were found for 1,3,4-oxadiazole thioethers, most of them bear also the sulfonamide group (Iqbal et al. 2006, Karabasanagouda et al. 2007, Zareef et al. 2007, Akhtar et al. 2008, Jansen et al. 2008, Letafat et al. 2008). 2-Thiomethyl-5-[(1-phenyl)-β-carbolino]-1,3,4-oxadiazole displays high selectivity and potent anticancer activity against ovarian cells (Formagio et al. 2008). 2-Thioalkyl-5-[1-methyl(4,6-dimethyl-2-oxo-1,2-dihydropyridine-3-carbonitrile)]-1,3,4-oxadiazole exhibits activity against Hepatitis B virus with low cytotoxicity and high selectivity (El-Essawy et al. 2008).

A potent anti-flammatory agent was prepared by Bhandari et al. (2008) in a result of reaction between 1-(3-methoxyphenyl)ethanone and 2-mercapto-5-[N-(2-aminobenz-1-yl)-(2,5-dichlorophenyl)]-1,3,4-oxadiazole. The reaction proceeded as S-acylation. The product exhibited higher anti-inflammatory activity than existing medications, but without their gastrointestinal toxic effects. Another thioether, 7-[5-(1-ethyl-1-hydroxypropyl)-1,3,4-oxadiazol-2-yl-thio]-4-(3-methylphenyl)quinoline-2-carbonitrile, was patented as an anti-inflammatory, anti-allergic, anti-asthmatic, anti-sclerotic, and cytoprotective drug (Ducharme et al. 2007).

1,3,4-Oxadiazol-2-one bearing 5-[4-(trifluoromethyl)phenyl]- and 3-[(5-chloro-2-hydroxyphenyl)methyl] groups is passing clinical evaluation of its neuroprotective properties (Romine et al. 2007). 2-Hydroxy-5-[2-(phenylthio)phenyl]-1,3,4-oxadiazole exhibits muscle relaxant activity comparable with that of Diazepam but without unwanted side effects that are peculiar to Diazepam (Almasirad et al. 2007). 2-Mercapto-1,3,4-oxadiazol-5-(5-chloro-3-methyl-1,2,4-triazolo[4,3-a]pyridyl-7-yl) is potent in the anxiety treatment showing 35 times Diazepam activity (Amr et al. 2008). 1,3,4-Oxadiazole bound at position 2 through sulfur with acetic acid (the rest of thioglycolic acid) and at position 5 bound with phenyl moiety, is active against human breast carcinoma (Sengupta et al. 2008).

Ethers of the 1,3,4-oxadiazole series also show biomedical activity. According to preclinical studies, 1,3,4-oxadiazoles containing 5-ethyl-2-ethoxy-4-phenyl—at position 2—and phenyl, or 4-methylphenyl, or 3-pyridyl—at position 5—inhibit breast cancer cell growth and induce apoptosis of these cells (Kumar et al. 2008). The ethers of 1,3,4-oxadiazole were stated as medications for the treatment addiction to dopamine-producing agents such as cocaine, opiates, morphine, amphetamines, nicotine, and alcohol (Zablocki et al. 2008). They can also inhibit inflammation (Rajak et al. 2007). Neurodiseases were proposed to treat with 2-aryl-5-ethers or thioethers of 1,3,4-oxadiazole (Lee et al. 2008a,b). Tranquilizing effects were observed from 2-[4-(β-D-glucopyranosyloxy)phenyl]-5-aryl-1,3,4-oxadiazoles (Li et al. 2007). 1,3,4-Oxadiazole substituted by the anisyl fragment at position 2 and by the (2-fluoro-4-chloro-3′-bromo-5′-chloro)diphenyl ethereal moiety at position 5, demonstrated activity against wild type human immunodeficiency (Aquino et al. 2008).

Oxadiazole amino derivatives are widely prepared and tested as diverse medications. 2-Amino-2-[2-(2-chlorobenzyloxy)phenyl]-1,3,4-oxadiazole (Zarghi et al.

2008) and 2-*N*-(2-chlorophenyl)amino-5-(1*H*-indol-3-yl)-1,3,4-oxadiazole (Siddiqui et al. 2008) are described as anticonvulsant agents. Derivatives of oxadiazole-2-methylammonium bromide regulate muscle activity (Bull et al. 2008). 2-Amino-*N*-(2,4,5-trifluorophenyl)-5-(benzimidazol-2-yl)-1,3,4-oxadiazole derivative was patented for the treatment of diabetes and obesity (Birch et al. 2007). 2-Amino-*N*-(4-chlorophenyl)-5-(2,6-dichlorobenzyl)-1,3,4-oxadiazole was active against human immunodeficiency virus (Barreiro et al. 2007).

1,3,4-Oxadiazoles bearing pyrrolidinyl- and piperidinyl-containing groups at positions 2 and 5 suppress appetite and obesity of humans (Alper et al. 2008). Thus, 2-{6-[4-(2-chloro-5-fluorophenoxy)piperidin-1-yl)pyridazazin-3-yl]}-5-methyl-1,3,4-oxadiazole demonstrates excellent activity in blocking fat accumulation (Liu et al. 2007b). Pyridyl-thienyl derivatives are patented as lymphocyte reducers (Bolli et al. 2007). 1,3,4-Oxadiazoles substituted with aryl(aryloxy)-oxazolines can be useful in the treatment of central nervous system diseases, the syndrome of autoimmune deficit, chronic obstructive pulmonary disease, psoriasis, dermatitis, autoimmune illness of intestinal tract (Crohn's disease), osteoarthritis, colitis, and other inflammatory diseases (Balachandran et al. 2008). 2-Pyridyl-5-*N*-arylamino derivatives of 1,3,4-oxadiazole were patented as inhibitors of conveying lymph, see Honda et al. (2008). Indazole-derived benzyloxadiazoles were claimed as modulators of serum glucocorticoid regulating protein kinase. The modulators play an important role in cell fission (Klein and Beier 2008a,b). Derivatives of 1,2-bis(1,3,4-oxadiazol-2-yl) ethane are active against both Gram positive and Gram negative bacteria at less than 6 mg/mL concentration (Holla et al. 2000). 2-Piperazino-5-pyrimidino-1,3,4-oxadiazole derivatives exhibit antiproliferative activity against human leukemia (Ishida et al. 2008). Shaharyar et al. (2007) found that 1-cyclopropyl-6-fluoro-3-[5-(4-nitrophenyl)-1,3,4-oxadiazol-2-yl]-7-piperazino-1,4-dihydro-4-quinolinone is encouraging for the treatment of human lung tumor cells.

1.4.4.2   Application to Environmental and Bioanalysis

The development of sensitive and selective fluorescent chemosensors for biologically and environmentally important ions is of intense current interest. Thus, Cu^{2+} has essential toxic nature. In this connection, the reaction between cupric cation and acylhydrazones (see Section 1.4.1) attracted attention (Li et al. 2009). The point is that acylhydrazones, as they are, emit extremely weak fluorescence. In the presence of Cu^{2+}, an instant response of acylhydrazones is observed by a dramatic enhancement of fluorescence. For acetonitrile solutions of acylhydrazones, the enhancement was by 1000 times. And this is despite the well-known fluorescence-quenching character of cupric ions! A linear response toward Cu^{2+} (in acetonitrile) was observed within the Cu^{2+} concentration range from 25×10^{-9} to 0.25×10^{-6} M, with a detection limit of 3.5×10^{-9} M. In a 20:80 mixture of acetonitrile and water (pH 7.2), the response was accumulated much slower and the detection limit was stepped up to 0.30×10^{-6} M. As seen, acetonitrile plays an important role in this kind of bioanalysis.

Section 1.4.1 has already considered oxidative cyclization of acylhydrazones into 1,3,4-oxadiazoles upon action of the cupric ions. The resulting oxadiazoles can be formed in the framework of copper complex. Importantly, the oxadiazole formation results in appearance of the molecules or ligands with high fluorescence. In acetonitrile, the oxidative ability of Cu^{2+} is enhanced because of coordination of the Cu^{1+} formed with this solvent. The presence of water lowers the acetonitrile coordination ability. Importantly, the stabilizing role of acetonitrile is significant for Cu^{1+} salts and remains insignificant for Cu^{2+} salts. The difference between Cu^{1+} and Cu^{2+} coordination with the solvent was thoroughly considered on pages 70 and 71 of the monograph by Todres (2009). The fluorescence enhancement described here results from acylhydrazone oxidation by Cu^{2+} into 1,3,4-oxadiazoles and is independent of counter anions such as NO_3^-, Cl^-, AcO^-, ClO_4^-, and SO_4^{2-}. No interference is observed from the other metal ions tested: Co^{2+}, Hg^{2+}, Mg^{2+}, Ca^{2+}, and Ba^{2+} (Li et al. 2009). Acylhydrazones can be recommended as chemidosimeters of Cu^{2+} in acetonitrile or in its mixture with water. The determination of Cu^{2+} is accessible at nanomolar levels.

Another challenge concerns determination of mercury in living cells. Mercuric ions can easily pass through biological membranes and cause serious damage to the central nervous and endocrine systems. Zhang et al. (2008) presented intracellular sensor of Scheme 1.159. Chemically, the sensor contains boron-dipyrrolomethane, a diphenylacetylenic bridge and leuco-rhodamine. The sensor is good at cell permeability. In its initial leuco form, the sensor emits fluorescence at 514 nm. Upon addition of Hg^{2+}, emission at 514 nm decreases, and a new emission band with maximum at 589 nm appears and gradually increases in intensity. The mercury ion desulfurizes the starting material. The 1,3,4-oxadiazole ring is formed whereas the leuco-rhodamine transforms into the quinoid state. The color of fluorescence changes from green to orange-red. This change of color can be observed by eye. The cells remained viable and no apparent toxicity or side effects were observed. The fluorescent probe of

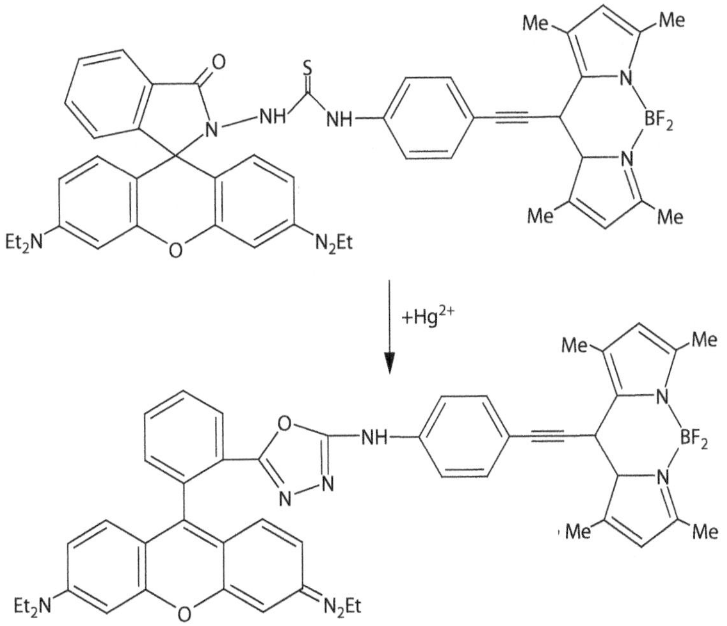

SCHEME 1.159

Scheme 1.159 showed excellent selectivity toward Hg^{2+} and did not give any observable response for Al^{3+}, Ba^{2+}, Ca^{2+}, Cd^{2+}, Cr^{2+}, Cu^{2+}, K^+, Mg^{2+}, Mn^{2+}, Na^+, Pb^{2+}, and Zn^{2+}. Only Ag^+ had a slight effect with a much longer equilibrium time (>40 min) than that of Hg^{2+} (<5 min). Both the leuco and quinoid forms of the probe had attainable fluorescent properties over wide pH span, from 4 to 10. The probe is suitable for application under physiological conditions.

Further, there is growing interest in recognition and detection of fluoride ions. These anions are associated with nerve gas, with uranium wastes, and with content of drinking water. Chemically, fluoride ion is an electron-donating species. On the other hand, 1,3,4-oxadiazole cycle is an electron deficient species. Based on these reasons, Kwak et al. (2007) proposed 2,5-bis(2-acetyl-4-methylphenyl)-1,3,4-oxadiazole as a sensor for fluoride ions. Indeed, addition of fluoride provides donor–acceptor interaction and the heterocyclic oxygen atom is enriched with electron density. Accordingly, ultraviolet absorption and fluorescent emission are changed. Among various halide and acetate ion tested, only fluoride anion induces a redshift of the absorption from 286 to 382 nm, so that a colorless solution of the diazole in dimethylformamide becomes yellow in the presence of fluoride. At the low concentration of fluoride anion (less than 20 equivalents), green fluorescent emission is exhibited, while the blue emission is developed in the higher fluoride concentration (more than 40 equivalents). This 2,5-bis(2-acetyl-4-methylphenyl)-1,3,4-oxadiazole can be used as a fluoride anion sensor in terms of naked-eye detection. The sensor is highly selective and its fluorescence depends on the concentration of fluoride ions.

2,5-Bis(2-hydroxy-5-bromophenyl)-1,3,4-oxadiazole was also proposed as fluoride sensor: Ultraviolet absorption and fluorescent spectra of the starting oxadiazole undergo significant changes upon addition of this anion (Kim et al. 2007).

1.4.4.3 Agricultural Protectors

1,3,4-Oxadiazoles bearing triazolemethylene and methylenecarbonylmethyl groups exhibit fungicide activity in plants (Yang and Zhai 2007). Having no cytotoxicity, derivatives of 1,3,4-oxadiazole-2-thiol accelerate wheat sprouting (Almajan et al. 2008) and show antibacterial activity against wheat powdery mildew (Liu et al. 2007c). Herbicidal activity was observed for 5-aryl-2-(3,5,6-trichloropyridyloxymethyl)-1,3,4-oxadiazoles (Chavan et al. 2006). 2-Alkylthio-5-(3,4,5-tribenzyloxyphenyl)-1,3,4-oxadiazoles exhibit marked activity against a series of fungi (Long et al. 2006). Thiooxomethyl-acetamide derivatives of 1,3,4-oxadiazole are active against the wheat scab fungus (inhibition rate were 90%, Sun and Qin 2006). 3-[6-(3-Chlorophenoxy)pyridine-3-yl]-5-isopropoxy-1,3,4-oxadiazol-2-one provides 100% control against *Piricularia oryzae* without damage on rice (Tajino et al. 2007).

1,3,4-Oxadiazole bearing 4-chlorophenyl in position 2 and 2-butyl-3(2*H*)-4-chloropyridazin-3-one joined through –O–CH$_2$– bridge with position 5, was found to be an effective insect growth regulator. It inhibits the larvae growth of the army worm, causing a prolonged larval period, smaller body size, and sluggish behavior (Huang et al. 2008). Army worms travel in large groups, ruining crops and grass. In contrast to many other insecticides, this insect-growth inhibitor differs in its photostability—the factor that promotes its high efficiency within fields. Being applied in mixtures with surfactants such as Triton X-100 and sodium sulfosuccinic acid, the inhibitor forms stable micelles. Micellization additionally enhances activity of the insect-growth inhibitor (Huang et al. 2007).

1.4.5 Technical Applications

1.4.5.1 Light Emitting

Of particular interest is aromatically substituted 1,3,4-oxadiazoles because they exhibit electron-transporting properties, high thermal and oxidative stability, strong fluorescence and high electroluminescence efficiency. The five-membered ring oxadiazole does not emit fluorescence. However, when the oxadiazole is attached to one or more homoaromatic or heteroaromatic rings, strong light absorption and phosphorescence emission appear (Hamada et al. 1992, Nakaya et al. 2007). Thus, 1,3-bis[2-(1,10-phenathroine-2-yl)-1,3,4-oxadiazole-5-yl]benzene were prepared and used in organic electroluminescent devices. The devices containing this oxadiazole showed high efficiency, high electric stability, and high durability (Miki et al. 2007).

2,5-Diaryl-1,3,4-oxadiazoles exhibit luminescent properties and are suitable candidates to use in optoelectronics. In this connection, iridium and platinum complexes should be mentioned. The complexes contain 2,5-diaryl-1,3,4-oxadiazoles as N-coordinated ligands (Inoue et al. 2007). European(III) complexes containing the electron-transporting 1,3,4-oxadiazole (non-coordinated) bridge in the polymeric core are good photoluminescent emitting materials (Xiang et al. 2006). Even side-chain location of non-coordinated 2,5-diaryl-1,3,4-oxadiazole together with tris(bipyridyl) units coordinated to ruthenium, does create conditions of the balanced charge mobility in the polymer network (Pefkianakis et al. 2008). Derivatives of 2,5-diaryl-1,3,4-oxadiazole are promising as materials for ordered-film or microbelt lasers and opto-electronic devices (Qu et al. 2011). For instance, 2-(4-biphenyl)-5-(4-tert-butylphenyl)-1,3,4-oxadiazole absorbs the pumped light and funnels the excitation to the doped molecule (Berggren et al. 1997). Another derivative, 2,2′-bis(1,3,4-oxadiazolyl-2-yl)biphenyl is an effective electron transport host for phosphorescent organic light-emitting diodes (Leung et al. 2007).

Light emission of 1,3,4-oxadiazole derivatives can be induced by neutron excitation. 2-(4-tert-Butylphenyl)-5-(4-biphenyl)-1,3,4-oxadiazole scintillates in the sample containing also trimethoxysilane, poly(ethylene glycol), and trimethylborate. (For the sample without oxadiazole, no neutron scintillation was observed.) The material was recommended for neutron detection (Koshimizu et al. 2008).

Molecules and polymers functionalized with the 1,3,4-oxadiazole ring constitute materials that are able of hole blocking and electron injecting (Shin et al. 2008). Ma et al. (2005) constructed a polymer light-emitting diode based on poly{[9,9-bis(6′-(N,N,N-trimethylammonium)hexyl iodide)fluorene-2,7-diyl]-alt-[2,5-bis(1,4-phenylene)-1,3,4-oxadiazole]}. This electrolyte was coated from a methanol solution on top of neutral conjugated polymer layers. No interfacial mixing or deterioration of underlying layers took place. Significant improvements in blue-, green-, and red-emitting devices were observed (Bazan 2007).

Introduction of the 1,3,4-oxadiazole ring in a macromolecule improves the polymer properties (Gong et al. 2009). Being incorporated into the main chain, these disubstituted rings contain no hydrogen atoms and do not participate in any rearrangements. They lack tension, have structural symmetry and are thermally unreactive. Specific properties determined by the electronic structure of oxadiazole ring, especially its electron-withdrawing character, initiated an intensive research aiming to use such polymers in microelectronics, optoelectronics, and other technical fields. Suffice it to enumerate representative recent patents and papers devoted to 1,3,4-oxadizole-containing polymers: Tamano and Takayama (2006), Ma et al. (2007), Murata et al. (2007), Nagata and Nishikawa (2007), Nishiji et al. (2007), Lin et al. (2008), Mikroyannidis et al. (2008), Tamura et al. (2008), Tao et al. (2008b).

1.4.5.2 Photoconductivity and Field-Effect Transition

Coupling bis(diazo)compound from 2,5-bis(4-aminophenyl)-1,3,4-oxadiazole with 3-hydroxy-N-(2-chlorophenyl)-2-naphthalenecarboxamide, Xue et al. (2007) obtained a bisazopigment. After repeated washing with water and then with dimethylformamide, the pigment acquires amorphous structure and shows good photoconductivity when being used as charge generation material. It is also good at xerographic application.

Lee et al. (2009) prepared oligomers containing the oxadiazole subunit, or the fluorenone, or the fumaronitrile segments and tested these different oligomers as field-effect transistors. Unexpectedly, it was oxadiazoles that occurred to be inferior at transition of the field effect. This must be noted for the sake of accuracy.

1.4.5.3 Formation of Liquid-Crystalline State

Liquid crystals, often referred to as the mesophase or mesomorphic phase, are a state in between anisotropic crystalline solids and isotropic liquids. Oxadiazole inserted in the molecular core seriously affects physicochemical properties of thermotropic liquid crystals. Symmetrically disubstituted 1,3,4-oxadiazole are characterized with biaxial nematic phase. These molecules are defined as boomerangs, because the exocyclic bond angle between C_2 and C_5 substituted positions is approximately $140°$ (Dingemans et al. 2006). Such an angle permits to construct 2,5-(phenylene ethynylene)-1,3,4-oxadiazoles in which phenylene moieties contain two para-located long-chain oxyalkyl groups. Such liquid crystals are biaxial, V-shaped, and shape persistent (Lehmann et al. 2008). Biaxiality imparts to the material new technically important properties.

The boomerangs have a large dipole moment of ca. 4D. The dipole moment is perpendicular to the molecular axis, a characteristic that influences the mesogenic properties. For the reasons, 1,3,4-oxadiazole derivatives are subjected to strong intermolecular association (Lai et al. 2002).

2-(4-Donor-phenyl)-5-(4′-acceptor-phenyl)-1,3,4-oxadiazoles dissolved in homogenously oriented liquid-crystalline polymers, can be included in films as components for light-emitting diodes (Wolarz et al. 2007). The first-encountered problem associated with the fabrication of multilayer organic light-emitting diodes with low-molecular liquid crystals, is the fluid nature of this kind of crystals. This may lead to interlayer mixing during the deposition of subsequent layer by one or another method. Usually, this problem is resolved by photochemically cross-polymerization of liquid-crystalline monomers themselves or with the polymeric film. Of course, the polymerizable functions should be included in the counterparts, such as in

2,5-bis-{4-[trans-(3,4-dioctyloxyphenyl)ethenyl]phenyl}-1,3,4-oxadiazole (Fang He et al. 2007). In this way, insoluble organic matrices are formed with needed orientation of the constituents.

Among the oxadiazole liquid crystals, those are the most attractive that show mesomorphic phase behavior below 100°C. Such liquid crystals are more stable as compared to the objects, for which the phase transition is observed at temperatures well above 150°C. The asymmetric derivatives with relatively low transition temperatures allow obtaining materials with exclusively nematic liquid-crystal phase behavior. For instance, this behavior was observed for 2-(4-undecyloxyphenyl)-5-(4-hydroxyphenyl)-1,3,4-oxadiazole, 2-{4-[4-octyloxy-2-(pent-4-enyloxy)carbonyloxy-phenyl]phenyl}-5-{[4-(4-undecyloxy)carbonyloxyphenyl]phenyl}-1,3,4-oxadiazole, and for 2-{4-[4-octyloxy-2-(5-(1,1,3,3,3-pentamethyldisiloxanyl)pentyloxyphenyl-carbonyloxy]phenyl}-5-{[4-(4-undecyloxy)carbonyloxyphenyl]phenyl}-1,3,4-oxadia-zole (Apreutesei and Mehl 2007).

With respect to applications in working devices, 1,3,4-oxadiazole metallocom-plexes (metallomesogens) and liquid crystalline polymers attract interest. Such a metallomesogen as bis[(3,4,5-trimethoxyphenyl)-1,3,4-oxadiazole]palladium dichlo-ride can be exemplified (Wen et al. 2005). From the family of liquid-crystalline polymers, such polyesters can be exemplified that contain 1,3,4-oxadiazole and bis(benzylidene) cycloalkanone units in the main chain (Balamurugan and Kannan 2008). Liquid-crystalline polymers containing mesogen-jacketed 1,3,4-oxadiazoles in side chains are also known (Yang et al. 2009). Homopolymers from 2,5-bis[(4-tert-butylphenyl)-1,3,4-oxadiazole]styrene and 2,5-bis[(4-alkoxyphenyl)-1,3,4-oxadia-zole]styrene can be mentioned. The comparison between these two homopolymers indicated that the flexibility of the side chain is crucial to determine the liquid-crys-talline structures. Namely, simply changing the chemical structures of small portion of the lateral group in the side chain greatly varies the molecular packing behavior and thus molecular shape in liquid-crystalline structures (Mochizuki et al. 2000, Chai et al. 2007, Xu et al. 2009). The styrene homopolymers mentioned occurred to be good blue light-emitting materials of acceptable solubility, high thermal stability, and fluorescence quantum yields over 60% (Wang et al. 2007e). These properties were further improved by introducing the carbazole pendant group in the oxadia-zole-containing side chains. Random bipolar copolymers were prepared using as monomers 2,5-bis[(alkyl or alkoxy)-1,3,4-oxadiazole]styrene and N-vinylcarbazole. That made the liquid-crystalline copolymers to be very appropriate as host materials for electrophosphorescent devices (Wang et al. 2008a,b,c).

1.4.5.4 Corrosion Inhibitors

Tu et al. (2008) synthesized 2,5-bis(2-undecyl)-1,3,4-oxadiazole and examined its inhibiting behavior for carbon steel in 1 and 2 mol/L aqueous hydrochloric acid. The compound behaves excellently as inhibitor at very low concentration and sup-presses both cathodic and anodic processes of steel corrosion in the acid solutions mentioned. The inhibition effect was explained by the oxadiazole chemisorption on steel surface. 2-Mercapto-5-(1-undecyl)- and 2-mercapto-5-(1-pentadecyl)-1,3,4-oxadiazoles are cathodic-type protectors preserving mild steel from corrosion in HCl (Rafiquee et al. 2007).

2-Mercapto-5-(phenylamino)-1,3,4-oxadiazole inhibits corrosion of ZA-27 alloy in 1 M hydrochloric or 1 M sulfuric acid with excellent performance, more than 90%. The alloy consists of zinc, aluminum, copper, and magnesium. The inhibitor covers the alloy surface with a protective layer thereby preventing acid corrosion (Pruthviraj 2008).

1.4.6 CONCLUSION

The main features of 1,3,4-oxadiazole are its electron-negative character and "half-aromaticity." Basicity of this heterocycle is weak. Conjugation of the π type through the 1,3,4-oxadiazole ring exists, but is not strong. Substitution reactions are rather limited but cycloaddition reactions are typical. Ring transformations proceed easily and with great diversity.

1,3,4-Oxadiazole derivatives find multipurpose biomedical applications. Aryl and heteroaryl derivatives emit fluorescence and are used in environmental and bioanalysis. Many compounds of the 1,3,4-oxadiazole series are efficient agriprotectors or active corrosion inhibitors.

A special and very important direction in technical application of 1,3,4-oxadiazoles is connected with new liquid crystals, polymeric and oligomeric materials. Owing to its unique structure, 1,3,4-oxadiazole is the most attractive component for such applications. Having high thermostability, oxidation resistance, and marked electron-withdrawing nature, this component is widely applied to organic light-emitting diodes, electrochromic cells, and xerographic devices. Being incorporated into the main chain, these disubstituted rings contain no hydrogen atoms and do not participate in any rearrangements. They lack tension, have structural symmetry, and are thermally unreactive.

Specific properties determined by the electronic structure of oxadiazole ring, especially its electron-withdrawing character are crucial for microelectronics, optoelectronics and other technical fields. Many devices containing this oxadiazole showed high efficiency, high electric stability, and high durability. Applications of 1,3,4-oxadiazole derivatives to neutron detection are also prospective.

REFERENCES TO SECTION 1.4

Akhtar, T.; Hameed, Sh.; Al-Masoudi, N.A.; Loddo, R.; la Colla, P. (2008) *Acta Pharm.* **58**, 135.

Al-Talib, M.; Tashtoush, H. (1999) *Indian J. Chem.* **38B**, 1374.

Almajan, G.L.; Barbuceanu, S.F.; Draghici, C. (2006) *Rev. Chem.* **57**, 1108.

Almajan, G.L.; Barbiceanu, S.F.; Saramet, I.; Dinu, M.; Doicin, C.V.; Draghichi, C. (2008) *Rev. Chim.* **59**, 395.

Almasirad, A.; Vousooghi, N.; Tabatabai, S.A.; Kebriaeezadeh, A.; Shafiee, A. (2007) *Acta Chim. Slovenica* **54**, 317.

Alper, Ph. B.; Lelais, G.; Epple, R.; Mutnick, D. (2008) WO Patent 109,702.

Amr, A.E.-G.E.; Mohamed, S.F.; Abdel-Hafez, N.A.; Abdalla, M.A. (2008) *Monatsh. Chem.* **139**, 1491.

Apreutesei, D.; Mehl, G.H. (2007) *J. Mater. Chem.* **17**, 4711.

Aquino, Ch. J.; Dickson, H.; Peat, A.J. (2008) WO Patent 157,330.

Augustine, J.K.; Vairaperumal, V.; Narasimhan, Sh.; Alagarsamy, P.; Radhakrishnan, A. (2009) *Tetrahedron* **65**, 9989.

Back, K.R.; Whaley, D.P. (2008) WO Patent 129,319.

Balachandran, S.; Gupta, N.; Karirnar, V.V.; Rundra, S.; Palle, V.P.; Dastiar, S.G. (2008) WO Patent 035,316.

Balamurugan, R.; Kannan, P. (2008) *J. Polym. Sci. A* **46**, 5760.

Barbuceanu, S.-F.; Almajan, G.L.; Saramet, I.; Draghici, C. (2008) *Rev. Chem.* **59**, 304.

Barradas, J.S.; Errea, M.I.; D'Accorso, N.B.; Sepulveda, C.S.; Talarico, L.B.; Damonte, E.B. (2008) *Carbohydr. Res.* **343**, 2468.

Barreiro, G.; Kim, J.T.; Guimaraes, C.R.W.; Bailey, Ch.M.; Domaoal, R.; Wang, L.; Anderson, K.S.; Jorgensen, W.L. (2007) *J. Med. Chem.* **50**, 5324.

Bazan, G.C. (2007) *J. Org. Chem.* **72**, 8615.

Bentiss, F.; Lagrenee, M. (1999) *J. Heterocycl. Chem.* **36**, 1029.

Berggren, M.; Dodabalapur, A.; Slusher, R.E.; Bao, Z. (1997) *Nature* **389**, 466.

Bhandari, Sh.V.; Bothara, K.G.; Raut, M.K.; Patil, A.A.; Sarkate, A.P.; Mokale, V.J. (2008) *Bioorg. Med. Chem.* **16**, 1822.

Bijev, A.T.; Prodanova, P. (2007) *Khim. Geterotsikl. Soed.* 383.

Birch, A.M.; Butlin, R.J.; Gill, A.L.; Groombridge, S.D.; Plowright, A.; Waring, M.J. (2009) WO Patent 024,821.

Birch, A.M.; Butlin, R.J.; Plowright, A. (2007) WO Patent 141, 517.

Bird, C.W. (1985) *Tetrahedron* **41**, 1409.

Bolli, M.; Lehmann, D.; Mathys, B.; Mueller, C.; Nayler, O.; Steiner, B.; Velker, J. (2007) WO Patent 086,001.

Boss, Ch.; Brisbare-Roch, C.; Jenck, F. (2009) *J. Med. Chem.* **52**, 891.

Boyd, G.V.; Dando, S.R. (1970) *J. Chem. Soc. C*, 1397.

Brain, C.T.; Paul, J.M.; Loong, Y.; Oakley, P.J. (1999) *Tetrahedron Lett.* **40**, 3275.

Bull, R.J.; Van der Heuvel, M.; Mete, A.; Nadin, A.J.; Ray, N.Ch. (2008) WO Patent 023,157.

Butlin, R.J.; Plowright, A. (2007) WO Patent 138,311.

Chai, Ch.-P.; Zhu, X.-Q.; Wang, P.; Ren, M.-Q.; Chen, X.-F.; Xu, Y.-D.; Fan, X.-H.; Ye, Ch.; Chen, E.-Q.; Zhou, Q.-F. (2007) *Macromolecules* **40**, 9361.

Chaudhary, A.; Dudhe, R.; Singh, P. (2007) *Orient. J. Chem.* **23**, 1101.

Chavan, V.P.; Sonawane, S.A.; Shingare, M.S.; Karale, B.K. (2006) *Khim. Geterotsikl. Soed.* **5**, 711.

Colquhoun, H.M.; Chan, Y.F.; Cardin, Ch.J.; Drew, M.G.B.; Can, Y.; Abd El Kader, K.; White, T.M. (2007) *Dalton Trans.* 3864.

Combs, A.P.; Yue, E.W. (2008) WO Patent 036,652.

Coppo, F.T.; Evans, K.A.; Graybill, T.L.; Burton, G. (2004) *Tetrahedron Lett.* **45**, 3257.

Cullen, M.D.; Deng, B.-L.; Hartman, T.L.; Watson, K.M.; Buckheit, R.W.; Pannecouque, Ch.; de Clerq, E.; Cushman, M. (2007) *J. Med. Chem.* **50**, 4854.

Dabiri, M; Salehi, P; Baghbanzadeh, M.; Bahramnejad, M. (2006) *Tetrahedron Lett.* **47**, 6983.

Dahl, B.H.; Peters, D.; Olsen, G.M.; Timmermann, D.B.; Joergensen, S. (2008) WO Patent 049,864.

Dingemans, T.J.; Madsen, L.A.; Zafiropoulos, N.A.; Lin, W.; Samulski, E.T. (2006) *Philos. Trans. R. Soc.* **364**, 2681.

Dobson, A.H.; Grundy, W. (2007) WO Patent 144,571.

Dolman, S.F.; Gosselin, F.; O'Shea, P.D.; Davies, I.W. (2006) *J. Org. Chem.* **71**, 9548.

Dong, Y.-B.; Sun, T.; Ma, J.-P.; Zhao, X.-X.; Huang, R.-Q. (2006) *Inorg. Chem.* **45**, 10613.

Du, M.; Li, Ch.-P.; You, Y.-P.; Jiang, X.-J.; Cai, H.; Wang, Q.; Guo, J.-H. (2007) *Inorg. Chim. Acta* **360**, 2169.

Du, M.; Zhang, Zh.-H.; Wang, X.-G.; Li, Ch.-P. (2006a) *J. Mol. Struct.* **791**, 131.

Du, M.; Zhang, Zh.-H.; Zhao, X.-J.; Xu, Q. (2006b) *Inorg. Chem.* **45**, 5785.

Ducharme, Y.; Wu, Y.-H.; Frenette, R. (2007) WO Patent 038,865.

El-Essawy, F.A.; El-Sayed, W.A.; El-Kafrawy, Sh.A.; Morshedy, A.S.; Abdel-Rahman, A.-H. (2008) *Zeitschr. Naturforsch. C* **63**, 667.

Emmerling, F.; Reck, G.; Kraus, W.; Orgzall, I.; Schultz, B. (2008) *Crystl. Res. Technol.* **43**, 99.

Erdem, S.S.; Ozpinar, G.A.; Sacan, M.T. (2005) *THEOCHEM* **726**, 233.

Fang He, C.; Richards, G.J.; Kelly, S.M.; Contoret, A.E.A.; O'Neill, M. (2007) *Liq. Crystl.* **34**, 1249.

Fard-Jahromi, M.J.S.; Morsali, A.; Zeller, M. (2008) *J. Coord. Chem.* **61**, 2227.

Feng, D.; Huang, Y.; Chen, R.; Yu, Y.; Song, H. (2007) *Synthesis*, 1779.

Formagio, A.S.N.; Tonin, L.T.D.; Foglio, M.A.; Madjarof, Ch.; de Carvalho, J.E.; da Costa, W.F.; Cardoso, F.P.; Sarragiotto, M.H. (2008) *Bioorg. Med. Chem.* **16**, 9660.

Franco, O.; Reck, G; Orgzall, I.; Schulz, B.W.; Schulz, B. (2003) *J. Mol. Struct.* **649**, 219.

Franzini, R.M.; Kool, E.T. (2008) *Org. Lett.* **10**, 2935.

Fujiwara, H.; Sugishima, Y.; Tsujimoto, K. (2008) *Tetrahedron Lett.* **49**, 7200.

Gavrilyuk, J.I.; Lough, A.J.; Batey, R.A. (2008) *Tetrahedron Lett.* **49**, 4746.

Gillis, B.T.; LaMontagne, M.P. (1967) *J. Org. Chem.* **32**, 3318.

Giovannoni, M.P.; Cesari, N.; Vergelli, C.; Graziano, A.; Biancalani, C.; Biagini, P.; Ghelardini, C.; Vivoli, E.; dal Piaz, V. (2007) *J. Med. Chem.* **50**, 3945.

Gong, X.; Yang, Y.; Xiao, S. (2009) *J. Phys. Chem. C* **113**, 7398.

Gopalakrishnan, M; Honore, M.P.; Lee, Ch.-H.; Malysz, J.; Fi, J.; Li, T.; Schrimpf, M.R.; Sippy, K.B.; Anderson, D.J. (2008) WO Patent 073,942.

Gosselin, F.; Britton, R.A.; Davies, I.W.; Dolman, S.J.; Gauvreau, D.; Hoerner, R.S.; Hughes, G.; Janey, J.; Lau, S.; Molinaro, C.; Nadeu, Ch.; O'Shea, P.D.; Palucki, M.; Sidler, R. (2010) *J. Org. Chem.* **75**, 4154.

Hamada, Y.; Adachi, Ch.; Tsutsuni, T.; Saito, S. (1992) *Optoelectron. Dev. Technol.* **10**, 397.

Hassan, N.A. (2007) *J. Heterocycl. Chem.* **44**, 933.

Hetzheim, A.; Moeckel, K. (1966) *Adv. Heterocycl. Chem.* **7**, 183.

Holla, B. Sh.; Gonsalves, R.; Shenoy, Sh. (2000) *Eur. J. Med. Chem.* **35**, 267.

Honda, T.; Fujisawa, K.; Aono, H.; Ban, M. (2008) WO Patent 093,677.

Huang, Q.; Kong, Y.; Liu, M.; Feng, J.; Liu, Y. (2008) *J. Insect. Sci.* **8**, 1.

Huang, Q.; Wu, J.; Kong, Y.; Liu, M.; Cao, S. (2007) *J. Environ. Sci. Health B* **42**, 305.

Hughes, T.V.; Xu, G.; Wetter, S.K.; Connoly, P.J.; Emanuel, S.L.; Karnachi, P.; Pollack, S.R.; Pandey, N.; Adams, M.; Moreno-Mazza, S.; Middleton, S.A.; Greenberger, L.M. (2008) *Bioorg. Med. Chem. Lett.* **18**, 4895.

Inoue, H.; Egawa, M.; Seo, S. (2007) US Patent 085,073.

Iqbal, R.; Zareef, M.; Ahmed, S.; Zaidi, J.H.; Arfan, M.; Shafique, M.; Al-Masoudi, N.A. (2006) *J. Chin. Chem. Soc.* **53**, 689.

Ishida, H.; Isami, Sh.; Matsumura, T.; Umehara, H.; Yamashita, Y.; Kajita, J.; Fuse, E.; Kiyoi, H.; Naoe, T.; Akinaga, Sh.; Shiotsu, Yu.; Arai, H. (2008) *Bioorg. Med. Chem. Lett.* **18**, 5472.

Isobe, T.; Ishikawa, T. (1999) *J. Org. Chem.* **64**, 6989.

Jansen, M.; Rabe, H.; Strehless, A.; Dieler, S.; Debus, F.; Dannhardt, G.; Akabas, M.H.; Lueddens, H. (2008) *J. Med. Chem.* **51**, 4430.

Johnstone, C.; Plowright, A. (2007) WO Patent 138,304.

Joshi, S.D.; Vagdevi, H.M.; Vaidya, V.P.; Gadaginamath, G.S. (2007) *Indian J. Heterocycl. Chem.* **17**, 165.

Joshi, S.D.; Vagdevi, H.M.; Vaidya, V.P.; Gadaginamath, G.S. (2008) *Eur. J. Med. Chem.* **43**, 1989.

Jursic, B.S.; Zdravkovski, Z. (1994) *Synth. Commun.* **24**, 1575.

Karabasanagouda, T.; Adhikari, A.V.; Shetty, N.S. (2007) *Phosphorus Sulfur* **182**, 2925.

Kawano, T.; Hirano, K.; Satoh, T.; Miura, M. (2010) *J. Am. Chem. Soc.* **132**, 6900.

Khatkar, A.; Singh, U.K.; Dewan, S.K.; Gagoria, J.; Sarita (2007) *Orient. J. Chem.* **23**, 623.

Kidwai, M.; Kumar, P.; Goel, Y.; Kumar, K. (1997) *Indian J. Chem.* **36B**, 175.

Kim, B. Ch.; Kim, K.-Y.; Lee, H.B.; Shin, H. (2008) *Org. Res. Develop.* **12**, 626.

Kim, T.H.; Lee, Ch.-H.; Kwak, Ch.G.; Choi, M.S.; Park, W.H.; Lee, T.S. (2007) *Mol. Crystl. Liq. Crystl.* **463**, 537.

Kizhnyaev, V.N.; Pokatilov, F.A.; Vereshchagin, L.I. (2008) *Vysokomol. Soed. C* **50**, 1296.

Klein, M.; Beier, N. (2008a) WO Patent 086,854.

Klein, M.; Beier, N. (2008b) Germ. Patent 102,007,002,717.

Koo, K.D.; Kim, M.J.; Kim, S.; Kim, K.-H.; Hong, S.Y.; Hur, G.-Ch.; Yim, H.J.; Kim, G.T.; Han, H.O.; Kwon, O.H.; Kwon, T.S.; Koh, J.S.; Lee, Ch.-S. (2007) *Bioorg. Med. Chem. Lett.* **17**, 4167.

Koshimizu, M.; Kitajima, H.; Iwai, T.; Asai, K. (2008) *Jpn. J. Appl. Phys.* **47**, 3717.

Kudelko, A.; Zielinski, W. (2009) *Tetrahedron* **65**, 1200.

Kumar, A.; D'Souza, S.S.; Gaonkar, S.L.; Rai, K.M.L.; Salimath, B.P. (2008) *Invest New Drugs* **26**, 425.

Kwak, Ch.K.; Lee, Ch.-H.; Lee, T.S. (2007) *Tetrahedron Lett.* **48**, 7788.

Lai, C.; Ke, Y.c.; Su, J.C.; Shen, C.; Li, W.R. (2002) *Liq. Crystl.* **29**, 915.

Lee, Ch.-H.; Ji, J.; Li, T.; Schrimpf, M.R.; Sippy, K.B.; Gopalakrishnan, M. (2008a) US Patent 255,203.

Lee, Ch.-H.; Ji, J.; Li, T.; Schrimpf, M.R.; Sippy, K.B.; Gopalakrishnan, M. (2008b) WO Patent 127,464.

Lee, T.; Landis, Ch.A.; Dhar, B.M.; Jung, B.J.; Sun, J.; Sarjeant, A.; Lee, H.-J.; Katz, H.E. (2009) *J. Am. Chem. Soc.* **131**, 1692.

Lehmann, M.; Koehn, Ch.; Kresse, H.; Vakhovskaya, Z. (2008) *Chem. Commun.* 1768.

Letafat, B.; Emami, S.; Alibadi, A.; Mohammadhosseini, N.; Moshafi, M.H.; Asadipour, A.; Shafiee, A.; Foroumadi, A. (2008) *Arch. Pharm.* **341**, 497.

Leung, M.-K.; Yang, Ch.-Ch.; Lee, J.-H.; Tsai, H.-H.; Lin, Ch.-F.; Huang, Ch.-Y.; Su, Y.O.; Chi, Ch.-F. (2007) *Org. Lett.* **9**, 235.

Li, A.-F.; He, H; Ruan, Y.-B.; Wen, Zh.-Ch.; Zhao, J.-S.; Jiang, Q.-J.; Jiang, Y.-B. (2009) *Org. Biol. Chem.* **7**, 193.

Li, J.; Yin, Sh.; Li, Y.; Fan, B. (2007) Chin. Patent 101,085,793.

Lin, Y.-H.; Wu, H.-H.; Wong, K.-T.; Hsieh, Ch.-Ch.; Lin, Y.-Ch.; Chou, P.T. (2008) *Org. Lett.* **10**, 3211.

Link, H. (1978) *Helv. Chim. Acta* **61**, 2419.

Liu, Ch.; Chi, H.; Cui, D.; Li, Zh. (2007c) Chin. Patent 1,927,859.

Liu, G.; Lynch, J.K.; Freeman, J.; Liu, B.; Xin, Zh.; Zhao, H.; Serby, M.D.; Kym, Ph.R.; Suhar, T.S.; Smith, H.T.; Cao, N.; Yang, R.; Janis, R.S.; Krauser, J.A.; Cepa, S.P.; Beno, D.W.A.; Sham, H.L.; Collins, Ch.A.; Surowy, T.K.; Camp, H.S. (2007b) *J. Med. Chem.* **50**, 3086.

Liu, Y.; Zong, L.; Zheng, L.; Wu, L.; Cheng, Y. (2007a) *Polymer* **48**, 6799.

Long, D.Q.; Li, D.J.; Cai, W.Q.; Chen, Sh.Sh. (2006) *Chin. Chem. Lett.* **17**, 1427.

Ma, B.; Kim, B.J.; Deng, L.; Poulsen, D.A.; Thompson, M.E.; Frechet, J.M.J. (2007) *Macromolecules* **40**, 8156.

Ma, W.L.; Iyer, P.K.; Gong, X.; Liu, B.; Moses, D.; Bazan, G.C.; Heeger, A.J. (2005) *Adv. Mater.* **17**, 274.

Mahmoudi, G.; Morsali, A. (2007) *Crystl. Eng. Commun.* **9**, 1062.

Malek, K.; Zborowsky, K.; Gebski, K.; Proniewicz, L.M.; Schroeder, G. (2008) *Mol. Phys.* **106**, 1823.

Mashraqui, S.H.; Subramanian, S.; Bhasikuttan, A.C. (2007) *Tetrahedron* **63**, 1680.

Mauro, M.; Panigati, M.; Donghi, D.; Mercandelli, P.; Mussini, P.; Sironi, A.; D'Alfonso, G. (2008) *Inorg. Chem.* **47**, 11154.

Mekuskiene, G.; Dodonova, J.; Burbuliene, M.M.; Vainilavicius, P. (2007) *Heterocycl. Commun.* **13**, 295.

Miki, T.; Yokoyama, N.; Hayashi, Sh.; Kusano, Sh.; Taniguchi, Y.; Ichikava, M. (2007) WO Patent 032,357.

Mikroyannidis, J.A.; Gibbons, K.M.; Kulkarni, A.; Jenekhe, S.A. (2008) *Macromolecules* **41**, 663.

Mochizuki, H.; Hasui, T.; Kawamoto, M.; Shiono, T.; Ikeda, T.; Adachi, Ch.; Taniguchi, Y.; Shirota, Y. (2000) *Chem. Commun.* 1923.

Murata, H.; Nakashima, H.; Kawakami, S.; Ohsawa, N.; Nomura, R.; Seo, S. (2007) US Patent 149,784.

Nagata, I.; Nishikawa, H. (2007) Jpn. Patent 246,672.

Nakaya, T.; Sato, M.; Kodera, T.; Takano, Sh.; Eto, N. (2007) Jpn. Patent 099,723.

Natera, J.; Otero, L.; Sereno, L.; Fungo, F.; Wang, N.-S.; Tsai, Y.-M.; Hwu, T.-Y.; Wong, K.-T. (2007) *Macromolecules* **40**, 4456.

Nishiji, A.; Kimura, K.; Yamazaki, Sh. (2007) Jpn. Patent 321,134.

Noguchi, Sh.; Koyama, T. (2008) WO Patent 156,159.

Ono, K.; Saito, K. (2008) Jpn. Patent 069,086.

Ono, K.; Wakida, M.; Hosokawa, R.; Saito, K.; Nishida, J.-i.; Yamashita, Y. (2007) *Heterocycles* **72**, 85.

Palmer, J.T.; Hirschbein, B.L.; Cheung, H.; McCarter, J.; Janc, J.W.; Yu, Z.W.; Wesolowski, G. (2006) *Bioorg. Med. Chem. Lett.* **16**, 2909.

Pefkianakis, E.K.; Tzanetos, N.P.; Kallitsis, J.K. (2008) *Chem. Mater.* **20**, 6254.

Peters, D.; Olsen, G.M.; Nielsen, E.O.; Timmermann, D.B.; Loechel, S.Ch. (2007a) WO Patent 138,038.

Peters, D.; Olsen, G.M.; Nielsen, E.O.; Timmermann, D.B.; Loechel, S.Ch.; Mikkelsen, J.D.; Hansen, H.B.; Redrobe, J.P.; Christensen, J.K.; Dyhring, T. (2007b) WO Patent 138,037.

Peters, D.; Timmermann, D.B.; Nielsen, E.O.; Olsen, G.M.; Dyhring, T.; Christensen, J.K. (2009) WO Patent 024,516.

Piatnitski-Chekler, E.L.; Elokdah, H.M.; Butera, J. (2008) *Tetrahedron Lett.* **49**, 6709.

Poitout, L.; Harnett, J.; Bigg, D.; Sackur, C.; Ferrandis, E. (2008) Fr. Patent 2,907,784.

Polshettiwar, V.; Varma, R.S. (2008) *Tetrahedron Lett.* **49**, 879.

Pruthviraj, R.D.; (2008) *Ultra Chem.* **4**, 1.

Qu, S.; Li, Y.; Wang, L.; Lu, Q.; Liu, X. (2011) *Chem. Commun.* 4207.

Rafiquee, M.Z.A.; Saxena, N.; Khan, S.; Quraishi, M.A. (2007) *Indian J. Chem. Technol.* **14**, 576.

Rajak, H.; Deshmukh, R.; Veerasamy, R.; Sharma, A.K.; Mishra, P.; Kharya, M.D. (2010) *Bioorg. Med. Chem. Lett.* **20**, 4168.

Rajak, H.; Kharya, M.D.; Mishra, P. (2007) *Yakugaku Zasshi* **127**, 1757.

Rajak, H.; Kharya, M.D.; Mishra, P. (2008) *Arch. Pharm.* **341**, 247.

Rastogi, N.; Varma, R.S.; Shukla, S.; Sethi, R. (2006a) *Indian J. Heterocycl. Chem.* **16**, 5.

Rastogi, N.; Varma, R.S.; Singh, A.P.; Kapil, A. (2006b) *Indian J. Heterocycl. Chem.* **15**, 339.

Reddy, P.S.N.; Reddy, P. (1987) *Indian J. Chem. B* **26**, 890.

Regel, E. (1977) *Liebigs Ann. Chem.*, **1977**, 159.

Reid, J.R.; Heindel, N.D. (1976) *J. Heterocycl. Chem.* **13**, 925.

Romine, J.L.; Martin, S.W.; Meanwell, N.A.; Gribkoff, V.K.; Boissard, Ch.G.; Dworetzky, S.I.; Natale, J.; Moon, S.; Ortiz, A.; Yeleswaram, S.; Pajor, L.; Gao, Q.; Starrett, J.E. (2007) *J. Med. Chem.* **50**, 528.

Rostamizadeh, Sh.; Ghamkhar, S. (2008) *Chin. Chem. Lett.* **19**, 639.

Saeed, A. (2007) *Khim. Geterotsikl. Soed.* 1264.

Saeed, A.; Mumtaz, A. (2008) *Chin. Chem. Lett.* **19**, 423.

Scopes, D.I.Ch.; Cheeseright, T.J.; Vinter, J.G. (2008) WO Patent 107,677.

Semenov, B.B.; Smushkevich, Yu.I. (2008) Russ. Patent 2,317,986.

Sengupta, P.; Dash, D.K.; Yeligar, V.C.; Murugesh, K.; Rajalingam, D.; Singh, J.; Maity, T.K. (2008) *Indian J. Chem.* **478**, 460.

Shaharyar, M.; Ali, M.A.; Abdullah, M.M. (2007) *Med. Chem. Res.* **16**, 292.

Shimizu, N.; Bartlett, P.D. (1978) *J. Am. Chem. Soc.* **100**, 4260.

Shin, D.Ch.; Jin, J.G.; Lee, Y.G. (2008) Kor. Patent 104,820.

Siddiqui, N.; Alam, M.Sh.; Ahsan, W. (2008) *Acta Pharm.* **58**, 445.

Singh, M.; Aggarwal, V.; Singh, U.P.; Singh, N.K. (2009a) *Polyhedron* **28**, 195.

Singh, M.; Butcher, R.J.; Singh, N.K. (2009b) *Polyhedron* **28**, 95.

Singh, H.; Yadav, L.D.S.; Bhattacharya, B.K. (1984) *J. Indian Chem. Soc.* **61**, 436.

Souldozi, A.; Slepokura, K.; Lis, T.; Ramazani, A. (2007) *Zeitschr. Natuforsch. B* **62**, 835.

Sridhar, D.; Arjun, M.; Jyoti, M.; Raviprasad, T.; Sarangapani, M. (2006) *Indian J. Heterocycl. Chem.* **16**, 61.

Summa, V.; Petrocchi, A.; Bonelli, F.; Crescenzi, B.; Donghi, M.; Ferrara, M.; Fiore, F.; Gardelli, C.; Gonzalez, Paz. O.; Hazuda, D.J.; Jones, Ph.; Kizel, O.; Laufer, R.; Monteagudo, E.; Muraglia, E.; Nizi, E.; Orvieto, J.; Pace, P.; Pescatore, G.; Scarpelli, R.; Stillmock, K.; Witmer, M.V.; Rowley, M. (2008) *J. Med. Chem.* **51**, 5843.

Sun, G.; Qin, Zh. (2006) *Nongyaoxue Xuebao* **8**, 272.

Tajino, H.; Kutsuma, S.; Kawaguchi, T. (2007) Jpn. Patent 137,834.

Tamano, M.; Takayama, M. (2006) Jpn. Patent 344,544.

Tamura, K.; Maeda, K.; Yashima, E. (2008) *Macromolecules* **41**, 5065.

Tandon, V.K.; Chhor, R.B. (2001) *Synth. Commun.* **31**, 1727.

Tang, X.-L.; Peng, X.-H.; Dou, W.; Mao, J.; Zheng, J.-R.; Qin, W.-W.; Liu, W.-Sh.; Chang, J.; Yao, X.-J. (2008) *Org. Lett.* **10**, 3653.

Tao, Y.; Ma, B.; Segalman, R.A. (2008b) *Macromolecules* **41**, 7152.

Tao, Y.; Wang, Q.; Shang, Y.; Yang, Ch.; Ao, L.; Qin, J.; Ma, D.; Shuai, Zh. (2009) *Chem. Commun.* 77.

Tao, Y.; Wang, Q.; Yang, Ch.; Wang, Q.; Zhang, Zh.; Zou, T.; Qin, J.; Ma, D. (2008a) *Angew. Chem. Intl. Ed.* **47**, 8104.

Todres, Z.V. (2009) *Ion-Radical Organic Chemistry.* CRC Press/Taylor & Francis, Boca Raton, FL.

Tu, J.; Fu, Ch.; Wang, L.; Liang, Y.; Zheng, J. (2008) *Zhongguo Fushi Yu Fanghu Xuebao* **28**, 44.

Tumosiene, I.; Beresnevicius, Z.I. (2007) *Khim. Geterotsikl. Soed.* 1353.

Vasil'ev, N.V.; Romanov, D.V.; Bazhenov, A.A.; Lyssenko, K.A.; Zatonsky, G.V. (2007) *J. Fluorine Chem.* **128**, 740.

Vasil'ev, N.V.; Romanov, D.V.; Truskanova, T.D.; Lyssenko, K.A.; Zatonsky, G.V. (2006) *Mendeleev Commun.* **3**, 186.

Vereshchagin, L.I.; Verkhozhina, O.N.; Pokatilov, F.A.; Strunevich, S.K.; Proidakov, A.G.; Kizhnyaev, V.N. (2007) *Zh. Org. Khim.* **43**, 1709.

Wang, P.; Chai, Ch.-P.; Chuai, Y.-T.; Wang, F.-Zh.; Chen, X.-F.; Fan, X.-H.; Xu, Y.-D.; Zou, D.-Ch.; Zhou, Q.-F. (2007e) *Polymer* **48**, 5889.

Wang, P.; Chai, Ch.-P.; Wang, F.-Zh.; Chuai, Y.-T.; Chen, X.-F.; Fan, X.-H.; Zou, D.-Ch.; Zhou, Q.-F. (2008a) *J. Polym. Sci. A* **46**, 1843.

Wang, P.; Chai, Y.-T.; Yang, Q.; Wang, F.-Zh.; Shen, Zh.-H.; Guo, H.-Q.; Chen, X.-F.; Fan, X.-H.; Zou, D.-Ch.; Zhou, Q.-F. (2008b) *J. Polym. Sci. A* **46**, 5452.

Wang, P.; Fin, H.; Liu, W.-L.; Chai, Ch.-P.; Shen, Zh.-H.; Guo, H.-Q.; Chen, X.-F.; Fan, X.-H.; Zou, D.-Ch.; Zhou, Q.-F. (2008c) *J. Polym. Sci. A* **46**, 7861.

Wang, Y.-T.; Tang, G.-M. (2007) *Inorg. Chem. Commun.* **10**, 53.

Wang, Y.-T.; Tang, G.-M.; Ma, W.-Y.; Wan, W.-Zh. (2007c) *Polyhedron* **26**, 782.

Wang, Y.-T.; Tang, G.-M.; Qiang, Zh.-W. (2007d) *Polyhedron* **26**, 4542.

Wang, Y.P.; Xie, W.F.; Li, B.; Li, W.L. (2007b) *Chin. Chem. Lett.* **18**, 1501.

Wang, Y.-G.; Xu, W.-M.; Huang, X. (2007a) *J. Combin. Chem.* **9**, 513.

Warkentin, J. (2009) *Acc. Chem. Res.* **42**, 205.

Warmus, J.S.; Flamme, C.; Zhang, L.Y.; Barrett, S.; Bridges, A.; Chen, H.; Gowan, R.; Kaufman, M.; Sebolt-Leopold, J.; Leopold, W.; Merriman, R.; Ohren, J.; Pavlovsky, A.; Przybranowski, S.; Tecle, H.; Valik, H.; Whitehead, Ch.; Zhang, E. (2008) *Bioorg. Med. Chem. Lett.* **18**, 6171.

Warrener, R.N.; Elsey, G.M.; Russell, R.A.; Tiekink, E.R.T. (1995) *Tetrahedron Lett.* **36**, 5275.

Warrener, R.N.; Margetic, D.; Tiekink, E.R.T.; Russell, R.A. (1977) *SYNLETT* **2**, 196.

Wen, Ch.-R.; Wang, Y.-J.; Wang, H.-Ch.; Sheu, H.-Sh.; Lee, G.-H.; Lai, Ch.K. (2005) *Chem. Mater.* **17**, 1646.

Werber, G.; Buccheri, F.; Noto, R.; Gentile, M. (1977) *J. Heterocycl. Chem.* **14**, 1385.

Werber, G.; Buccheri, F.; Vivona, N.; Bianchini, R. (1979) *J. Heterocycl. Chem.* **16**, 145.

Wolarz, E.; Chrzumnicka, E.; Fisher, T.; Stumpe, J. (2007) *Dyes Pigments* **75**, 753.

Wolkenberg, S.E.; Boger, D.L. (2002) *J. Org. Chem.* **67**, 7361.

Xiang, N.J.; Leung, L.M.; So, Sh.K.; Wang, J.; Su, Q.; Gong, M.L. (2006) *Mater. Lett.* **60**, 2909.

Xu, Y.-D.; Yang, Q.; Shen, Zh.-H.; Chen, X.-F.; Fan, X.-H.; Zhou, Q.-F. (2009) *Macromolecules* **42**, 2542.

Xue, J.; Wang, Sh.; Zhang, L.; Li, X. (2007) *Dyes Pigments* **75**, 369.

Yang, Q.; Jin, H.; Xu, Y.; Wang, P.; Liang, X.; Shen, Zh.; Chen, X.; Zou, D.; Fan, X.; Zhou, Q. (2009) *Macromolecules* **72**, 1037.

Yang, Ch.; Tao, Y.; Qin, J. (2008) Chin. Patent 101,274,916.

Yang, Zh.; Yang, Sh.; Zhang, J. (2007) *J. Phys. Chem. A* **111**, 6354.

Yang, Sh.-h.; Zhai, Zh.-w. (2007) *Shandong Huang* **36**, 1.

Young, J.B.; Eid, R.; Turner, Ch.; de Vita, R.J.; Kurtz, M.M.; Tsao, K.-L.C.; Chicchi, G.G.; Wheeldon, A.; Carlson, E.; Mills, S.G. (2007) *Bioorg. Med. Chem. Lett.* **17**, 5310.

Zablocki, J.; Abelman, M.; Organ, M.; Bilokin, Ya.; Jiang, R.; Elzein, E.; Kobayashi, T.; Kalla, R.; Perry, Th.; Li, X.; Diamond, I.; Yao, L.; Fan, P.; Arolfo, M.P.; Jian, Zh. (2008) US Patent 207,610.

Zareef, M.; Igbal, R.; Gamboa de Domingues, N.; Rodrigues, J.; Zaidi, J.H.; Arfan, M.; Supuran, C.T. (2007) *J. Enzyme Inhib. Med. Chem.* **22**, 301.

Zarghi, A.; Hajimahdi, Z.; Mohebbi, Sh.; Rashidi, H.; Mozaffari, S.; Sarraf, S.; Faizi, M.; Tabatabaee, S.A.; Shafiee, A. (2008) *Chem. Pharm. Bull.* **56**, 509.

Zarudnitski, E.V.; Pervak, I.I.; Merkulov, A.S.; Yurchenko, A.A.; Tolmachev, A.A. (2008) *Tetrahedron* **64**, 10431.

Zhang, R.; Jordan, R.; Nuyken, O. (2003) *Macromol. Rapid Commun.* **24**, 246.

Zhang, X.; Xiao, Y.; Qian, X. (2008) *Angew. Chem. Intl. Ed.* **47**, 8025.

Zhang, Zh.-H.; Tian, Y.-L.; Guo, Y.-M. (2007) *Inorg. Chim. Acta* **360**, 2783.

LOOKING OUT OVER CHAPTER 1: COMPARISON OF OXADIAZOLE ISOMERS

The most notable feature of oxadiazoles is their difference in the conjugation conditions. The base-catalyzed hydrogen-deuterium isotope exchange illustrates such a difference for isomeric dimethyl oxadiazoles (Scheme 1.160) (Brown and Ghosh 1969). In the case of 3,5-dimethyl-1,2,4-oxadiazole, the 5-methyl group is selectively involved in this kind of exchange. In the case of 2,5-dimethyl-1,3,4-oxadiazole,

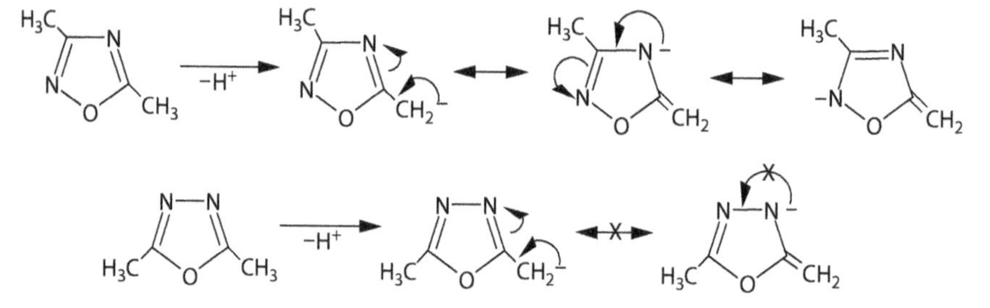

SCHEME 1.160

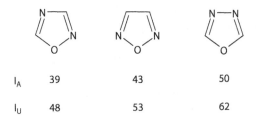

I_A	39	43	50
I_U	48	53	62

SCHEME 1.161

no isotopic exchange takes place under the same conditions. The first step in the exchange reaction obviously consists in the proton detachment from a methyl group with localization of the negative charge at the residual methylene group. The very possibility of the isotope exchange depends on stability of the anions formed. In its turn, the anion stability depends on delocalization of the negative charge within the different isomers of oxadiazoles. As seen from Scheme 1.160, this delocalization proceeds more effectively in 1,2,4 isomer than in its 1,3,4 counterpart. Correspondingly, H/D exchange does take place in 3,5-dimethyl-1,2,4-oxadiazole and does not in 2,5-dimethyl-1,3,4-oxadiazole.

It is interesting to compare the aromaticity indexes, I_A, calculated by Bird (1985, 1992) on the basis of bond lengths in oxadiazole isomers. Applicability of this index was compared with other ones accepted in heterocyclic chemistry (see Balaban et al. 2004). For the aims of comparison of the isomeric oxadiazoles, I_A's are useful enough and they correlate to the available chemical data. For isomeric oxadiazoles, Scheme 1.161 juxtaposes the Bird's indexes I_A's calculated on the basis of bond lengths in 1985 and the Bird's united heteroaromaticity indexes, I_U's, calculated on the basis of formation heat in 1992 (Scheme 1.161).

$I_A = 50$ ($I_U = 62$) for 1,3,4-oxadiazole is higher than $I_A = 39$ ($I_U = 48$) for 1,2,4-oxadiazole. 1,2,5-Oxadiazole with its $I_A = 43$ ($I_U = 53$) occupies the middle position. (Benzene with its $I_A = I_U = 100$ is a reference compound.) Decreased aromaticity of oxadiazoles gives rise to instability of their anion-radicals. The unstable ones undergo ring cleavage.

A pertinent difference should be noted between N-oxides of 1,2,4- and 1,2,5-oxadiazoles. Upon reduction with zinc in acetic acid, 1,2,4-oxadiazole-4-oxide gives rise to 1,2,4-oxadiazole. Under the same conditions, 1,2,5-oxadiazole-5-oxide forms glyoxime. It is the equilibrium between the ring-closed and ring-opened forms that is responsible for the different behavior of 1,2,5-oxadiazole-5-oxide in relation to 1,2,4-oxadiazole-4-oxide.

Ring-to-open-chain transferability is the intrinsic and crucial property of 1,2,5-oxadiazole-2(or 5)-oxides. This transformation leads to nitroso compounds as intermediates. The latter can recyclize, but isomeric oxides are formed. This endows furoxan chemistry with definite specificity and with ability of NO-releasing in biosystems.

Being reach in nitrogen and oxygen, oxadiazoles are highly energetic heterocycles. Derivatives of 1,2,5-oxadiazole N-oxide are the most attractive explosive systems exhibiting acceptable exploitation properties. Certainly, the warning statement should be repeated: It is both dangerous and illegal to participate in unauthorized experimentation with explosives.

1,2,4-Oxadiazole oxides differ in the fragility of their heterocyclic rings: The rings are capable of transforming into the nitrolic acid and nitrosocarbonyl species upon photoexcitation or thermal initiation.

In majority, 1,2,3-oxadidiazoles exist in the open diazoketone tautomeric form. Data on sydnones and sydnone imines were subjects of long-time discussions. At present, it is accepted that these compounds are not delocalized systems, but remain substituted oxadiazolium salts or zwitterions. Many sydnones and sydnone imines are pharmacologically active. Their bioactivity is frequently associated with ring cleavage in metabolic chains. This ring opening is an intermediary step and the final reaction leads to liberation off in vivo active nitrogen monoxide. In this respect, N-oxides of 1,2,3- and 1,2,5-oxadiazoles are similar.

REFERENCES TO CHAPTER 1: OVERLOOK

Balaban, A.T.; Oniciu, D.C.; Katritzky, A.R. (2004) *Chem. Rev.* **104**, 2777.
Bird, C.W. (1985) *Tetrahedron* **41**, 1409.
Bird, C.W. (1992) *Tetrahedron* **48**, 335.
Brown, D.J.; Ghosh, P.B. (1969) *J. Chem. Soc. B*, 270.

2 Thiadiazoles

2.1 1,2,3-THIADIAZOLES

2.1.1 FORMATION

2.1.1.1 From Hydrazones

Interaction of hydrazones with thionyl chloride is known as the Hurd–Mori reaction (1955). Thionyl chloride is used as a reactant and as a solvent or as a reactant dissolved in organic solvents. Yields are usually good. For instance, 4-phenyl- and 4-*tert*-butyl-1,2,3-thiadiazoles were obtained from the corresponding hydrazones with yields of 83% and 89%, respectively. Acetyl hydrazone, however, reacts with thionyl chloride unexpectedly forming 5-chloro-4-methyl-1,2,3-thiadiazole. This obstacle is probably caused by an in situ halogenation of the starting material (Sandrinelli et al. 2006).

2.1.1.2 From Semicarbazones

Scheme 2.1 describes the reaction between a semicarbazone with thionyl chloride. Namely, furan and thiophene fused with cyclohexanone semicarbazones were transformed to the corresponding 1,2,3-thiadiazoles upon treatment with thionyl chloride. The reaction includes an oxidation step and eventually leads to a benzene fused with three 1,2,3-thiadiazole rings, see the central compound of Scheme 2.1 (Shekarchi et al. 2003). This case of oxidation is a challenge from the mechanistic point of view.

In reactions of semicarbazones from acetones with thionyl chloride, both the methylene and methyl groups of the former substituted acetone could participate in the cyclization. Scheme 2.2 shows that two isomers can be formed. As noted by Zimmer and Meier (1981) then Morzherin et al. (2001), the cyclization proceeds with high regioselectivity: Only the methylene group takes part to form 4-methyl substituted thiadiazole. Its non-methylated counterpart is not formed at all. This feature is in accordance with results of C-nitrosation or bromination of the considered acetone moiety as noted by Rukavishnikov and Tkachev (1992) or Kolesnik et al. (1955). In all of the reactions, just the methylene group rather than the methyl function was involved. Notably, this methylene group is activated by α-effect due to the adjacent C=N fragment. When two different ethylene groups are adjacent to the C=N fragment, a more acidic ethylene will cyclize with higher regioselectivity (Fujita et al. 1993). Bulky substituents at one of the methylene groups blocked its participation in the cyclization (Hanold et al. 1986, L'abbe et al. 1995a). The initial semicarbazone can be used as a mixture of Z and E isomers. However, it is the E configuration that predominantly participates in this kind of cyclization (Zimmer and Meier 1981).

Of course, the regioselectivity of the cyclization changes depending on the R substituent. Thus, the non-methyl isomer of Scheme 2.2 is produced almost exclusively

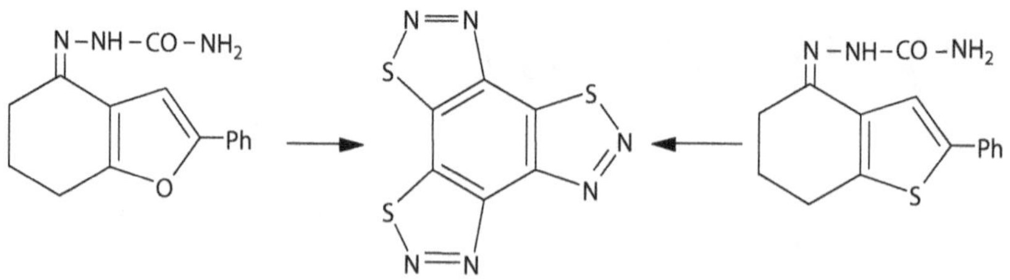

SCHEME 2.1

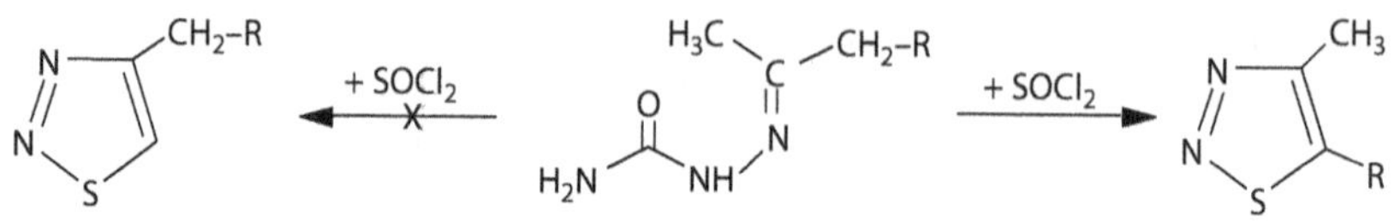

SCHEME 2.2

when $R = NMe_2$ (Fujita et al. 1993). Naturally, the methyl group is also involved in the cyclization when the α-carbon atom mentioned above bears no hydrogen. This case is illustrated by Scheme 2.3 (Attanasi et al. 2003). The reaction kinetics depends on the size of the leaving group (R). Yields of 60% are reached in 1.2 h when the leaving group is *tert*-butoxy and in 168.0 h when the amino group is the leaving one. N-ethoxycarbonylhydrazones are used in some special cases because ethoxycarbonyl is a good leaving group (Lardies et al. 2007, Petrov et al. 2007, Cerrada et al. 2009).

One interesting case is represented by the bis(1,2,3-thiadiazole-4-yl)ferrocene formation according to Scheme 2.4. The yield of the product is good. The product is stable and quantitatively gives rise to molecular ion upon electron impact (Al-Smadi and Ratrout 2004). The thiadiazolyl rings formed get along together and, seemingly, each of them is coplanar with its own cyclopentadienyl nucleus.

The character of thionyl chloride interaction with semicarbazones is sometimes very sensitive to the reaction temperature. Saravanan et al. (2007) gave examples where semicarbazones broke down to give their parent ketones under thionyl chloride action at 0°C. In these cases, a temperature of −15°C is optimal.

The (semicarbazone-thionyl chloride) reaction can be catalyzed by sodium hydrogen sulfate supported on silica gel under microwave irradiation (Gopalakrishnan et al. 2008).

Another application of an oxidation reaction (with hydrogen peroxide as an oxidant) was used to transform the hydrogen-bonded thionophenylhydrazone of Scheme 2.5 into the corresponding 1,2,3-thiadiazole (Zaleska et al. 2003).

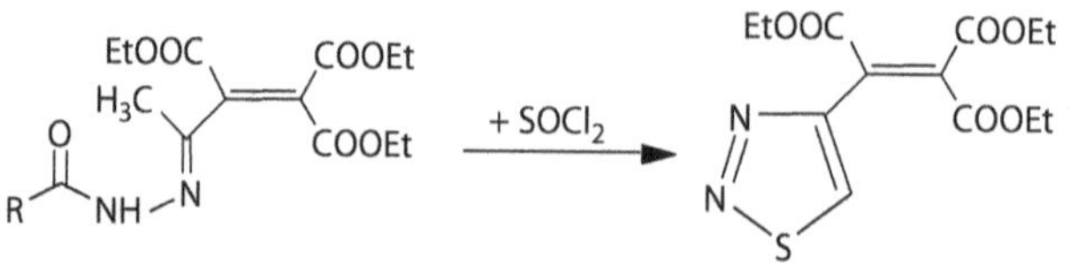

SCHEME 2.3

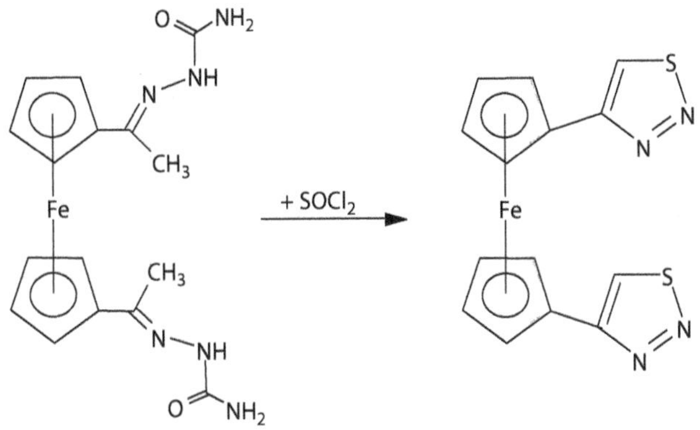

SCHEME 2.4

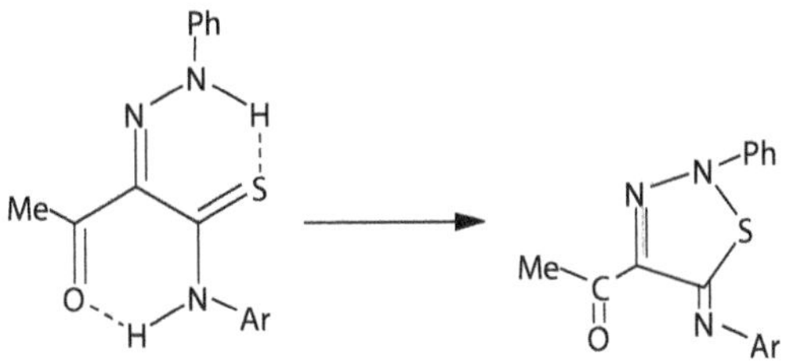

SCHEME 2.5

As mentioned, the formation of the 1,2,3-thiadiazole cycle from thionophenyl-hydrazones requires an oxidant (Weber et al. 1977). In attempts to prepare alkyl- or aromatic derivatives of 2*H*-[1,2,3]-thiadiazolo[5,4-b]indole, Kondratieva et al. (2007) used the following oxidants: bromine, iodine, and *N*-chlorosuccinimide. The oxidation with bromine (in acetic acid) or with *N*-chlorosuccinimide (in ethyl acetate) was turned out to be the method of choice. The cyclization is preceded by S-halogenation so that the final products are formed as hydrochlorides (see Scheme 2.6).

2.1.1.3 From Diazocompounds

The simplest and most common process for the synthesis of benzo-1,2,3-thiadiazoles consists in the diazotization 2-aminobenzenethiols. Thus, the diazo method was applied to the thiobenzyl aniline derivative in Scheme 2.7. The corresponding thiadiazole derivative was obtained with 90% yield. This compound exhibits the

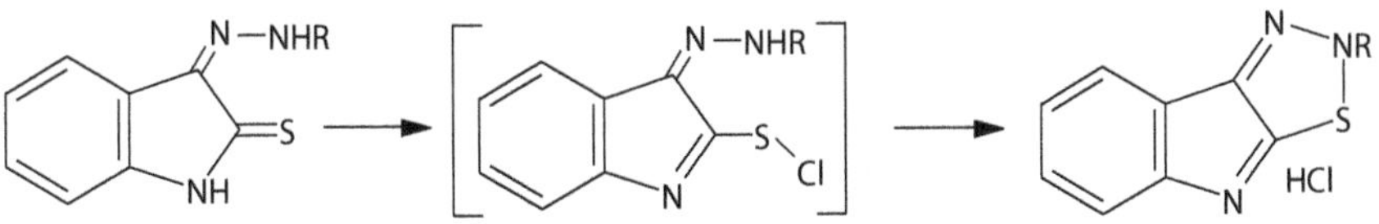

SCHEME 2.6

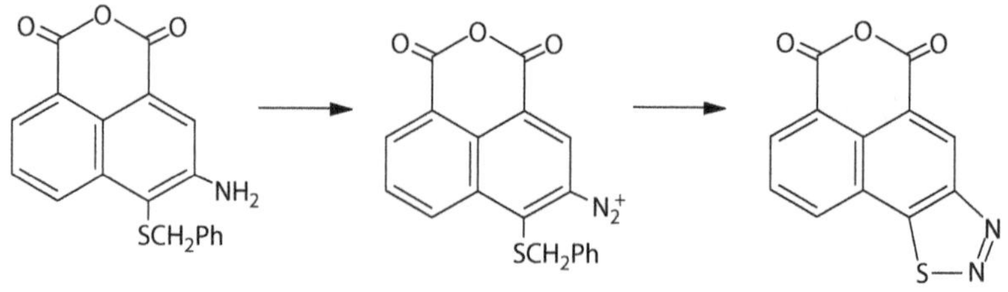

SCHEME 2.7

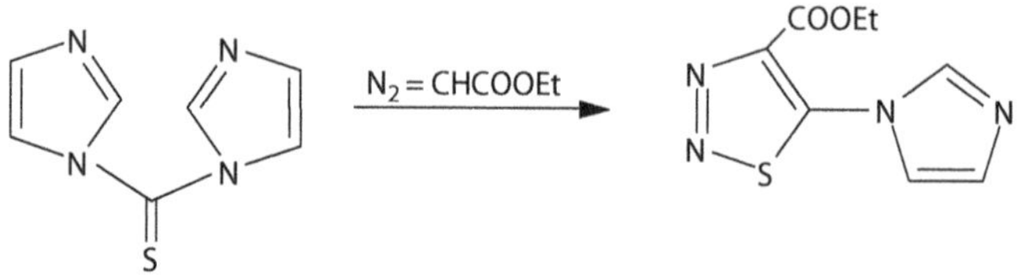

SCHEME 2.8

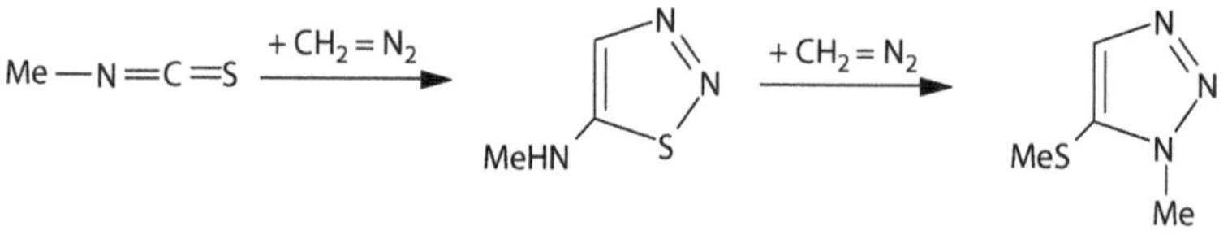

SCHEME 2.9

photonuclease properties. It intercalates into DNA efficiently and damages DNA at as low as $10\,\mu$M concentration under photochemical conditions (Li et al. 2003).

One understandable variation of the diazo method is the reaction of aromatic diazo oxide with phosphorus pentasulfide. For instance, diazotization of naptho-1-amino-2-naphthol leads to 1,2-naphthoquinone-1-diazide that reacts with freshly prepared phosphorus pentasulfide in dry benzene while the reaction temperature is gradually raised from 25°C to 70°C. The overall result is the formation of naphtho-[2,1-d]-[1,2,3]-thiadiazole (Bamberger et al. 1923).

The reaction of ethyldiazoacetate with 1,1′-thiocarbonyl-bis(imidazole) proceeds with high yield and regioselectivity: Only one of the imidazolyl moieties takes part in the reaction depicted in Scheme 2.8 (Aoyoma et al. 1986).

The reaction between diazomethane and methylisothiocyanate leads to 5-methyl-amino-1,2,3-thiadiazole, which can be converted into 1-methyl-5-methylthio-1,2,3-triazole upon action of the same reactant (Scheme 2.9) (Hoff and Block 1974).

Treatment of diazodiketones with Lawesson's reagent gives excellent yields of 4,5-disubstituted 1,2,3-thiadiazoles according to Scheme 2.10 (Caron 1986). [Lawesson's reagent is 2,4-bis(4-methoxyphenyl)-1,3-dithia-2,4-diphosphetane 2,4-disulfide.]

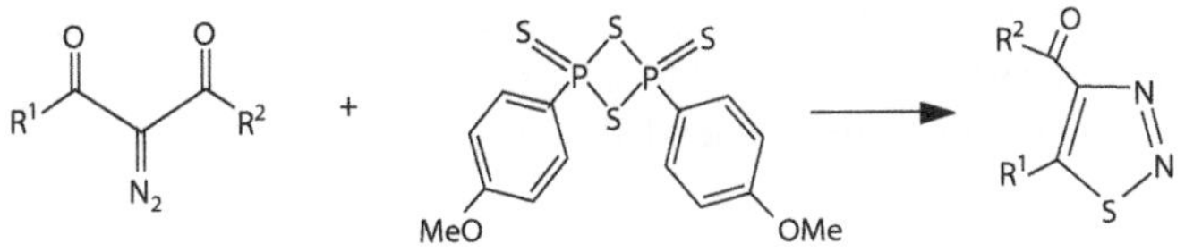

SCHEME 2.10

2.1.1.4 From Compounds with Activated Methylene Fragments

Benzimidazole 1-amino-2-methylene hydroxide gives benzimidazole-[1,2-*c*]-[1,2,5]-thiadiazole chloride under treatment with thionyl chloride. The sequence of the transformations depicted by Scheme 2.11 was proposed (Tumkevicius et al. 2003). At the first step, thionyl chloride transforms the starting hydroxide into the chloride derivative. The next intermediary step consists of the formation of the thiadiazole S-oxide fragment. The latter undergoes a Pummerer-like rearrangement. The rearrangement is initiated by the attack of thionyl chloride followed by subsequent elimination of sulfurous acid and hydrochloric acid during treatment with aqueous sodium bicarbonate solution.

Wu et al. (2007) proposed another way to prepare 4,5-diaryl-1,2,3-thiadiazole, namely, condensation of the correspondingly substituted deoxybenzoin with toluene sulfonylhydrazide and cyclization upon action of thionyl chloride (Scheme 2.12).

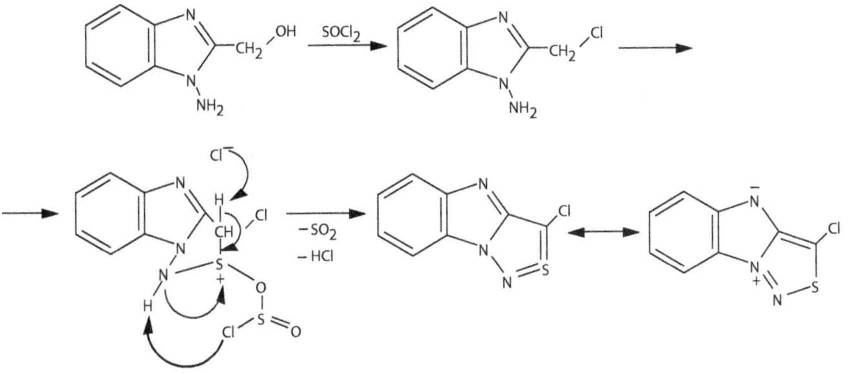

SCHEME 2.11

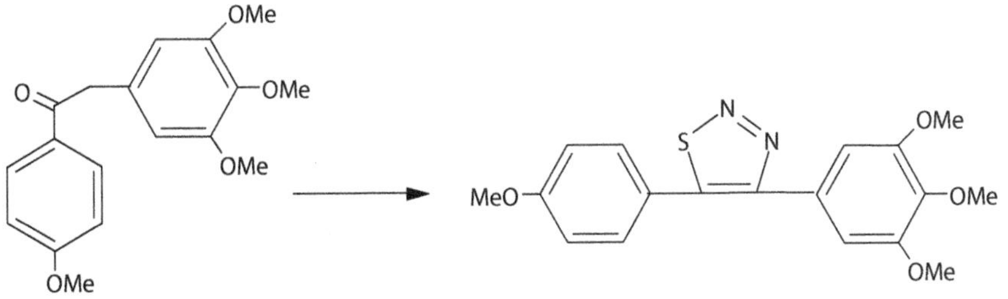

SCHEME 2.12

For readers interested in preparation of 1,2,3-thiadiazole variable derivatives, authoritative reviews by Bakulev and Mokrushin (1986) and, especially, by Bakulev and Dehaen (2004) can be recommended; the review of 2004 combines data up to 2002.

2.1.1.5　From Other Heterocycles

This method is exemplified by Scheme 2.13. The thiadiazole formed was then hydrogenated to prepare saturated "pyridine" part. Such a hydrogenated product is biomedically important (Girard et al. 1989).

Treatment of 2-mercaptotriazole derivatives with hydrochloric acid leads to derivatives of 5-amino-1,2,3-thiadiazole (Scheme 2.14) (L'abbe and Vanderstede 1989). The method is based on the equilibrium that exists between the initial and final materials. (Reverse transformation takes place under treatment with metal hydroxides.)

Upon passing of hydrogen sulfide gas through ethanolic aqueous solution of 1,2,3-oxadiazoles with ammonium hydrosulfide, the corresponding derivatives of 1,2,3-thiadiazoles were obtained (Wolf 1904).

Treatment of 5-aroyl-3[2H]-isothiazolones with arylhydrazones produces N-substituted 1,2,3-thiadiazolines (Tsolomitis and Sandris 1984). Scheme 2.15 illustrates another approach to benzo-1,2,3-thiadiazoles via diazotization of aminobenzoisothiazole and subsequent recyclization in the presence of cuprous chloride.

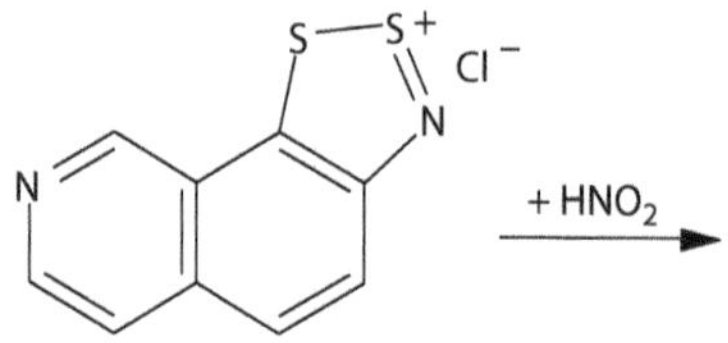

SCHEME 2.13

SCHEME 2.14

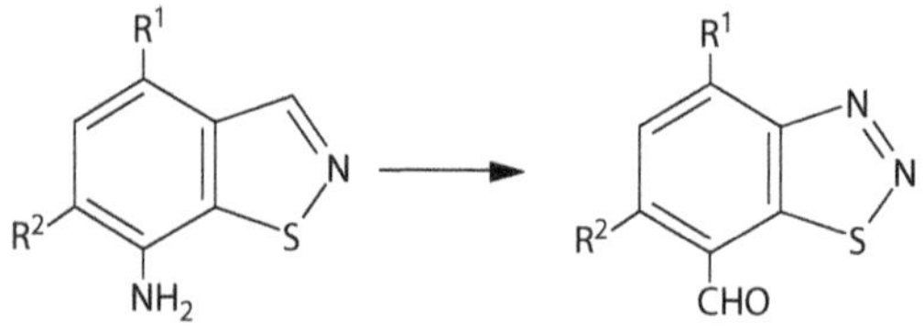

SCHEME 2.15

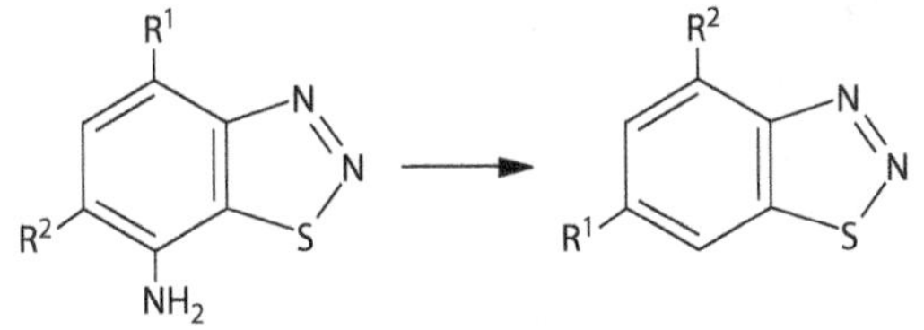

SCHEME 2.16

The releasing methyne group is stabilized by transformation into the formyl one (Haddock et al. 1971).

Interestingly, diazotization of substituted α-aminobenzo-1,2,3-thiadiazoles and removing the diazonium group by phosphorous acid leads to benzo-1,2,3-thiadiazoles containing a different orientation of substituents as compared to the starting material. Scheme 2.16 illustrates such a rearrangement (Haddock et al. 1970).

Obushak et al. (2007) proposed an interesting method of assembly of the 1,2,3-thiadiazole moiety from 4-amino-4H-1,2,4-triazole-3-thiol and quinoline-4-carboxylic acid in the presence of phosphoryl chloride. The assemblage proceeds at the expense of the amino and thiol groups of the triazole derivative and one of its nitrogen at the participation of the carboxylic group belonging to quinoline.

2.1.2 FINE STRUCTURE

Calculations of aromaticity indexes (I_A) based on bond lengths, give $I_A = 100$ for benzene and $I_A = 54$ for 1,2,3-thiadiazole (Bird 1985). Bakulev and Dehaen (2004) compiled the x-ray crystallographic data of many 1,2,3-thiadiazole derivatives. All the thiadiazoles are completely planar with deviation less than 0.02° from the plane. Normally, 4,5-disubstituted and 4-monosubstituted 1,2,3-thiadiazoles display full stability either to reduction or to oxidation. In particular, they are unreactive even upon action of such strong reagents as chromic acid (L'abbe et al. 1992a) or lithium aluminum hydride (Bakulev and Mokrushin 1986). Two specific cases of oxidation with cytochrome or peracetic acid are mentioned in Section 2.1.3, together with other methods of ring cleavage. Molecular ions of 1,2,3-thiadiazoles (i.e., cation-radicals) are also stable enough and give highly intense peaks in mass spectra Bakulev and Dehaen (2004). These data testify for aromaticity of 1,2,3-thiadiazole ring. However, degree of aromaticity is not very high because both the N=N and C=C edges of the cycle can, in any event, be involved in cycloaddition reactions (see Section 2.1.3).

From the aromaticity point of view, data of diazotization of amino-1,2,3-thiadiazoles are of interest. The diazo compounds of this class are unstable. However, like the aromatic azo compounds, the presence of an electron-withdrawing group in the 4-position of 5-amino-1,2,3-thiadiazole exerts a marked stabilizing effect on the resulting diazonium salt (Pain and Slack 1965, Goerdeler and Gnad 1966).

2.1.3 REACTIVITY

It is worth considering separately the materials on reactivity of 1,2,3-thiadiazole itself and its derivatives fused with aromatic fragments.

2.1.3.1 Monocyclic 1,2,3-Thiadiazoles

2.1.3.1.1 Salt Formation

Upon action of trimethyl oxonium tetrafluoroborate, 1,2,3-thiadiazole gives rise to 3-methyl-1,2,3-thiadiazolium tetrafluoroborate with 96% yield. In the case of 5-chloro-1,2,3-thiadiazole, methylation on the N_3 position is also takes place. The phenyl or *tert*-butyl group in position 4 exerts a shielding effect on position 3 and methylation occurs preferentially at N_2 (yield is 81% or 87%, respectively). Methylation of 4-(methoxycarbonyl)-1,2,3-thiadiazole is special and interesting. Although 4-(methoxycarbonyl) is undoubtedly shielding a substituent, the compound is methylated at N_3 for 92%. The N_2 isomer is formed with the yield of 8% only (L'abbe et al. 1993). L'abbe et al. (1991) proposed the following explanation for the outside effect of the 4-(methoxycarbonyl) group: Methylation first occurs at the ester function, and the transient product then isomerizes to 3-methyl-1,2,3-thiadiazolium tetrafluoroborate. Scheme 2.17 summarizes this assumption.

Alkylation on position 2 was also observed in the case of Scheme 2.18. Namely, methylation of phenylhydrazone derived 8-oxo-8*H*-indeno[1,2-*d*]-[1,2,3]-thiadiazole with Meerwein's reagent furnishes two methylated products. Single crystal x-ray analyses of these methylated products reveal no significant S···NPh interaction, thus excluding a thiapentalene structure (L'abbe et al. 1992b). Stabilization depicted in Scheme 2.18 is possible due to reorganization of the conjugated bond system.

2.1.3.1.2 Oxidation

Two special cases of oxidation is represented by the reaction of 1,2,3-thiadiazoles with cytochrome P450 and oxygen or with peracetic acid. Babu and Vaz (1997) proposed Scheme 2.19 for the thiadiazole transformations upon oxidative action of cytochrome. The sequence is important to understand metabolism of 1,2,3-thiadiazole biomolecules.

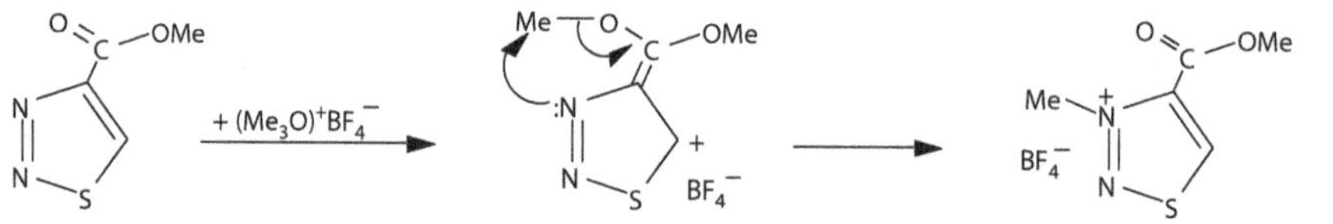

SCHEME 2.17

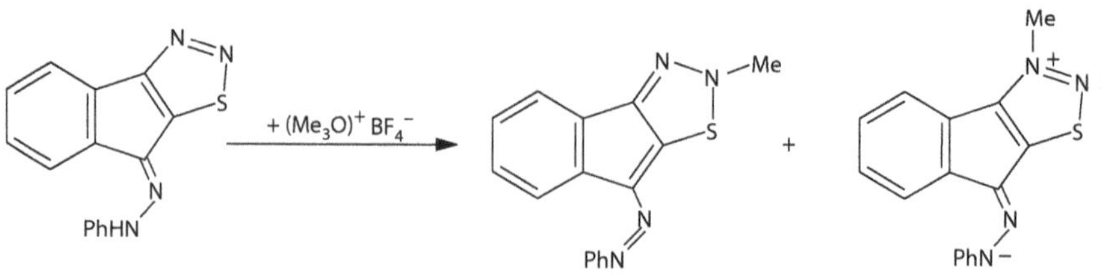

SCHEME 2.18

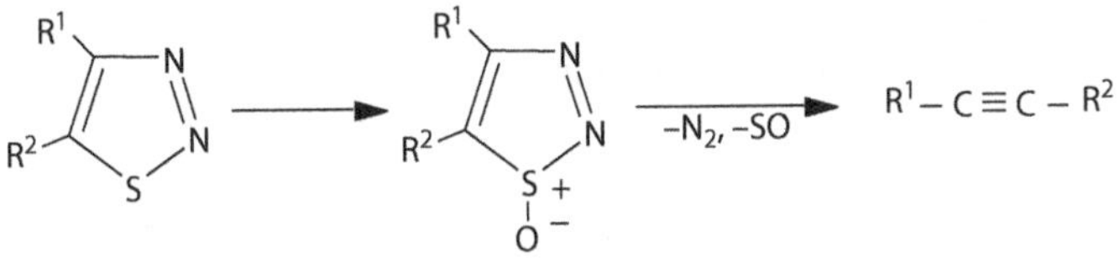

SCHEME 2.19

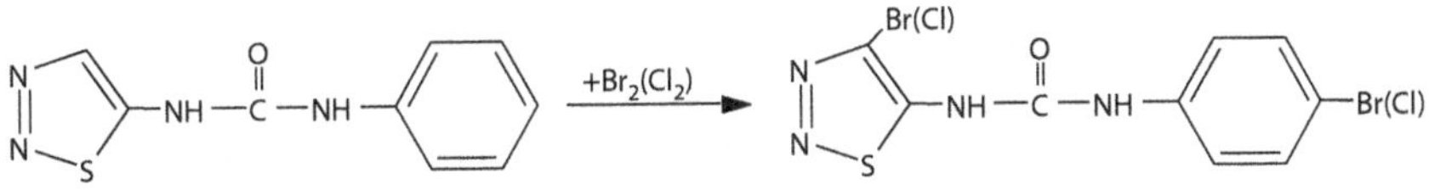

SCHEME 2.20

Trickers et al. (1977) and Winter et al. (1978) performed oxidation of 1,2,3-thiadiazoles with peracetic acid. At first, 3-oxides were formed. The oxides can be isolated, but upon further action of peracetic acid they underwent oxidation at sulfur. The final (and relatively stable) products were 1,1,3-trioxides (see also Gericke et al. 1975). Irradiation of the latter gave dioxazolones and nitriles (major products) as well as alkynes (minor products).

2.1.3.1.3 Electrophilic Substitution

Halogenation of 5-(2-N-penylureido)-1,2,3-thiadiazole touches both the thiadiazole and phenyl rings leading to dichloro or dibromo derivative as it is depicted in Scheme 2.20 (Shafran et al. 1993).

As seen from Scheme 2.20, activity of both the rings in electrophilic substitution is rather equal, and the carboxamidic moiety assists to this reaction.

Nitration, Mannich, and Vilsmeier reactions were performed with 3-aryl-1,2,3-thiadiazolium containing the enolate substituent (Scheme 2.21) (Masuda et al. 1979).

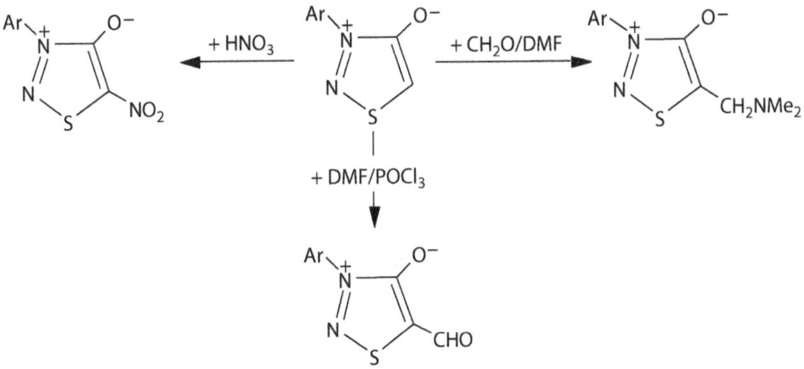

SCHEME 2.21

The enolate activating effect prevails over the weakening influence of the positive charge in the electrophilic thiadiazole ring.

2.1.3.1.4 Nucleophilic Substitution

There are few clear examples of such reactions. Ring hydrogen in 1,2,3-thiadiazole can undergo rapid deuterium exchange under basic conditions (Elvidge et al. 1974). The chlorine atom in 5-chloro-1,3,3-thiadiazole can be displaced by the methoxy group (Modarai et al. 1974). Upon action of aniline, 4-*N*-carbamoyl-5-chloro-1,2,3-thiadiazole undergoes substitution of *N*-phenylamino for chlorine at the same place; all of other fragments remain their integrity (Morzherin et al. 1994).

Nucleophilic substitution of ethylenediamine for chlorine in 5-chloro-4-carboethoxy-1,2,3-thiadiazole is followed with the Dimroth-type rearrangement of the five-membered cycles and the formation of the seven-membered ring (Scheme 2.22) (Volkova et al. 1999). The reaction leads to the bis {[1,2,3]-triazolo-[1,5*b*; 5′,1′*f*]-[1,3,6]-thiadiazepine di(ethoxycarbonyl) derivative, which is of special interest for pharmacology.

One interesting case of nucleophilic substitution was described by Katritsky et al. (2001): 4-Methylthiophenol displaces the benzotriazole moiety, rather than the thiadiazole one. In other words, fugacity of the benzotriazolyl fragment is stronger than that of the thiadiazolyl moiety (Scheme 2.23).

2.1.3.1.5 Ring Cleavage

Treatment of 4,5-diphenyl-1,2,3-thiadiazole with butyl lithium at −60°C initiates nitrogen gas evolution and extrusion of sulfur. Perfectly clean 1,2-diphenylacetylene is formed (Bakulev and Mokrushin 1986). This is preparatively promised. Raap and Micetich (1968) produced proofs that ring disclosure in 4,5-diphenyl-1,2,3-thiadiazole upon action of butyl lithium begins from the attack on the sulfur atom.

A large-scale manufacturing can be possible in the framework of electrochemical reduction, such an opportunity was demonstrated by Abramov et al. (1992) by the preparation of alkynethiolates upon electrochemical reduction of 4-alkyl-1,2,3-thiadiazoles.

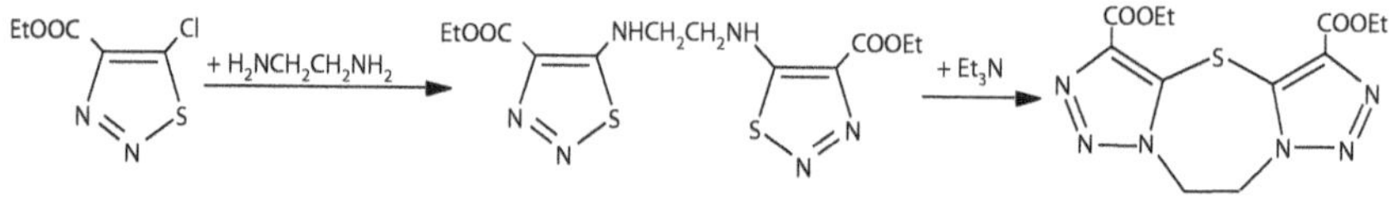

SCHEME 2.22

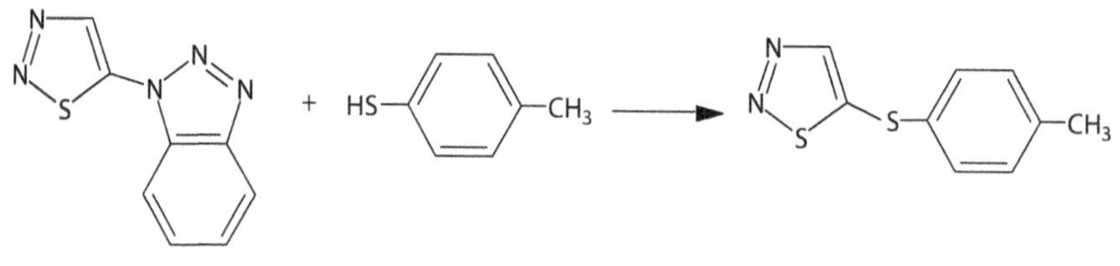

SCHEME 2.23

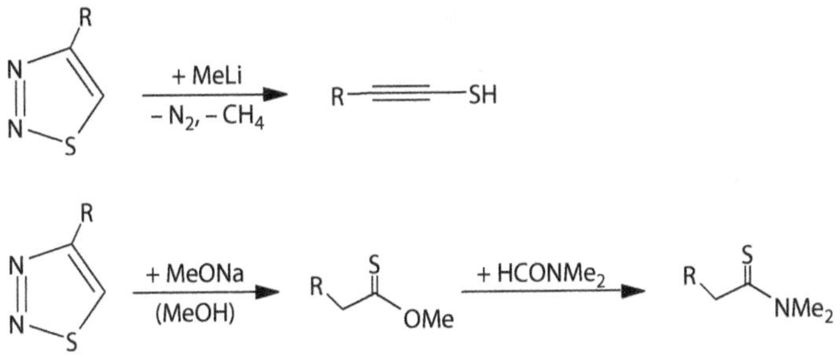

SCHEME 2.24

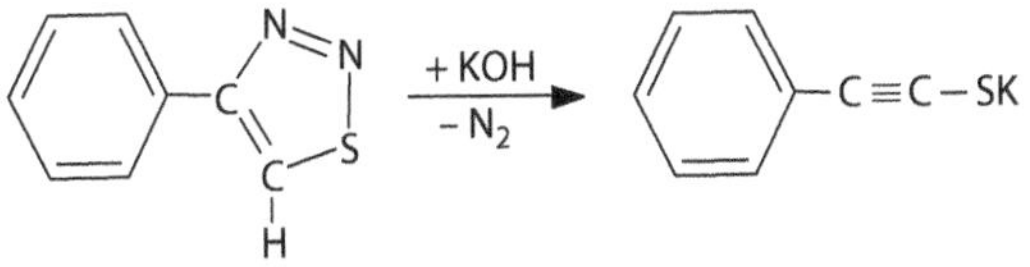

SCHEME 2.25

Upon treatment with strong bases, 1,2,3-thiadiazoles transform into alkynethiolates or thiolates according to Scheme 2.24 (Sandrinelli et al. 2006, Androsov et al. 2007).

The treatment with potassium hydroxide in dioxane allows obtaining potassium 2-phenylethynyl thiolate with quantitative yield (Scheme 2.25) (Zachinyaev and Orlov 1980).

Synthesis of dendrimers starting from nitrogen-elimination ring cleavage of 1,2,3-thiadiazole derivatives was described by Al-Smadi (2007a,b). Namely, [(1,2,3-thiadiazole-4-yl)phenoxymethyl]benzene was treated with potassium *tert*-butoxide and then with Frechet[G-3]-Br dendrone proposed by Hawker and Frechet (1990). The resulting polymer contains alkynethio core with branched layers. The volume inside this dendrimer is shielded from relatively large molecules while still remaining accessible to small ones. (This is a specific case of host-guest recognition.)

In 1977, Malek-Yazdi and Yalpani (1977) proposed their own way to substitute thioamides starting from 1,2,3-thiadiazole derivatives (see Scheme 2.26). Depending on the substituent nature, yields reach 70%–100%.

Scheme 2.27 depicts the decomposition of 4-(2-chloro-5-nitrophenyl)-1,2,3-thiadiazole (Androsov and Neckers 2007). 2-(2-Chloro-5-nitrophenyl)ethynethiolate is formed after ring cleavage with sodium carbonate in water. Reacting as its

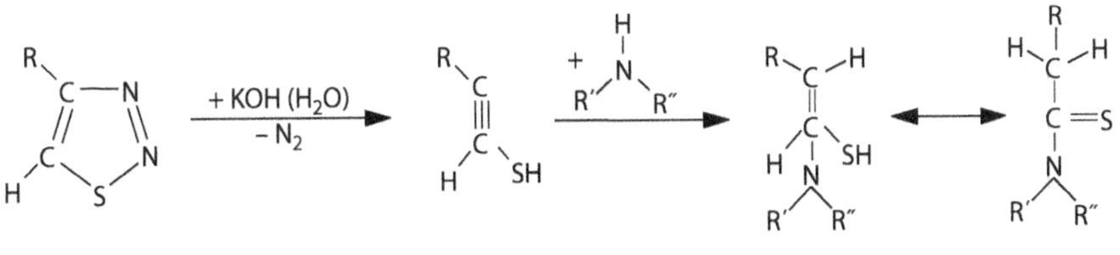

SCHEME 2.26

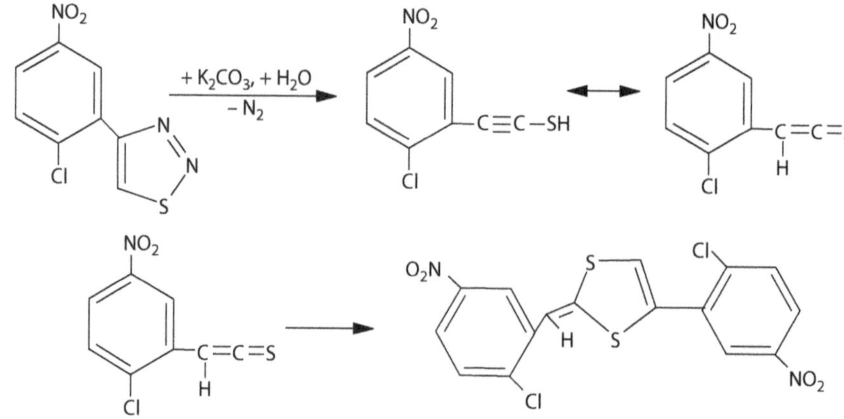

SCHEME 2.27

thioketene valence isomer, this thiolate eventually gives the dithiafulvalene derivative in Scheme 2.27.

In the presence of butylamine, 1-butyl-2-mercaptoindole is formed as depicted
in Scheme 2.28. The nitro group plays a crucial role in the reaction: This group
makes the thiadiazole ring susceptible for proton abstraction and activates the chlorine substituent to intramolecular nucleophilic attack by the amide nitrogen as noted
by Scheme 2.28 (Androsov and Neckers 2007).

1,3,5-Tris-(1,2,3-thiadiazol-4-yl)benzene reacts with potassium *tert*-butoxide
and forms tripotassium 1,3,5-tris(ethynothiolate). Coupling of tripotassium
1,3,5-tris(ethynothiolate) with a dendron bearing the methylene bromide terminal

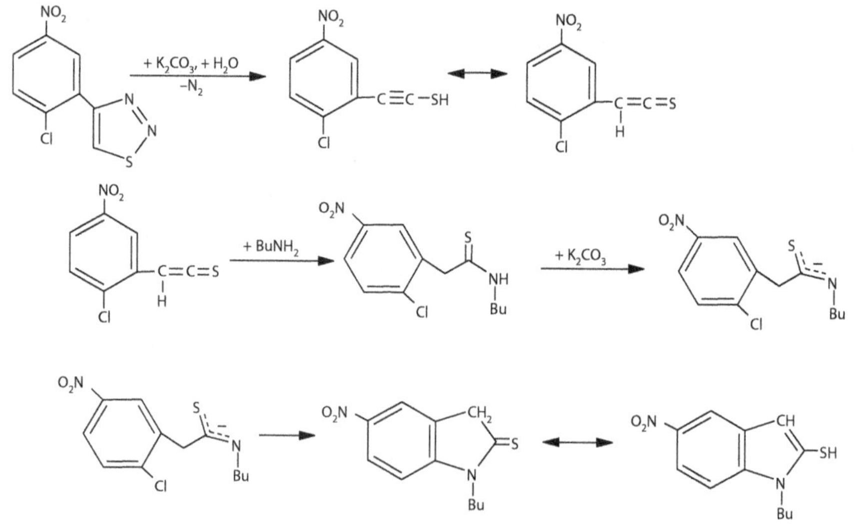

SCHEME 2.28

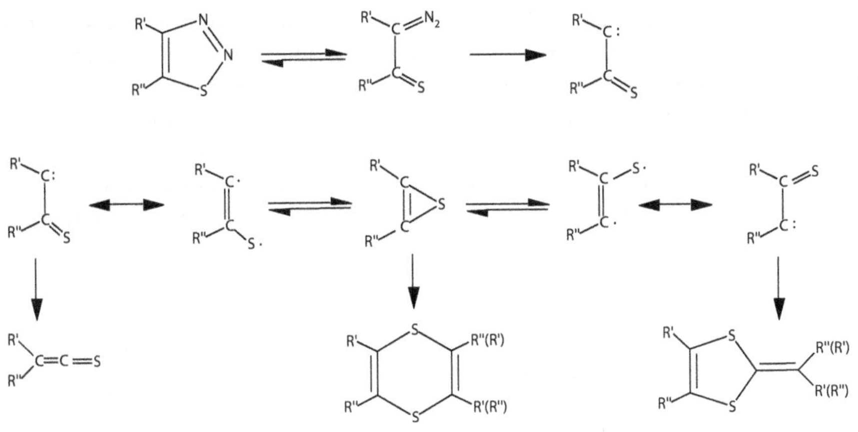

SCHEME 2.29

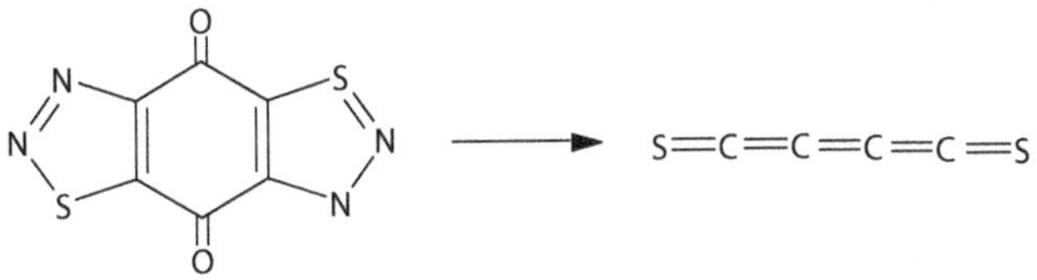

SCHEME 2.30

function leads to a tris(thiodendrimer) of technically useful properties (L'abbe et al. 1995b).

2.1.3.1.6 Photolysis

1,2,3-Thiadiazoles are somewhat light sensitive and experiments with them are usually performed under dark conditions. Upon irradiation at 230 nm, extrusion of nitrogen takes place and the intermediates formed give rise to the gamma of products depicted in Scheme 2.29 (Kirmse and Horner 1958).

Photolysis of the 1,4-benzoquinone fused at its two edges with the 1,2,3-thiadiazole rings at 366 nm proceeds according to Scheme 2.30 (Maier et al. 1991) and leads to 1,2,3-butatriene-1,4-dithione. Analogously, photolysis of benzotris(thiadiazole) of Scheme 2.1 at 245 nm leads to 1,2,3,4-pentatetraene-1,5-dithione, $S{=}C{=}C{=}C{=}C{=}S$ (Maier et al. 1990).

2.1.3.1.7 Thermolysis

Having been heated, 1,2,3-thiadiazoles evolve nitrogen molecule and give a range of products derived from intermediates generated in this way. Generally speaking, 1,2,3-thiadiazoles attract a lot of attention because they are the only representatives of the isomeric thiadiazole family where loss of nitrogen molecule can readily occur. One of the most useful applications of this thermally induced decomposition is the synthesis of thioketenes. Thermolysis at 220°C–230°C leads to thioketenes, dithiins, and 1,4-dithiafulvenes. Scheme 2.29 explains the formation of these products (Seybold and Heibl 1975, Thomas 1996). Due to the equilibriums depicted in Scheme 2.29, thermolysis of 4-methyl-5-phenyl- or 5-methyl-4-phenyl-1,2,3-thiadiazole in the presence of methyl phenyl acetylene leads to the same mixture of dimethyldiphenyl thiophenes irrespective of the starting regioisomer mentioned (Schaumann et al. 1979).

Flash-vacuum pyrolysis of benzotris(thiadiazole) of Scheme 2.1 results in the formation of 1,2,3,4-pentatetraene-1,5-thione, $S{=}C{=}C{=}C{=}C{=}S$ (Maier et al. 1990).

2.1.3.1.8 Cycloaddition

Such reactions take place with 1,2,3-thiadiazoles if they have special features in their structure. When 3-methyl-4,5-diaryl-1,2,3-thiadiazoles are treated with cesium fluoride, ylides of Scheme 2.31 are formed. Being formed in the presence of alkynes, the ylides develop pyrazolo-thiadiazole compounds (Butler et al. 1999).

As to the carbon–carbon edge of 1,2,3-thiadiazole, Scheme 2.32 shows its participation in the cycloaddition reaction with 4-phenyl-4H-1,2,4-triazole-3,5-dione (Hanold et al. 1986).

2.1.3.1.9 Ring Transformation

Transformations like Cornforth and Dimroth rearrangements allow to obtain 1,2,3-triazoles from 1,2,3-thiadiazoles (see Masuda et al. 1979, Abbe et al. 1991, 1993). Thus, Schiff bases formed from 1,2,3-thiadiazole formyl derivatives rearrange to 1,2,3-triazole carbothioamides. When 1,2,3-thiadiazole contains the hydroxyl, methoxyl, or picrylamino substituent, the reaction is reversible (see Scheme 2.33) (Glukhareva et al. 2000, 2004, 2005). The reversibility is based on the initial equilibrium of Scheme 2.29 followed by cyclization of the diazo group involving either sulfur or nitrogen atom.

Transformation of a 1,2,3-thiadiazole derivative to a derivative of tetrathiafulvalene deserves special mention because it leads to an interesting organic metal.

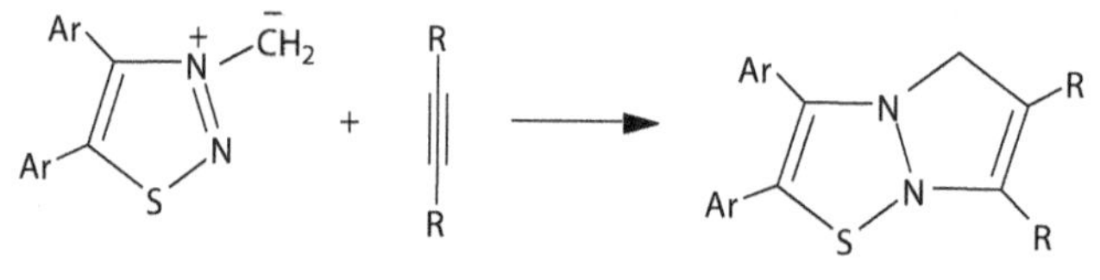

SCHEME 2.31

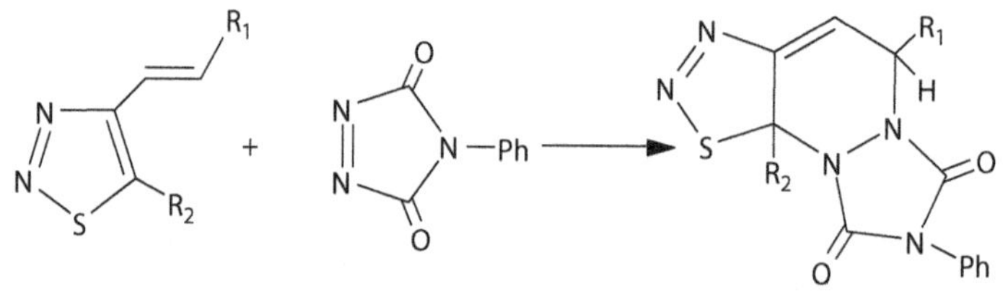

SCHEME 2.32

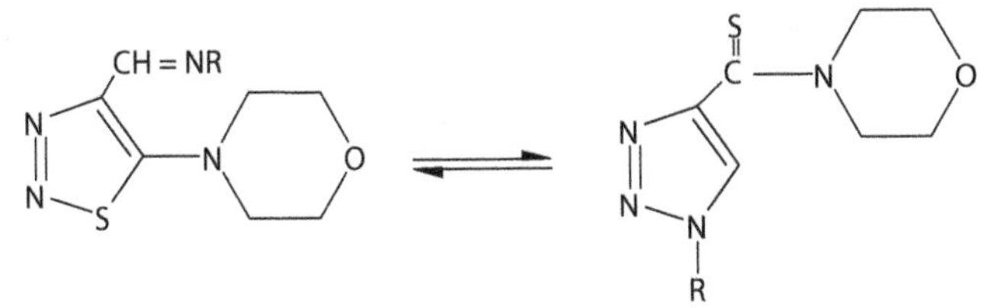

SCHEME 2.33

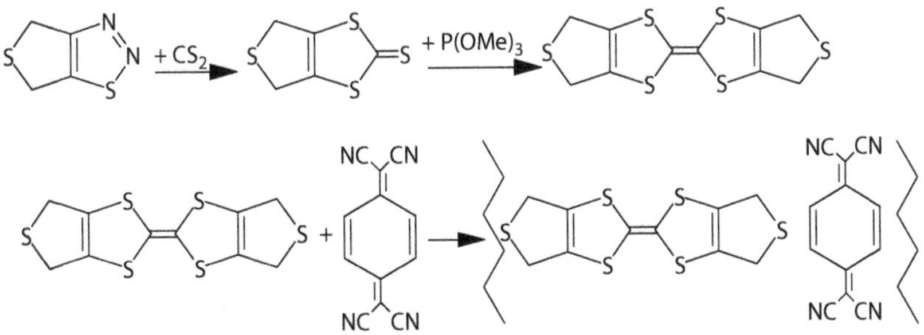

SCHEME 2.34

Scheme 2.34 shows this transformation. The final product of Scheme 2.34 (namely, the charge-transfer complex between the donor, the derivative of tetrathiafulvalene, and the acceptor, the 1,4-quinodimethane tetracyano derivative) showed room-temperature electric conductivity of 1.3×10^2 S/cm (Rovira et al. 1994). The latter value is very attractive for technical applications.

4-(2-Hydroxyphenyl)-1,2,3-thiadiazole forms 2-thiomethylbenzofuran upon treatment with base and methyl halide (Scheme 2.35) (D'hooge et al. 1997).

The analogous reaction with 4-(2-aminophenyl)-1,2,3-thiadiazole leads to 2-thiomethylindole (Abramov et al. 2000).

Heating solutions of 5-phenoxy-1,2,3-thiadiazole in dimethylformamide at 100°C in the presence of excess sodium hydride results in the formation of 1,4-benzoxathiin (Scheme 2.36) (Katritzky et al. 2002).

Treatment of methyl 1,2,3-thiadiazole-4-carboxylate with powdered samarium and iodine in methanol at 0°C results in the formation of a mixture 1,2,5-trithiepanes according to Scheme 2.37 (Miyawaki et al. 2004).

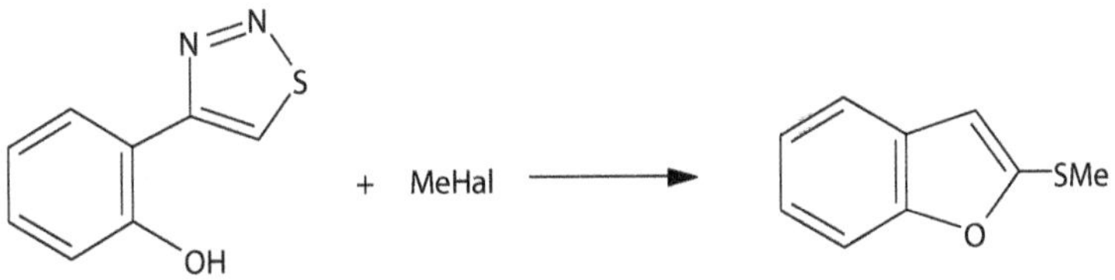

SCHEME 2.35

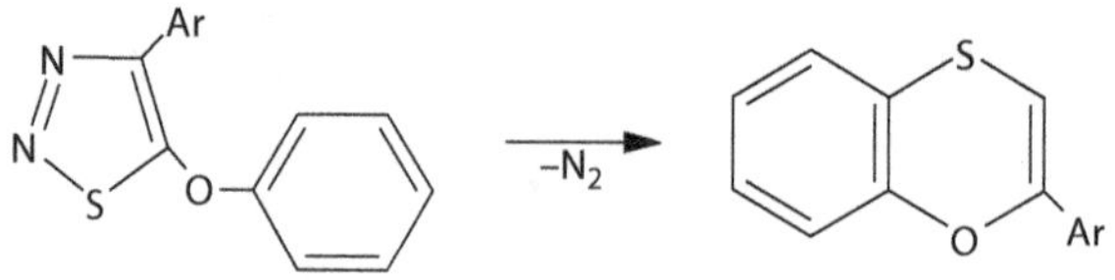

SCHEME 2.36

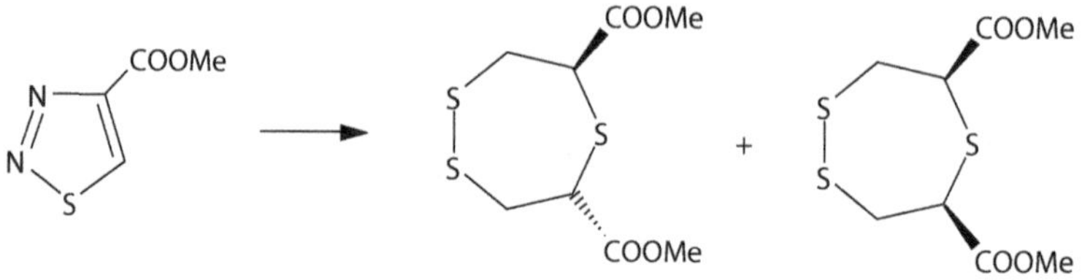

SCHEME 2.37

2.1.3.1.10 Coordination

Coordination of 1,2,3-thiadiazoles to iron proceeds in accordance with Scheme 2.29: In the complex, the ligand appears as diazothioketone (Pannell et al. 1983b). On the contrary, complexes with manganese, chromium, molybdenum, and tungsten contains the unopened 1,2,3-thiadiazoles, and the metal is coordinated with nitrogen atom in position 2 of the thiadiazole ring (Baetzel and Boese 1981, Pannell et al. 1983a, Mayr et al. 1991). Reasons of so different directions of coordination remain unclear.

2.1.3.2 Fused 1,2,3-Thiadiazoles

As distinct from 1,2,3-thiadiazole, its benzo-fused counterpart shows a high degree of stability under a variety of conditions.

2.1.3.2.1 Electrophilic or Nucleophilic Substitution

Benzo-1,2,3-thiadiazole undergoes electrophilic substitution at the α-position, whereas a chlorine atom in the β-position is displaced by a variety of nucleophiles (Kurzer 1975).

2.1.3.2.2 Oxidation

Benzo-[1,2-c]-[1,2,3]-thiadiazole does not change upon oxidation by potassium permanganate or ferricyanide, chromic, or nitric (dilute) acid. However, upon action of 30% hydrogen peroxide in methanolic acetic acid, this fused compound gives S-monoxide after 45 day endurance (Naghipur et al. 1990).

2.1.3.2.3 Ring Cleavage

The reaction of benzo-1,2,3-thiadiazoles with bis(tert-butyl)peroxide does not proceed as oxidation, but as ring cleavage so as tert-butoxy radical adds to the sulfur atom with the nitrogen evolution and the formation of sulphenyl ester of Scheme 2.38 (Albertazzi et al. 1984).

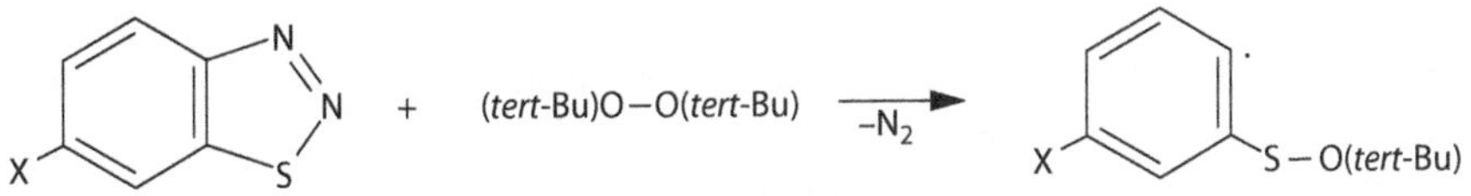

SCHEME 2.38

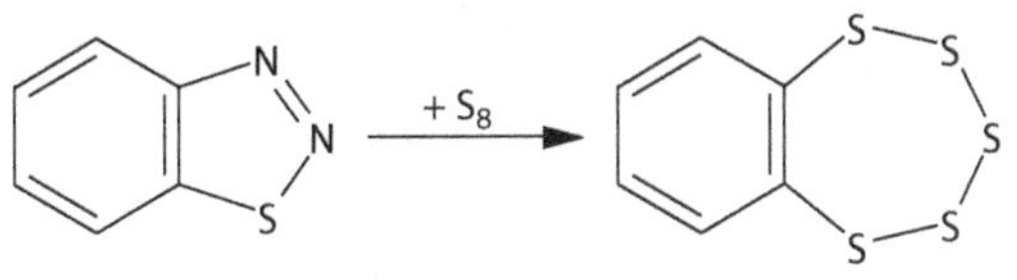

SCHEME 2.39

2.1.3.2.4 Thermolysis

Unlike simple 1,2,3-thiadiazoles, the transformation of benzothiadiazole into thioketene on thermolysis does not take place. However, thermolysis in the presence of sulfur gives rise to 1,2,3,4,5-benzopentathiepine (Scheme 2.39) (Chenard and Miller 1984).

2.1.3.2.5 Coordination

Coordination of benzo-[1,2-c]-[1,2,3]-thiadiazole to manganese proceeds at the expense of nitrogen in position 2 of the heteroring (Mayr et al. 1991). To explain the preference of position 2 over position 3 in this coordination, total energies of 2- and 3-protonated benzo[1,2-c]-[1,2,3]-thiadiazoles were calculated on ab initio basis. The difference in energy was found to be in favor of the 2-protonated benzothiadiazole (Mayr et al. 1991).

In contrast to the complexation with metals, the reaction of this benzothiadiazole with arsenic pentafluoride was directed to the nitrogen atom in position 3 of the thiadiazole ring (Apblett et al. 1986). It is a challenge to find reasons for such different directions of coordination.

2.1.4 BIOMEDICAL IMPORTANCE

2.1.4.1 Medical Significance

Compounds containing 1,2,3-thiadiazole moieties are widely used to improve the presently existing drug repertoire. Thus, introduction of such components in antitumor drugs of the pyrazolone series improves their metabolic stability and enhances their oral bioavailability (Tripathy et al. 2007). The thiadiazole derivative of Scheme 2.19 was also tested against cancer. This medication exhibits excellent antitumor effect in vivo and relatively low cytotoxicity (Wu et al. 2007). A series of 1,2,3-thiadiazoles was found to be potent necroptosis inhibitors (Teng et al. 2007). They can be used in novel therapeutic intervention strategies for the treatment of maladies where necrosis is known to play a prominent role, including stroke, trauma, and possibly some forms of neurodegeneration.

4,5-Bis(4-methoxyphenyl)-1,2,3-thiadiazole inhibits platelet aggregation in humans (Thomas et al. 1985). 4,5-Diphenyl-1,2,3-thiadiazole was claimed as an anti-inflammatory agent (Lau 1997). 4-Methylene(1-hydroxy-2-naphthyl)-1,2,3-thiadiazole is useful for the treatment of airway disorders, against allergic asthma, bronchitis, inflammation, rheumatism, thrombosis, ischemia, angina pectoris, arteriosclerosis, and skin diseases. It is also active as a cytoprotective agent for gastrointestinal tracts (Kobori et al. 1995).

Derivatives of 1,2,3-thiadiazole manifest antifungal activity at low toxicity. For example, 8-sulfamoyl-4,5-dihydronaphtho[1,2-d][1,2,3]thiadiazole is effective against *Cryptococcus neoformans*. The toxicity test on brine shrimps demonstrated that the compound exhibits low cytotoxicity (Jalilian et al. 2003).

1,2,3-Thiadiazole thioacetanilides were forwarded as a novel class of potent HIV-1 non-nucleoside reverse transcriptase inhibitors (Zhan et al. 2008). Liu and Zhan (2009) have taken a patent for 2-[(4-phenyl-1,2,3-thiadiazol-5-yl)thio]-*N*-(2-chlorophenyl)acetamide as an agent that exhibits inhibitory activity against HIV-1 with EC_{50} value of $26.11 \pm 5.28\,\mu M$.

2.1.4.2   Imaging Agents

1,2,3-Thiadiazole compounds are objects in search of imaging agents. Among them, 5-chloro-6,7,8,9-tetrahydro[1,2,3]thiadiazolo[5,4-h]isoquinoline should be distinguished. Rajagopalan et al. (2007) took out a patent for targeting and photoactivation at target sites.

2.1.4.3   Agricultural Protectors

The diester formed from 1,2,3-thiadiazole-5-carboxylic acid and ethyleneglycol was patented as a protector from agricultural and horticultural diseases (Tsubata et al. 1998). *N*-(3-Chloro-4-methylphenyl)-[4-methyl-1,2,3-thiadiazole-5-carboxamide] is an excellent plant activator (Tsubata et al. 1999). Recently, novel 1,2,3-thiadiazole-4-acetamide derivatives were described (Dong et al. 2007). These derivatives are active in plant-growth regulation and exhibit high fungicidal activity.

N-Phenyl-*N*′-(1,2,3-thiadiazol-5-yl)urea is a cotton defoliant. It also regulates the plant growth and, in this sense, exceeds commonly used compounds (Krueger et al. 1983). This thiadiazolyl–urea compound enhances seed germination in lettuce (Babiker et al. 1992) and turnip (Murthy et al. 1998), accelerates bud break of apple trees (Wang et al. 1986). It promotes sprouting in potato (Ji and Wang 1988) and growth of pumpkin cotyledons (Baskakov et al. 1981). It also increases grape-berry weights (Reinolds et al. 1992), enhances cropping and ripening of kiwi fruits (Famiani et al. 1999) as well as induces shoot regeneration in pigeon peas (Eapen et al. 1998). Being introduced in cultural mixtures, this compound assists in regeneration of cacao plants (Li et al. 1998).

S-Methyl ester of benzo-1,2,3-thiadiazole-7-thiocarboxylic acid is a plant activator that protects plants from diseases (see, e.g., Shimono et al. 2007). The compound was patented by Maetzke (1996) and described in detail by Kunz et al. (1997). This ester is also used as an elicitor on aphid growth and development in wheat, shortening the development duration of the aphid, decreases its adult weight, and prevents plant juice from being suctioned out (Zhu and Zhao 2006). The compound induces disease resistance in wheat (Gorlach et al. 1996), tobacco (Friedrich et al. 1996), melons (Huang et al. 2000, Wang et al. 2008), watermelons (Tao et al. 2009), maize (Morris et al. 1998), strawberry (Terry and Joyce 2000), rice (Chen et al. 2007), and pea (Dann and Deverall 2000).

S-Methyl ester of benzo-1,2,3-thiadiazole-7-thiocarboxylic acid is used to help pepper fruit to rise upon ripe state, protecting the unripe fruit from *Colletotrichum gloeosporioides* (Lee et al. 2009). As a result, the resistance of unripe and ripe fruits is similar. Grey mould post-harvest infection caused by *Botrytis cinera*, one of the main causes decay in harvested tomato, was also proposed to be controlled with the help of this ester (Iriti et al. 2007).

It also is applicable to improve low-quality coffee beans in cases of bad plant harvest (Iriti and Faro 2003, de Nardi et al. 2006) and develops storability of post-harvest loquat (Zhu et al. 2007b) and banana (Zhu and Ma 2007b) fruits.

In connection with the comparison between properties of isomeric thiadiazoles, it is interesting to note that the replacement of the 1,2,3-thiadiazole ring in S-methyl ester of benzo-1,2,3-thiadiazole-7-thiocarboxylic acid with the 1,2,5-thiadiazole moiety results in the loss of bioactivity (Kunz et al. 1997).

2.1.5 CONCLUSION

As it follows from the materials presented, 1,2,3-thiadiazoles are interesting, first of all, due to their reactivity. They are easily transformed into alkynes, alkynethiols, and rearranged into other heterocyclic compounds. Technically, this class of thiadiazoles still awaits its application, but its biomedical and agrichemical importance is well documented.

REFERENCES TO SECTION 2.1

Abbe, G.L.; Vanderstede, E.; Dehaen, W.; Delbeke, P.; Toppet, S. (1991) *J. Chem. Soc. Perkin Trans. 1*, 607.

Abbe, G.L.; Verbeke, M.; Dehaen, W.; Toppet, S. (1993) *J. Chem. Soc. Perkin Trans. 1*, 1719.

Abramov, M.A.; Dehaen, W.; D'hooge, B.; Petrov, M.L.; Smeets, S.; Toppet, S.; Voets, M. (2000) *Tetrahedron* **56**, 3993.

Abramov, M.A.; Niyazymbetov, M.E.; Petrov, M.L. (1992) *Zh. Obshch. Khim.* **62**, 2138.

Albertazzi, A.; Leardini, R.; Pedulli, G.F.; Tundo, A.; Zanardi, G. (1984) *J. Org. Chem.* **49**, 4482.

Al-Smadi, M. (2007a) *J. Heterocycl. Chem.* **44**, 915.

Al-Smadi, M. (2007b) *Asian J. Chem.* **19**, 1783.

Al-Smadi, M.; Ratrout, S. (2004) *Molecules* **9**, 957.

Androsov, D.A.; Neckers, D.C. (2007) *J. Org. Chem.* **72**, 5368.

Androsov, D.A.; Petrov, M.L.; Shchipalkin, A.A. (2007) *Zh. Org. Khim.* **43**, 1863.

Aoyoma, T.; Iwamoto, Y.; Shiori, T. (1986) *Heterocycles* **24**, 589.

Apblett, A.; Chivers, T.; Richardson, J.E. (1986) *Can. J. Chem.* **64**, 849.

Attanasi, O.A.; De Crescenitini, L.; Favi, G.; Filippone, P.; Giorgi, G.; Mantellini, F.; Santeusanio, S. (2003) *J. Org. Chem.* **68**, 1947.

Babiker, A.G.T.; Parker, C.; Suttle, J.C. (1992) *Weed Res.* **32**, 243.

Babu, B.R.; Vaz, A.D.N. (1997) *Biochemistry* **36**, 7209.

Baetzel, V.; Boese, R. (1981) *Z. Naturforsch.* **36B**, 172.
Bakulev, V.A.; Dehaen, W. (2004) *The Chemistry of 1,2,3-Thiadiazoles.* Wiley, Hoboken, NJ.
Bakulev, V.A.; Mokrushin, V.S. (1986) *Khim. Geterotsikl. Soed.* 1011.
Bamberger, E.; Baum, M; Schlein, L. (1923) *J. Prakt. Chem.* **105**, 266.
Baskakov, Yu.A.; Shapovalov, A.A.; Zhirmunskaya, N.M. (1981) *Dokl. AN SSSR* **267**, 1514.
Bird, C.W. (1985) *Tetrahedron* **41**, 1409.
Butler, R.N.; Cloonan, M.O.; McArdle, P.; Cunnigham, D. (1999) *J. Chem. Soc. Perkin Trans. 1*, 1415.
Caron, M. (1986) *J. Org. Chem.* **51**, 4075.
Cerrada, E.; Laguno, M.; Lardies, N. (2009) *Eur. J. Inorg. Chem.*, 137.
Chen, J.; Zhang, W.; Song, F.; Zheng, Zh. (2007) *Protoplasma* **230**, 13.
Chenard, B.L.; Miller, T.J. (1984) *J. Org. Chem.* **49**, 1221.
D'hooge, B.; Smeets, S.; Toppet, S.; Dehaen, W. (1997) *J. Chem. Soc. Chem. Commun.* 1851.
Dann, E.K.; Deverall, B.J. (2000) *Plant Pathol.* **49**, 324.
de Nardi, B.; Dreos, R.; Del Terra, L.; Martellossi, Ch.; Asquini, E.; Tornoncasa, P.; Gasperini, D.; Pacchioni, B.; Rathinavelu, R.; Pallavicini, A.; Graziosi, G. (2006) *Genome* **9**, 1594.
Dong, W.-L.; Yao, H.-W.; Wang, F.-L.; Li, Zh.-M.; Shen, L.-L.; Qiuan, Yu-M.; Zhao, W.-G. (2007) *Gaodeng Xuexiao Huaxue Xuebao* **28**, 1671.
Eapen, S.; Tivarekar, S.; George, L. (1998) *Plant Cell Tissue Organ Cult.* **53**, 217.
Elvidge, J.A.; Jones, J.R.; O'Bien, C.; Evans, E.A.; Sheppard, H.C. (1974) *Adv. Heterocycl. Chem.* **16**, 10.
Famiani, F.; Battistelli, A.; Moscatello, S.; Boco, M.; Antognozzi, E. (1999) *J. Horticult. Sci. Biotechnol.* **74**, 375.
Friedrich, L.; Lawton, K.; Ruess, W.; Masner, P.; Specker, N.; Gut-Rella, M.; Meier, B.; Dincher, S.; Staub, T.; Uknes, S.; Metraux, J.-P.; Kessmann, H.; Ryals, J. (1996) *Plant J.* **10**, 61.
Fujita, M.; Kobori, T.; Hijama, T.; Kondo, K. (1993) *Heterocycles* **36**, 33.
Gericke, R.; Rogalski, W.; Bergman, R.; Wahlig, H.; Hameister, W. (1975) Germ. Patent 2,345,402.
Girard, G.R.; Bondinell, W.E.; Hillegas, L.M.; Holden, K.G.; Pendleton, R.G.; Uzinskas, I. (1989) *J. Med. Chem.* **32**, 1566.
Glukhareva, T.V.; Dyudya, L.V.; Pospelova, T.A.; Bakulev, V.A. (2005) *Khim. Geterotsikl. Soed.* 631.
Glukhareva, T.V.; Morzherin, Yu.Yu.; Dyudya, L.V.; Malysheva, K.V.; Tkachev, A.V.; Padva, A.; Bakulev, V.A. (2004) *Izv. RAN Ser. Khim.* 1258.
Glukhareva, T.V.; Morzherin, Yu.Yu.; Mokrushin, V.S.; Tkachev, A.V.; Bakulev, V.A. (2000) *Khim. Geterotsikl. Soed.* 707.
Goerdeler, J.; Gnad, G. (1966) *Chem. Ber.* **99**, 1618.
Gopalakrishnan, M.; Thanusu, J.; Kangarajan, V. (2008) *J. Sulfur Chem.* **29**, 179.
Gorlach, J.; Volrath, S.; Knauf-Beiter, G.; Hengy, G.; Beckhove, U.; Kogel, K.-H.; Oostendorp, M.; Staub, T.; Ward, E.; Kessmann, H.; Ryals, J. (1996) *Plant Cell* **8**, 629.
Haddock, E.; Kirby, P.; Jonson, A.W. (1970) *J. Chem. Soc. C* 2514.
Haddock, E.; Kirby, P.; Jonson, A.W. (1971) *J. Chem. Soc. C* 3994.
Hanold, N.; Kalbitz, H.; Zimmer, O.; Meier, H. (1986) *Liebigs Ann.* **1986**, 1344.
Hawker, C.J.; Frechet, J.M.J. (1990) *J. Am. Chem. Soc.* **112**, 7638.
Hoff, S.; Block, A.P. (1974) *Rec. Trav. Chim.* **93**, 317.
Huang, Y.; Deverall, B.J.; Tang, W.H.; Wu, F.W. (2000) *J. Plant Biol.* **106**, 651.
Hurd, C.D.; Mori, R.I. (1955) *J. Am. Chem. Soc.* **77**, 5359.
Iriti, M.; Faoro, F. (2003) *J. Plant Pathol.* **85**, 265.
Iriti, M.; Mapelli, S.; Faoro, F. (2007) *Food Chem.* **105**, 1040.
Jalilian, A.R.; Sattari, S.; Bineshmarvasti, M.; Daneshtalab, M.; Shafiee, A. (2003) *Farmaco* **58**, 63.

Ji, Z.L.; Wang, S.Y. (1988) *Plant Growth Regul.* **7**, 37.

Katritzky, A.R.; Nikonov, G.N.; Tymoshenko, D.O.; Moyano, E.L.; Steel, P.J. (2002) *Heterocycles* **56**, 483.

Katritzky, A.R.; Tymoshenko, D.O.; Nikonov, G.N. (2001) *J. Org. Chem.* **66**, 4045.

Kirmse, W.; Horner, L. (1958) *Liebigs Ann.* **614**, 4.

Kobori, T.; Fujita, M.; Kondo, S. (1995) Jpn. Patent 043,258.

Kolesnik, V.D.; Rukavishnikov, A.V.; Tkachev, A.V. (1955) *Mendeleev Commun.* 179.

Kondratieva, M.L.; Pepeleva, A.V.; Bel'skaya, N.P.; Koksharov, A.V.; Groundwater, P.V.; Robeyns, K.; van Meervelt, L.; Dehaen, W.; Fan, Zh.-j.; Bakulev, V.A. (2007) *Tetrahedron* **63**, 3042.

Krueger, H.R.; Arndt, F.; Rush, R. (1983) Germ. Patent 3,139,506.

Kunz, W.; Schurter, R.; Maetzke, T. (1997) *Pesticide Sci.* **50**, 275.

Kurzer, F. (1975) *Org. Compds. Sulfur Selenium Tellurium* **3**, 670.

L'abbe, G.; Bastin, L.; Dehaen, W.; Delbecke, P.; Toppet, S. (1992a) *J. Chem. Soc. Perkin Trans. 1*, 1755.

L'abbe, G.; Dehaen, W.; Bastin, L.; Declercq, J.-P.; Feneau-Dupont, J. (1992b) *J. Heterocycl. Chem.* **29**, 461.

L'abbe, G.; Dehaen, W.; Haelterman, B.; Vangeneugden, D. (1995b) *Acros Org. Acta* **1**, 61.

L'abbe, G.; Delbecke, P.; Bastin, L.; Dehaen, W.; Toppet, S. (1993) *J. Heterocycl. Chem.* **30**, 301.

L'abbe, G.; Frederix, A.; Toppet, S.; Deeler, J.P. (1991) *J. Heterocycl. Chem.* **28**, 477.

L'abbe, G.; Vanderstede, E. (1989) *J. Heterocycl. Chem.* **26**, 1811.

L'abbe, G.; Vossen, P.; Dehaen, W.; Toppet, S. (1995a) *J. Chem. Soc. Perkin Trans. 1*, 2079.

Lardies, N.; Romeo, I.; Cerrada, E.; Laguna, M.; Skabara, P.J. (2007) *Dalton Trans.* 5329.

Lau, C.K. (1997) US Patent 5,677,318.

Lee, S.-H.; Hong, J.-Ch.; Jeon, W.-B. (2009) *Plant Cell Rep.* **28**, 1573.

Li, Y.; Xu, Y.; Qian, X.; Qu, B. (2003) *Bioorg. Med. Chem. Lett.* **13**, 3513.

Li, Z.; Traore, A.; Maximova, S.; Guiltinan, M.J. (1998) *In Vitro Cell Develop. Biol. Plant* **34**, 293.

Liu, X.; Zhan, P. (2009) Chin. Patent 101,362,737.

Maetzke, T. (1996) Eur. Patent 690,061.

Maier, G.; Schrot, J.; Reisenauer, H.P.; Janoschek, R. (1990) *Chem. Ber.* **123**, 1753.

Maier, G.; Schrot, J.; Reisenauer, H.P.; Janoschek, R. (1991) *Chem. Ber.* **124**, 2617.

Malek-Yazdi, F.; Yalpani, M. (1977) *Synthesis* 328.

Masuda, K.; Akai, Y.; Itoh, M. (1979) *Synthesis* 470.

Mayr, A.J.; Carrasco-Flores, B.; Cervantes-Lee, F.; Pannell, K.H. (1991) *J. Organometal. Chem.* **405**, 309.

Miyawaki, K.; Suzuki, H.; Morikawa, H. (2004) *Org. Biomol. Chem.* **2**, 2870.

Modarai, B.; Ghanderhari, M.H.; Masscumi, H.; Shafiee, A.; Lalerazi, I.; Badali, A. (1974) *J. Heterocycl. Chem.* **11**, 343.

Morris, S.W.; Vernooij, B.; Titatarn, S.; Starrett, M.; Thomas, S.; Wiltse, C.C.; Frederiksen, R.A.; Hulbert, S.; Ukness, S. (1998) *Mol. Plant Microbe Interact.* **11**, 643.

Morzherin, Yu.Yu.; Glukhareva, T.V.; Mokrushin, V.S.; Tkachev, A.V.; Bakulev, V.A. (2001) *Heterocycl. Commun.* **7**, 173.

Morzherin, Yu.Yu.; Tarasov, E.V.; Bakulev, V.A. (1994) *Khim. Geterotsikl. Soed.* 554.

Murthy, B.N.S.; Murch, S.J.; Saxena, P.K. (1998) *In Vitro Cell Dev. Biol. Plant* **34**, 267.

Naghipur, A.; Reszka, K.; Lown, J.W.; Sapse, A.-M. (1990) *Can. J. Chem.* **68**, 1950.

Obushak, N.D.; Pokhodylo, N.; Krupa, I.I.; Matiichuk, V.S. (2007) *Russ. J. Org. Chem.* **43**, 1223.

Pain, D.L.; Slack, R. (1965) *J. Chem. Soc.* 5166.

Pannell, K.H.; Mayr, A.J.; Haggard, R.; McKennis, J.S.; Dawson, J.C. (1983a) *Chem. Ber.* **116**, 230.

Pannell, K.H.; Mayr, A.J.; van Derveer, D. (1983b) *J. Am. Chem. Soc.* **105**, 6186.

Petrov, M.L.; Shchipalkin, A.A.; Kuznetsov, V.A. (2007) *Zh. Org. Khim.* **43**, 631.

Raap, R.; Micetich, R.G. (1968) *Can. J. Chem.* **46**, 1057.

Rajagopalan, R.; Jacobs, F.; Dorshow, R.B. (2007) WO Patent 103,250.

Reinolds, A.G.; Wardle, D.A.; Zurowski, C. (1992) *J. Am. Soc. Horticult. Sci.* **117**, 85.

Rovira, C.; Veciana, J.; Santalo, N.; Tarres, J.; Cirujeda, J.; Molina, E.; Llorca, J.; Espinosa, E. (1994) *J. Org. Chem.* **59**, 3307.

Rukavishnikov, A.V.; Tkachev, A.V. (1992) *Synth. Commun.* **22**, 1049.

Sandrinelli, F.; Boudou, C.; Caupene, C.; Averbuch-Pouchot, M.-Th.; Perrio, S.; Metzner, P. (2006) *SYNLETT*, 3289.

Saravanan, S.; Muthsubramanian, Sh.; Vasantha, S.; Sivakolunthu, S.; Raghavaiah, P. (2007) *J. Sulfur Chem.* **28**, 181.

Schaumann, E.; Ehlers, J.; Foerster, W.R.; Adiwidjaja, G. (1979) *Chem. Ber.* **112**, 1769.

Seybold, G.; Heibl, C. (1975) *Angew. Chem. Intl. Ed.* **14**, 248.

Shafran, Y.M.; Bakulev, V.A.; Sheverin, V.A.; Kolobov, M.Y. (1993) *Khim. Geterotsikl. Soed.* 840.

Shekarchi, M.; Ellahiyan, F.; Akbarzadeh, T.; Shafiee, A. (2003) *J. Heterocycl. Chem.* **40**, 427.

Shimono, M.; Sugano, Sh.; Nakayama, A.; Jiang, Ch.-J.; Ono, K.; Toki, S.; Takatsuji, H. (2007) *Plant Cell* **19**, 2064.

Tao, Y.; Bin, P.; Wu, H.J. (2009) *China Cucurbitis Vegetables* **22**, 4.

Teng, X.; Keys, H.; Jeevanandam, A.; Porco, J.A.; Degterev, A.; Yuan, J.; Cuny, G.D. (2007) *Bioorg. Med. Chem. Lett.* **17**, 6836.

Terry, L.A.; Joyce, D.C. (2000) *Pest Manag. Sci.* **56**, 989.

Thomas, E.W. (1996) *Comprehensive Heterocycl. Chem.* **4**, 289.

Thomas, E.W.; Nishizawa, E.E.; Zimmermann, D.S.; Williams, D. (1985) *J. Med. Chem.* **28**, 442.

Trickers, G.; Braun, H.P.; Meir, H. (1977) *Libigs Ann.* **1977**, 1347.

Tripathy, R.; Ghose, A.; Singh, J.; Bacon, E.R.; Angeles, Th.S.; Yang, Sh.X.; Albom, M.S.; Aimone, L.S.; Herman, J.L.; Mallamo, J.P. (2007) *Bioorg. Med. Chem. Lett.* **17**, 1793.

Tsolomitis, A.; Sandris, C. (1984) *J. Heterocycl. Chem.* **21**, 1679.

Tsubata, K.; Sanpei, O.; Takagi, K.; Umetani, K.; Uchikurohane, T.; Tajima, S. (1999) WO Patent 9,923,084.

Tsubata, K.; Shimaoka, T.; Takagi, K.; Baba, K.; Tajima, S. (1998) Jpn. Patent 278949.

Tumkevicius, S.; Labanauskas, L.; Bucinskaite, V.; Brukstus, A.; Urbelis, G. (2003) *Tetrahedron Lett.* **44**, 6635.

Volkova, N.N.; Tarasov, E.V.; Dehaen, W.; Bakulev, V.A. (1999) *Chem. Commun.* 2273.

Wang, S.Y.; Steffens, G.L.; Faust, M. (1986) *Phytochemistry* **25**, 311.

Wang, Y.; Li, X.; Bi, Y.; Ge, Y.-h.; Li, Y.-c.; Xie, F. (2008) *Agricult. Sci. China* **7**, 217.

Weber, G.; Buccheri, F.; Gentile, M.; Librici, L. (1977) *J. Heterocycl. Chem.* **14**, 853.

Winter, W.; Pluecken, U.; Meier, H. (1978) *Z. Naturforsch. B* **33**, 316.

Wolf, L. (1904) *Liebigs Ann.* **333**, 1.

Wu, M.; Sun, Q.; Yang, Ch.; Chen, D.; Ding, J.; Chen, Y.; Lin, L.; Xue, Y. (2007) *Bioorg. Med. Chem. Lett.* **17**, 869.

Zachinyaev, Ya.V.; Orlov, D.S. (1980) *Khim. Promyshl. Ser. React. Osobo Chistye Veshestva* **6**, 51.

Zaleska, B.; Trzewik, B.; Grochowski, J.; Serda, P. (2003) *Synthesis*, 2559.

Zhan, P.; Liu, X.; Cao, Y.; Wang, Y.; Pannecouque, Ch.; de Clerq, E. (2008) *Bioorg. Med. Chem. Lett.* **18**, 5368.

Zhu, Ch.-sh.; Zhao, H.-y. (2006) *Yingyoung Shengtai Xuebao* **17**, 668.

Zhu, S.J.; Ma, B.-C. (2007) *J. Horticult. Sci. Biotechnol.* **82**, 500.

Zhu, S.J.; Zhang, Z.W.; Xu, J.W.; Ma, L.Y.; Tang, W.L.; Liu, D.J. (2007) *Acta Horticult.* **750**, 445.

Zimmer, O.; Meier, H. (1981) *Chem. Ber.* **114**, 2938.

2.2   1,2,4-THIADIAZOLES

2.2.1   FORMATION

2.2.1.1   From Thioamides, Aminothioureas, or N-Arylthioureas

Oxidation of thioamides with a variety of oxidizing agents is a widely used method for the synthesis of 3,5-disubstituted 1,2,4-thiadiazoles. The most common oxidizing agents are halogens, hydrogen peroxide, and nitrous acids (see, e.g., Podolesov and Jordanovska 1985, Balya et al. 2006) and α-arylsulfonyl-α-bromoacetophenons (Shafiee et al. 1999). When phenyliododiacetate is used as an oxidant, the reaction includes steps of dimerization and cyclization according to Scheme 2.40 (Mamaeva and Bakibaev 2003).

Thiobenzamides can react with each other giving rise to 3,5-diaryl-1,2,4-thiadiazoles. Thus, a solution of 4-methoxythiobenzamide in dimethylsulfoxide was heated for 8 h at 38°C with 3-hydroxythiobenzamide in the presence of concentrated hydrochloric acid. The crude product forms bis(hydroxyphenyl)thiadiazole of Scheme 2.41 after treatment with boron tribromide in methylene chloride at −78°C during 18 h (Bey et al. 2008).

One wrongly forgotten method is the oxidation of thiobenzamide into 3,5-diphenyl-1,2,4-thiadiazole. Alcoholic iodine, ammonium persulfate, or nitric-sulfuric acid mixture was used as an oxidant (Friedman et al. 1937 and references therein).

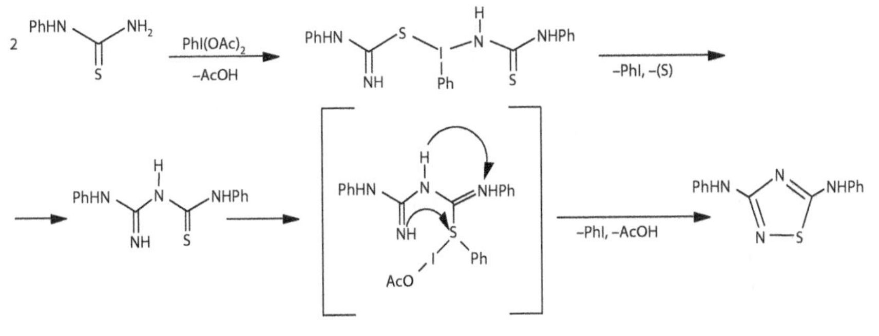

SCHEME 2.40

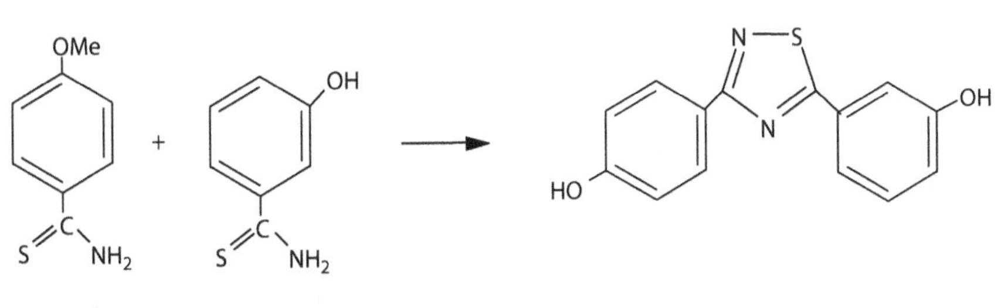

SCHEME 2.41

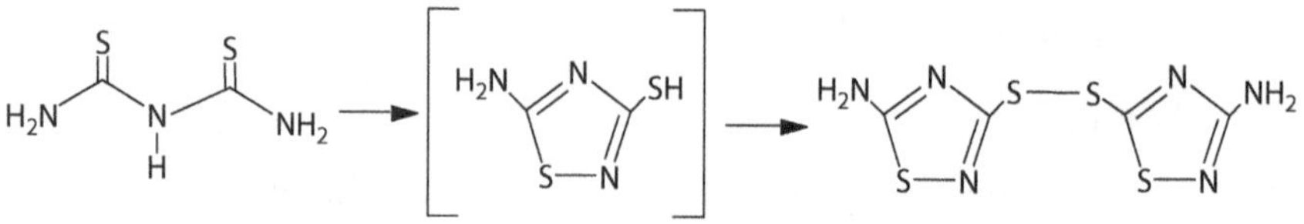

SCHEME 2.42

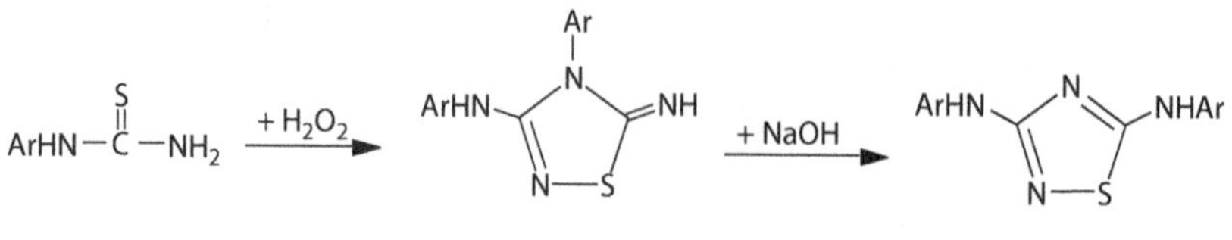

SCHEME 2.43

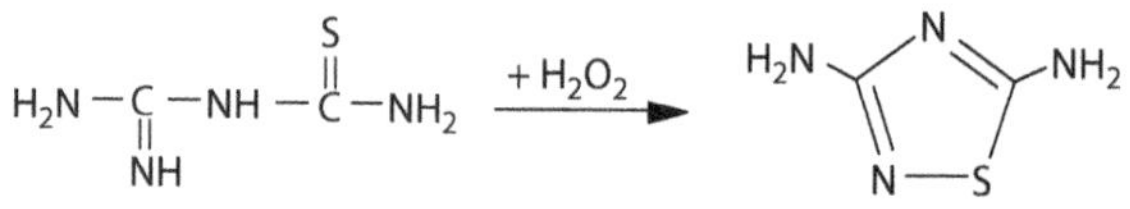

SCHEME 2.44

Under the action of hydrogen peroxide in aqueous sodium hydroxide solution, dithiobiuret gives rise to 2-amino-3-mercapto-1,2,4-thiadiazole that dimerizes according to Scheme 2.42 (Cho et al. 2003).

Oxidation of *N*-arylthioureas with hydrogen peroxide leads to 3-(*N*-arylamino)-4-aryl-5-imino-1,2,4-thiadiazolines. The thiadiazolines isomerize to 3,5-(*N*,*N*′-diarylamino)-1,2,4-thiadiazoles upon action of sodium hydroxide (Scheme 2.43). The products formed were found to have anticonvulsant activity from tests of the maximal electroshock-induced seizures (Gupta et al. 2009).

Danilova et al. (2009) patented process for preparation of 3,5-diamino-1,2,4-thiadiazole by intramolecular cyclocondensation of 2-imino-4-thiobiuret upon action of hydrogen peroxide. The advantage of the process is high yield (90%) of the product and contracted the reaction time (2 h instead of usual 60 h). The diamine is needed for synthesis of medicines, macroheterocyclic compounds, various biomolecules, etc. Scheme 2.44 shows the starting materials and the final product.

2.2.1.2 From Hydroxylamines, Amidoximes, Amidines, Isothiocyanates, Isocyanates, or Thiosemicarbazides

Reaction of 2-hydroxylamino-4,5-dihydroimidazolium-*O*-sulfonate with carbon disulfide in the presence of triethylamine leads to 6,7-dihydro-5*H*-imidazolo-[2,1-*c*]-[1,2,4]-thiadiazole (Scheme 2.45) (Saczewski et al. 2003).

The amidoxime cyclization with aryl isothiocyanate generates 5-(arylamino)-1,2,4-thiadiazoles. This is the classical Tiemann's synthesis; a variation of such a synthesis (also proposed by Tiemann 1891) consists of the amidoxime cyclization with carbon disulfide in the presence of elemental sulfur and sodium methoxide

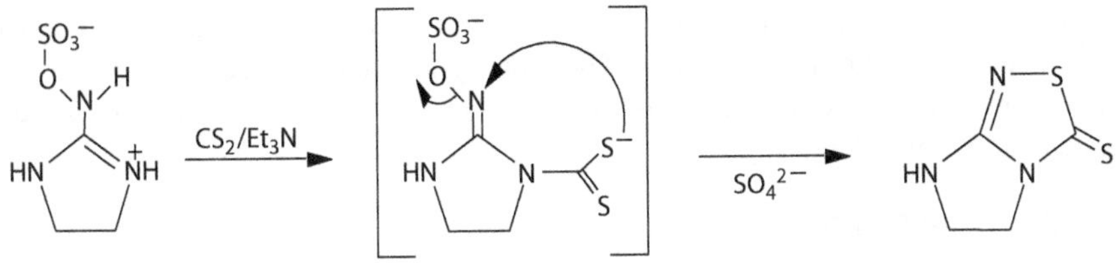

SCHEME 2.45

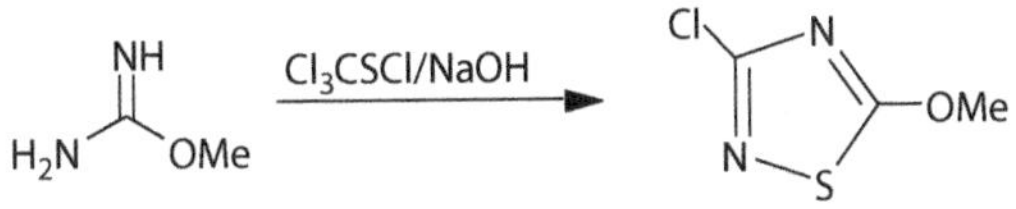

SCHEME 2.46

to obtain 5-mercapto-1,2,4-thiadiazoles. A detailed and simplified procedure was claimed by Tegler and Shoger (1988).

The amidine cyclization was performed using trichloromethylsulfenyl chloride in basic medium (see Scheme 2.46) (Toung et al. 2003, Keith et al. 2008).

Reaction of thionicotinamide with dimethylformamide dimethyl acetal leads to the formation of 2-(1,2,4-thiadiazol-5-yl)pyridine if the intermediary thioacylamidine is intercepted by hydroxylamine-*O*-sulfonic acid according to Scheme 2.47 (Richardson and Steel 2007).

Formidoyl isothiocyanates was transformed into 5-chloro-1,2,4-thiadiazolium chlorides by treatment with twofold excess of methyl sulfenyl chloride. Scheme 2.48 gives a sample of such reactions. The choice of the sample is explained with the ability of the thiadiazoline formed to eliminate the *tert*-butyl group (Morel et al. 2003).

Nasim and Crooks (2009) identified *N*-chlorosuccinimide as a convenient and safe oxidant for the co-joint condensation of isothiocyanate and isocyanates to afford 1,2,4-thiadiazolidine-3,5-diones. Wu and Zhang (2008) proposed one-pot synthesis

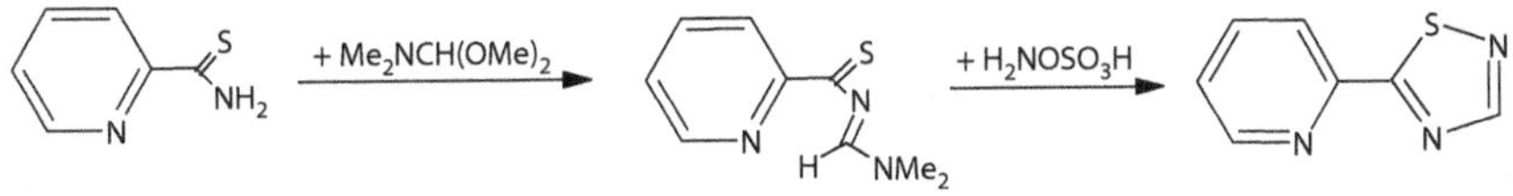

SCHEME 2.47

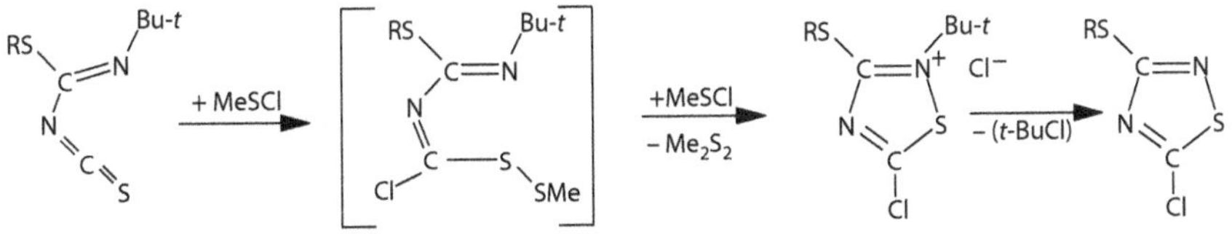

SCHEME 2.48

of 3-substituted 5-amino-1,2,4-thiadiazoles based on amidines together with iso-thiocyanates in dimethylformamide. Di(isopropyl)ethylamine (Huenig's base) and di(isopropyl) azo dicarboxylate was used as reagents. Yields of thiadiazoles were reported to be excellent.

2.2.1.3 From Nitriles

Aromatic nitriles form symmetrically substituted 3,5-diaryl-1,2,4-thiadiazole when reacted with S_8 in a sealed tube in the presence of trioctylamine (Mack 1967) or with sulfur monochloride/dichloride in the presence of aluminum or ferric chloride (Komatsu et al. 1983).

2.2.1.4 From Other Heterocycles

Raficul et al. (2004) described the thermolysis of 1,3,5-oxathiazine-3-oxides into 1,2,4-oxathiazoles followed by transformation into 1,2,4-thiadiazoles upon action of silica gel (Scheme 2.49).

3,5-Disubstituted 1,2,4-thiadiazoles were prepared by ultrasound heating of the corresponding 3,5-disubstituted 1,2,4-dithiazolium phosphates (Zarafu et al. 2008).

Glinka et al. (2003) described the formation of 1,2,4-thiadiazole diester from 3-amino-1,2-oxazole upon cooperative action of potassium thiocyanate, methyl-chloroformate, acetic acid, methanol, and thionyl chloride. The overall reaction is depicted by Scheme 2.50.

Makhova et al. (2004) discovered that furoxan derivatives can be transformed into derivative of 1,2,4-thiadiazole upon treatment with ethoxycarbonyl isothiocya-nate, EtOOCNCS. Scheme 2.51 represents such a reaction.

Vivona et al. (1977) performed rearrangements of 3-amino-1,2,4-oxadiazoles into 3-(acylamino)-5-(phenylamino)-1,2,4-thiadiazoles upon action of phenyl isothiocya-nate. Scheme 2.52 shows that the final product is formed from the intermediary thiourea. This reaction usually gives high yields of 5-amino derivatives as synthons for further synthesis.

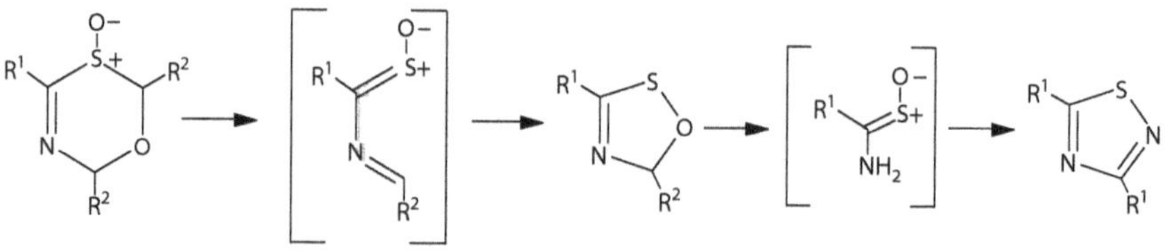

SCHEME 2.49

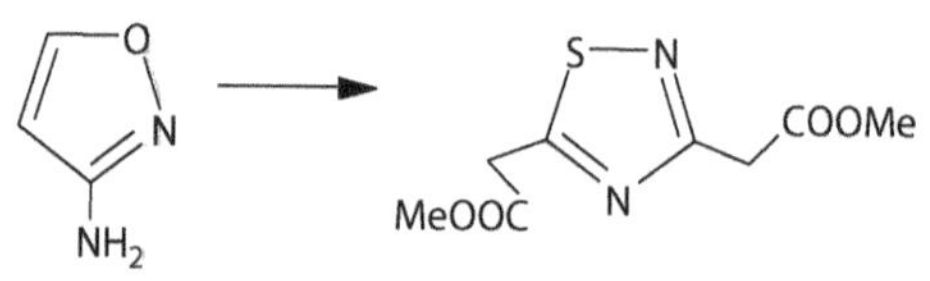

SCHEME 2.50

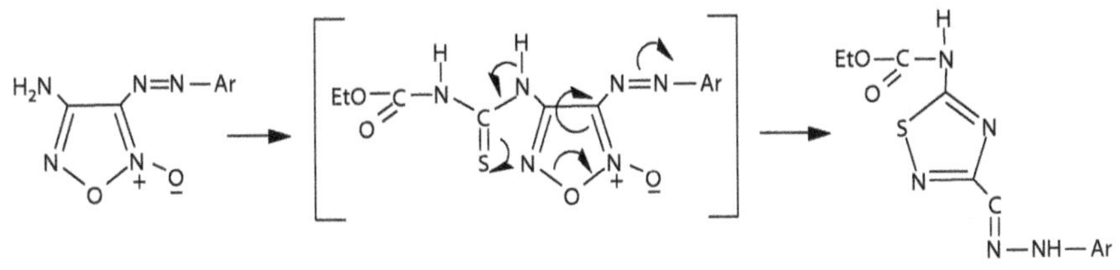

SCHEME 2.51

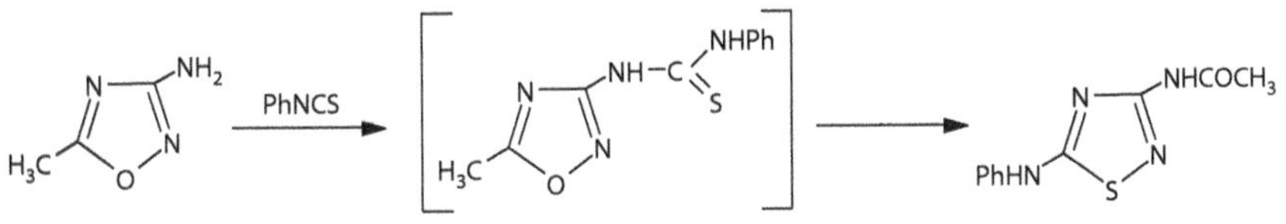

SCHEME 2.52

Pelter and Suemengen (1977) conducted thermal rearrangement of 1,2,4-oxadia-zol-5-thione into 1,2,4-thiadiazol-5-one (Scheme 2.53).

Reuman et al. (2007) used ethoxycarbonyl cyanate to transform 3-[4-(trifluo-romethyl)phenyl]-1,2,4-thiaoxazol-5-one into ethyl 3-[4-(trifluoromethyl)phenyl]-1,2,4-thiadiazole-5-carboxylate as a synthon for further synthesis of a compound regulating lipid/cholesterol metabolism (see Shen et al. 2007, 2008). Scheme 2.54 illustrates the transformation mentioned.

2.2.2 FINE STRUCTURE

Based on bond lengths in cyclic systems, Bird (1985) derived their aromaticity indexes, I_A. According to this author, I_A for 1,2,4-thiadiazole comes to 72, whereas for benzene I_A is equal to 100.

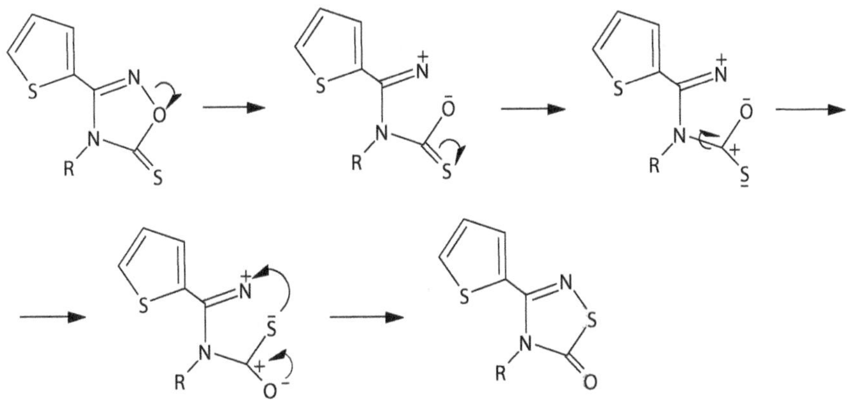

SCHEME 2.53

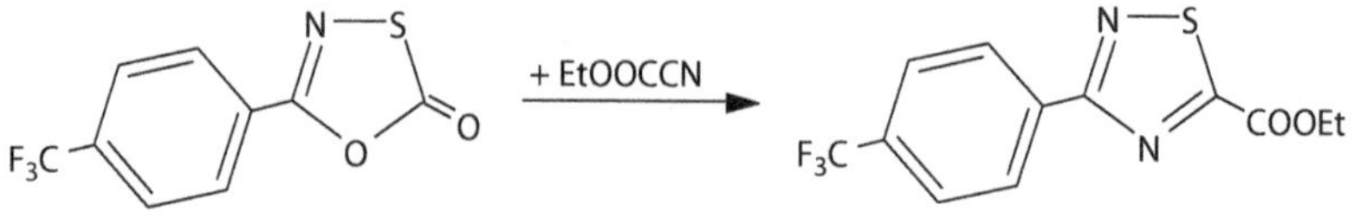

SCHEME 2.54

2.2.3 Reactivity

2.2.3.1 Reduction

Reduction of the 1,2,4-thiadiazole ring proceeds easily unless there are substituents at the carbon atoms of the ring (Kurzer 1965). One particular case to be noted concerned selective reduction of N^+–S bond in a 1,2,4-thiadiazolium derivative of Scheme 2.55 (Zhang et al. 2008). As reducing agents, glutathione, cysteine, ascorbic acid, and thioethanol were used. The reduction also occurred in Sprague-Dawley rat and Yorkshire swine plasma, suggesting that thiol-containing biomolecules existing in the plasma are mainly responsible for this reaction. These data are important for development of 1,2,4-thiadiazolium pharmaceuticals that could be effective in non-enzymatic processes.

2.2.3.2 Oxidation

1,2,4-Thiadiazole is quite sensitive to oxidation. Substituents at positions 3 and 5, however, stabilize the ring toward oxidizing agents (Kurzer 1965).

2.2.3.3 Nucleophilic Substitution

In respect of nucleophilic substitution, accessible positions 3 and 5 differ in their reactivity: Halogen substituents at the 5 position may be displayed by various nucleophiles, whereas halogen substituents in the position 3 are inert toward most nucleophilic reactants (Franz and Dingra 1984).

2.2.3.4 Ring Cleavage

One principally important case of ring cleavage includes the formation of a bicyclic intermediate with a three center-four electron bond at S atom (Scheme 2.56) (Akiba and Yamamoto 2007). In Scheme 2.56, N^* stands for ^{15}N. The intermediate differs

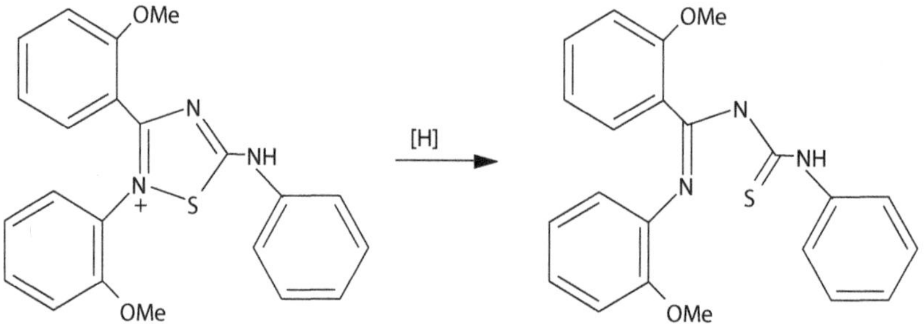

SCHEME 2.55

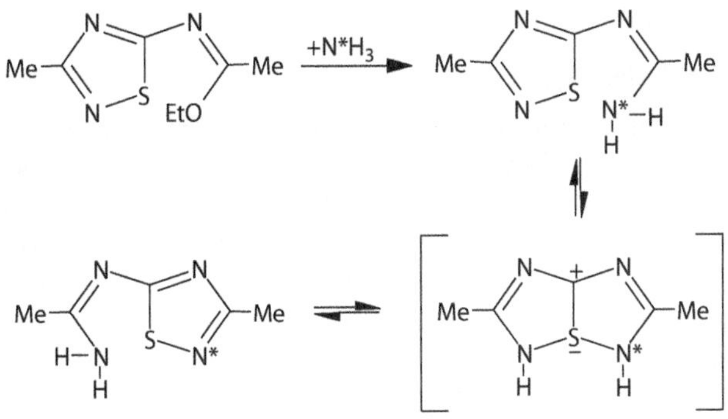

SCHEME 2.56

with an expanded valence shell at the sulfur atom. Notably, the oxygen analog is not capable of this valence-shell expansion and the ring transformation mentioned in Scheme 2.56 does not take place with 1,2,4-oxadiazoles.

2.2.3.5 Photolysis

Photoirradiation of 5-phenyl-1,2,4-thiadiazole results in the formation of benzonitrile (Pavlik et al. 2003).

2.2.3.6 Thermolysis

1,2,4-Thiadiazoles are stable when heated. This stability was ascribed to the aromatic nature of the ring system (Kurzer 1965).

2.2.3.7 Ring Transformation

Zyabrev et al. (2002, 2003) described rearrangements involving the 1,2,4-S,N,N skeleton. Scheme 2.57 presents two actual cases.

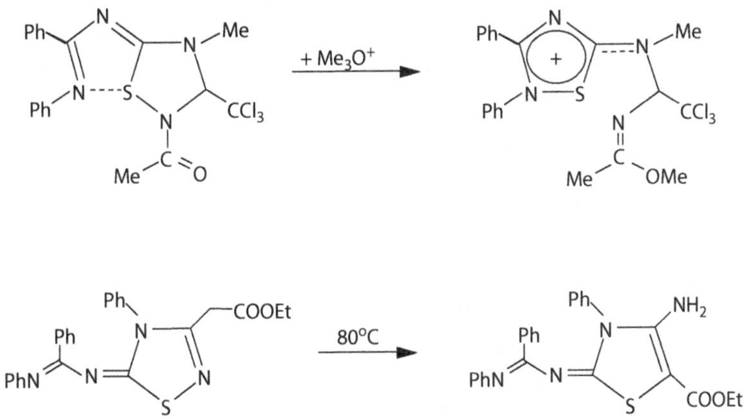

SCHEME 2.57

2.2.4 Biomedical Importance

2.2.4.1 Medical Significance

Besides the existing repertoire of 1,2,4-thiadiazole drugs, new compounds are claimed as medically significant. O-Derivatives of 3-hydroxy-5-phenyl-1,2,4-thiadiazole exerts cardioprotective action due to the inhibition of voltage-gated Na^+ channels (Hartmann et al. 1998). There is a claim that 3,5-diamino-1,2,4-thiadiazole derivatives are useful in the treatment of hypertension (Cohnen and Armah 1982). Hydrochlorides of 3-aryl-2-(propylphenyl)-5-imino-1,2,4-thiadiazoles exhibit remarkable antiplatelet and anticoagulant activities (Rehse and Martens 1993).

The product of hydroxylamine addition to 7-cyano-2-(3-methyl-1,2,4-thiadiazol-5-yl)-heptanoic acid phenylamide is active in treating cancer (Joel and Marson 2008).

1,2,4-Thiadiazolyl nitrones are capable of direct trap and stabilize oxygen, carbon, and sulfur-centered free radicals. Accordingly, they exhibit neuroprotective activity. (The central nervous system is more sensitive to radical damage compared to liver and heart.) The neuroprotective activity of the nitrone pharmacophore depends in great part on the nature of substituents on the nitrone group. Such groups as thiadiazoles (1,2,4 and 1,2,3) as well as furoxans stabilize the spin-adducts resulting from addition of a radical species to the nitrone function. The thiadiazolyl nitrones show excellent free-radical scavenging capacity and good neuroprotective effects without cellular toxicity (Porcal et al. 2008).

2-(Arylalkylpyrazolyl)-5-(*N*-arylamino)-1,2,4-thiadiazoles are patented as medicaments for the prevention and treatment of eye disorders, inflammatory processes, airway pathologies, ischemic cardiac diseases, and disorders in the central nervous system (Bolea et al. 2009). 5-[*N*-(2-pyridyl)amino]-1,2,4-thiadiazole derivatives were claimed to cure diabetes or obesity (Aicher et al. 2008, Hashimoto et al. 2008). Aminoamido-1,2,4-thiadiazole compounds were patented for treatment infection (Huang 2008a–d, 2009) and ulcerative (Muchowski et al. 2008) diseases. Other 1,2,4-thiadiazole amino derivatives are effective in the treatment of tropical (Gallagher and Gahman 2008) or neurodegenerative (Griffioen et al. 2007, 2008, 2009, Macdonald et al. 2008) diseases. Thus, 3-methyl-1-[3-(4-methylbenzyl)-1,2,4-thiadiazol-5-yl]-4-(4-methoxyphenyl)sulfonyl piperazine inhibits Parkinson's disease with activity of 93% (Griffioen et al. 2008). 2-{4-[3-(3,4-Dichlorophenyl)-[1,2,4]-thiadiazol-5-yl]-2-mentylphenoxy}-2-methyl propionic acid is a real candidate for treatment of metabolic syndromes in the human bodies (Shen et al. 2008).

2.2.4.2 Agricultural Protectors

The (2,4-dichlorophenoxy)propanamide derivative of 1,2,4-thiadiazolopyrimidine was found to have herbicidal activity (Liu et al. 2007, 2008). A number of 1,2,4-thiadiazole aryloxyphenylamidines (Kunz et al. 2007, 2008) and thiocarbamate (Takyo and Ihara 2008) derivatives were successfully tested as insectofungicides.

5-Ethoxy-3-(trichloromethyl)-1,2,4-thiadiazole is effective as a soil fungicide for the prevention and control of damping-off, root rot, and stem disease. It is also an inhibitor of nitrification of paddy fields (Carrasco et al. 2004).

2.2.5 TECHNICAL APPLICATIONS

1,2,4-Thiadiazolotriazines were patented as components of optical compensation films for liquid crystal displays. They differ by good control of wavelength dispersion (Nagata and Nishikawa 2007).

2.2.6 CONCLUSION

Compounds containing the 1,2,4-thiadiazole or 1,2,4-thiadiazolidine ring attract much attention due to their biomedical importance. Many of them are of interest as candidates for pharmaceuticals, agricultural improvers, or bioanalytical reagents. This is the reason of diversity in methods of their synthesis and wideness of studies on their reactivity. In the 1,2,4-thiadiazole reactions, specific intermediary states are turned to be possible that are distinguished by expansion of the sulfur valence shell. Technical application is at present scarce, but will undoubtedly develop in the future.

REFERENCES TO SECTION 2.2

Aicher, Th.D.; Boyd, S.A.; Chicarelli, M.F.; Condroski, K.R.; Carrey, R.F.; Hinklin, R.J.; Singh, A.; Turner, T.M. (2008) WO Patent 091,770.

Akiba, K.-Y.; Yamamoto, Y. (2007) *Heteroat. Chem.* **18**, 161.

Balya, A.G.; Belyuga, A. G.; Brovarets, V.S.; Drach, B.S. (2006) *Zh. Org. Farm. Khim.* **4**, 49.

Bey, E.; Marchais-Oberwinkler, S.; Werth, R.; Negri, M.; Al-Soud, Ya.A.; Kruchten, P.; Oster, A.; Frotscher, M.; Birk, B.; Hartmann, R.W. (2008) *J. Med. Chem.* **51**, 6725.

Bird, C.W. (1985) *Tetrahedron* **41**, 1409.

Bolea, Ch.; Celanire, S.; le Poul, E.; Gagliardi, S.; Rencurosi, A.; Farina, M. (2009) WO Patent 010,871.

Carrasco, D.; Fernandez-Valiente, E.; Quesada, A. (2004) *Biol. Fertil. Soils* **39**, 186.

Cho, N.S.; Kim, Y.H.; Park, M.S.; Kim, E.H.; Kang, S.K.; Park, S. (2003) *Heterocycles* **60**, 1401.

Cohnen, E.; Armah, B. (1982) Eur. Patent 044,266.

Danilova, E.A.; Melenchuk, T.V.; Islyaykin, M.K.; Sud'ina, E.E. (2009) Russ. Patent 2,348,623.

Franz, J.E.; Dingra, O.P. (1984) *Comprehens. Heterocycl. Chem.* **6**, 464.

Friedman, B.S.; Sparks, M.; Adams, R. (1937) *J. Am. Chem. Soc.* **59**, 2262.

Gallagher, J.L.; Gahman, T.C. (2008) WO Patent 086,176.

Glinka, T.; Huie, K.; Cho, A.; Ludwikow, M.; Blais, J.; Griffith, D.; Hecker, S.; Dudley, M. (2003) *Bioorg. Med. Chem.* **11**, 591.

Griffioen, G.; Coupet, K.M.E.; Duhamel, H.R.; Wera, S.; Gomme, E.; van Damme, N.; van der Auwera, I.; Lox, M.; van Dooren, T.; Decruy, T. (2007) WO Patent 090,617.

Griffioen, G.; Coupet, K.M.E.; Duhamel, H.R.; Wera, S.; Gomme, E.; van Damme, N.; van der Auwera, I.; Lox, M.; van Dooren, T.; Decruy, T. (2009) US Patent 054,410.

Griffioen, G.; Wera, S.; Duhamel, H.R.; van Damme, N.; Gomme, E. (2008) WO Pat. 061,781.

Gupta, A.; Mishra, P.; Pandeya, S.N.; Kashaw, S.K.; Stables, J.P. (2009) *Eur. J. Med. Chem.* **44**, 1100.

Hartmann, M.; Decking, U.K.M.: Schrader, J. (1998) *Naunyn-Schmeidelberg's Arch. Pharmacol.* **358**, 554.

Hashimoto, N.; Sagara, Yu.; Asai, M.; Nishimura, T. (2008) WO Patent 044,777.

Huang, Zh. (2008a) Chin. Patent 101,153,028.

Huang, Zh. (2008b) Chin. Patent 101,210,020.

Huang, Zh. (2008c) Chin. Patent 101,210,023.

Huang, Zh. (2008d) Chin. Patent 101,230,069.

Huang, Zh. (2009) Chin. Patent 101,357,229.

Iizawa, Y.; Okonogi, K.; Hayashi, R.; Iwahi, T.; Yamazaki, T.; Imada, A. (1993) *Antimicrob. Agents Chemother.* **37**, 100.

Joel, S.P.; Marson, Ch.M. (2008) WO Patent 047,138.

Keith, J.M.; Apodaca, R.; Xiao, W.; Sierstad, M.; Pattabiraman, K.; Wu, J.; Webb, M.; Karbarz, M.J.; Brown, S.; Wilson, S.; Scott, B.; Tham, Ch.-S.; Luo, L.; Palmer, J.; Wennerholm, M.; Chaplan, S.; Breitenbucher, J.G. (2008) *Bioorg. Med. Chem. Lett.* **18**, 4838.

Komatsu, M.; Shibata, J.; Ohshiro, Y.; Agawa, T. (1983) *Bull. Chem. Soc. Jpn.* **56**, 180.

Kunz, K.; Dunkel, R.; Greul, J.; Guth, O.; Benoit, H.; Ilg, K.; Moradi, W.; Seitz, Th.; Ebbert, R.; Dahmen, P.; Voerste, A.; Wachendorf-Neumann, U.; Franken, E.-M.; Malsam, O.; Tietjen, K. (2008) WO Patent 128,639.

Kunz, K.; Greul, J.; Guth, O.; Benoit, H.; Hartmann, B.; Ilg, K.; Moradi, W.; Seitz, Th.; Dahmen, P.; Voerste, A.; Wachendorf-Neumann, U.; Dunkel, R.; Ebbert, R.; Franken, E.-M.; Malsam, O. (2007) WO Patent 031,513.

Kurzer, F. (1965) *Adv. Heterocycl. Chem.* **5**, 285.

Liu, G.-H.; Xue, Yu.-N.; Lu, X.-Q.; Liu, M.-M.; Wang, W.-Y.; Yang, L.-Zh. (2008) *Pest. Manag. Sci.* **64**, 556.

Liu, G.-H.; Xue, Yu.-N.; Yao, M.; Fang, H.-B.; Yu, H. (2007) *Jiegou Huaxue* **26**, 450.

Macdonald, G.J.; Bartolome-Nebreda, J.M.; van Gool, M.L.M. (2008) WO Patent 128,996.

Mack, W. (1967) *Angew. Chem. Internl. Ed.* **6**, 1084.

Makhova, N.N.; Ovchinnikov, I.V.; Kulikov, A.C.; Moltov, S.I.; Baryshnikova, E.L. (2004) *Pure Appl. Chem.* **76**, 1691.

Mamaeva, E.A.; Bakibaev, A.A. (2003) *Tetrahedron* **59**, 7521.

Morel, G.; Marchand, E.; Sinbandhit, S.; Toupet, L. (2003) *Heteroat. Chem.* **14**, 95.

Muchowski, P.J.; Muchowski, J.M.; Schwarcz, R.; Guidetti, P. (2008) WO Patent 022,286.

Nagata, I.; Nishikawa, H. (2007) Jpn. Patent 246,672.

Nasim, Sh.; Crooks, P.A. (2009) *Tetrahedron Lett.* **50**, 257.

Pavlik, J.W.; Changton, C.; Tserikas, V.M. (2003) *J. Org. Chem.* **68**, 4855.

Pelter, A.; Suemengen, D. (1977) *Tetrahedron Lett.* **22**, 1945.

Podolesov, B.D.; Jordanovska, V.B. (1985) *J. Serb. Chem. Soc.* **50**, 119.

Porcal, W.; Hernandes, P.; Gonzalez, M.; Ferreira, A.; OLea-Azar, C.; Cerecetto, H.; Castro, A. (2008) *J. Med. Chem.* **51**, 6150.

Raficul, I.M.; Shimada, K.; Aoyagi, S.; Takikawa, Y.; Kabuto, C. (2004) *Heteroat. Chem.* **15**, 175.

Rehse, K.; Martens, A. (1993) *Arch. Pharm.* **326**, 399.

Reuman, M.; Hu, Zh.; Kuo, G.-H.; Li, X.; Russell, R.; Shen, L.; Youelles, S.; Zhang, Y. (2007) *Org. Res. Develop.* **11**, 1010.

Richardson, Ch.; Steel, P.J. (2007) *Tetrahedron Lett.* **48**, 4553.

Saczewski, F.; Saczewski, J.; Gdaniec, M. (2003) *J. Org. Chem.* **68**, 4791.

Shafiee, A.; Ebrahimzadeh, M.A.; Maleki, A. (1999) *J. Heterocycl. Chem.* **36**, 901.

Shen, L.; Zhang, Ya.; Wang, A.; Sieber-McMaster, E.; Chen, X.; Pelton, P.; Xu, J.Z.; Yang, M.; Zhu, P.; Zhou, L.; Reuman, M.; Hu, Zh.; Russell, R.; Gibbs, A.C.; Ross, H.; Demarest, K.; Murray, W.V.; Kuo, G-H. (2007) *J. Med. Chem.* **50**, 3954.

Shen, L.; Zhang, Ya.; Wang, A.; Sieber-McMaster, E.; Chen, X.; Pelton, P.; Xu, J.Z.; Yang, M.; Zhu, P.; Zhou, L.; Reuman, M.; Hu, Zh.; Russell, R.; Gibbs, A.C.; Ross, H.; Demarest, K.; Murray, W.V.; Kuo, G-H. (2008) *Bioorg. Med. Chem.* **16**, 3321.

Takyo, H.; Ihara, H. (2008) WO Patent 032,858.

Tegler, J.J; Shoger, K.D. (1988) US Patent 4,758,578.

Tiemann, F. (1891) *Chem. Ber.* **24**, 369.

Toung, R.L.; Wodzinska, J.; LI, W.; Lowrie, J.; Kurkrea, R.; Desilets, D.; Karimian, K.; Tam, T.F. (2003) *Bioorg. Med. Chem. Lett.* **13**, 5529.

Vivona, N.; Gusmano, G.; Macaluso, G. (1977) *J. Chem. Soc. Perkin Trans. 1,* 1616.

Wu, Y.-J.; Zhang, Yu. (2008) *Tetrahedron Lett.* **49**, 2869.

Zarafu, I.; Ivan, L.V.; Harasim, I. (2008) *Rev. Chim.* **59**, 101.

Zhang, F.; Estavillo, C.; Mohler, M.; Cai, J. (2008) *Bioorg. Med. Chem. Lett.* **18**, 2172.

Zyabrev, V.S.; Rensky, M.A.; Drach, B.S. (2002) *Zh. Obshch. Khim.* **72**, 1402.

Zyabrev, V.S.; Rensky, M.A.; Rusanov, E.B.; Drach, B.S. (2003) *Heteroat. Chem.* **14**, 474.

2.3   1,2,5-THIADIAZOLES

2.3.1   FORMATION

In contrast to rubricating in other sections, the rubricating in Section 2.3 is intentionally performed according to the reactants introducing sulfur. Such an approach gives a chance to show diversity of the sulfur reactants transforming the organic nitrogen-containing substrates into 1,2,5-thiadiazoles.

2.3.1.1   With Thionyl Chloride

Reaction of vicinal diamines with thionyl chloride is a routine method of the 1,2,5-thiadiazole preparation, especially of its benzo-fused derivatives. As established, *ortho*-phenylene diamines react with thionyl chloride forming *ortho*-sulfenamidoanilines and *ortho*-bis(sulfenamido)benzenes. After that, cyclization takes place and derivatives of benzothiadiazole (piazothiol) are formed. The reaction sequence was pointed out by Pesin and Muravnik (1965), then Wucherpfenning (1968). An improved and near quantitative synthesis of *ortho*-bis(sulfenamido)benzene was completed by its x-ray crystal structure. This amide has a planar Z,Z-conformation. On heating or on standing in moist air, it is converted into piazothiol with the almost quantitative yield (Bagryanskaya et al. 2001). Scheme 2.58 illustrates the transformation.

As mentioned, intermediary *ortho*-sulfenamidoaniline then transforms into piazothiol. Accordingly, the ready-made *ortho*-sulfenamidoaniline can be successfully used avoiding the use of thionyl chloride (Pesin and Muravnik 1965, Philipp et al. 2004).

2.3.1.2   With Disulfur Dichloride

Disulfur dichloride reacts with heminal aminonitriles to form 1,2,5-thiadiazole chloroderivatives according to Scheme 2.59 (Jung et al. 2003).

SCHEME 2.58

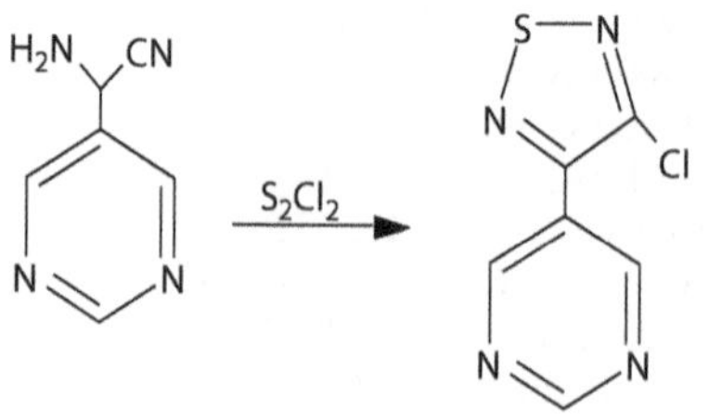

SCHEME 2.59

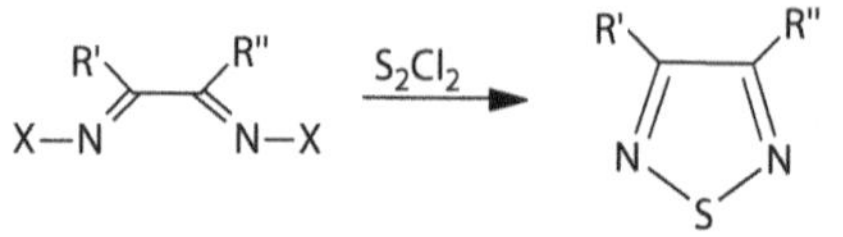

SCHEME 2.60

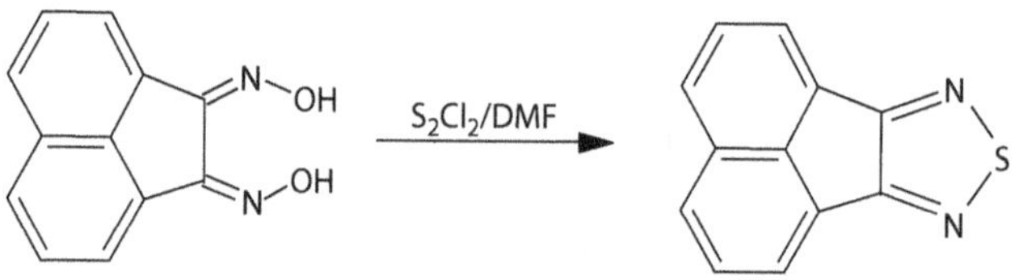

SCHEME 2.61

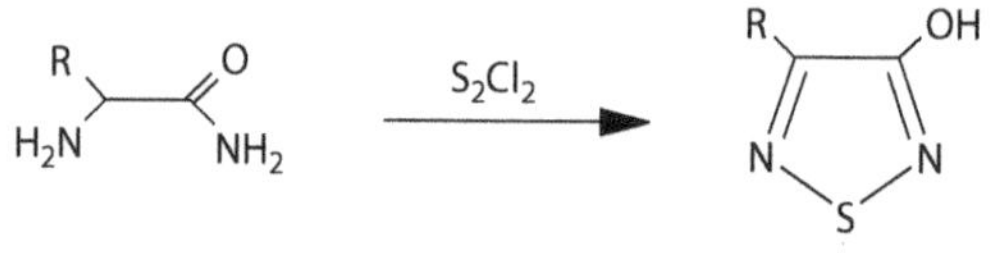

SCHEME 2.62

1,2-Diimines (Weinstock et al. 1967, Buchwald and Ruehlmann 1979) as well as 1,2-dioximes (Weinstock et al. 1967) react with disulfur dichloride to afford the corresponding 1,2,5-thiadiazoles (Scheme 2.60).

Aromatic dioximes also react according to Scheme 2.60 as exemplified by the case of acenaphthoquinone dioxime (Scheme 2.61) (Pilgram 1970).

2-Aminoacylamide with disulfur dichloride gives 4-substituted 3-hydroxy-1,2,5-thiadiazoles (Scheme 2.62) (Naito et al. 1968).

2-Aminoacetamidine hydrobromide reacts with disulfur dichloride to afford 3-amino-1,2,5-thiadiazole in good yield (Scheme 2.63) (Weinstock et al. 1967).

Scheme 2.64 represents one specific case of the disulfur dichloride reaction with N-(trimethylsilyl)pyrid-2-yl ketimide. The product is the fused thiadiazolium chloride with the positive charge on sulfur and the pyridyl nitrogen as a part of 1,2,5-thiadizole system (Bacon et al. 2008).

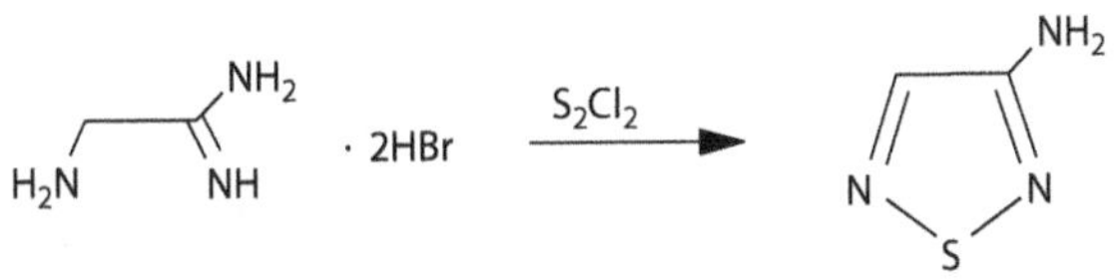

SCHEME 2.63

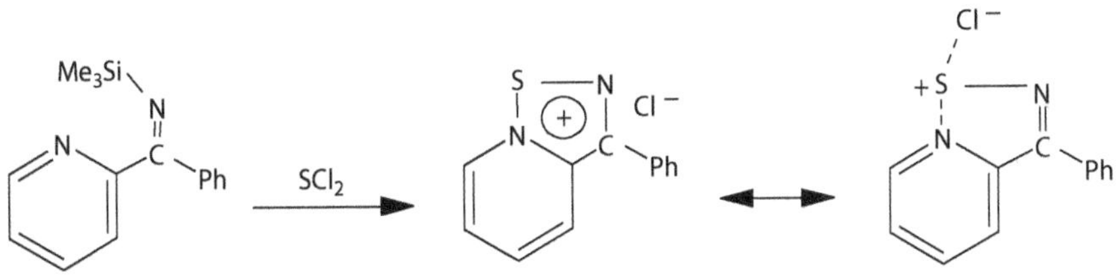

SCHEME 2.64

2.3.1.3 With Sulfamides

Colyer et al. (2010) described the sulfanilide reaction with 4-cyano-1,2-phenylene diamine, in which an amino group in position 1 was protected by the *p*-methoxybenzyl fragment. The reaction proceeds in diglyme at 160°C and leads to the corresponding dihydro dioxide with 60% yield (Scheme 2.65).

2.3.1.4 With Cyclotrithiazyl Trichloride, (NSCl)₃

Such a synthon as 3-amino-1,2,5-thiadiazole is needed for preparation of bioactive compounds. Usually, it is obtained from aminoacetamidine and disulfur dichloride (Weinstock et al. 1967). However, the synthesis of the amidine requires four steps and the disulfur dichloride reaction can give problems. Duan et al. (1997) have proposed the reaction of (NSCl)₃ with commercially available *N*-vinylphthalimide (Scheme 2.66). This is a two-step process and the overall yield compares favorably with the method of Weinstock et al. (1967).

The reaction depicted in Scheme 2.66 was recommended to be performed in the presence of molecular sieves. The sieves remove traces of water from the solvent

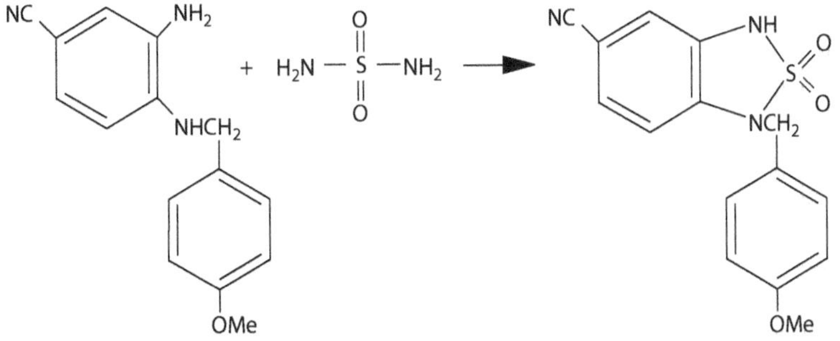

SCHEME 2.65

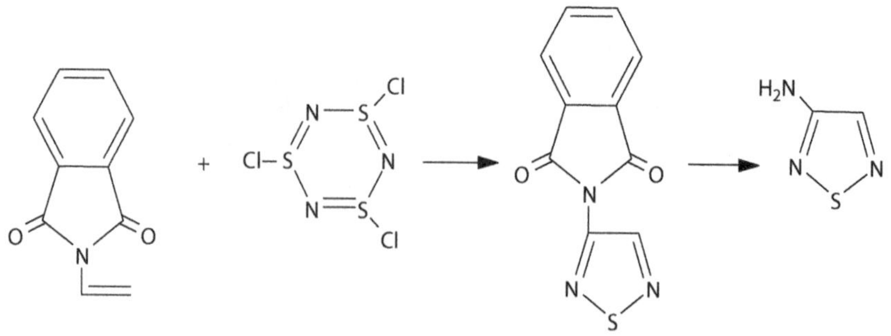

SCHEME 2.66

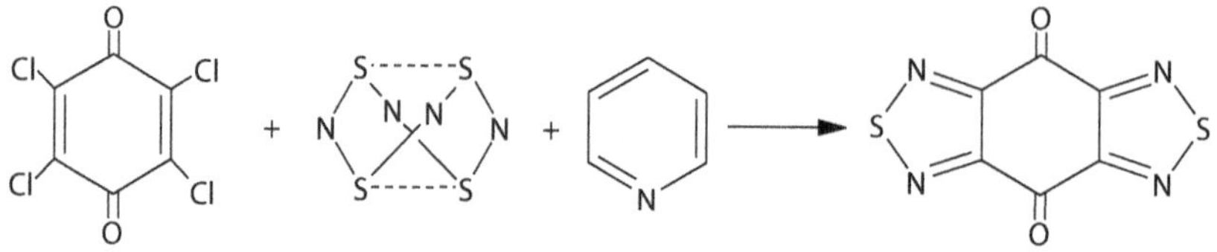

SCHEME 2.67

(tetrahydrofuran) and thus minimize undesirable hydrolysis. They also serve to absorb the hydrogen chloride formed in the reaction. The phthalyl fragment group is readily eliminated upon action of methylhydrazine.

Cyclotrithiazyl trichloride, (NSCl)$_3$, was also used to transform triazoles into 1,2,5-thiadiazoles. The reaction is assumed to proceed by initial ring opening of the triazoles with the formation of their diazoimines (Rees and Yue 2001).

2.3.1.5   With Tetrasulfur Tetranitride, (NS)$_4$

Preparation of electric conductors containing thiadiazole and p-quinone fragments is an actual task. A convenient one-pot synthesis of dithiazolo-p-benzoquinone consists of the reaction between chloranil and (NS)$_4$ with the participation of pyridine (Scheme 2.67) (Shi et al. 1995).

2.3.1.6   With Sulfurous Anhydride

This method of cyclization was proposed by Philipp et al. (2004) and is presented by Scheme 2.68. Although the reaction mechanism was not discovered, the authors suppose that amino groups of the starting material are converted to a bis(trimethylsilyl) moiety; the intermediate, upon exposure to liquid SO$_2$, produces the thiadiazole product.

Sulfurous anhydride and potassium cyanide were used to assemble a 1,2,5-thiadiazole derivative. Namely, reacting in anhydrous acetonitrile or in absolute ethanol, sulfurous anhydride and potassium cyanide produce potassium pyrosulfite and 3-cyano-4-hydroxy-1,2,5-thiadiazole. After the reaction with concentrated hydrochloric acid, the organic product was obtained with excellent yield (Ross and Smith 1964).

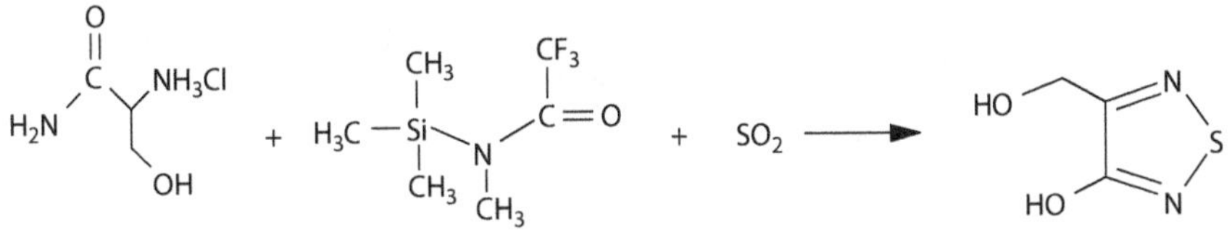

SCHEME 2.68

2.3.1.7 With Sulfur Diimides

As a modification of Scheme 2.68, the reaction of perfluoronaphthalene with sulfur diimide was proposed. Scheme 2.69 gives the step sequence. The product, perfluoro-naphtho-[1,2-c]-[1,2,5]-thiadiazole was prepared with 70% yield (Lork et al. 2002).

Analogous synthesis had been proposed by Bagryanskaya et al. (1994). Using *N,N'*-bis(trimethylsilyl) sulfur diimide in the presence of cesium fluoride, Makarov et al. (2005) synthesized [1,2,5]-thiadiazolo-[3,4-c]-[1,2,5]-thiadiazolyl, according to Scheme 2.70. (Upon one-electron reduction, this compound forms an anion-radical. The potassium salt of the anion-radical is long-lived even in the crystalline state.)

2.3.1.8 From Other Heterocycles

The use of tetrasulfur tetranitride, $(SN)_4$, has been already described by Scheme 2.67. The complex of this tetranitride with antimony pentachloride was used to transform substituted isoxazoles into carbonyl derivatives of 1,2,5-thiadiazole (Scheme 2.71) (Kim and Kim 2007). The product of this reaction was used to prepare 3,3':4,3''-*tert*-1,2,5-thiadiazole, a useful oligoheterocyclic compound.

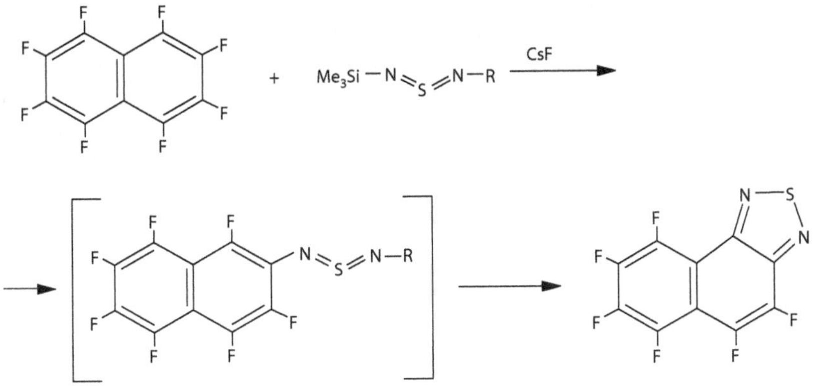

SCHEME 2.69

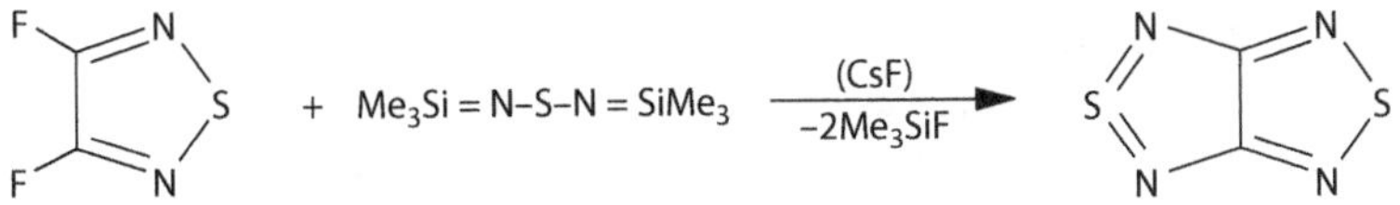

SCHEME 2.70

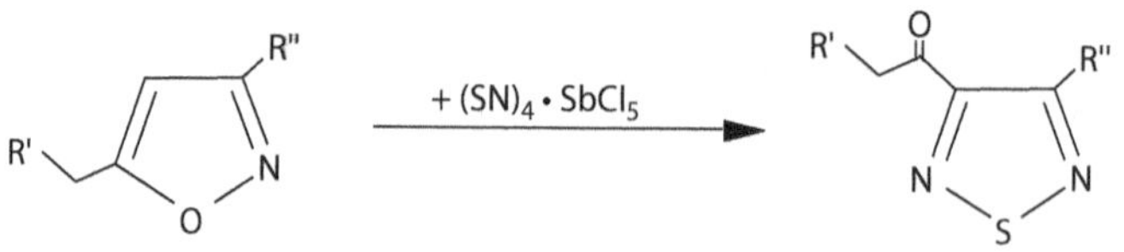

SCHEME 2.71

Very easy replacement of selenium by sulfur with formation of the 1,2,5-thiadiazole ring was observed in the case of iron(II) complex of (1,2,5-selenadiazolo)hexaphenyl porphyrazine. The transformation proceeds merely upon hydrogen sulfide saturation of the complex in chloroform–pyridine solution (Ul-Haq et al. 2007).

2.3.2 Fine Structure

1,2,5-Thiadiazole is a planar system with C_{2v} symmetry (Momany and Bonham 1964). Its dipole moment is equal to 1.58D (Stiefvater 1978). Structural and quantum mechanical studies established extensive π-electron delocalization in 1,2,5-thiadiazole (see Salmond 1968). Its Bird aromaticity index is 84 (100 for benzene). These indexes were derived from bond lengths in the corresponding molecules (Bird 1985). Calculations in the framework of Density Functional Theory (DFT) (Glossman-Mitnik 2001) also point out to aromaticity of 1,2,5-thiadiazole (it is more aromatic than thiophene).

Gaberkorn et al. (2006), Stuzhin et al. (2007), and Zhou (2007) confronted 1,2,5-thiadiazole with benzene as participants fused to the porphyrazine system. The authors concluded that π-electrons in the 1,2,5-thiadiazole rings are conjugated with the porphyrazine macroring π-system to a weaker extent, as compared to the benzene rings. This is the result of π-deficient character of the 1,2,5-thiadiazole ring, its general electron–acceptor property. Unsymmetrical annulation of the benzene donors and 1,2,5-thiadiazole acceptors to porphyrazines provides the molecular systems with the push–pull character and leads to second-order nonlinear optical (NLO) properties (Kudrik et al. 2001). Nevertheless, progressive introduction of the thiadiazole rings in the macrocycle disfavors the nonlinear optical limiting performance (Donzello et al. 2003).

In the fully symmetrical tetrakis(thiadiazolyl)porphyrazine, the 1,2,5-thiadiazole units form an extended π-electron system that permeates the entire molecular frame and this seriously enhanced stability of the whole macrocyclic molecule (Cai et al. 2007, Donzello et al. 2007). Such an extension of the π-electron system results in unprecedented stability of the corresponding anion-radical that was prepared within a film of tetrakis(thiadiazolyl)porphyrazine: The ESR signal of this anion-radical was persistent even in air (Miyoshi et al. 2007).

π-Delocalization is also well-preserved in benzo-[1,2c]-[1,2,5]-thiadiazole (piazothiol), although the benzenoid ring of this compound keeps the o-quinonoid character (Fedin et al. 1967, Breier et al. 1969, Gul'maliev et al. 1973, Palmer and Kennedy 1978). In piazothiol and its derivatives, the sulfur–nitrogen bond distances are middle between the ordinary and double S–N bond lengths. Lengths of the carbon–carbon bonds in the fused six-membered ring are close to that in o-benzoquinone (Luzzati 1951, Gieren et al. 1980, Huebner et al. 1984, Suzuki et al. 2001).

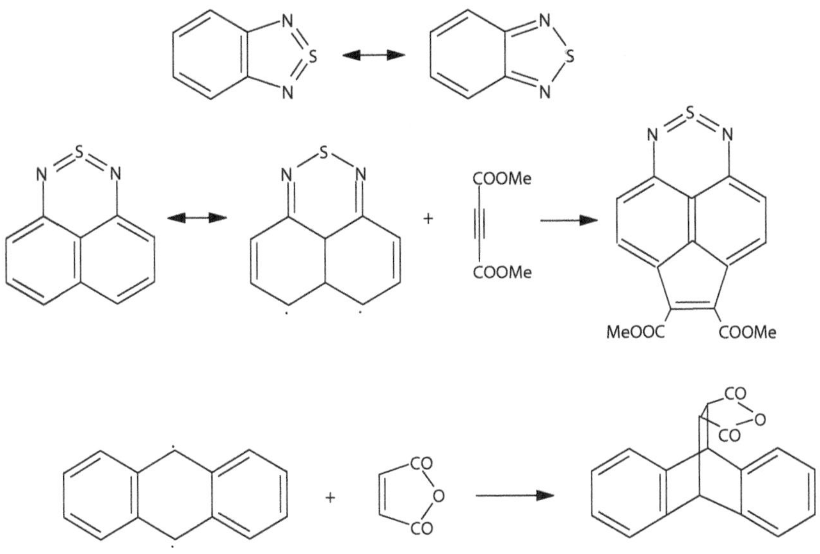

SCHEME 2.72

In the literature, two resonance structures of piazothiol are considered. The first structure has the *o*-quinonoid six-membered ring and the two-coordinated sulfur. The second structure has homoaromatic ring and the four-coordinated sulfur (see Scheme 2.72). From this point of view, naphtho-[1,8-*cd*]-[1,2,5]-thiadiazine represents a related compound to consider the resonance problem. For this compound, the structure with the four-coordinated sulfur and with the completely aromatic naphthalene moiety seems to be reasonable. Nevertheless, it readily undergoes cycloaddition with dimethyl acetylenedicarboxylate (Scheme 2.72) (Gait et al. 1975). The latter reaction makes this naphthothiadiazine close to anthracene. Anthracene also has radical unsaturation at the positions active in cycloaddition (see Scheme 2.72).

Naphtho-[2.3-*c*]-[1,2,5]-thiadiazole (depicted in Scheme 2.73) is totally aromatic according to calculations of the individual canonical or localized π molecular orbitals (Miao et al. 2007). Nevertheless, naphtho-[2.3-*c*]-[1,2,5]-thiadiazole reacts with maleic anhydride in refluxing benzene solution to give a product of cycloaddition, (Scheme 2.73). Alike anthracene, naphtho-[2.3-*c*]-[1,2,5]-thiadiazole is readily oxidized to dione (see Scheme 2.73). On the other side, the thiadiazole of Scheme 2.73 is much less reactive than anthracene: A competitive reaction using *N*-phenylmaleimide and equimolecular amounts of the thiadiazole and anthracene leads to the single adduct of anthracene with *N*-phenylmaleimide. No detectable amount of the thiadiazole adduct was obtained. This difference in reactivity is kinetic rather than thermodynamic in nature: Anthracene does not react with the thiadiazole adduct despite prolonged refluxing in benzene (Cava and Schlessinger 1964).

Cycloaddition of *N*-phenylmaleimide to benzo-[1,2-*c*:4,5-*c'*]-bis{[1,2,5]-thiadiazole} also takes place at the expense of meso positions of this three-ring compound. As calculations show, just the meso positions have the largest atomic orbital

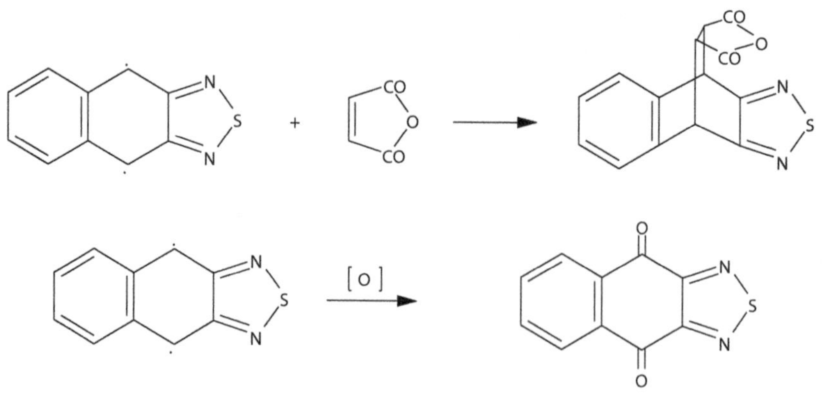

SCHEME 2.73

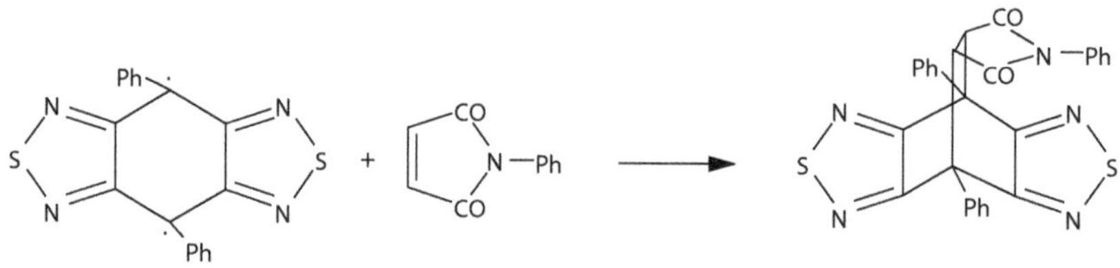

SCHEME 2.74

coefficients in the HOMO and lowest unoccupied molecular orbital (LUMO). The cycloaddition proceeds even though these meso positions hold the bulky phenyl groups (see Scheme 2.74) (Yamashita et al. 1997).

Spin-delocalization encompasses the whole molecular contour in ion-radicals of 1,2,5-thiadiazoles fused with aromatic fragments (Strom and Russell 1965, Kursanov and Todres 1967, Solodovnikov and Todres 1967, 1968, Todres et al. 1968, 1969, Kwan et al. 1976). It means that such ion-radicals undergo complete π-electron delocalization, which characterizes compounds of the aromatic family. According to calculations (Gul'maliev et al. 1975, Konchenko et al. 2009), single occupied molecular orbital in anion-radical of benzo-[1,2-c]-[1,2,5]-thiadiazole has π character.

2.3.3 Reactivity

It is worth to consider separately the materials on reactivity of 1,2,5-thiadiazole itself and its derivatives fused with aromatic fragments.

2.3.3.1 Monocyclic 1,2,5-Thiadiazoles

2.3.3.1.1 Salt Formation

Nitrogen atoms in 1,2,5-thiadiazole bear lone electron pairs, but, according to electrophilicity of the heterocycle as a whole, the compound is a weak base with $pK_a = -4.9$ (Kurita and Takayama 1997). Accordingly, N-alkylation does not come

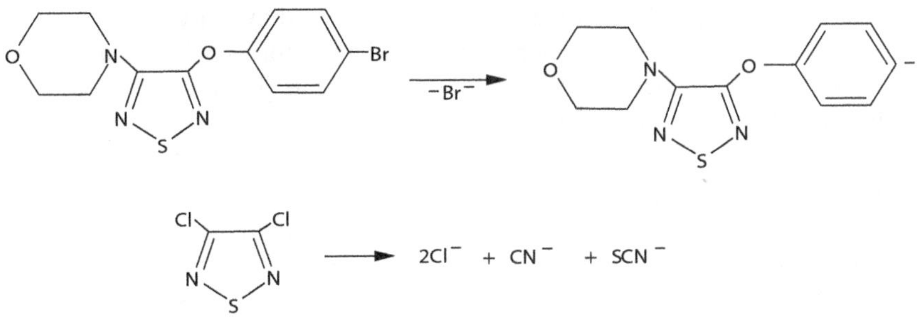

SCHEME 2.75

about readily. Strong alkylating agents such as triethyloxonium tetrafluoroborate are needed (Masuda et al. 1981).

2.3.3.1.2 Reduction

Despite its strong electrophilic character, the thiadiazole ring is not reduced if there are competitive reducing groups bound to the ring. Thus, thiadiazole acyl chlorides are reduced to the alcohols with sodium borohydride in dioxane without degradation of the thiadiazole ring (Mulvey and Weinstock 1967). Interesting is the difference in behavior of the following two 1,2,5-thiadiazole halo derivatives upon cathodic reduction (see Scheme 2.75) (Nastapova et al. 2002).

2.3.3.1.3 Oxidation

Oxidation agents of moderate strength such as oxirane or 3-chloroperoxybenzoic acid transform 1,2,5-thiadiazoles into non-aromatic S-oxides and S,S-dioxides. For instance, 3,4-dimethoxy-1,2,5-thiadiazole gives 3,4-dimethoxy-1,2,5-thiadiazole S-oxide when 1 equivalent of 3-chloroperoxybenzoic acid was used or 3,4-dimethoxy-1,2,5-thiadiazole S,S-dioxide when 2 equivalents of 3-chloroperoxybenzoic acid were used (Algieri et al. 1982). The methyl groups do not oxidize.

2.3.3.1.4 Electrophilic Substitution

Usually, non-fused 1,2,5-thiadiazoles are resistant to electrophilic substitution. Nevertheless, boiling of 1,2,5-thiadiazole, hydrogen chloride, formaldehyde, and acetic acid leads to the formation of 3,4-bis(chloromethyl)-1,2,5-thiadiazole (Cookson and Richards 1974). Containing an electron donor substituent, 3-amino-1,2,5-thiadiazole undergoes halogenation with no ring destruction to give 4-halo derivatives of the starting material. The process was patented by Menzl (1964).

2.3.3.1.5 Nucleophilic Substitution

Being treated with sodium deuteroxide in d_2-water, 1,2,5-thiadiazoles undergo hydrogen–deuterium isotope exchange (Bertini and de Munno 1967). 3-Chloro-4-morpholino-1,2,5-thiadiazole transforms into 4-morpholino-1,2,5-thiadiazol-3-ol (95%) during refluxing with sodium hydroxide in dimethylsulfoxide. The reaction 3,4-dichloro-1,2,3-thiadiazole with morpholine affords

3-chloro-4-morpholino-1,2,5-thiadiazole with 97% yield (Weinstock et al. 1976). With potassium fluoride, 3,4-dichloro-1,2,5-thiadiazole forms 3,4-difluoro derivative as the major product (73%). The minor product is 3-chloro-4-fluoro-1,2,5-thiadiazole (Geisel and Mews 1982).

2.3.3.1.6   Ring Cleavage

The reactions of 1,2,5-thiadiazoles with Grignard (de Manno et al. 1986), alkyllithium (Kouvetakis et al. 1994), or with metal-amide (Komin and Carmack 1976) reagents lead mostly to products of ring-opening. The reactions begin from an attack on the sulfur atom. Merschaert et al. (2006) found that sterically shielded metal amides allow control of ring opening and help avoid the subsequent decomposition of the ring-opened product. These authors performed large-scale transformation of 3,4-dichloro-1,2,5-thiadiazole into 3-amino-4-(propylthio)-1,2,5-thiadiazole by means of ring cleavage and then by recyclization of a stable intermediary product. The reaction was carried out at −78°C in a heptane–ether mixture (see Scheme 2.76).

The product of ring cleavage (see Scheme 2.76), was obtained with 90% yield. The product of recyclization was obtained with 90% yield. The reaction is prospective for manufacture of pharmaceuticals. The starting material is commercially available, and the ring-opened product was prepared in kg scale. It is isolable and no detectable degradation has been observed after its storage for several years at −20°C. The process has been patented (Borgese et al. 2006).

2.3.3.1.7   Photolysis

Usually, photolysis of aromatic compounds does not proceed readily. Although 1,2,5-thiadiazoles are more or less aromatic in their nature, 3,4-diphenyl-1,2,5-thiadiazole suffers slow photochemical degradation giving benzonitrile and sulfur (Cantrell and Haller 1968).

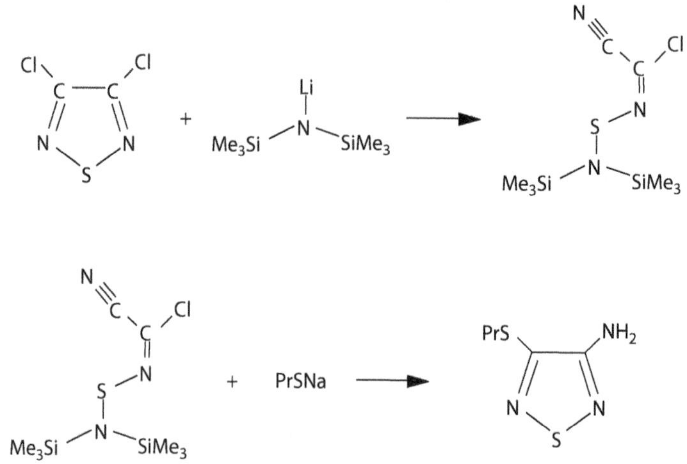

SCHEME 2.76

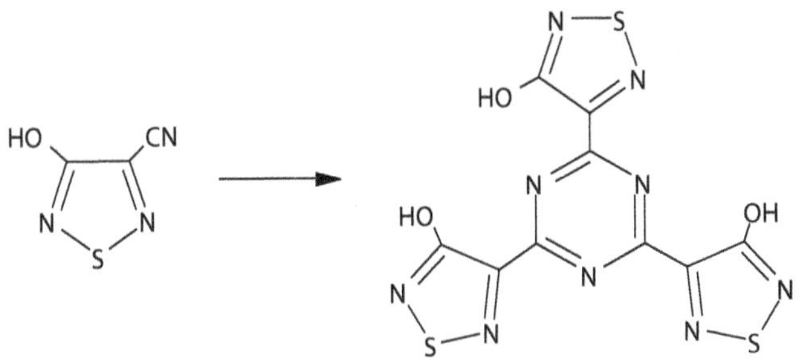

SCHEME 2.77

2.3.3.1.8 Thermolysis

The parent 1,2,5-thiadiazole shows thermal stability, up to 220°C (Cantrell and Haller 1968). The thiadiazole thermal stability is caused by the relatively high degree of aromaticity. Being heated above 162°C, 4-hydroxy-1,2,5-thiadiazole-3-carbonitrile melts, but rapidly crystallizes to give the trimer 2,4,6-tris(4-hydroxy-1,2,5-thiadiazol-3-yl)-1,3,5-triazine (Scheme 2.77). This triazine is thermally stable up to 360°C (Ross and Smith 1964).

Rather unexpected result was obtained from thermolysis (250°C) of 1,2,5-thiadiazole-1-oxide. Instead of destruction, this S-oxide deoxygenated smoothly to yield parent 1,2,5-thiadiazole (Dunn and Rees 1989).

2.3.3.1.9 Coordination

Coordination of 1,2,5-tiadiazole-3.4-dicarboxylate (L) to a metal (M) was reported by Li et al. (2007) (M stands for cobalt or nickel). The metal is bound with both ring-nitrogen atoms and both carboxylate groups. The complex structures are established as $[M_2(L)_2(H_2O)_2]_n$ (Li et al. 2007). 1,2,5-Thiadiazole itself also coordinates to two tungsten pentacarbonyl fragments at the expense of two nitrogens of the thiadiazole ring (Bock et al. 1988). The coordination increases stability of the 1,2,5-thiadiazole anion-radical, which is instable in the non-coordinated state. (The lower the molecular orbital energy, the more probable the spin-density localization is in the framework of this molecular orbital. Meanwhile, coordination with a metal often decreases the energy level of the molecular orbital encompassing an organic ligand. Accordingly, this increases the stability of the ligand in the anion-radical form.)

2.3.3.1.10 Ring Transformation

Although the reactions of 1,2,5-thiadiazoles with Grignard reagents lead mostly to products of ring cleavage through an attack on the sulfur atom, treatment of compounds intermediately formed with tellurium tetrachloride leads to the corresponding 1,2,5-telluradiazole. In this case, experimental condition must be followed carefully. The best solvent is tetrahydrofuran. 1,2,5-Thiadiazole was added to ethyl magnesium bromide, the mixture was stirred at room temperature and cooled to −60°C, followed by addition of TeCl$_4$. The resulting solution was treated with

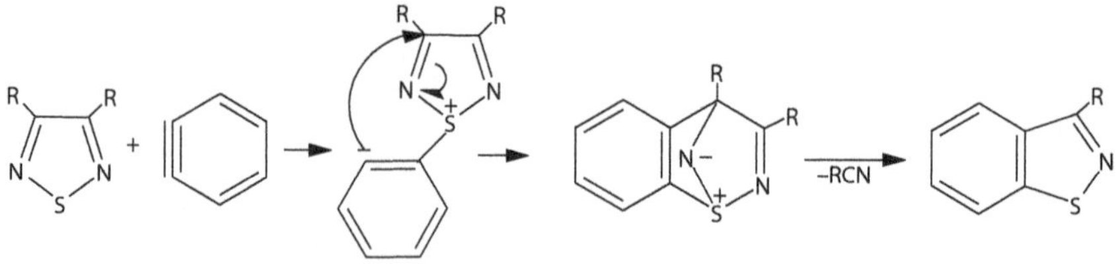

SCHEME 2.78

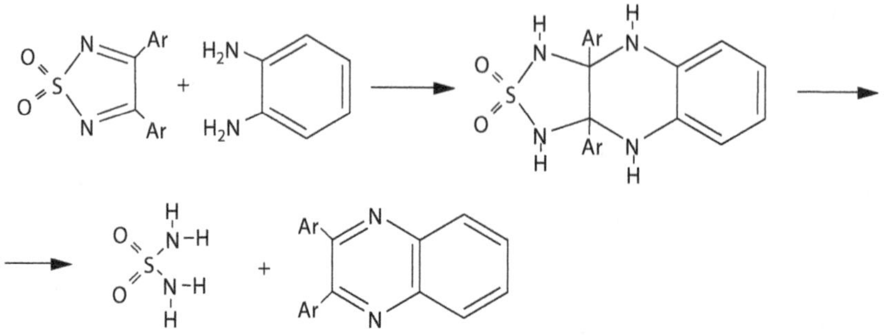

SCHEME 2.79

triethylamine and stirred at room temperature to give 1,2,5-telluradiazole (Bertini et al. 1982).

Compounds of the 1,2,5-thiadiazole group can be transformed into derivatives of benzoisothiazole. Scheme 2.78 illustrates this transition. Benzyne was generated from benzenediazonium-2-carboxylate in one pot with a thiadiazole. The reactions proceed in boiling tetrahydrofuran and lead to isothiazole derivatives with moderate to good yields (Bryce et al. 1988).

Transition from 3,4-diaryl-1,2,5-thiadiazole-1,1-dioxides to 2,3-diarylquinoxalines was performed according to Scheme 2.79, the reaction proceeds in dimethylformamide at room temperature, the yields achieve 80%–95% (Mirifico et al. 2008).

Pyrimidino-[3,4-*d*]-[1,2,5]-thiadiazole 2-oxide gives pyrimidino-[3,4-*d*]-[1,2,5]-oxadiazole upon oxidation of sodium hypochlorite. Yields are 35%–40% (Yavolovsky et al. 2000).

2.3.3.2   Fused 1,2,5-Thiadiazoles

2.3.3.2.1   Salt Formation

Benzo-fusion does little to aid *N*-alkylation, which still requires reactive alkylating agents.

2.3.3.2.2   Reduction

The fused 1,2,5-thiadiazoles are insensitive to mild reducing agents, but more powerful ones give rise to ring cleavage and desulfurization with the formation of

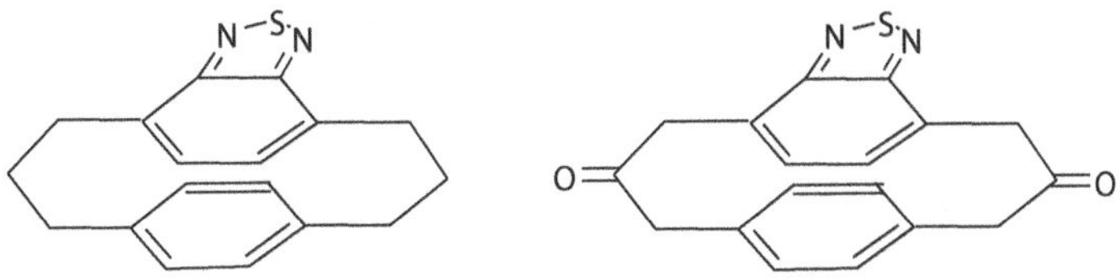

SCHEME 2.80

1,2-diamino compounds. As this takes place, sensitive functional groups such as bromo, chloro, cyano, or carboxyl bound with the aryl-fused ring can be tolerated through the use of magnesium in methanol, to give synthetically important aryl-1,2-diamines (Hatta et al. 1991) or, through the use of sodium borohydride, to give industrially significant 1,2-diamino-4,5-phthalodinitrile (Burmester and Faust 2008). The same method was used to reduce piazothiol-α-sulfonyl chloride to 1,2-phenylenediamine-3-sulfonyl chloride with the yield lower than 5%. However, zinc in acetic acid allows obtaining the diamine with a yield of 84% while the sulfonamide moiety remains to be intact (Rosen et al. 2009). This diamino sulfonyl chloride is a starting material for the synthesis of cholecystokinin-2 receptor. Importantly, piazothiol-α-sulfonyl chloride is commercially available.

Practically important aminopiazothiols can be easily obtained via reduction with iron shavings in diluted acetic acid; the nitro group is reduced earlier than the thiadiazole ring. This fact can be correlated with the paradigm about localization of an unpaired electron on the nitro group in anion-radicals of nitroaromatic heterocyclic compounds such as nitropiazothiols, nitropiazoselenols (Todres et al. 1968, Todres 2009), or nitrobenzothiazoles (Ciminale 2004).

The nitro groups in piazothiol are also converted into the amino functions by *E. coli* cells immobilized in carragheen gel. The heterocyclic ring, despite its electrophilicity, keeps is integrity during this bioreaction, too (Davidenko and Romanovskaya 1989). This is important from the point of metabolism. The authors noted that yields of the amines are high and the gel works efficiently and reusably, in the column regime.

Watanabe et al. (2010) prepared paracyclophanes containing piazothiol and benzene (Scheme 2.80). In the cyclophanes, the piazothiol and benzene rings are completely overlapped in an almost parallel fashion with the tilt of 1.5° and an average transannular distance of 0.32 nm. Both compounds of Scheme 2.80 were compared with respect to their ultraviolet and fluorescent spectra. Surprisingly, these spectra occurred to be similar regardless of the presence of the $-CH_2COCH_2-$ or $-CH_2CH_2CH_2-$ bridges. This indicates that charge transfer from benzene to piazothiol takes place via through-space interaction. The direction of the charge transfer is from benzene to piazothiol: Piazothiol not only has a high electron affinity but also an ionization potential similar to that of benzene (Watanabe et al. 2010 and references therein). So, upon photoexcitation, piazothiol is reduced (partially) at the expense of benzene, which is its through-space counterpart in the paracyclophane.

SCHEME 2.81

2.3.3.2.3 Oxidation

Having heterocycles of pronounced electrophilicity, aryl-annulated 1,2,5-thiadiazoles are hardly oxidized. Strong oxidizing agents can convert piazothiol into 1,2,5-thiadiazole-3,4-dicarboxylic acid. Chromic acid affords this carboxylic acid in moderate yield (Pesin et al. 1964a). Potassium permanganate leads to over-oxidation. If the benzenoid ring of piazothiol bears electron-withdrawing substituents, the potassium permanganate oxidation results in 1,2,5-thiadiazole-3,4-dicarboxylic acid with good yields. The oxidation of α-nitropiazothiol is an example. Carmack et al. (1961) patented the method (see Scheme 2.81).

The presence of an annulated acceptor ring also assists in oxidation. Thus, pyrimidino-[3,4-d]-[1,2,5]-thiadiazole 2-oxide gives 6-amino-5-nitrouracyl upon oxidation of hydrogen peroxide with yields up to 70% (Yavolovsky et al. 2000).

Electrochemical oxidation of aryl-annulated 1,2,5-thiadiazoles requires high anodic potentials. Nevertheless, inclusion of piazothiol in the high-molecular chain with fluorene does not prevent anodic oxidation of the polymer. Oxidation of poly[(9,9-dioctylfluorene)-alt-(α,α′-piazothiol)] in the presence of tri(n-propyl) amine leads to electrochemically generated chemiluminescence (Chang et al. 2008). The polymer (further denoted as R) was tested as nanoparticles of sizes lower than 25 nm. [The measurement technique was proposed and described by Palacios et al. (2006).] The chemiluminescence is a result of redox reactions and annihilation of the ion-radicals formed, see p.2 of the supporting information to the communication by Chang et al. (2008).* The whole process may involve the following steps:

1. *Redox reactions in the system under consideration*
   Polymer oxidation: $R - e \rightarrow R^{(+\cdot)}$,
   Hole transport (conduction): $R^{(+\cdot)}$ (double layer at electrode) $\rightarrow R^{(+\cdot)}$ (surface of nanoparticle),
   Oxidation reaction of $(n\text{-}Pr)_3N$ followed by deprotonation of $R^{(+\cdot)}$:

$$Pr_2NCH_2CH_2CH_3 - e \rightarrow Pr_2NCH_2CH_2CH_3^{(+\cdot)},$$

$$Pr_2NCH_2CH_2CH_3^{(+\cdot)} \rightarrow Pr_2NC^{(\cdot)}HCH_2CH_3 + H^+,$$

---

* Chang, Ya.-L., Palacios, R.E., Fan, F.-R.F., Bard, A.J., and Barbara, P.F., Electrogenerated chemiluminescence of single conjugated polymer nanoparticles, *J. Am. Chem. Soc.*, 130(28), 8906, 2008. With permission.

One-electron reduction of $R^{(+\cdot)}$:

$$R^{(+\cdot)} + Pr_2NCH_2CH_2CH_3 \rightarrow R + Pr_2NCH_2CH_2CH_3^{(+\cdot)},$$

$$R^{(+\cdot)} + Pr_2NC^{(\cdot)}HCH_2CH_3 \rightarrow R + Pr_2NC^{(+)}HCH_2CH_3.$$

2. *Possible schemes for electrochemically generated luminescence*

a. Excited state of polymer (R*) is formed via electron transfer from $Pr_2NC^{(\cdot)}HCH_2CH_3$:

$$R^{(+\cdot)} + Pr_2NC^{(\cdot)}HCH_2CH_3 \rightarrow R^* + Pr_2NC^{(+)}HCH_2CH_3, R^* \rightarrow R + h\gamma.$$

b. Neutral R is reduced by $Pr_2NC^{(\cdot)}HCH_2CH_3$ and polymer anion-radical $R^{(-\cdot)}$ is formed:

$$R + Pr_2NC^{(\cdot)}HCH_2CH_3 \rightarrow R^{(-)} + Pr_2NC^{(+)}HCH_2CH_3.$$

Then, the excited state of polymer, R^*, is formed by annihilation of $R^{(+\cdot)}$ and $R^{(-\cdot)}$:

$$R^{(+\cdot)} + R^{(-)} \rightarrow R^*R + h\gamma.$$

The high number of detected photons from individual nanoparticle (1500 photons during 100 s) highlights the potential of this phenomenon as a very sensitive analytical method. It also allows studying the heterogeneous electron-transfer kinetics at the single particle level.

2.3.3.2.4 *Electrophilic Substitution*

Piazothiol itself is readily nitrated in α-position (Efros and Levit 1953.) Pesin and Belen'kaya (1967) published interesting results on nitration of α-hydroxypiazothiol and α-ethoxypiazothiol. The hydroxy compound gives β-nitro-α-hydroxypiazothiol and not α′-nitro-α-hydroxypiazothiol. By contrast, the ethoxy compound gives an equal-amount mixture of β-nitro-α-ethoxypiazothiol and α′-nitro-α-ethoxypiazothiol. It is clear, that regioselectivity in nitration of α-hydroxypiazothiol is dictated by stability of the β-nitro-α-hydroxypiazothiol formed. In this regioisomer, the hydrogen bonding between the available hydroxyl group and the entering nitro group provides a uniformity of nitration.

Similar to α-ethoxypiazothiol, α-methylpiazothiol forms a mixture of β-nitro-α-methylpiazothiol (12%) and α-methyl-α′-nitropiazothiol (53%) upon nitration (Pilgram 1974). From the two nitro isomers obtained, only α-methyl-α′-nitropiazothiol is required as a synthon: Its reduction affords α-methyl-α′-aminopiazothiol as a starting material for syntheses of interesting biologically active compounds. The low yield as

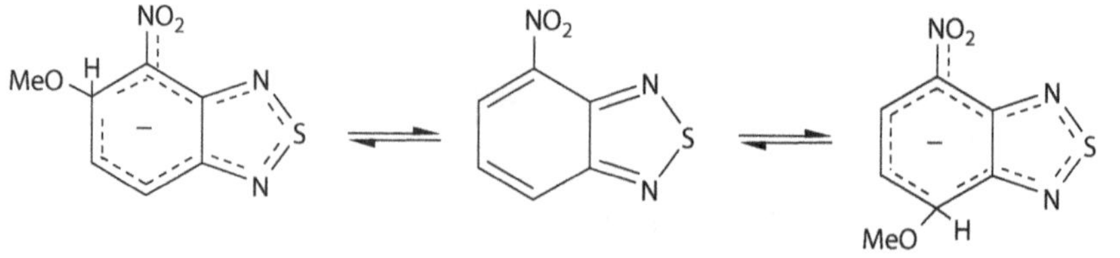

SCHEME 2.82

well as the poor regioselectivity made this route to the aimed amino compound unattractive for large-scale manufacture. Another limitation is the safety concerns associated with nitration on a large scale. To circumvent all of the problems, Liu et al. (2003) has devised an alternative method based on the possibility of α-methylpiazothiol to give α-methyl-α′-bromopiazothiol selectively (90% yield) upon bromination. The latter was transformed into α-methyl-α′-aminopiazothiol, using benzophenone imine as an ammonia equivalent and tris(dibenzylideneacetone) dipalladium as a catalyst (this is the Buchwald methodology). The scale-up synthesis was able to produce 14 kg of α-methyl-α′-aminopiazothiol in high yield (Liu et al. 2003).

2.3.3.2.5 Nucleophilic Substitution

The reaction of α-nitropiazothiol with methoxide in methanolic dimethylsulfoxide gives rise to the isomeric Meisenheimer complexes depicted in Scheme 2.82. This nucleophilic reaction is reversible, but the rate of decomposition is different for the isomers: The complex with the methoxy group in ortho position to the nitro group decomposes 700 times faster than the complex containing the methoxy group in para position to the nitro group (Deicha and Terrier 1981). The higher stability of the para adduct relative to the ortho isomer may reasonably be attributed to the fact that the nitro group can better accommodate the negative charge when it occupies a position para, rather than ortho to the point of attachment of methoxide. Furthermore, the para adduct may take advantage of a larger extent of electron delocalization because of more remoteness of the methoxy group from the nitro group. This is obvious from comparison of structures for the isomeric Meisenheimer complexes in Scheme 2.82.

Mononitro derivatives of piazothiol give nitramines with 85%–95% yields upon treatment by hydroxylamine hydrochloride. As this takes place, the amino group occurs in the vicinal position with respect to the nitro group so that β-nitropiazothiol forms β-nitro-α-aminopiazothiol and α-nitropiazothiol gives α-nitro-β-aminopiazothiol (Pesin et al. 1964b, Cillo and Lash 2004). In the last case, the entering amino group occupies the position adjacent to the α-nitro group despite the fact that α′ position is free. It is the amino-nitro mutual disposition that occurs to be preferential thermodynamically due to combination of the orientation and field effects. Seemingly, hydrogen bonding between the already existing nitro and entering amino groups also makes this regioselectivity in line with thermodynamic requirements.

Let us compare the described nucleophilic amination of isomeric nitropiazothiols with the literature data on electrophilic nitration of monoacetamidopiazothiol isomers. The electrophilic reaction proceeds in the normal way: β-acetamidopiazothiol gives α-nitro-β-acetamidopiazothiol and α-acetamidopiazothiol forms

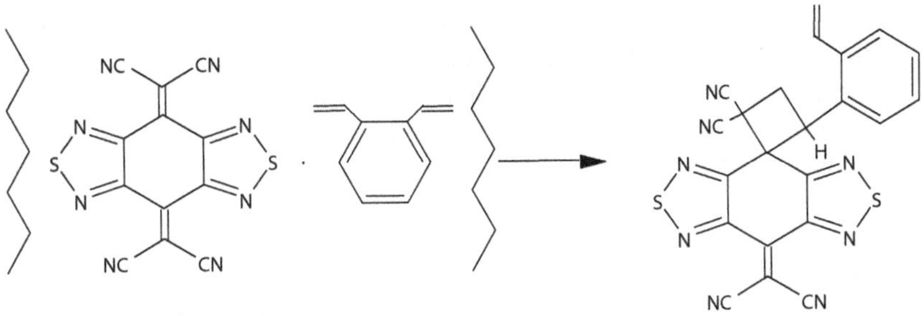

SCHEME 2.83

α-acetamido-α′-nitropiazothiol (Pesin et al. 1962). Note: Unsubstituted piazothiol is nitrated at α-position (Efros and Levit 1953).

2.3.3.2.6 Cycloaddition

Schemes 2.72 through 2.74 should be supplemented with Scheme 2.83 that introduces reaction between bis{[1,2,5]-thiadiazolo}tetracyanoquinodimethane and 1,2-divinyl-benzene. The reaction proceeds in a confined environment, that is, within the chiral crystal of the charge-transfer complex, upon light irradiation. The resulting cycloadduct is optical active and formed in 95% *e,e* purity (Suzuki et al. 1994).

2.3.3.2.7 Ring Cleavage

Reduction with tin dichloride is a typical way to cleave 1,2,5-thiadiazole ring resulting in the 1,2-diamine formation. The 1,2-phenylene diamines and tin tetrachloride often form stable complexes. To liberate the diamines from the complexes, tin has to be deposited with hydrogen sulfide. Chloro derivatives of benzothiadiazoles are reduced with difficulty. The more chlorine substituents are in the benzene ring, the more difficultly in proceeding with this reduction (Pesin and Sergeev 1968).

2.3.3.2.8 Thermolysis

The regularities mentioned for thermolysis of non-fused 1,2,5-thiadiazoles are also applicable to aryl-fused compounds of 1,2,5-thiadiazole group. One specific effect deserves to be distinguished for blends of polymers with the donor nature and polymers containing piazothiol moieties as acceptor fragments. Thermal treatment of such blends improves charge mobility within the blends due to the formation of ordered nanostructures, basically at the expense of the piazothiol-containing constituents (Liao et al. 2007). This opens a way to enhance light harvesting in photovoltaic devices. The role of thiadiazole fragments in polymer components of solar cells is considered in Section 2.4.5.

2.3.3.2.9 Coordination

Benzo-[1,2-*c*]-[1,2,5]-thiadiazole forms 1:1 adduct with arsenic pentafluoride and the complexation accomplished by the nitrogen atom (Apblett et al. 1986). Coordination to iridium, osmium, platinum, and ruthenium proceeds in the same manner whereas copper, chromium, molybdenum, and tungsten form 1:2

complexes, binding both the ring nitrogens (Kuyper and Vrieze 1975, Meij et al. 1977, Bel'skii et al. 1984, Kaim and Kohlmann 1985, Herberhold and Hill 1989). Being incorporated into the osmium complex containing also the carbonyl, chlorine, and p-tolyl ligands, benzo-[1,2-c]-[1,2,5]-thiadiazole is easily displaced by carbon monoxide (Herberhold and Hill 1989). Consequently, the thiadiazole binding in the complex is not strong.

One feature deserves to be mentioned concerning complexation of the benzo-[1,2-c]-[1,2,5]-thiadiazole anion-radical with two M(CO)$_5$ fragments, M=Cr, Mo, or W. Being free from coordination, this anion-radical is stable. Coordination does not change the stability, but does change the electron spin resonance spectra of the anion-radical (Bock et al. 1988). This means that the metal participates in spin delocalization.

Bis-{[1,2,5]-thiadiazolo}tetracyanoquinodimethane of Scheme 2.83 coordinates to ruthenium through the cyano groups, only. In the complex, electron transfer from the metals to the ligand takes place, resulting in stabilizing of the organic dianion (Miyasaka et al. 2011).

Two phenomena make possible complexation between α,α'-bis(4-pyridyl)piazothiol and 2,6-dichoro-2,5-dihydroxy-1,4-benzoquinone: Charge-transfer and hydrogen bonding. As established by x-ray analysis of the crystalline complex, the two hydroxyl protons of the quinone are transferred to the pyridine ring. The remaining benzoquinone bisphenolate, obviously, shows up as a donor relative to the electrophylic piazothiol counterpart (Akhtaruzzaman et al. 2004).

Interesting cases of self-coordination were found for the α,α'-bis(ethynyltrimethylsilyl) derivative of benzo-[1,2-c]-[1,2,5]-thiadiazole (Boudebous et al. 2008) as well as for bis{[1,2,5]-thiadiazolo}-[3,4-b:3,4-e]-pyrazine (Yamashita et al. 1988a) and bis[[1,2,5]-thiadiazolo}-[1,2-c:4,5-c']-benzene (Yamashita et al. 1997). These compounds were isolated as stable crystalline solids. X-ray analysis revealed the tape-like molecular network for those solids as exemplified by Scheme 2.84.

Theoretical calculations of the single bis(thiadiazolo)benzene pointed out that the sulfur atom is positively charged, while the nitrogen atoms are negatively charged (Yamashita et al. 1997). Therefore, the efficient intermolecular interaction in Scheme 2.84 is theoretically explainable.

In the presence of dithiafulvalene groups connected with the quinonoid homocyclic ring in the bis(thiadiazolo)derivative of Scheme 2.84, the type of self-coordination is cardinally changed. The coordination takes place at the expense of only the 1,3-dithiole sulfur atoms according to Scheme 2.85 (Huang and Kertesz 2005 and references therein).

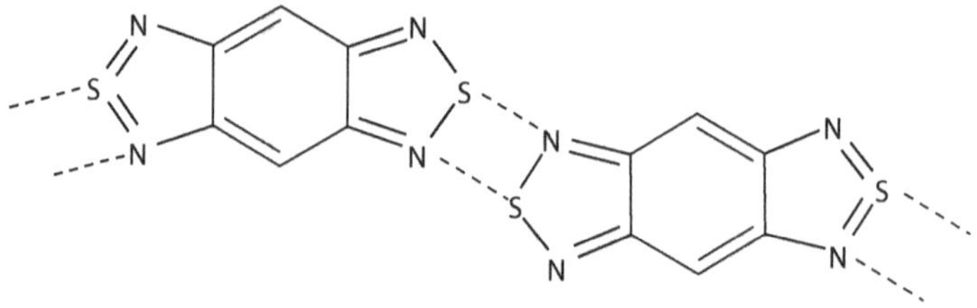

SCHEME 2.84

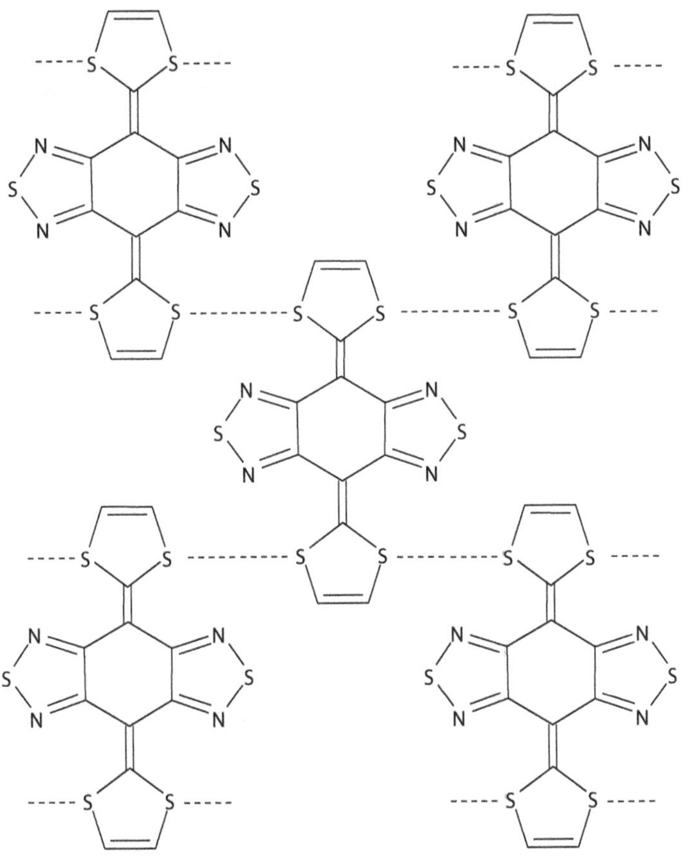

SCHEME 2.85

Additionally, there are data on the formation of stable hydrogen-bonded complexes between 1,2,5-thiadiazole- and aminopyrimidine-containing pharmaceuticals (Giusepetti et al.1994, Caira et al. 2003).

2.3.4 BIOMEDICAL IMPORTANCE

2.3.4.1 Medical Significance

Morpholino-4-(3-*tert*-butylamino-2-hydroxypropoxy)-1,2,5-thiadiazole forms a salt with maleic acid. The salt is active in the treatment of ocular and blood-vessel hypertension, heart pain and attacks (Mulvey and Tull 1976).

4-Amino-*N*-(4-methoxy-1,2,5-thiadiazol-3-yl)benzenesulfonamide exhibits antibacterial action and activity in therapy for chancroid in men (Plummer et al. 1983). Chancroid is a bacterial sexually transmitted disease characterized by painful sores on the genitalia. The thiadiazole compound enhances its activity in the presence of 5-[(3,4,5-trimethoxyphenyl)methyl]pyrimidine-2,4-diamine. Probably, high efficiency of the mix is caused by hydrogen bonding between both components resulting in the formation of an intricate, highly stabilized framework (Giusepetti et al. 1994).

Metabolic transformation may involve just this complex as it is. 5-{[3,5-Dimethoxy-4-(methoxyethoxy)phenyl]methyl}pyrimidine-2,4-diamine also forms the intricate, highly stabilized hydrogen-bonded complex with 4-amino-N-(4-methoxy-1,2,5-thiadiazol-3-yl)benzenesulfonamide (Caira et al. 2003).

3-(4-Hexyloxy-1,2,5-thiadiazol-3-yl)-1-methyl-5,6-dihydro-2H-pyridine is a drug that has recently been recommended to use for treatment of schizophrenia symptoms and for robust improvements in verbal learning as well as in short-term memory (Shekhar et al. 2008).

Some drugs of the piazothiole class were recently patented. Thus, cyclopropylmethyl-{7-(β,β'-dimethylpiazothiol-α-yl)-2,5,6-trimethyl-7H-pyrrolo-[2,3-d]-pyrimidin-4yl}-propylamine was claimed as a pharmaceutical against diseases induced or facilitated by corticotrophin releasing factor (Schoeffter 2004). N-{2-(Benzo-[1,2-c]-[1,2,5]-thiadiazol-7-yl-amino)-6-(2,6-dichlorophenyl)pyrido-[2,3-d]-pyrimidin-7-yl}-N'-(1,1-dimethyl)urea was patented as a drug against leukemia, in particular, against its myeloid version (Bourrie and Casellas 2008). Other piazothiol derivatives have been tested against malaria: 3-carboxy-1,2,5-thiadiazoles with hydroxyl, or 4-methoxy, or 4-amino-group in position 4 (Cameron et al. 2004) as well as a sulfonamide that contains piazothiol, cyanotetrahydropyrimidine and imidazole fragments (Bendale et al. 2007). These compounds were designed as future second-generation drugs. The second generation drugs are needed because of widespread resistance to well-established agents of the first generation.

Nitro derivatives of piazothiol possess insectofungicidial activity (Belen'kaya et al.1970). S,S-Dioxides of 2-aryl-N,N-dihydropiazothiol have advanced into development as pain relievers and as possible drugs to treat vasomotor symptoms (O'Neil et al. 2010). Porphyrin derivatives with four piazothiol chromophores at the meso positions were proposed to be developed for photodynamic therapy initiated by two-photon excitation (Ishi-i et al. 2007).

2.3.4.2   Applications to Environmental and Bioanalysis

Maisonneuve et al. (2008) synthesized a benzo-[1,2-c]-[1,2,5]-thiadizol-6-yl-N-triazolyl β-cyclodextrin as a fluorescent chemosensor that exhibits a high selectivity to nickel(2+), among a series of cations in acetonitrile solution. The sensor forms a stable complex with nickel and its fluorescence is quenched. As an industrial pollutant, nickel is a toxic element causing lung injury, allergy, and carcinogenesis. The method proposed is applicable to monitoring of Ni^{2+} in industrial, environmental, and food samples.

Liu et al. (2008) published a description of a novel colorimetric and fluorimetric chemosensor for biologically dangerous traces of mercury ions in natural waters. The chemosensor is poly(p-phenylene ethylene) that contains α,α'-piazothiol in the backbone. A highly Hg^{2+}-selective fluorescence quenching property in conjunction with a visible colorimetric change from yellow to violet can be observed. So, this polymer can serve as a naked-eye indicator for the presence of mercuric cation in human-available water sources.

Pu and Liu (2008) synthesized a cationic polyfluorene derivative with piazothiol content. The high charge density induced by cationic oligo(ethylene) and oligo(ethylene oxide) side chains containing trimethylammonium end groups results in good water solubility. When negatively charged highly sulfated glucosaminogly-can was added into aqueous polymer solution, it induced polymer aggregation, giving rise to enhanced energy transfer from fluorene segments to the piazothiol units. By increasing the sulfoglycan concentration, the orange piazothiol emission intensity progressively increased at the expense of the blue fluorene emission. In contrast, addition of hyaluronic acid, an analog of the sulfoglycan, enhanced the piazothiol emission insignificantly. This is the basis to distinguish the sulfoglycan from hyaluronic acid. The sulfoglycan can be quantitatively determined because the piazothiol emission intensity linearly increases with the sulfoglycan concentration (Pu and Liu 2008).

Piazothiol derivatives were proposed for use as sensitive fluorophores for selective DNA detection (Neto et al. 2007). One "light-up" probe is based on piazothiol proposed, the quinonoid ring of piazothiol bearing the p-anisyl substituent at α-position and the ethenylphenyl substituent at α'-position. The presence of the acceptor piazothiol fragment and the donor anisyl group makes intramolecular charge transfer possible. Moreover, the insertion of the anisyl group on the molecular architecture of the piazothiol marker increases its thermal and excited-state stability. The planar ethenylphenyl moiety is ideal to be inserted between two consecutive base pairs of DNA. Such a probe shows a significant increase of the fluorescence intensity (hyperchromic effect) and a marked redshift of the long-wavelength emission maximum. The emission does not interfere with absorption of the DNA bases. The piazothiol marker permits quantitative detection of DNA at the highest level of sensitivity.

A mixture of poly[(9,9-dioctylfluorenyl-2,7-diyl)-co-(α,α'-piazothiol)] with poly(styrene-co-maleic anhydride) was proposed for fluorometric determination of Fe^{2+} and Cu^{2+} at their joint presence in drinking water or in cell cultures (Chan et al. 2011).

To perform fluorescent analysis of DNA, Li et al. (2008a) used water-soluble cationic polyelectrolyte poly[9,9'-bis(6''-trimethylammonium)hexylfluorene-alt-α,α'-piazothiol dibromide]. The piazothiol polymer exhibits more stable and readily quantifiable optical signals.

Ishi-i et al. (2008) proposed to use a complex of piazothiol bearing (phenyl-4-formyl) groups at α- and α'-position with two zinc(2+) cations as a fluorescence chemosensor in studies of enzyme activities and inhibition of enzymatic reactions.

1,2,5-Thiadiazole is also patented as an additive to organic solvents to solubilize biomolecules (e.g., nucleic acids, proteins, peptides, amino acids). This helps to separate the biomolecules by electrophoresis or liquid chromatography (Tatsuta and Yamaguchi 2009). Supposedly, 1,2,5-thiadiazole forms van der Waals or charge-transfer complexes with the biomolecules and thereby assists in their solubilization. (The analogous effect is known for inclusion complexes with biomolecules, see Todres 2006, 2009.)

2.3.5 Technical Applications

2.3.5.1 Microelectronics

New electronic materials containing fused 1,2,5-thiadiazole rings differ in the strong intermolecular contacts as well as in high polarizability and in extended π-conjugation that decreases electrostatic repulsion. Compounds of this class were studied as radicals, charge-transfer complexes, radical-ions, and dianions.

Crystalline samples of neutral π-electron radicals containing 1,2,5-thiadiazole and dithiazonyl fragments are able to conduct electric current at room temperature (Barclay et al. 1999). (These radicals are stable. They are generated by reduction of dithiazolium salts with sodium dithionite, zinc–copper couple, or triphenylstibine.) The radicals are of particular interest for applications in magneto-thermal switching and information-storage devices. In crystals, the radicals form π-stacks and exhibit magnetic bistability. Namely, a paramagnetic phase (in non-dimerized π-stacks) coexists over a specific temperature range with a diamagnetic phase (in π-dimer stacks). Importantly, the region of this bistability exists in the range from 230 to 320 K (Brusso et al. 2004). This region encompasses room temperatures.

The tetrathiafulvalene fused with 1,2,5-thiadiazole ring forms charge-transfer 3:2 complex with tetracyanoquinodimethane that shows electric conductivity of 5.6×10^{-2} S/cm in compaction pellets (Yamashita et al. 1992). This conductivity is thought to be high.

The hexafluorophosphate salt of the cation-radical from the just mentioned tetrathiafulvalene fused with 1,2,5-thiadiazole ring shows electric conductivity of 1.1×10^2 S/cm and metallic temperature dependence down to 100 K (Yamashita et al. 1992). Cation-radical salts of naphtho-[2,3-c]-[1,2,5]-thiadiazole-4,9-dione of Scheme 2.73 exhibit high electric conductivity (Yamashita et al. 1994).

Disodium salt of the dianion from the tetracyanoquinodimethane fused with the two 1,2,5-thiadiazole rings (at the both back sides of quinodimethane) also manifests electric conductivity (Yamashita et al. 1987, 1988b).

Bis-α,α'-[5-(5-hexylthiophene-2-yl)thiophene-2-yl]benzo-[1,2-c]-[1,2,5]-thiadiazole is soluble in common solvents and solution-processed devices were prepared by spin coating. The organic thin films exhibit very low band gap, high whole mobility coupled with solution processability and ambient stability (Sonar et al. 2008). These properties make the materials of this type excellent candidates for application in organic electronics, see also Shin et al. (2009).

2.3.5.2 Light Emitters and Solar Cells

(4,8-Dithien-2-yl-2$\lambda^4\delta^2$-benzo-[1,2-c:4,5-c']-bis{[1,2,5]-thiadiazole} (an acceptor) forms a copolymer with N-[3,4,5-tris(dodecyloxy)phenyl]dithieno-[3,2-b:2',3'-d]-pyrrole (a donor). The copolymer shows the very low optical band gap (0.5–0.6 eV). Four differently colored redox states of the polymer can be accessed at moderate potentials and each of the states has good stability. This opens a way to apply the polymer to electrochromic devices (Steckler et al. 2009).

Design of solar cells containing organic light-emitting diodes flows with various derivatives of 1,2,5-thiadiazole (see, e.g., patents claimed by Takada and Amano 2006, Takashima and Funabashi 2008). Such diodes provide the cells with twist response time, high efficiency and color purity, low-voltage driving requirements, simplicity, and cheapness. A desired requirement in all the diodes is the simultaneous supply of electrons and holes to the active light-emitting polymer layer sandwiched between two electrodes. The experimental data (see, e.g., Admassie et al. 2006) show that high photoluminescence quantum efficiency as well as balance of oppositely charged carriers is needed for a good performance of single-layer polymer light-emitting diodes. For realization of full-color, efficient emitting materials of all three primary colors (red, green, blue) are required. Green and blue emitting materials with reasonable color purity are mentioned in reviews by Blouin and Leclerc (2008) and Grimsdale et al. (2009). However, in comparison with green and blue emitting materials, red ones remain a challenge. Benzothiadiazole (piazothiol) as a component of the main chain in the conjugated polymers exhibits red-light emitting properties and is applicable in electroluminescent devices. In such polymers, the noncovalent intramolecular interactions take place between electron donor and electron acceptor moieties that alternate throughout the polymer backbone. This narrows the gap between HOMO and LUMO (Ozen et al. 2007). The energy band gap decreases (see, for example, Thomas et al. 2008). Importantly, a good balance is maintained in electron and hole transport within the polymers.

One of the designed red emitters of high intensity is α-(N,N-diphenylamino)-α'-(2,2-dicyanoethyleno)piazothiol (Ju et al. 2008). In this short π-conjugated asymmetric structure, the piazothiol center is an electron acceptor with respect to the N,N-diphenyl amine donor and the 2,2-dicyanoethylene is the electron withdrawing-transporting group. The thiadiazole acceptor ability can be enforced by incorporation of benzo-[1,2-c:4,5-c']-bis([1,2,5]-thiadiazole) into the polymer chain (Li et al. 2008b) or by using nitro derivatives of piazothiol as a part of the alternating polymer sequence (He et al. 2009). These approaches additionally decrease the band gap, enhance the luminescence efficiency-intensity, and elongate the service life.

Relevant examples are copolymer of thiophenes with piazothiols (He et al. 2009, Lin et al. 2011, Liu et al. 2011, Steinberger et al. 2011), poly[(9,9-dioctylfluorene)-alt-piazothiol] (Donley et al. 2005, van Vooren et al. 2008, Bolognesi et al. 2009), and poly[2,6-(4,4-bis-(2-alkyl)-4H-cyclopenta[2,1-b;3,4-b']dithiophene)-alt-piazothiol] (Zhang et al. 2007, Hwang et al. 2008). Piazothiol bearing (phenyldinapthyl) (hexylthienyl) fragments in α,α'-positions is a pure-red molecular emitter for non-doped diodes (Huang et al. 2008).

A series of naphtho-[2,3-c]-[1,2,5]-thiadiazole derivatives presents both carrier transporting property and high fluorescence quantum yield. Thus, 4,9-bis{(diphenyl-4-yl)naphtho-[2,3-c]-1,2,5]-thiadiazole} and 4,9-bis{(naphtho-2-yl)naphtho-[2,3-c]-[1,2,5]-thiadiazole} manifest ambipolar transporting property with almost identical hole and electron mobilities. Due to the absence of a strong donor group in the molecular structure, intermolecular dipole–dipole interaction and concentration quenching are effectively suppressed (Wei et al. 2008). The high and ambipolar transporting property, good-film forming ability and bulky molecular structure of these naphthothiadiazoles make them high-performance components of non-doped red-emitting diodes.

For solar cells, polymer gel electrolyte was proposed on the basis of cyanoacrylic acid bound with a sequence from thiophene-piazothiol-thiophene-benzene-bis(fluorenylamine). With this gel, more than 6% power conversion efficiency was achieved (Kim et al. 2008).

Experimental results (Sandanayaka et al. 2004, Clarke et al. 2009) as well as quantum chemical calculations (Loboda et al. 2009) show that fullerene derivatives, acting as electron acceptors in ground state, significantly enhance the average static hyperpolarization. Fullerenes are also used as blends with 1,2,5-thiadiazole copolymers. Fullerene C_{60} is most often used. However, there are data on fullerenes

C_{70} (Clarke et al. 2009, Dante et al. 2009) and C_{80} as more efficient constituents (Zoombelt et al. 2009).

Polymers with electroluminescent properties can be prepared by attaching piazo-thiol derivatives to the side or main chain. Thus, attachment of a piazothiol derivative to polyfluorene side chain leads to the formation of a highly efficient red emitting material (Liu et al. 2008b). Especially promising is the main-chain poly[N-9'-alkyl-2,7-carbazole-*alt*-5,5-(α,α'-di-2-thienylpiazothiol)]: It combines high glass transition temperature, good solubility, relatively high molecular weight, and air stability. In the crystalline state, the thiadiazole units display supramolecular interactions that are observable even in oligomers (Anant et al. 2008). Photovoltaic cells based on piazothiol-containing polymers blended with methyl [6,6]-phenylfullerene-C_{60} butyrate (PCBM) exhibit power conversion efficiency as high as 2.2% (Zhou et al. 2008), 3.6% (Blouin et al. 2007, Lu et al. 2008), or even 5.1% (Hou et al. 2008). Power conversion efficiency of 2.4% was also claimed for photodiodes based on PCBM blend with alternating polymer containing 9-(2-ethylhexyl)-9-hexylfluorene, N-hexylpyrrole, and piazothiol (Svensson et al. 2003). Prospective are also PCBM blends with the thiophene copolymers based on benzo-bis-(thiadiazole) (see Scheme 2.84) (Bundgaard and Krebs 2007, Bundgaard et al. 2007, Qian et al. 2008).

What is more (in the sense of "additionally important") effective is also a bulk-heterojunction solar cell based on the intimate blend between 4,7-bis{2-[1-(2-ethylhexyl)-4,5-dicyanoimidazol-2-yl]vinyl}piazothiol as the acceptor material and poly[N-(2'-decyclodecyl)carbazole]-2,7-diyl as the donor counterpart (Ooi et al. 2008). Then, the blend of poly(9,9-dioctylfluorene-*alt*- α,α'-piazothiol) and poly[9,9-dioctyl-fluorene-*alt*-N-(4-butylphenyl)diphenylamine] is characterized with the charge-trans-fer character and strong luminescence emission (Ramon and Bittner 2007).

Cationic polyelectrolyte poly[9,9'-bis(6''-trimethylammoniumhexylfluorene)-*alt*-α,α'-piazothiol dibromide] is a proper material for light-emitting diodes (Liu and Bazan 2004). The bromide ions in this material can be exchanged with bulkier anions. Increasing the counteranion size decreases inter-chain contacts and self-aggregation. As a result, the bulk-emission efficiency substantially grows (Yang et al. 2006, Bazan 2007). The anionic component in conjugated polyelectrolyte thus provides a versatile structural handle to fine-tune properties relevant to optoelectronic applications (a fact that has not been widely recognized). Furthermore, the volume of the anion dominates the hygroscopicity of the material, with bromide being the most moistening absor-bent. The bulky tetra[3,5-di(trifluoromethyl)phenyl] borate retains very little water. The presence of water significantly increases the ionic conductivity of polyelectrolytes and decreases the device efficiency (Ortony et al. 2008). The higher piazothiol content in the backbone, the lower is humidification of the material (Wang and Bazan 2006).

2.3.6 CONCLUSION

1,2,5-Thiadiazoles are compounds of pronounced chemical stability, diverse reactiv-ity, and temperate electron affinity. These properties make them attractive for aca-demia as well as for industry. From the academic point of view, the key sulfur atom is closer to be ambivalent. Sulfur of such a kind does not preclude ring conjugation. The ring as a whole is more or less aromatic. Arylene-annulated 1,2,5-thiadiazoles react in a peculiar manner that differs from usual aromatic compounds. Polyacene-annulated 1,2,5-thiadiazoles combine both acceptor and donor properties in the same molecule. From the practical point of view, 1,2,5-thiadiazoles have undoubtedly biological importance. As pointed out, chemical stability of benzothiadiazole (piazothiol) can vary, but not in metabolic chains (Rosen et al. 2008). This makes piazothiol derivatives useful for therapeutic practice.

1,2,5-Thiadiazoles attract attention as components of solar cell, which are near-future alternative sources of energy. There are various examples of α,α'-disubstituted benzo-[1,2-c]-[1,2,5]-thiadiazole derivatives as efficient fluorophores. Many of them constitute liquid crystals. In particular, liquid crystals containing benzo-[1,2-c]-[1,2,5]-thiadiazole α,α'-connected with ethynyl fragments are very interesting materials exhibiting large birefringence and luminescence properties (Vieira et al. 2008).

REFERENCES TO SECTION 2.3

Admassie, Sh.; Yakob, Z.; Zhang, F; Mammo, W.; Yohannes, T.; Solomon, Th. (2006) *Bull. Chem. Soc. Ethiopia* **20**, 309.

Akhtaruzzaman, M.; Tomura, M.; Nishida, J.-I.; Yamashita, Y. (2004) *J. Org. Chem.* **69**, 2953.

Algieri, A.A.; Luke, G.M.; Standridge, R.T.; Brown, M.; Partyka, R.A.; Grenshaw, R.R. (1982) *J. Med. Chem.* **25**, 210.

Anant, P.; Lucas, N.T.; Jacob, J. (2008) *Org. Lett.* **10**, 5533.

Apblett, A.; Chivers, T.; Richardson, J.F. (1986) *Can. J. Chem.* **64**, 849.

Bacon, C.E.; Eisler, D.J.; Melen, R.L.; Rawson, J.M. (2008) *Chem. Commun.* 4924.

Bagryanskaya, I.Yu.; Gatilov, Yu.V.; Makarov, A.Yu.; Shakirov, M.M.; Shuvaev, K.V.; Zibarev, A.V. (2001) *Russ. J. Gen. Chem.* **71**, 1050.

Bagryanskaya, I.Yu.; Gatilov, Yu.V.; Miller, A.O.; Shakirov, M.M.; Zibarev, A.V. (1994) *Heteroatom. Chem.* **5**, 561.

Barclay, T.M.; Cordes, A.W.; Haddon, R.C.; Itkis, M.E.; Oakley, R.T.; Reed, R.W.; Zhang, H. (1999) *J. Am. Chem. Soc.* **121**, 969.

Bazan, G.C. (2007) *J. Org. Chem.* **72**, 8615.

Bel'skii, V.K.; Ellert, D.G.; Seifulina, Z.M.; Novotortsev, V.M.; Tsveniashvili, V.Sh.; Garnovskii, A.D. (1984) *Izv. AN SSSR Ser. Khim.* 1914.

Belen'kaya, I.A.; Chizhov, N.P.; Chigareva, N.G. (1970) *Khim.-Farm. Zh.* **5**(12), 66.

Bendale, P.; Olepu, S.; Suryadevara, K.; Bulbule, V.; Rivas, K.; Nallan, L.; Smart, B.; Yokoyama, K.; Ankala, S.; Pendyala, P.R.; Floyd, D.; Lombardo, L.J.; Williams, D.K.; Buckner, F.S.; Chakrabarti, D.; Verlinde, Ch.L.M.J.; van Voorhis, W.C.; Gelb, M.H. (2007) *J. Med. Chem.* **50**, 4585.

Bertini, V.; de Munno, A. (1967) *Gazz. Chim. Ital.* **97**, 1614.

Bertini, V.; Lucchesini, F.; de Munno, A. (1982) *Synthesis* 681.

Bird, C.W. (1985) *Tetrahedron* **41**, 1409.

Blouin, N.; Leclerc, M. (2008) *Acc. Chem. Res.* **41**, 1110.

Blouin, N.; Michaud, A.; Leclerc, M. (2007) *Adv. Mater.* **19**, 2295.

Bock, H.; Haenel, P.; Niedlein, R. (1988) *Phosphorus Sulfur* **39**, 235.

Bolognesi, A.; Betti, P.; Botta, C.; Destri, S.; Giovannela, U.; Moreau, J.; Pasini, M.; Porzio, W. (2009) *Macromolecules* **42**, 1107.

Borghese, A.; Mancuso, V.; Merschaert, A. (2006) WO Patent 068,821.

Boudebous, A.; Constable, E.C.; Housecroft, C.E.; Neuburger, M.; Schaffner, S. (2008) *Aust. J. Chem.* **61**, 755.

Bourrie, B.; Casellas, P. (2008) Fr. Patent 2,910,813.

Breier, L.; Petrovskii, P.V.; Todres, Z.V.; Fedin, E.I. (1969) *Khim. Geterotsikl. Soed.* 62.

Brusso, J.L.; Clements, O.P.; Haddon, R.C.; Itkis, M.E.; Leitch, A.A.; Oakley, R.T.; Reed, R.W.; Richardson, J.F. (2004) *J. Am. Chem. Soc.* **126**, 8256.

Bryce, M.R.; Dransfield, T.A.; Kandeel, K.A.; Vernon, J.M. (1988) *J. Chem. Soc. Perkin Trans. 1*, 2141.

Buchwald, H.; Ruehlmann, K. (1979) *J. Organomet. Chem.* **166**, 25.

Bundgaard, E.; Krebs, F.C. (2007) *Solar Energy Mater. Solar Cells* **91**, 1019.

Bundgaard, E.; Shaheen, S.E.; Krebs, F.C.; Ginley, D.S. (2007) *Solar Energy Mater. Solar Cells* **91**, 1631.

Burmester, Ch.; Faust, R. (2008) *Synthesis* 1179.

Cai, X.; Zhang, Y.; Zhang, X.; Jiang, J. (2007) *TEOCHEM* **812**, 63.

Caira, M.R.; Bettinetti, G.; Sorrenti, M.; Cattenacci, L. (2003) *J. Pharm. Sci.* **92**, 2164.

Cameron, A.; Read, J.; Tranter, R.; Winter, V.J.; Sessions, R.B.; Brady, R.L.; Vivas, L.; Easton, A.; Kendrik, H.; Croft, S.L.; Barros, D.; Lavandera, J.L.; Martin, J.J.; Risco, F.; Garcia-Ochoa, S.; Gamo, F.J.; Sanz, L.; Leon, L.; Ruiz, J.R.; Gabarro, R.; Mallo, A.; Gomes de las Heras, F. (2004) *J. Biol. Chem.* **279**, 31429.

Cantrell, T.S.; Haller, W.S. (1968) *J. Chem. Soc. Chem. Commun.* 977.

Carmack, M.; Shew, D.; Weinstock, L.M. (1961) US Patent 2,980,687.

Cava, M.P.; Schlessinger, R.H. (1964) *Tetrahedron Lett.* **5**, 3815.

Chan, Y.H.; Jin, Y.; Wu, Ch.; Chiu, D.T. (2011) *Chem. Commun.* **47**, 2820.

Chang, Ya.-L.; Palacios, R.E.; Fan, F.-R.F.; Bard, A.J.; Barbara, P.F. (2008) *J. Am. Chem. Soc.* **130**(28), 8906.

Cillo, C.M.; Lash, T.D. (2004) *J. Heterocycl. Chem.* **41**, 955.

Ciminale, F. (2004) *Tetrahedron Lett.* **45**, 5849.

Clarke, T.; Ballantyne, A.; Jamieson, F.; Brabec, Ch.; Nelson, J.; Durrant, J. (2009) *Chem. Commun.* 89.

Colyer, D.E.; Nortcliffe, A.; Wheeler, S. (2010) *Tetrahedron Lett.* **51**, 5306.

Cookson, R.F.; Richards, A.C. (1974) *J. Chem. Soc. Chem. Commun.* 585.

Dante, M.; Garcia, A.; Nguen, Th.-Q. (2009) *J. Phys. Chem. C* **113**, 1596.

Davidenko, T.I.; Romanovskaya, I.I. (1989) *Khim.-Farm. Zh.* **23**(4), 473.

de Manno, A.; Bertini, V.; Picci, N. (1986) *Heterocycles* **24**, 1131.

Deicha, C.; Terrier, F. (1981) *J. Chem. Res.* 312.

Donley, C.L.; Zaumseil, J.; Andreasen, J.W.; Nielsen, M.M.; Sirringhaus, H.; Friend, R.H.; Kim, J.-S. (2005) *J. Am. Chem. Soc.* **127**, 12890.

Donzello, M.P.; Ercolani, C.; Gaberkorn, A.A.; Kudrik, E.V.; Meneghetti, M.; Marcolongo, G.; Rizzoli, C.; Stuzhin, P.A. (2003) *Chem. Eur. J.* **9**, 4009.

Donzello, M.-P.; Ercolani, C.; Kadish, K.M.; Ricardi, G.; Rosa, A.; Stuzhin, P.A. (2007) *Inorg. Chem.* **46**, 4145.

Duan, X.-G.; Duan, X.-L.; Rees, Ch.W.; Yue, T.-Y. (1997) *J. Chem. Soc. Perkin Trans. 1*, 2597.

Dunn, P.J.; Rees, C.W. (1989) *J. Chem. Soc. Perkin Trans. 1*, 2485.

Efros, L.S.; Levit, R.M. (1953) *Zh. Obshch. Khim.* **23**, 1552.

Fedin, E.I.; Todres, Z.V.; Efros, L.S. (1967) *Khim. Geterotsikl. Soed.* 297.

Gaberkorn, A.A.; Donzello, M.-P.; Stuzhin, P.A. (2006) *Zh. Org. Khim.* **42**, 946.

Gait, S.F.; Rance, M.J.; Rees, C.W.; Stephenson, R.W.; Storr, R.C. (1975) *J. Chem. Soc. Perkin Trans. 1*, 556.

Geisel, M.; Mews, R. (1982) *Chem. Ber.* **115**, 2135.

Gieren, A.; Lamm, V.; Haddon, R.C.; Kaplan, M.L. (1980) *J. Am. Chem. Soc.* **102**, 5070.

Giusepetti, G.; Tadini, C.; Bettinetti, G.P. (1994) *Acta Crystallogr.* **C50**, 1289.

Glossman-Mitnik, D. (2001) *TEOCHEM* **549**, 285.

Grimsdale, A.C.; Chan, K.L.; Martin, R.E.; Jokisz, P.G.; Holmes, A.B. (2009) *Chem. Rev.* **109**, 897.

Gul'maliev, A.M.; Stankevich, I.V.; Todres, Z.V. (1973) *Khim. Geterotsikl. Soed.* 1473.

Gul'maliev, A.M.; Stankevich, I.V.; Todres, Z.V. (1975) *Khim. Geterotsikl. Soed.* 1055.

Hatta, T.; Mataka, S.; Tashiro, M.; Numano, K.; Suzuki, H.; Torii, A. (1991) *J. Heterocycl. Chem.* **28**, 55.

He, Y.; Wang, X.; Zhang, J.; Li, Y. (2009) *Macromol. Rapid Commun.* **30**, 45.

Herberhold, M.; Hill, A.F. (1989) *J. Organometal. Chem.* **377**, 151.

Hou, J.; Chen, H.-Y.; Zhang, Sh.; Li, G.; Yang, Y. (2008) *J. Am. Chem. Soc.* **130**, 16144.

Huang, J.; Kertesz, M. (2005) *J. Phys. Chem. B* **109**,12891.

Huang, J.; Quiao, X.; Xia, Y.; Zhu, X.; Ma, D.; Cao, Y.; Roncali, J. (2008) *Adv. Mater.* **20**,4172.

Huebner, T.; Lamm, V.; Neidlein, R.; Droste, D. (1984) *Z. Naturforsch. B* **39**, 485.

Hwang, I.-W.; Cho, Sh.; Kim, J.Y.; Lee, K.; Coates, N.E.; Moses, D.; Heeger, A.J. (2008) *J. Appl. Phys.* **104**, 033706.

Ishi-I, T.; Nakamura, N.; Esaki, N.; Amemori, Sh. (2008) *Chem. Lett.* **37**, 1166.

Ishi-I, T.; Taguri, Y.; Kato, Sh.-i.; Shigeiwa, M.; Gohohmaru, H.; Maeda, Sh.; Mataka, Sh. (2007) *J. Mater. Chem.* **17**, 3341.

Ju, J.U.; Jung, S.O.; Zhao, Q.H.; Kim, Y.H.; Je, J.T.; Kwon, S.K. (2008) *Bull. Korean Chem. Soc.* **29**, 335.

Jung, M.H.; Park, J.G.; Park, W.K. (2003) *Arch. Pharm. Pharm Med. Chem.* **336**, 230.

Kaim, W.; Kohlmann, S. (1985) *Inorg. Chim. Acta* **101**, L21.

Kim, J.-J.; Choi, H.; Lee, J.-W.; Kang, M.-S.; Song, K.; Kang, S.O.; Ko, J. (2008) *J. Mater. Chem.* **18**, 5223.

Kim, K.J.; Kim, K. (2007) *Tetrahedron* **63**, 5014.

Komin, A.P.; Carmack, M. (1976) *J. Heterocyclic. Chem.* **13**, 13.

Konchenko, S.N.; Gritsan, N.P.; Lonchakov, A.V.; Radius, U.; Zibarev, A.V. (2009) *Mendeleev Commun.* **19**, 7.

Kouvetakis, J.; Grotjahn, D.; Becker, P.; Moore, S.; Dupon, R. (1994) *Chem. Mater.* **6**, 636.

Kudrik, E.V.; Bauer, E.M.; Ercolani, C.; Chiesi-Villa, A.; Rizzoli, C.; Gaberkorn, A.; Stuzhin, P.A. (2001) *Mendeleev Commun.* 45.

Kurita, Y.; Takayama, C. (1997) *J. Phys. Chem. A* **101**, 5593.

Kursanov, D.N.; Todres, Z.V. (1967) *Dokl. AN SSSR* **172**, 1086.

Kuyper, J.; Vrieze, K. (1975) *J. Organometal. Chem.* **86**, 127.

Kwan, C.L.; Carmack, M.; Kochi, J.K. (1976) *J. Phys. Chem.* **80**, 1786.

Li, H.; Yang, R.; Bazan, G.C. (2008a) *Macromolecules* **41**, 1531.

Li, J.-R.; Yu, Q.; Tao, Y.; Bu, X.-H.; Ribas, J.; Batten, S.R. (2007) *Chem. Commun.* 2290.

Li, X.; Liu, A.; Xun, Sh.; Qiao, W.; Wun, X.; Wang, Zh.Yu. (2008b) *Org. Lett.* **10**, 3785.

Liao, L.; Dai, L.; Smith, A.; Durstock, M.; Lu, J.; Ding, J.; Tao, Y. (2007) *Macromolecules* **40**, 9406.

Lim, E.; Lee, S.; Lee, K.K. (2011) *Chem. Commun.* **47**, 914.

Liu, B.; Bazan, G.C. (2004) *J. Am. Chem. Soc.* **126**, 1942.

Liu, J.; Chen, L.; Shao, Sh.; Xie, Zh.; Cheng, Y.; Cheng, Y.; Wang, L.; Jing, X.; Wang, F. (2008b) *J. Water Chem.* **18**, 319.

Liu, Q.; Jiang, R.; Guan, B.; Tang, Zh.; Pei, J.; Song, Y. (2011) *Chem. Commun.* **47**, 740.

Liu, Sh.-J.; Fang, Ch.; Zhao, Q.; Fan, Q.-L.; Huang, W. (2008a) *Macromol. Rapid Commun.* **29**, 1212.

Liu, Yu.; Prashad, M.; Repic, O.; Blacklock, Th.J. (2003) *J. Heterocycl. Chem.* **40**, 713.

Loboda, O.; Zalesny, R.; Avramopoulos, A.; Luis, J.-M.; Kirtman, B.; Tagmatarchis, N.; Reis, H.; Papadopoulus, M.G. (2009) *J. Phys. Chem. A* **113**, 1159.

Lork, E.; Mews, R.; Shakirov, M.M.; Watson, P.G.; Zibarev, A.V. (2002) *J. Fluorine Chem.* **115**, 165.

Lu, J.; Liang, F.; Drolet, N.; Ding, J.; Tao, Y.; Movileanu, R. (2008) *Chem. Commun.* 5315.

Luzzati, V. (1951) *Acta Crystallogr.* **4**, 193.

Maisonneuve, S.; Fang, Q.; Xie, J. (2008) *Tetrahedron* **64**, 8716.

Makarov, A.Yu.; Irtegova, I.G.; Vasilieva, N.V.; Bagryanskaya, I.Yu.; Borrmann, T.; Gatilov, Yu. V.; Lork, E.; Mews, R.; Stohrer, W.-D.; Zibarev, A.V. (2005) *Inorg. Chem.* **44**, 7194.

Masuda, K.; Adachi, J.; Nomura, K. (1981) *J. Chem. Soc. Perkin Trans. 1*, 1033.

Meij, R.; Kaandorp, T.A.M.; Stufkens, D.J.; Vrieze, K. (1977) *J. Organometal. Chem.* **128**, 203.

Menzl, K. (1964) Germ. Patent 1,175,683.

Merschaert, A.; Boquel, P.; Gorissen, H.; van Hoeck, J.-P.; Borghese, A.; Antoine, L.; Mancuso, V.; Mocker, A.; Vanmarsenille, M. (2006) *Tetrahedron Lett.* **47**, 8285.

Miao, Sh.; Schleyer, P.v.R.; Wu, J.I.; Hardcastle, K.I.; Bunz, U.H.F. (2007) *Org. Lett.* **9**, 1073.

Mirifico, M.V.; Caram, J.A.; Vasini, E.J.; Piro, O.E.; Castellano, E.E. (2008) *J. Phys. Org. Chem.* **22**, 163.

Miyasaka, H.; Morita, T.; Yamashita, M. (2011) *Chem. Commun.* **47**, 271.

Miyoshi, Ya.; Kubo, M.; Fujinawa, T.; Suzuki, Y.; Yoshikawa, H. (2007) *Angew. Chem. Intl. Ed.* **46**, 5532.

Momany, F.A.; Bonham, R.A. (1964) *J. Am. Chem. Soc.* **86**, 162.

Mulvey, D.M.; Tull, R. (1976) *J. Org. Chem.* **41**, 3121.

Mulvey, D.M.; Weinstock, L.M. (1967) *J. Heterocycl. Chem.* **4**, 445.

Naito, T.; Okumura, J.; Kasai, K.-I. (1968) *Chem. Pharm. Bull.* **16**, 544.

Nastapova, N.V.; Yanilkin, V.V.; Morozov, V.I.; Eliseenkova, R.M.; Bredikhina, Z.A.; Bredikhin, A.A.; Buzykin, B.I. (2002) *Proceedings—Electrochem. Soc. (Org. Electrochem.)*, p. 154.

Neto, B.A.D.; Lapis, A.A.M.; Mancilha, F.S.; Vasconcelos, I.B.; Thum, C.; Basso, L.A.; Santos, D.S.; Dupont, J. (2007) *Org. Lett.* **9**, 4001.

O'Neil, D.J.; Adeldoyin, A.; Alfinito, P.D.; Bray, J.A.; Cosmi, S.; Deecher, D.C.; Fensome, A.; Harrison, J.; Leventhal, L.; Mann, Ch.; McComas, C.; Sullivan, N.L.; Spangler, T.B.; Uveges, A.J.; Trybulski, E.J.; Whiteside, G.T.; Zhang, P. (2010) *J. Med. Chem.* **53**, 4511.

Ooi, Z.E.; Tam, T.L.; Shin, R.Y.Ch.; Chen, Zh. K.; Kietzke, Th.; Sellinger, A.; Baumgarten, M.; Muellen, K.; de Mello, J.C. (2008) *J. Mater. Chem.* **18**, 4619.

Ortony, J.H.; Yang, R.; Brzezinski, Ja.Z.; Edman, L.; Nguen, Th.-Q.; Bazan, G.C. (2008) *Adv. Mater.* **20**, 298.

Ozen, A.S.; Atilgan, C.; Sonmez, G. (2007) *J. Phys. Chem. C*, **111**, 16362.

Palacios, R.E.; Fan, F.-R.F.; Bard, A.J.; Barbara, P.F. (2006) *J. Am. Chem. Soc.* **128**, 9028.

Palmer, M.H.; Kennedy, S.M.F. (1978) *J. Mol. Struct.* **43**, 33.

Pesin, V.G.; Belen'kaya, I.A. (1967) *Khim. Geterotsikl. Soed.* 289.

Pesin, V.G.; Muravnik, R.S. (1965) *Latvias PSR Zinatnu Akad. Vest. Kim. Ser.* 233.

Pesin, V.G.; Sergeev, V.A. (1968) *Khim. Geterotsikl. Soed.* 1001.

Pesin, V.G.; Sergeev, V.A.; Khaletskii, A.M. (1964a) *Zh. Obshch. Khim.* **34**, 3753.

Pesin, V.G.; Khaletski, A.M.; Sergeev, V.A. (1962) *Zh. Obshch. Khim.* **32**, 181.

Pesin, V.G.; Khaletski, A.M.; Sergeev, V.A. (1964b) *Zh. Obshch. Khim.* **34**, 261.

Philipp, D.M.; Muller, R.; Goddard, W.A.; Abboud, Kh. A.; Mullins, M.J.; Snelgrove, R.V.; Athey, Ph.S. (2004) *Tetrahedron Lett.* **45**, 5441.

Pilgram, K. (1970) *J. Org. Chem.* **35**, 1165.

Pilgram, K. (1974) *J. Heterocycl. Chem.* **11**, 835.

Plummer, F.A.; Nsanze, H.; D'Costa, L.J.; Karasira, P.; Maclean, I.W. (1983) *New Engl. J. Med.* **309**, 67.

Pu, K.-Y.; Liu, B. (2008) *Macromolecules* **41**, 6636.

Qian, G.; Dai, B.; Luo, M.; Yu, D.; Zhan, J.; Zhang, Zh.; Ma, D.; Wang, Zh.Yu. (2008) *Chem. Mater.* **20**, 6208.

Ramon, J.G.S.; Bittner, E.R. (2007) *J. Chem. Phys.* **126**, 181101.

Rees, C.W.; Yue, T.-Y. (2001) *J. Chem. Soc. Perkin Trans. 1*, 662.

Rosen, M.D.; Hack, M.D.; Allison, B.D.; Phuong, V.K.; Woods, C.R.; Morton, M.F.; Prendergast, C.E.; Barrett, T.D.; Schubert, C.; Li, L.; Wu, X.; Wu, J; Freedman, J.M.; Shankley, N.P.; Rabinovitz, M.H. (2008) *Bioorg. Med. Chem.* **16**, 3917.

Rosen, M.D.; Simon, Z.M.; Tarantino, K.T.; Zhao, L.X.; Rabinovitz, M.H. (2009) *Tetrahedron Lett.* **50**, 1219.

Ross, J.M.; Smith, W.C. (1964) *J. Am. Chem. Soc.* **86**, 2861.

Salmond, W.G. (1968) *Quart. Rev.* **22**, 253.

Sandanayaka, A.S.D.; Matsukawa, K.; Ishi-I, T.; Mataka, Sh.; Araki, Ya.; Ito, O. (2004) *J. Phys. Chem. B* **108**, 19995.

Schoeffter, Ph. (2004) WO Patent 069,257.

Shekhar, A.; Potter, W.Z.; Lightfoot, J.; Lienemann, J.; Dube, S.; Mallincrodt, F.P.B.; McKinzie, D.L.; Felder, Ch.C. (2008) *Am. J. Psychiatry* **165**, 1033.

Shi, Sh.; Katz, Th. J.; Yang, B.Y.; Liu, L. (1995) *J. Org. Chem.* **60**, 1285.

Shin, R.Y.C.; Sonar, P.; Siew, P.S.; Chen, Zh.-K.; Sellinger, A. (2009) *J. Org. Chem.* **74**, 3293.

Solodovnikov, S.P.; Todres, Z.V. (1967) *Khim. Geterotsikl. Soed.* 811.

Solodovnikov, S.P.; Todres, Z.V. (1968) *Khim. Geterotsikl. Soed.* 360.

Sonar, P.; Singh, S.P.; Sandhakar, S.; Dodabalapur, A.; Sellinger, A. (2008) *Chem. Mater.* **20**, 3184.

Steckler, T.T.; Zhang, X.; Hwang, J.; Honeyager, R.; Ohira, Sh.; Znang, X.-H.; Grant, A.; Ellinger, S.; Odom, S.A.; Sweat, D.; Tanner, D.B.; Rinzler, A.G.; Barlow, S.; Bredas, J.-L.; Kippelen, B.; Marder, S.; Reynolds, J.R. (2009) *J. Am. Chem. Soc.* **131**, 2824.

Steinberger, S.; Mishra, A.; Reinold, E.; Leuichkov, F.; Uhrich, Ch.; Pfeifer, M.; Baeuerle, P. (2011) *Chem. Commun.* **47**, 1982.

Stiefvater, O.L. (1978) *Z. Naturforsch. A* **33**, 1518.

Strom, E.T.; Russell, G.A. (1965) *J. Am. Chem. Soc.* **87**, 3326.

Stuzhin, P.A.; Pimkov, I.V.; Ul-Khaq, A.; Ivanova, S.S.; Popkova, I.A.; Volkovich, D.I.; Kuz'mitskii, V.A.; Donzello, M.-P. (2007) *Russ. J. Org. Chem.* **43**, 1854.

Suzuki, T.; Fukushima, T.; Yamashita, Y.; Miyashi, T. (1994) *J. Am. Chem. Soc.* **116**, 2793.

Suzuki, T.; Tsuji, T.; Okubo, T.; Okada, T.; Obana, Y.; Fukushima, T.; Miyashi, T., Yamashita, Y. (2001) *J. Org. Chem.* **66**, 8954.

Svensson, M.; Zhang, F.; Inganas, O.; Andersson, M.R. (2003) *Synth. Met.* **135–136**, 137.

Takada, Y.; Amano, S. (2006) Jpn. Patent 045,143.

Takashima, Y.; Funabashi, M. (2008) Jpn. Patent 162,921.

Tatsuta, K.; Yamaguchi, Y. (2009) WO Patent 028,692.

Thomas, K.R.J.; Huang, T.-H.; Lin, J.T.; Pu, Sh.-Ch.; Cheng, Y.M.; Hsieh, Ch.-Ch., Tai, Ch.P. (2008) *Chem. Eur. J.* **14**, 11231.

Todres, Z.V. (2006) *Organic Mechanochemistry and Its Practical Applications*, CRC Press/Taylor & Francis, Boca Raton, FL.

Todres, Z.V. (2009) *Ion-Radical Organic Chemistry. Principles and Applications*, CRC Press/Taylor & Francis, Boca Raton, FL.

Todres, Z.V.; Lyakhovetskii, Yu.I.; Kursanov, D.N. (1969) *Izv. AN SSSR Ser. Khim.* 1455.

Todres, Z.V.; Tsveniashvili, V. Sh.; Zhdanov, S.I.; Kursanov, D.N. (1968) *Dokl. AN SSSR* **181**, 906.

Ul-Haq, A.; Donzello, M.-P.; Stuzhin, P.A. (2007) *Medeleev Commun.* **17**, 337.

van Vooren, A.; Kim, J.-S.; Cornil, J. (2008) *ChemPhysChem* **9**, 989.

Vieira, A.A.; Cristiano, R.; Bortoluzzi, A.J.; Gallardo, H. (2008) *J. Mol. Struct.* **875**, 364.

Wang, F.; Bazan, G.C. (2006) *J. Am. Chem. Soc.* **126**, 15786.

Watanabe, M.; Goto, K.; Shibahara, M.; Shinmyozu, T. (2010) *J. Org. Chem.* **75**, 6104.

Wei, P.; Duan, L.; Zhang, D.; Qiao, J.; Wang, L.; Wang, R.; Dong, G.; Qui, Y. (2008) *J. Mater. Chem.* **18**, 806.

Weinstock, L.M.; Davies, P.; Handelsman, B.; Tull, R. (1967) *J. Org. Chem.* **32**, 2823.

Weinstock, L.M.; Mulvey, D.M.; Tull, R. (1976) *J. Org. Chem.* **41**, 3121.

Wucherpfenning, W. (1968) *Chem. Ber.* **101**, 371.

Yamashita, Y.; Hagiya, K.; Saito, G.; Mukai, T. (1988b) *Bull. Chem. Soc. Jpn.* **61**, 271.

Yamashita, Y.; Ono, K.; Tanaka, Sh.; Imaeda, K.; Inokuchi, H. (1994) *Adv. Mater.* **6**, 295.

Yamashita, Y.; Ono, K.; Tomura, M.; Tanaka, Sh. (1997) *Tetrahedron* **53**, 10169.

Yamashita, Y.; Sato, K.; Suzuki, T.; Kabuto, C.; Mukai, T.; Miyashi, T. (1988a) *Angew. Chem. Intl. Ed.* **27**, 434.

Yamashita, Y.; Suzuki, T.; Mukai, T. (1987) *J. Chem. Soc. Chem. Commun.* 1184.

Yamashita, Y.; Tanaka, S.; Imaeda, K.; Inokuchi, H.; Sano, M. (1992) *Chem. Lett.* 419.

Yang, R.; Garcia, A.; Korystov, D.; Mikhailovsky, A.; Bazan, G.C.; Nguen, Th.-Q. (2006) *J. Am. Chem. Soc.* **128**, 16532.

Yavolovsky, A.A.; Ivanov, E.I.; Ivanova, R.Yu. (2000) *Khim. Geterotsikl. Soed.* 1571.

Zhang, M.; Tsao, H.N.; Pisula, W.; Yang, Ch.; Mishra, A.K.; Muellen, K. (2007) *J. Am. Chem. Soc.* **129**, 3472.

Zhou, E.; Nakamura, M.; Nishizawa, T.; Zhang, Yu.; Wei, Q.; Tajima, K.; Yang, Ch.; Hashimoto, K. (2008) *Macromolecules* **41**, 8302.

Zhou, H. (2007) *Wuji Huaxue Xuebao* **23**, 778.

Zoombelt, A.P.; Fonrodona, M.; Wienk, M.M.; Sieval, A.B.; Hummelen, J.C.; Janssen, R.A. (2009) *Org. Lett.* **11**, 903.

2.4 1,3,4-THIADIAZOLES

2.4.1 FORMATION

2.4.1.1 From Thiosemicarbazides

Heating (80°C–90°C) thiosemicarbazides with fatty acids in concentrated sulfuric acid affords 2-amino-5-alkyl-1,3,4-thiadiazoles in good yields (Funatsukuri and Veda 1996, Jatav et al. 2006).

The formation of 1,3,4-thiadiazoles can be achieved via reaction of a thiosemicarbazide with phosphorus oxychloride (Raslan and Khalil 2003) or, as in Scheme 2.86, with diethylchlorophosphate and dimethylformamide (Yarovenko et al. 2004). Cyclization of the starting material can also be performed by cold sulfuric acid (Rostom et al. 2003) or acyl chlorides (Thomasco et al. 2003).

Scheme 2.87 represents one-pot procedure to obtain 2-amino-5-(E)-styryl-1,3,4-thiadiazole (Song and Tan 2008). Phosphorus oxychloride was used to perform dehydration-cyclization of thiosemicarbazide and trans-cynnamic acid. The product (75% yield) served as a starting material in preparation of bioactive aroylurea derivatives containing 1,3,4-thiadiazole.

Reaction of thiosemicarbazides with dinitriles is another way to 1,3,4-thiadiazoles, see, for example, Sancak et al. (2007). Chauviere et al. (2003) reported a large-scale applicable method based on thiosemicarbazide and 1-methyl-2-nitroimidazole-5-carbonitrile (Scheme 2.88).

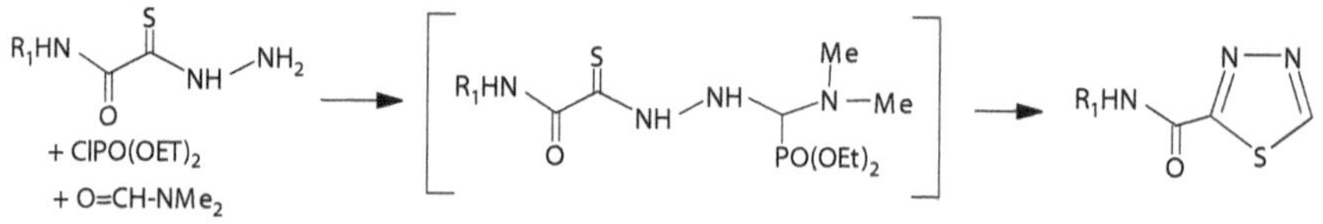

SCHEME 2.86

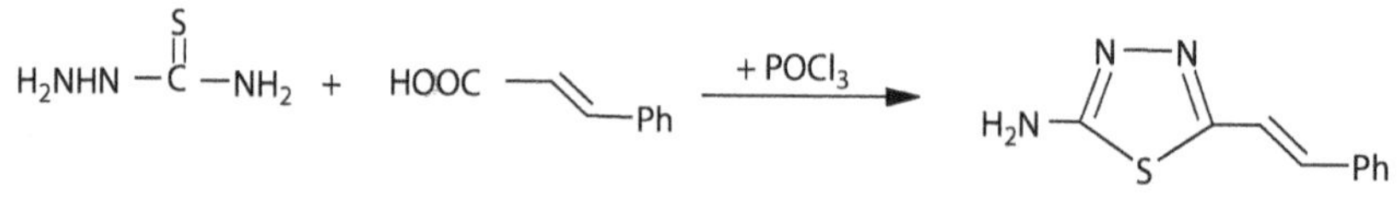

SCHEME 2.87

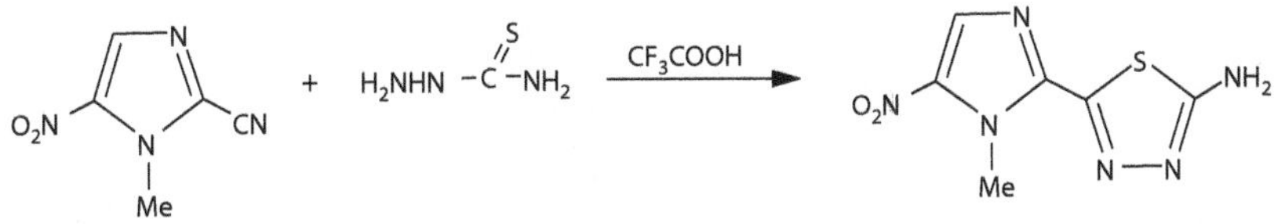

SCHEME 2.88

Condensation of thiosemicarbazide with enones leads to thiosemicarbazones. Being heated in the pyridine–acetic anhydride mixture, the thiosemicarbazones transform into 1,3,4-thiadiazolines, examples can be found in the work by Anis'kov et al. (2009).

When thiosemicarbazide and 1,1-cyclopropane dicarboxylic acid are treated with phosphorus oxychloride, 1,1-bis(2-amino-1,3,4-thiadiazol-5-yl)cyclopropane is formed in good yield (Scheme 2.89) (Sharba et al. 2005a).

In Scheme 2.89, the carbon atom that is needed to build the five-membered cycle, originates from the carboxylic function. However, the literature gives an example when locking the cycle proceeds with preservation of the carboxylic group (Scheme 2.90) (Abdel-Rahman 2006).

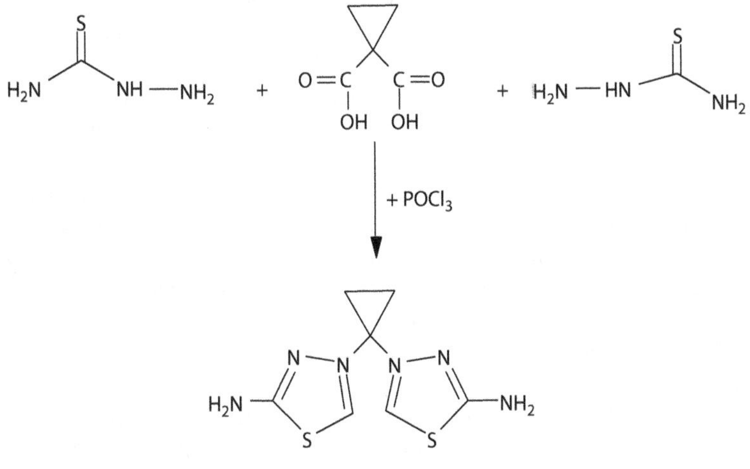

SCHEME 2.89

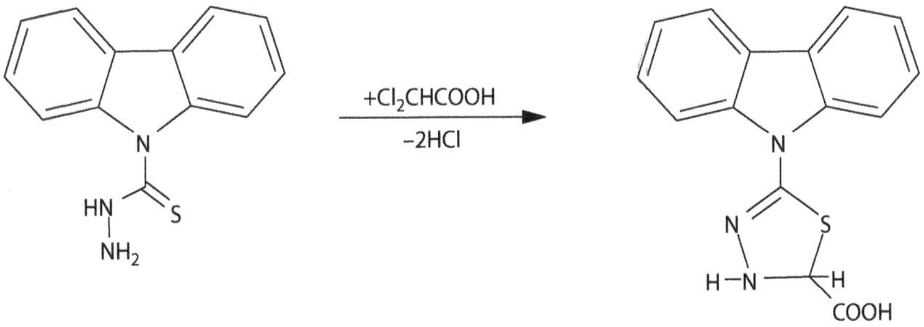

SCHEME 2.90

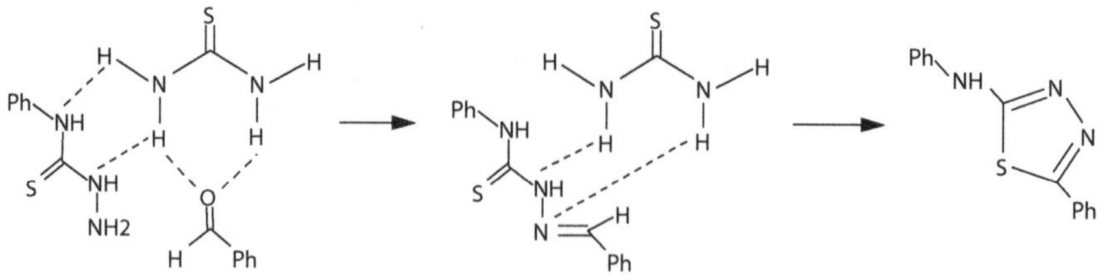

SCHEME 2.91

A microwave variant of the thiosemicarbazide reaction with aryl and alkyl carboxylic acids (in POCl₃) has also been reported (Yuye 2007). Phosphorus oxychloride acts as a solvent, a dehydrating agent and, perhaps, as a reactant to transform the
carboxylic acid into the corresponding acylchloride (cf., Ferreira da Silva et al. 2008).

Rostamizadeh et al. (2010) described an efficient synthesis of 1,3,4-thiadiazoles
from thiosemicarbazides and aldehydes in acetonitrile solution using thiourea as
organic catalyst. Thiourea gathers the reactants in a hydrogen-bond complex. Within
the complex, transformation of a thiosemicarbazide and aldehyde into a semicarbazone takes place. Transition of the thiosemicarbazone into the final thiadiazole is
also assisted by hydrogen bonding with thiourea (Scheme 2.91).

Formation of 1,3,4-oxadiazoles from thiosemicarbazides can be performed in
solutions, using various oxidants such as ferric chloride (Jung et al. 2004, Jatav et al.
2006, Sharma et al. 2008a), ferric sulfate (Chauviere et al. 2003), ferric ammonium
sulfate (Foroumadi et al. 2003, Mohammadhosseini et al. 2008). Concentrated sulfuric acid is also able to cyclize a thiosemicarbazide (Pintilie et al. 2007). The best
results were obtained when the reaction was performed at 0°C–5°C (Saha et al. 2010).
In cases of persistence, polyphosphoric acid is recommended: It plays the triple role
as a solvent, as a catalyst and as a water scavenger (Mashelkar and Audi 2006).

Solid-phase synthesis (oxidation of thiosemicarbazide derivatives on an amide
resin with ferric chloride) was reported by Kilburn et al. (2003) and Kappel et al.
(2004). The thiadiazole is formed on the resin and its elimination proceeds upon
treatment with trifluoroacetic acid.

Oxidation of 1,1-dimethyl-4-phenyl-thiosemicarbazide by cupric chloride in
methanol leads to two products. One of them is a salt containing [CuCl₄]²⁻ anion and
two positively charged 3-methyl-5-(N-phenylamino)-1,3,4-thiadiazolium cations.
Another product is 5-(2,2-dimethylhydrazino)-4-phenyl-3-(phenylimino)-2-isopro-
pyl-1,2,4-thiadiazole. The formation of the both products is depicted in Scheme 2.92
(Lopez-Torres et al. 2007). The authors suppose that the key intermediate in this

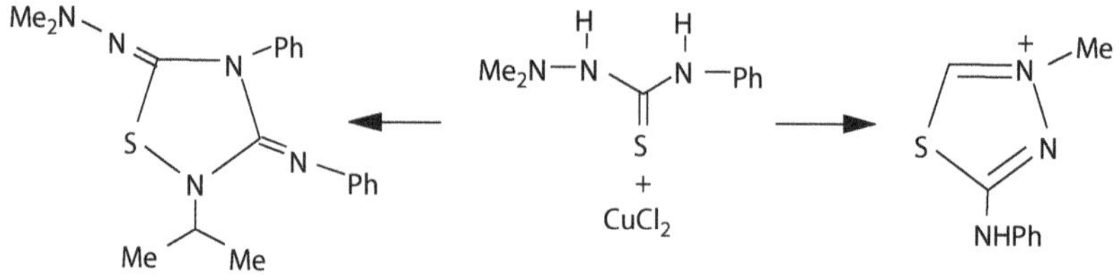

SCHEME 2.92

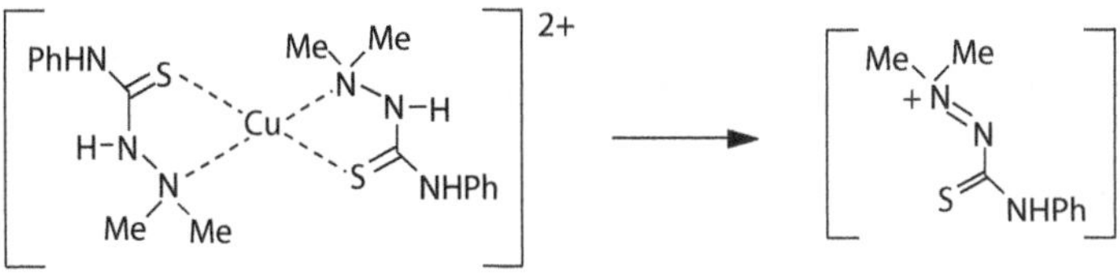

SCHEME 2.93

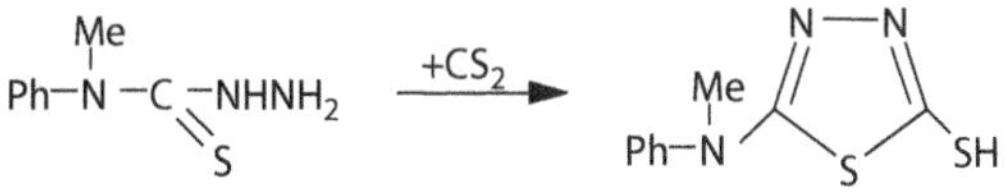

SCHEME 2.94

reaction is the diazenium cation formed from the copper complex of the starting 1,1-dimethyl-4-phenyl-thiosemicarbazide (Scheme 2.92).

The diazenium cation of Scheme 2.93 cyclizes to give the 3-methyl-1,3,4-thiadiazolium or couples with elimination of sulfur to give the 1,2,4-thiadiazole derivative as it is implied by Scheme 2.92.

Scheme 2.94 gives one additional method to prepare 1,3,4-thiadiazoles, from thiosemicarbazide and carbon disulfide (Spalinska et al. 2006).

Liu et al. (2005a) patented a method for producing 2-mercapto-5-phenyl-1,3,4-thiadiazole from alkali dithiocarbazide and phenylamidine hydrochloride. Another patent of Liu et al. (2005b) claims the synthesis of 2-mercapto-5-(2-pyridyl)-1,3,4-thiadiazole from carbon disulfide and 2-pyridyl (Py) amidrazone (Scheme 2.95). The methods have the advantages of wide reactant resources, mild reaction conditions, high product purity, and excellent yields.

2.4.1.2   From Thiosemicarbazones

Scheme 2.96 describes the ferric chloride oxidation of thiosemicarbazones prepared from thiosemicarbazides and aldehydes to obtain derivatives of 2-amino-1,3,4-thiadiazole (Werber et al. 1977, Shawali and Sayed 2007).

Ammonium ferric sulfate is also used for oxidation of thiosemicarbazones (Hadizadeh and Vosooghi 2008). Efficient is a method of the thiosemicarbazone

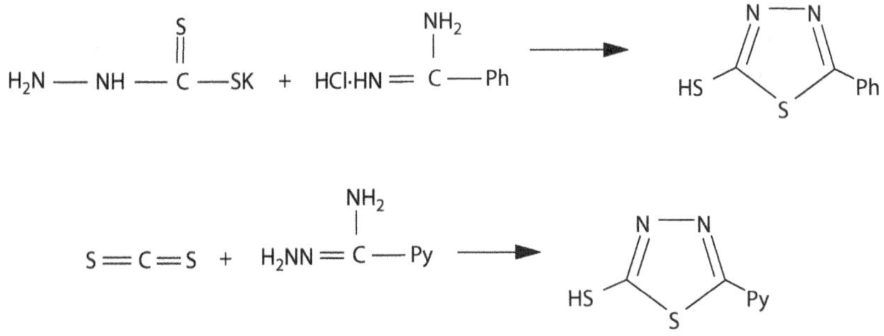

SCHEME 2.95

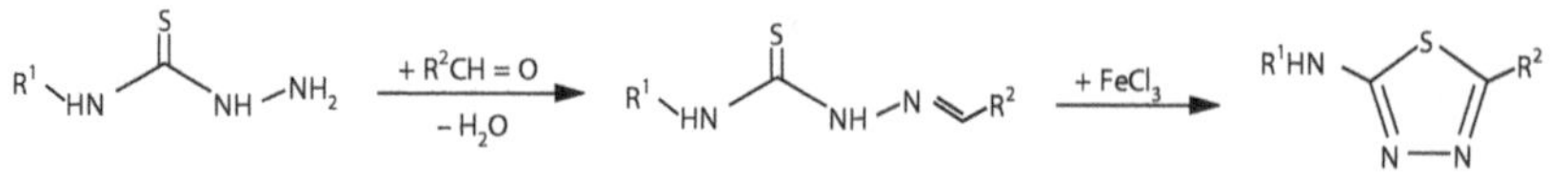

SCHEME 2.96

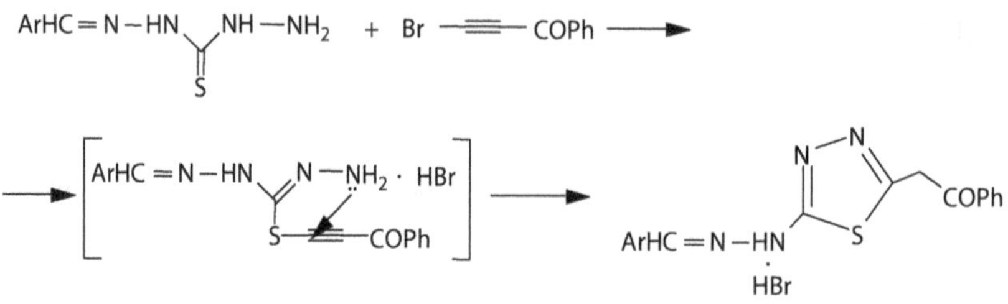

SCHEME 2.97

cyclization by means of silica-supported dichlorophosphate upon microwave irradiation (Li et al. 2008a).

Thiosemicarbazones of Scheme 2.97 react with 3-bromo-1-phenylprop-2-yn-1-one (Glotova et al. 2008). The proposed mechanism involves nucleophilic replacement of the bromine atom at the triple-bonded carbon atom. Benzoyl-ethynyl sulfides are intermediary formed. At the next step, intramolecular cyclization takes place via addition of the amino group at the electron-deficient β-carbon atom of the triple bond to form the final product.

Another way to obtain amino derivatives of 1,3,4-thiadiazole is oxidation of aldehyde thiosemicarbazones using cupric perchlorate (Scheme 2.98) (le Meo et al. 1999).

Heravi et al. (2004) described a one-pot synthesis of [1,3,4]-thiadiazolo-[2,3-c]-triazin-4-one. The synthesis started from 6-methyl-4-amino-3-thiono-[1,2,4]-triazine-2-one (a cyclic analog of thiosemicarbazone), phenylacetic acid or derivatives of benzoic acid and was promoted with sulfuric acid adsorbed on silica gel. Scheme 2.99 shows the total equation; yields of the final products were 80%–90%.

It should be mentioned that treatment of thiosemicarbazones with acetic anhydride in the presence of pyridine (at the bath temperature of 145°C) leads to the formation of 1,3,4-thiadiazoline-4-acetyl compounds (Somogyi et al. 2008).

2.4.1.3 From Dithiourea

The patent by Danilova et al. (2007) claims preparation of 2,5-diamino-1,3,4-thiadiazole by heating dithiourea with excess of hydrogen peroxide. The cyclocondensation

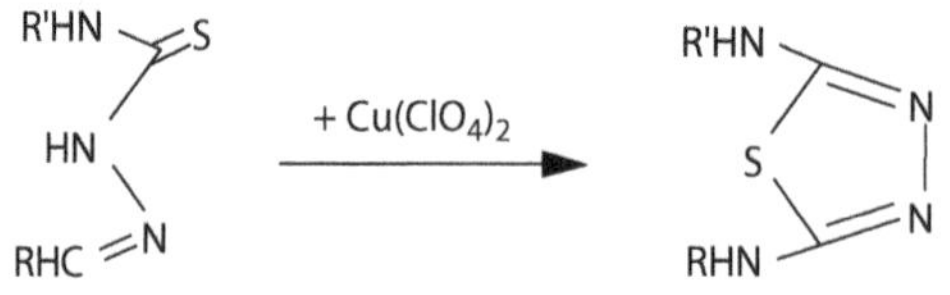

SCHEME 2.98

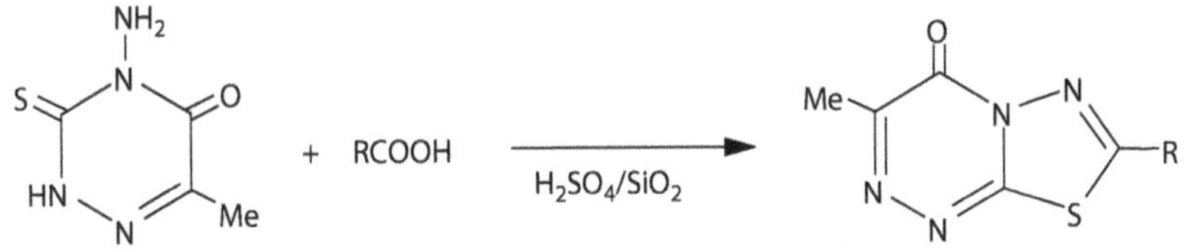

SCHEME 2.99

proceeds almost quantitatively without formation of colored by-products, see also Melenchuk et al. (2008).

2.4.1.4 From Isothiocyanates

Rostamizadeh et al. (2008) proposed condensation of substituted phenylisothiocyanates with hydrazine hydrate followed by the addition of benzaldehyde derivatives. The reaction was performed in the 1-butyl-3-methylimidazolium tetrafluoroborate ionic liquid and led to 1,3,4-thiadiazoles in excellent yields (Scheme 2.100). The ionic liquid plays a dual role: It is a solvent and a catalyst. It is easily recovered and can be repeatedly used.

A facile one-pot synthesis of 2,5-disubstituted 1,3,4-thiadiazoles was achieved by ultrasonic irradiation of a mixture containing ammonium isothiocyanate, N-arylglycine hydrazides, 1-naphthylacetyl chloride, poly(ethylene glycol), and dichloromethane. The method requires short reaction time and results in high-yields of the thiadiazole formation (Li et al. 2007a).

2.4.1.5 From Sulfides and Hydrazides or Amides

This route to 1,3,4-thiadiazoles consists of treatment of acylhydrazides with diphosphorus pentasulfide, cf. early works by Stolle and Stevens (1904), Stolle and Johannissien (1904) and nowadays publication by Shafiee et al. (1995). The use of P_2S_5 leads to unstable products of low yields, whereas the use of Lawesson's reagent gives rise to excellent yields (Seed 2007). Scheme 2.101 depicts the reaction of diacylhydrazide with Lawesson's reagent (here, An stands for 4-anisyl). A solvent-free protocol has also been recently reported (Kiryanov et al. 2001).

Propylphosphonic acid cyclic anhydride (trade mark T3P) is generally used as a coupling agent and water scavenger. Augustine et al. (2009) applied T3P, phosphorus

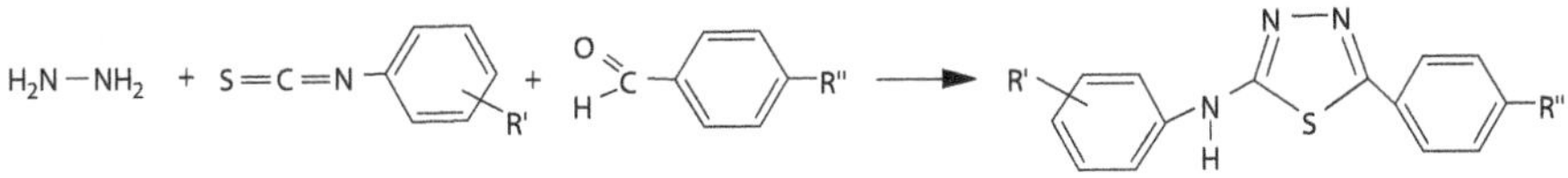

SCHEME 2.100

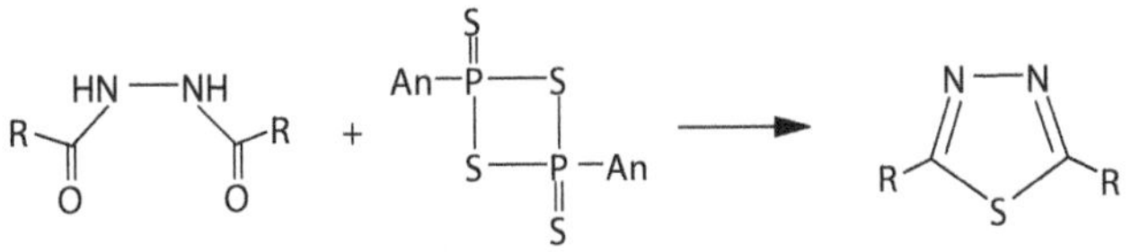

SCHEME 2.101

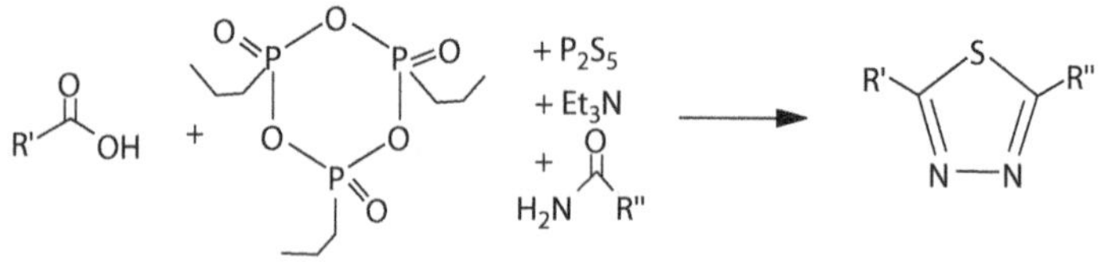

SCHEME 2.102

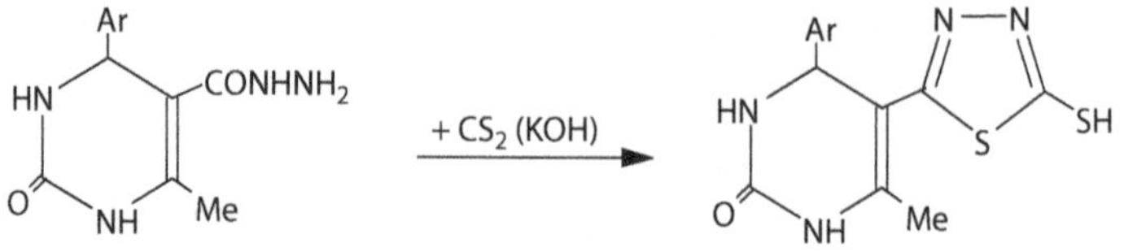

SCHEME 2.103

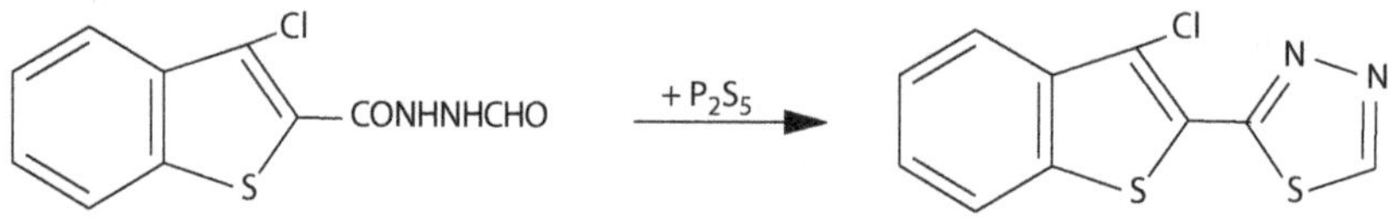

SCHEME 2.104

pentasulfide, triethylamine, and carboxamides for the one-pot synthesis of 1,3,4-thiadiazoles directly from carboxylic acids. Yields were excellent. Scheme 2.102 shows the overall transformation, the nature of functional groups R' and R" has no effect on the results. All of the by-products are water-soluble and easily washed away from the aimed compounds.

The use of diphosphorus pentasulfide distributed in alumina opens a way to microwave-assisted one-pot and solvent-free syntheses of 1,3,4-oxadiazoles from various hydrazides and triethylorthoalkanates. Yields are from moderate to good, up to 70% (Polshettiwar and Varma 2008).

Scheme 2.103 represents the reaction between a carbohydrazide and carbon disulfide. The reaction proceeds in the presence of potassium hydroxide (Singh et al. 2004). The corresponding dithiocarbazate is intermediately formed. The latter transforms into the thiadiazole upon action of hydrazine hydrate (Sharma and Fernandes 2007).

Cyclization of hetaryl (*N*-formylhydrazides) with phosphorus pentasulfide is exemplified by Scheme 2.104 (Sharba et al. 2005b).

2.4.1.6 From vic-Mercapto Amines or vic-Thiono Amines and Carboxylic Acids or Ketones

vic-Mercapto amines react with carboxylic acids giving 1,3,4-thiadiazole derivatives. The carbon atom of the acid locks the cycle. Condensation proceeds in phosphorus oxychloride (Scheme 2.105) (Tan et al. 2007, compare also Matiychuk et al. 2007, Prasad et al. 2009). In this reaction, phosphorus oxychloride acts as a solvent and as a dehydrating agent. When a mercapto amine and a carboxylic acid are poorly

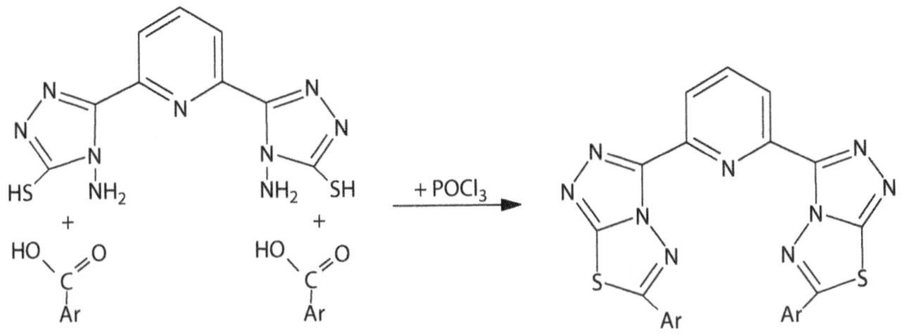

SCHEME 2.105

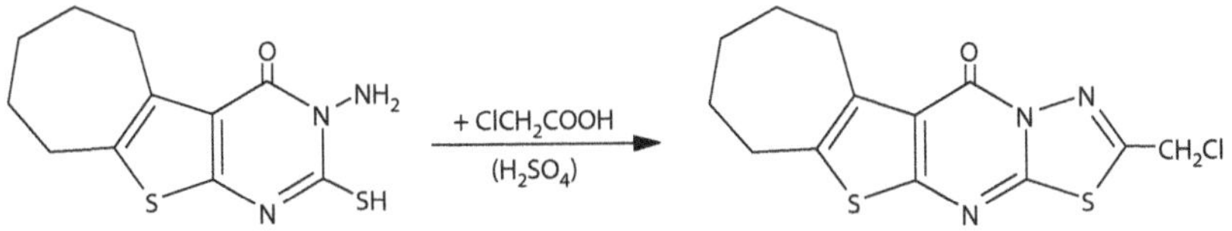

SCHEME 2.106

soluble in POCl$_3$, a phase transfer catalyst is used, for example, tetrabutylammonium iodide (Li and Fu 2007).

Scheme 2.106 describes the reaction that was performed under microwave irradiation in the presence of catalytic amounts of sulfuric acid (Prasad et al. 2007).

Scheme 2.107 illustrates reaction of a mercapto amine with 1,1′-diacetylferrocene. The condensation was performed refluxing pyridine for 2 h (El-Shiekh et al. 2006).

Schemes 2.108 represents reaction of an aminopyrimidine thione with acetic anhydride (Kovalenko et al. 2007). In some cases, the 1,3,4-thiadiazole cycle can be formed when amino thiones are treated with carboxylic acids in phosphorus oxychloride (Mekuskiene et al. 2007).

2.4.1.7 From Elemental Sulfur or Sulfur-Containing Compounds and Hydrazones

Symmetrical dialkyl (Hagen et al. 1980) and diaryl (Mazzone et al. 1983) 1,3,4-thiadiazoles were synthesized in a one-pot procedure. Scheme 2.109 represents this reaction in the top line and its modification (Moss and Taylor 1982) in the bottom line.

Scheme 2.110 gives another typical example that consists of the reaction between C-acetyl-N-(5-penylpyrazol-3-yl)hydrazonoyl chloride and potassium thiocyanate (Abdelhamid et al. 2001).

The treatment of hydrazonoyl chlorides with triethylamine leads to nitrile imines. In the presence of 1,3-thiazole-2(3H)-thiones, this reaction leads to the formation of the 1,3,4-thiadiazole ring spiro-connected with the 1,3-thiazole counterpart (Scheme 2.111) (Budarina et al. 2007).

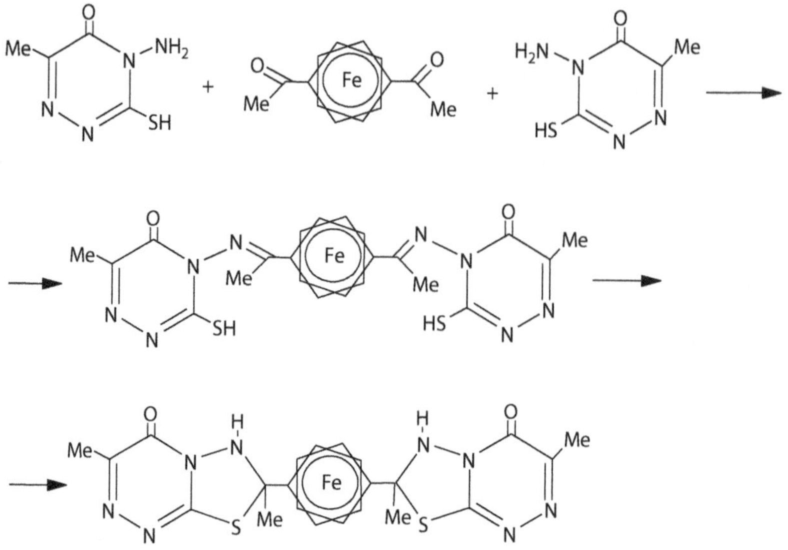

SCHEME 2.107

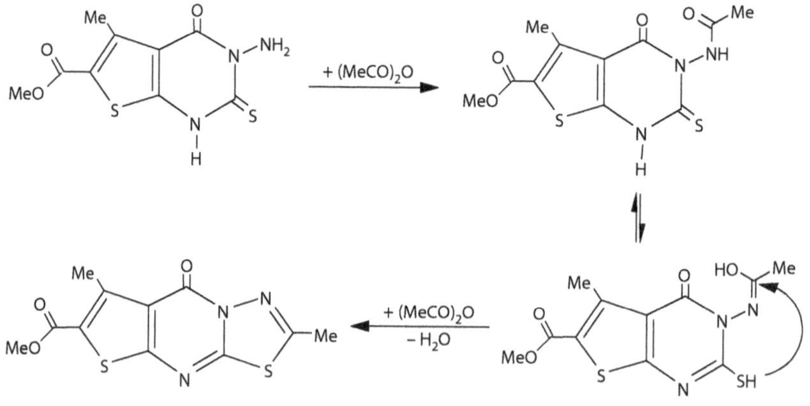

SCHEME 2.108

Using thioxanthene-9-thione as a substrate, Hafez et al. (2008) have studied regioselectivity of spirocondensation with nitrile imines. As seen from Scheme 2.112, the electron-deficient carbon of the nitrile imine adds to the electron-enriched thione sulfur whereas the electron-rich nitrogen of the nitrile imine adds to the electron-stripped thione carbon. The product of the first route mentioned is obtained exclusively and in high yield. The alternative regioisomer is not formed, at all.

Some modification of the reaction depicted in Schemes 2.111 and 2.112, leads to spiro-connected 1,3,4-thiadiazole-indolinone (see Scheme 2.113). Having been high-throughput screened, the spiro-products of this class showed their potential utility as selective therapeutics to treat osteoarthritis (Bursavich et al. 2007).

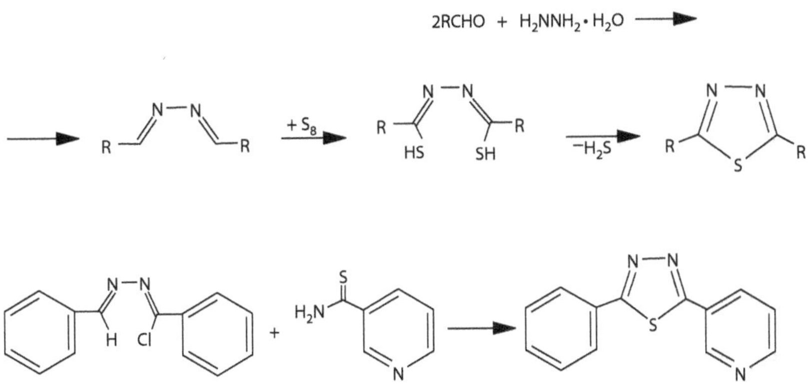

SCHEME 2.109

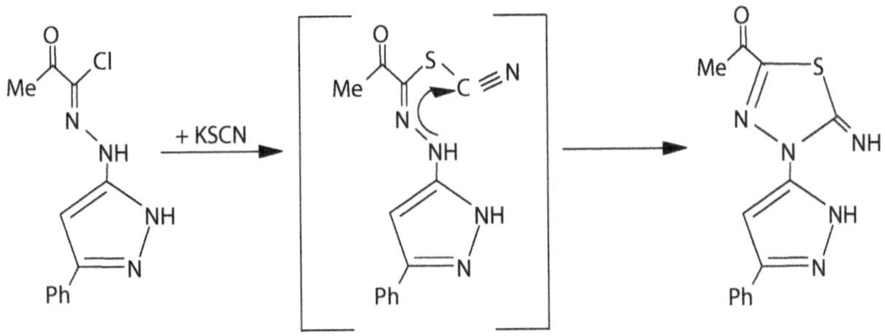

SCHEME 2.110

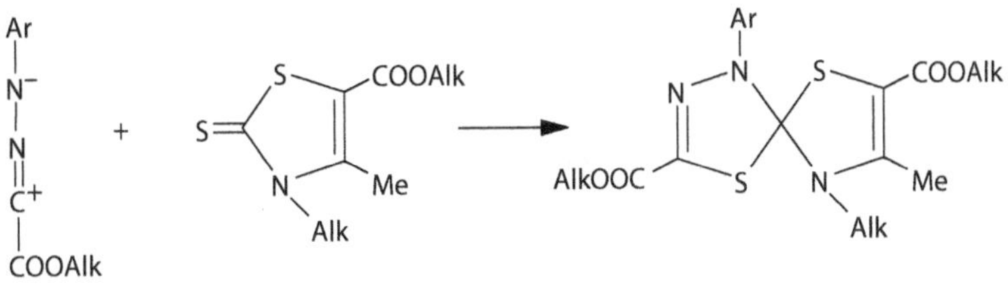

SCHEME 2.111

2.4.1.8 From Elemental Sulfur, Malodinitrile, and Nitrogen Heterocycles

This type of the 1,3,4-thiadiazole formation can be exemplified by transformation of melatonin in refluxing ethanolic triethylamine solution (Scheme 2.114) (Doss et al. 2003). The transformation was used to enhance the melatonin antioxidant activity (Elmegeed et al. 2008).

2.4.1.9 From Other Heterocycles

2,5-Diaryl-1,3,4-oxadiazoles give 2,5-diaryl-1,3,4-thiadiazoles upon treatment with thiourea, although heating in a sealed tube is necessary to obtain acceptable yields. Urea is formed as the by-product (Scheme 2.115) (Linganna and Lokanatha Rai 1998).

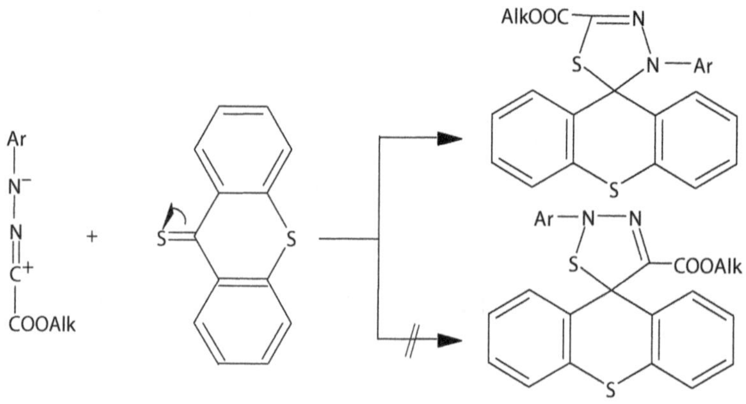

SCHEME 2.112

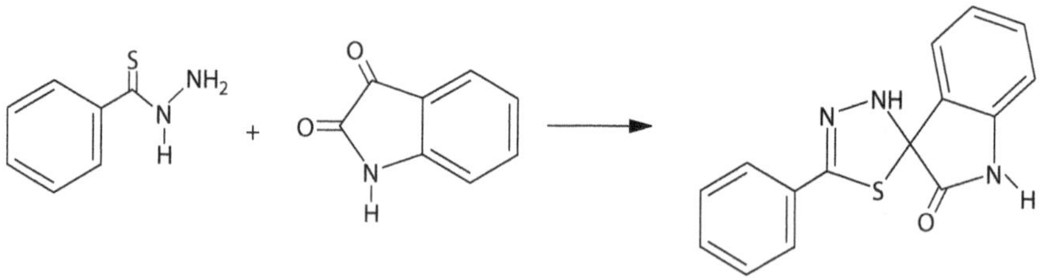

SCHEME 2.113

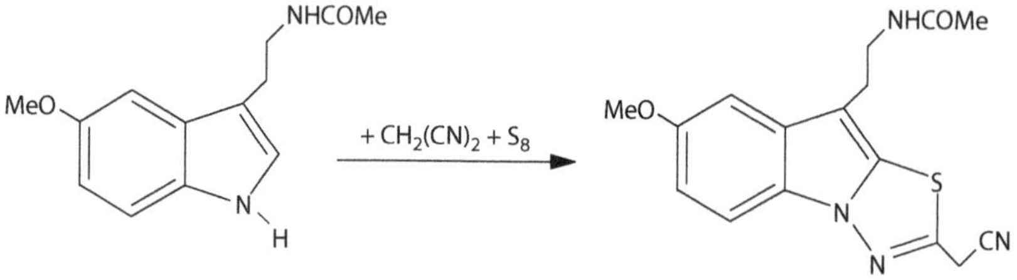

SCHEME 2.114

Elhazazi et al. (2003) proposed somewhat curious transition from the diazepine ring to the 1,3,4-thiadiazole one. Scheme 2.116 represents this transition.

Scheme 2.117 depicts another ring contraction, when 1,3,4-thiadiazine derivatives transform into 1,3,4-thiadiazole ones (Fleischhauer et al. 2008).

2.4.2 FINE STRUCTURE

Molecule of 1,3,4-thiadiazole is planar with point group C_{2v} (Hegelund et al. 2008). The carbon–sulfur bonds of the ring have some double character resulting from

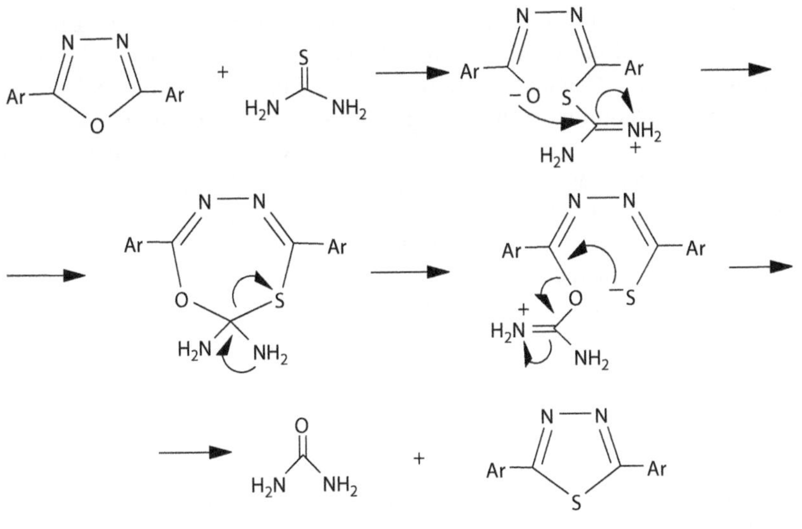

SCHEME 2.115

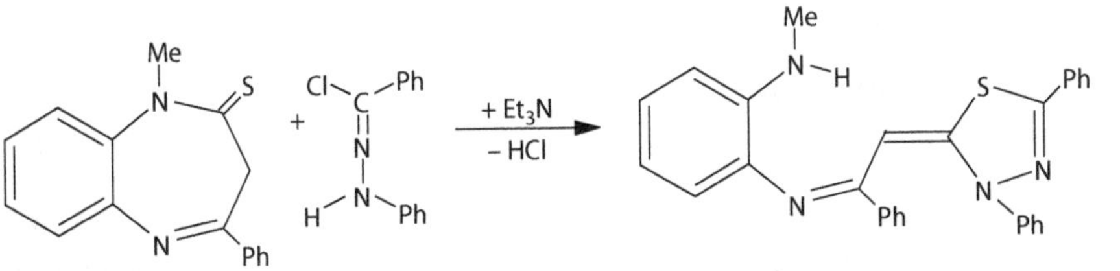

SCHEME 2.116

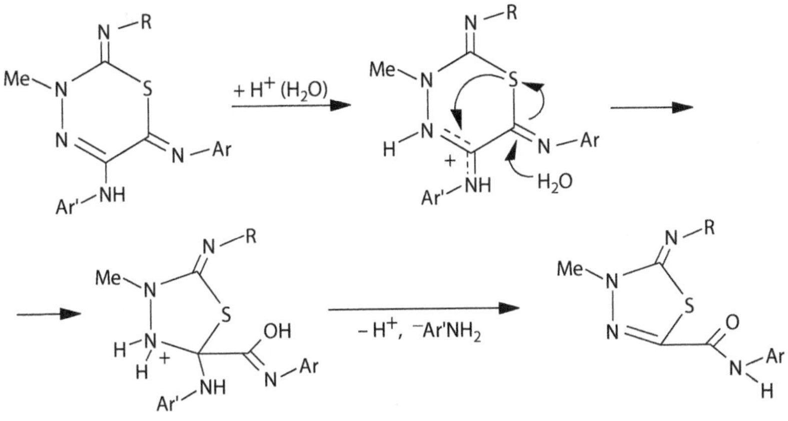

SCHEME 2.117

cyclic delocalization of the π electrons (Mathew and Palenik 1974). Bird (1985) characterizes 1,3,4-thiadiazole with aromaticity index $I_A = 63$, whereas $I_A = 100$ correspond to benzene. (Calculations were based on bond lengths in the corresponding molecules.) According to such a degree of aromaticity, α-amino derivatives of 1,3,4-thiadiazole form stable diazo salts that can be involved in Sandmeyer reactions (Cowdrey and Davies 1952, Rao and Srinvasan 1970, Heindl et al. 1975) and azocoupling (Pandey et al. 2003, Seferoglu et al. 2008). At the same time, the 1,4-diazadiene character of 1,3,4-thiadiazole was revealed by thermal cycloaddition with dimethyl acetylenedicarboxylate (Moriarty and Chin 1972) or with norbornene under high pressure (Warrener et al. 1997).

With respect to an aromatic ring, the 1,3,4-thiadiazolyl substituent simultaneously acts as meta and para–ortho orientant. This heterocycle is a strong acceptor and a moderate donor. Treatment of (1,3,4-thiadiazol-2-yl)-benzene with classical nitration combination ($HNO_3 + H_2SO_4$) leads to 1-(1,3,4-thiadiazol-2-yl)-3-nitrobenzene together with 1-(1,3,4-thiadiazol-2-yl)-4-nitrobenzene and 1-(1,3,4-thiadiazol-2-yl)-2-nitrobenzene (Ohta et al. 1953). The amount of the meta-product is equal to the combined amount of the ortho and para products. Similarly, chlorobenzene also produces a mixture of ortho, meta, and para nitrochlorobenzenes under treatment with classical nitration combination ($HNO_3 + H_2SO_4$).

2.4.3 REACTIVITY

2.4.3.1 Salt Formation

1,3,4-Thiadiazoles are readily alkylated and acylated at N_3 or N_4 position. N-Alkylation can be achieved using alkyl sulfates (Mayer and Lauerer 1970), alkyl halides (Takamizawa et al. 1980), and trialkyloxonium tetrafluoroborates (Curphey and Prasad 1972). By stepwise treatment with trialkyloxonium tetrafluoroborates, both ring nitrogen atoms may be alkylated, in sequence.

N-Acylation is accompanied with migration of the entered acyl group with the formation of the carbon-substituted product according to Scheme 2.118 (Alemagna and Bacchetti 1972).

Scheme 2.119 represents one unusual case of intramolecular quaternization (Eldavy et al. 1979).

For 2-amino-1,3,4-thiadiazole, the amino and imino forms coexist, although the amino tautomer is favored (Hough and Jones 1984). Quaternization with benzyl chloride shifts the equilibrium to the imino form, so that the product is 2-imino-3-benzyl-1,3,4-thiadiazoline) (Scheme 2.120) (Wheeler and Blanchard 1992).

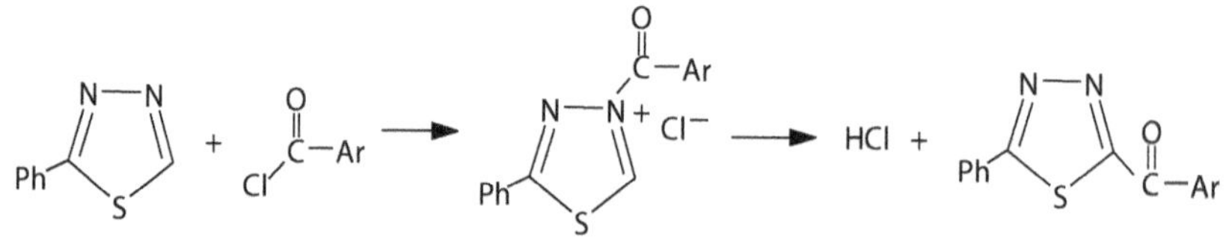

SCHEME 2.118

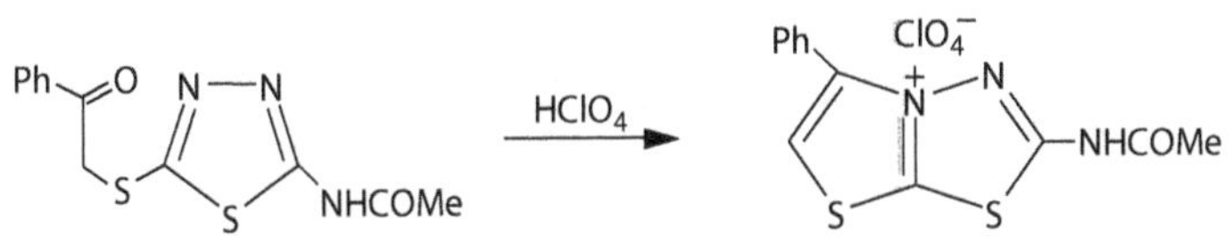

SCHEME 2.119

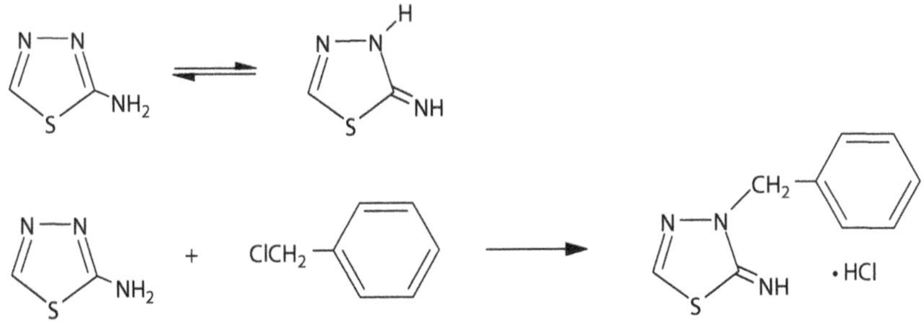

SCHEME 2.120

2.4.3.2 Electrophilic Substitution

Due to low electron density of the carbon atoms in 1,3,4-thiadiazoles, standard electrophilic substitution hardly proceeds. Thus, nitration of 2-phenyl-1,3,4-thiadiazole leads to substitution in the phenyl group, only (Ohta et al. 1953), see also Section 2.4.2. Halogenation was observed only for 2-amino-1,3,4-thiadiazole, with substitution of bromine for hydrogen at position 5 (Werber et al. 1977).

2.4.3.3 Nucleophilic Substitution

1,3,4-Thiadiazole ring is electrophilic in its nature. Therefore, halogen substituents in α position are highly activated and can be displaced by a wide range of nucleophiles such as alcoholates, arylthiolates, or primary amines (Hagen et al. 1980). Upon treatment with hydroxylamine, 2-aryl-1,3,4-thiadiazoles form 5-amino-2-aryl-1,3,4-thiadiazoles (Rao and Srinivasan 1970). Considered as a nucleophilic substitution, this reaction, however, may proceed as nucleophilic addition (see Scheme 2.121).

2.4.3.4 Ring Cleavage

Strong bases evoke ring fission in 1,3,4-thiadiazoles. For instance, 2-phenyl-1,3,4-thiadiazole decomposes giving benzonitrile and thiocyanate upon treatment with alkali (Alemagna and Bacchetti 1972). Reaction between 2-(4-metoxyphenyl)-1,3,4-thiadiazole and *n*-butyl lithium at −78°C leads to lithium thiocyanate and 4-methoxybenzonitrile in near quantitative yields (Scheme 2.122) (Seed 2007).

2.4.3.5 Photolysis

Hipler et al. (2002) and Rostkowska et al. (2010) studied photolysis of 1,3,4-thiadiazole-2,5-dithiol in argon matrices. In the argon matrix, the substrate is stabilized in the thiolothione form. Upon ultraviolet irradiation, this "mixed" form undergoes

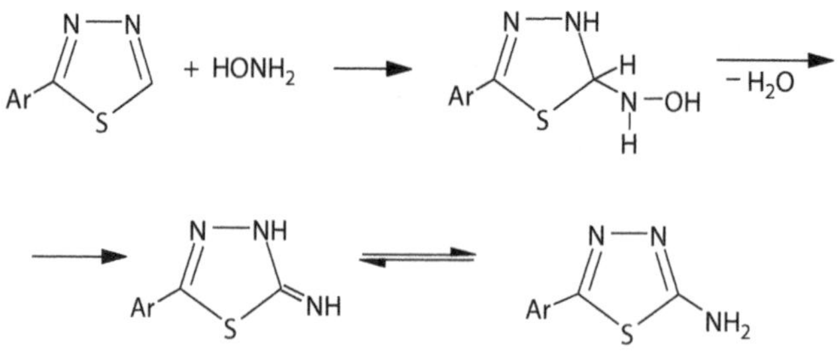

SCHEME 2.121

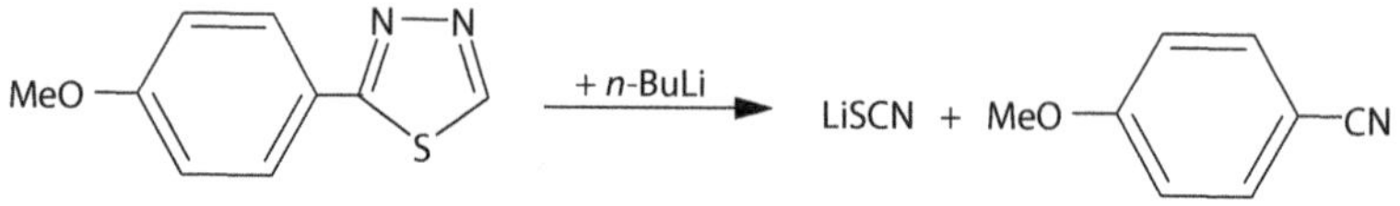

SCHEME 2.122

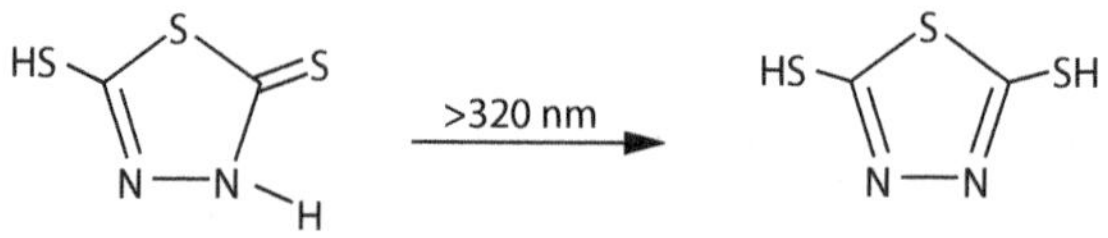

SCHEME 2.123

a hydrogen atom transfer from the endocyclic nitrogen atom to the thione sulfur atom so that the dithiol form is fixed (Scheme 1.123). The analogous structures were fixed for 2-mercapto-5-methyl- and 2-mercapto-5-thiomethyl-1,3,4-thiadiazoles as the products of photoirradiation at >295 nm (Rostkowska et al 2010).

2.4.3.6 Thermolysis

As has already been mentioned, in argon matrix 1,3,4-thiadiazole-2,5-dithiol exists in the thiol-thione form. Upon thermal treatment, this form undergoes decomposition eliminating carbon disulfide (Hipler et al 2002).

2.4.3.7 Coordination

With metals of binary coordinative ability, 1,3,4-thiadiazole-2,5-dithiol forms polymeric systems, in which each metal binds two mercapto sulfur atoms of two different ligand molecules along with both nitrogen atoms of the same ligand ring (Ortega et al. 1996, 1997, Li et al. 2008b). The specific case of the 1,3,4-thiadiazole-2,5-dithiol coordination to rhenium pentacarbonyle triflate is depicted in Scheme 2.124. The structure of the complex formed was established by x-ray diffraction analysis. The ring sulfur remains out of coordination, but both dithiol sulfur atoms are involved in bonding with rhenium. One of the ring nitrogens coordinatively adds rhenium pentacarbonyl. Besides, an interesting π-coordination of triflate with one of the carbon–nitrogen bonds is observed (Tseng et al. 2007).

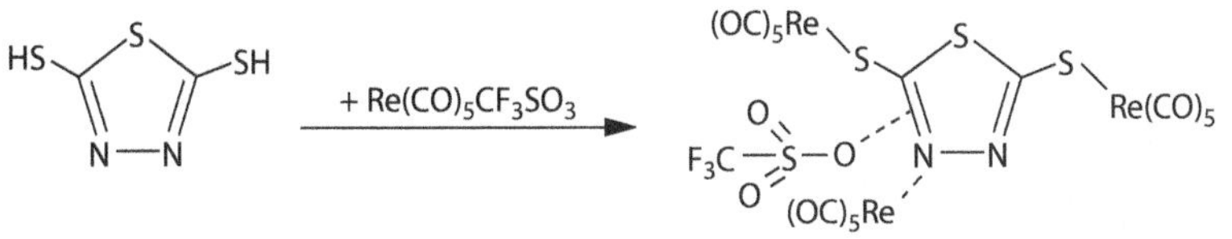

SCHEME 2.124

Coordination between 1,3,4-thiadiazole-2,5-dithiol and palladium dichloride involves one of the thiadiazole nitrogen atoms, and one of the thiol group displaces one of the palladium chlorine. The palladium atoms in the main polymer chain are coordinatively bound through the chlorine atoms remaining after the other one has been substituted by the thiolate sulfur (El-Shekeil and Al-Shuja'a 2007). In 1,3,4-thiadiazole-2,5-dithiol, the thiol group does not prevent the metal coordination with the heteroring. When these thiol groups are blocked, [1,3,4-thiadiazole-2,5-bis(thioacetic acid)], coordination to metal proceeds only at the expense of the carboxylic functions, and the heteroring in the polymeric complex formed remains free. Such a type of coordination to rare earth metal centers has been proven by x-ray crystallographic analysis (Wang et al. 2007). Whereas 1,3,4-thiadiazole-2,5-dithiol is not choosy to metal cations, blockade of its two thiol groups with picolinyl moieties makes the modified ligand selective to Cu^{2+} and Pb^{2+} ions (Lu et al. 2008).

In metal complexes of 2,5-bis(4-pyridyl)-1,3,4-thiadiazole, the following nitrogens available for coordination: Two nitrogens of the two pyridine substituents and two nitrogens of each thiadiazole rings. Many complexes with this ligand are mono- or bifunctionally bonded through pyridyl N atoms, but the trifunctionally bonded complexes are also known. In the latter, metal occurs to be bound with two pyridyl N atoms and with one of the two thiadiazole N atoms (Ettorre et al. 2006, 2008). As a multidentate ligand, N,N'-bis(5-ethyl-1,3,4-thiadiazol-2-yl)-2,6-pyridinedicarboxamide has an interesting pocket-like structure (see Scheme 2.125). With two-coordinating metals, this ligand forms polynuclear polymeric complexes that are depicted in Scheme 2.125 (Shen et al. 2008).

IR spectroscopic and x-ray diffraction studies showed that the coordination sphere of a complex between 2-amino-5-ethyl-1,3,4-thiadiazole and copper dichloride includes four molecules of the heterocyclic ligand and copper is bound with one nitrogen of each thiadizole ring and with one of the two chlorides. The copper center was five-coordinated. As for the second chloride of $CuCl_2$, this ion is in the outer sphere. Notably, the amino group substituted the thiadiazole ring remains free from the coordination (Kadirova et al. 2007).

With diorganotin dichloride, 2-mercapto-5-methyl-1,3,4-thiadiazole reacts in its thione tautomeric form (see Scheme 2.126). To elucidate the structure of the tin complex, x-ray diffraction study has been performed. As occurred, tin is bound with the nitrogen of the NH moiety covalently whereas the remaining thiadiazole nitrogen atom is bound with tin coordinatively. Surprisingly, neither the cyclic nor the thione sulfur is involved in coordination with tin, despite tin's intrinsic affinity to sulfur (Ma et al. 2006).

Coupling of the diazocompound from 2-amino-5-mercapto-1,3,4-thiadiazole with phenol or 2-naphthol results in the formation of the azo ligand. Three molecules of

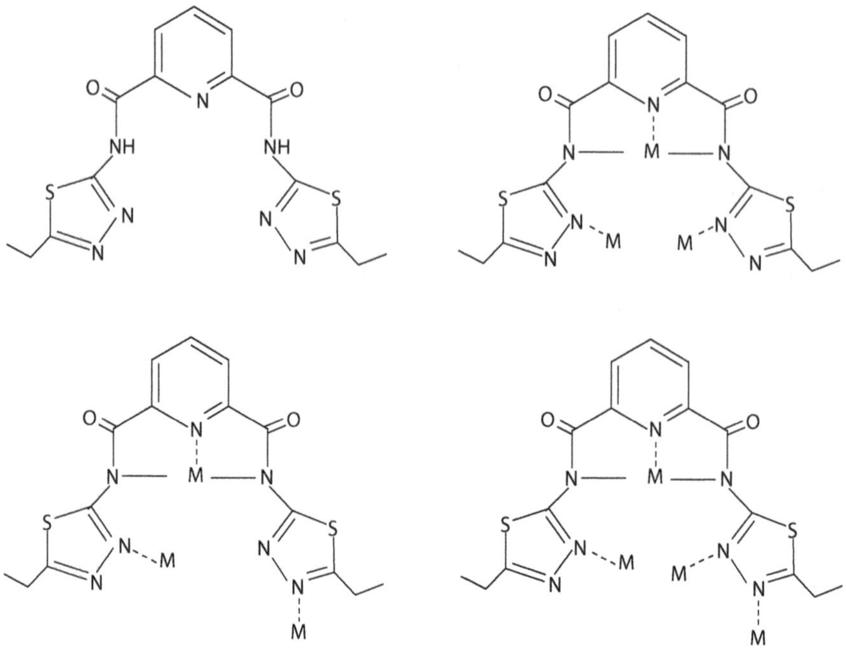

SCHEME 2.125

SCHEME 2.126

this ligand bind with one aluminum exclusively through the mercapto sulfur atoms (Ortega-Luoni et al. 2007). In contrast, a Schiff base from 2-amino-5-mercapto-1,3,4-thiadiazole and 2-formylthiophene gives with dimethyltin dichloride a cluster, in which the both thiadiazole nitrogens occurred to be coordinated whereas the azomethyne nitrogen, the thienyl and mercaptane sulfurs remain free (Ma et al. 2007).

In the case of 2-amino-5-methyl-1,3,4-thiadiazole, coordination to the metals mentioned proceeds through the nitrogen atom of the amino group and, in some complexes, also through one of the two ring-nitrogen atoms (Fabretti et al. 1980, 1985). It is the ring-nitrogen atom that is vicinal to the amino but not to the methyl group. Weakening the amino group basicity does not change the situation: 5-Ethyl-1,3,4-thiadiazole-2-(N-benzenesulfonyl)amine is bound with cupric ion through 3 but not 4 nitrogen of the heterocyclic moiety (Hangan et al. 2007).

If a ligand contains another nitrogen moiety of stronger basicity than that of the thiadiazole nitrogens, the latter do not participate in coordination, and the sulfur atom can form a coordinative bond with a metal. Such a case is just depicted by

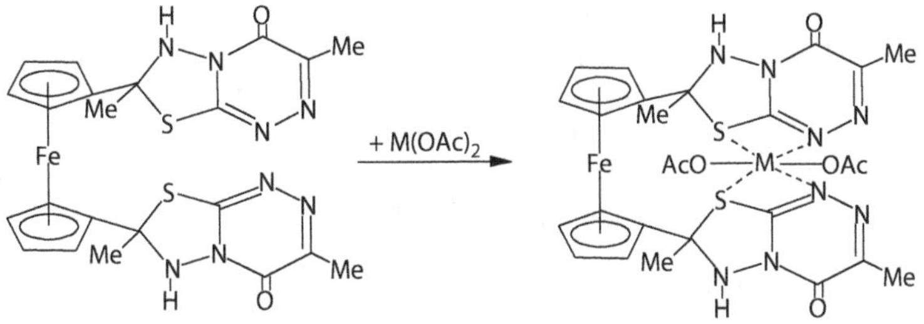

SCHEME 2.127

Scheme 2.127 (El-Shiekh et al. 2006). The ligand in this coordinative reaction is the same that has already been mentioned in Scheme 2.107.

Thiadiazol hydrazines increase the silver-photographic speed (Chen et al. 1993a). It was demonstrated that this effect is caused by adsorption of the compounds on the silver halide grains. Essentially, adsorption does not have the physical nature, but just consists in coordination of the hydrazines to the silver ion by means of the thiadiazole chemisorption, that is, coordination (Chen et al. 1993b).

2-Amino-5-mercapto-1,3,4-thiadiazole is used for the conservation of copper-tin bronzes due to formation of insoluble complexes (Bastidas and Otero 1996). The ligand metal adsorption is an initial step in the formation of such complexes. Yang et al. (2008) investigated the complexation of this ligand during the formation of its monolayer on silver electrode. Based on surface-enhanced Raman scattering and electrochemical observations, adsorption–complexation modes were elucidated. These modes depend on pH of aqueous solution, in which the thiadiazole is dissolved. Under pH 1, the thiadiazole was protonated. The authors assume that protonation does not prevent participation of the nitrogen atoms in adsorption on the silver surface. The sulfur atoms also participate in this adsorption-complexation. All of the sulfur and nitrogen constituents are bound with silver. Near pH 7, the ligand stands at the silver surface in a tilted mode and occurred to be bound through as the thiadiazole as the mercapto sulfur atom. From the basic solution, under pH 13, the thiadiazole attaches the silver surface and is fixed by the two sulfur atoms and only one nitrogen atom belonging to the amino group.

2.4.4 Biomedical Importance

2.4.4.1 Medical Significance

Derivatives of 1,3,4-thiadiazole become a fragment of many therapeutically active compounds. As resistance to antifungal and antimicrobial drugs is widespread, there is an increasing need in new ones with diminished side effects. Taking several drugs in combinations produces a harmful drug–drug interaction. The latter problem is especially important for senior patients and new drugs can help to avoid this serious obstacle.

Antifungal activity of 1,3,4-thiadiazole derivatives were reported by Chen et al. (2007). 2-Substituted 5-(nitroaryl)-1,3,4-thiadiazoles exhibit activity against *Helicobacter pylori*. According to in vitro tests, this activity is higher than that of standard antibacterial drugs (Mohammadhosseini et al. 2008, Mirzaei 2008, Foroumadi et al. 2008). Yoshida et al. (2000) performed synthesis of cepheme derivatives bearing 1,3,4-thiadiazolyl fragments. Introduction of this fragment dramatically improves the activity against *H. pylori*. In particular, 7-β-(2-phenylacetamido)-3-(5-methyl-1,3,4-thiadiazolyl-2-yl)thio-3-cephem-4-carboxylic acid was found to be very active (in vitro) and produces practically no side effects alike diarrhea. 5-(4-Hydroxy-2′,4′-difluoro-1,1′-biphenyl-5-yl)-2-(2-propenyl)-1,3,4-thiadiazole exhibits 73% of anti-inflammatory activity whereas the commonly used drugs show the activity about 25% (Kuecuekguezel et al. 2007).

The fused 1,3,4-thiadiazolo-1,2,4-triazole compound bearing 4-aminophenyl and 3,4-dichlorophenyl substituents at the thiadiazole and triazole rings, respectively, exhibits activity against human acute lymphoblastic leukemia (Al-Soud et al. 2008). The fused compound bearing 2,3-dichlorophenyl in the thiadiazole moiety and 4-chlorophenyl in the triazole fragment exhibits high activity and specificity against cervical and oral cancers (Kundu et al. 2007). *N*-Substituted 2-amino-5-(ethylphenyl)- or 5-(2,4-dihydroxyphenyl)-1,3,4-thiadiazoles inhibit proliferation of tumor cells derived from cancer of nervous system and peripheral cancers (Matysiak 2007, 2008, Rzeski et al. 2007, Thomas and Toledo-Sherman 2007, Juszczak et al. 2008). At the same time, these compounds are not toxic to normal cells. Moreover, they possess a prominent neuroprotective activity. 1,3,4-Thiadiazole-2-(*N*-piperidinoamine) derivatives were claimed as medicines for treating or preventing central nervous system disorders (e.g., schizophrenia) and without motor side effects (Macdonald et al. 2008). 5-Substituted 1,3,4-thiadiazoles bearing arylethenylquinazolones at position 2 exhibit sedative-hypnotic and central nervous system depressant activities (Jatav et al. 2008a,b).

5-Amino-1,3,4-thiadiazole-2-thiol derivatives exhibit antidepressant activity close to that of standards. The newly proposed antidepressants have passed neurotoxicity tests (Yusuf et al. 2008).

2-(*N*-Acetylamino)-5-sulfamido-1,3,4-thiadiazole is active in treating glaucoma, altitude sickness, cystinuria, and for increasing the secretion and flow of urine (see overviews by Maren 1967 and Supuran 2007). As an anticonvulsant agent, it is also used to treat epilepsy. Anti-convulsion activity has also been reported for 1,3,4-thiadiazoles bearing aminomethylene quinazolones or methylene indoles fragments (Singh et al. 2007a,b). Many other *N*-(1,3,4-thiadiazol-2-yl)sulfonamides were proposed to treat/prevent disorders of fatty-acid metabolism and glucose utilization as well as disorders, in which insulin resistance is involved (Keil et al. 2007, Schoenafinger et al. 2007a,b).

Chagas' disease is one of the most important parasitic infections in Latin America: 16–18 million people are currently infected by *Trypanosoma cruzi* and 120 million people are at risk. The first compound of Scheme 2.128 treats Chagas' disease, but toxicity and mutagenicity are its side effects. Brazilian chemists designed and patented (*N*-aryl)hydrazinoderivatives of 5-(1-methyl-5-nitroimidazol-2-yl)-1,3,4-thiadiazole (Ferreira da Silva et al. 2006). The most active candidates were found to be *N*-(1,2-dihydroxypenyl)hydrazinoderivative (Carvalho et al. 2004) and *N*-(1,2-dihydroxy-3-methoxypenyl)hydrazinoderivative (Carvalho et al. 2008). These are the top and bottom compounds, respectively, depicted on the right part of Scheme 2.128.

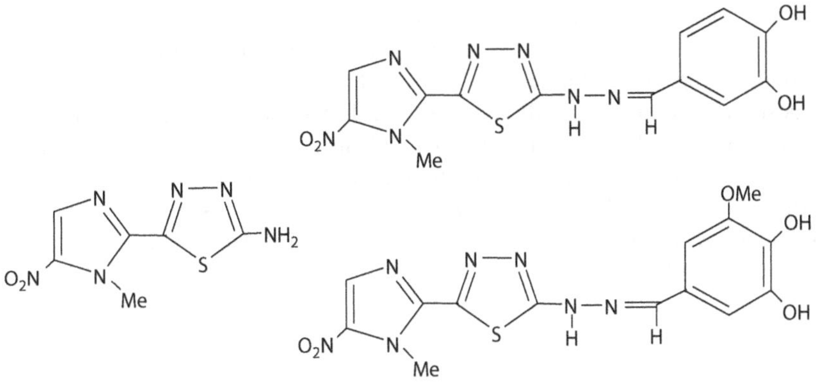

SCHEME 2.128

Recently, a new series of mesoionic 1,3,4-thiadiazolium-2-phenylamine chlorides were prepared. These compounds are derivatives or analogs of the natural amide piperine and are also prospective for the treatment of Chagas' disease. From the compounds tested, 4-phenyl-5-[4-(3,4-methylenedioxaphenyl)-1(*E*),3(*E*)-butadienyl]-1,3,4-thiadiazolium-2-phenylamine chloride showed the best activity profile (Ferreira da Silva et al. 2008).

Leishmaniasis is a disease caused by parasites belonging to the genus *Leishmania*. Leishmaniasis affects the skin, mucous membranes, and internal organs. Leishmaniasis is also called sand-fly disease, Dum-Dum fever, espundia, and kala azar, which is Hindi for "black fever." It is relatively unknown in the developed world but affects many poor countries. 1,3,4-Thiadiazoles (Hemmateenejad et al. 2007) and 1,3,4-thiadiazolium chlorides (Rodrigues et al. 2007, Soares-Bezzera et al. 2008) were found to have the leishmanicidal activity.

The leishmanicidal activity of various 1,3,4-thiadiazole derivatives was correlated by quantitative structure–activity relationship (QSAR) method. A QSAR is a mathematical relationship between biological activity of a molecular system and its geometric and chemical characteristics. All the derivatives analyzed had 1,3,4-thiadiazole as a central part, in which position 2 bore the (5-nitrofur-2-yl), or (5-nitrothien-2-yl), or (5-nitro-1-methyl-1*H*-imidazol-2-yl) moieties. Position 5 of these derivatives bore substituents of morpholyl, or piperidyl, or piperazinyl series. For the compounds considered, energies of the LUMOs were calculated. It was obtained from QSAR analysis that LUMO energy represents significant impact on the leishmanicidal activity. This means that the 1,3,4-thiadiazole derivatives act on the leishmania parasites through an electron transfer reaction. This reaction leads to transformation of nitro substituents to the corresponding nitroso groups (Hemmateenejad et al. 2007). The question remains whether the leishmanicidal activity is caused by electron transfer itself or by action of the nitroso compounds formed (or by either phenomena). In vitro tests, 5-nitrofuran derivatives occurred to be more active than the corresponding 5-nitrothiophene analogs (Behrouzi-Fardmoghadam et al. 2008).

Aryl and hetaryl substituted 1,3,4-thiadiazoles can regulate the enzyme activation (Honda et al. 2008, Lachance et al. 2008, De et al. 2009) or act as anti-inflammatory and analgesic agents (Kumar et al. 2007). Analgesic activity exceeding conventional analgesics was observed for sulfonylurea chloride derived from 2-amino-5-mercapto-1,3,4-thiadiazole (Sharma et al. 2008b) and from 1,2,4-triazolo-[3,4-b]-[1,3,4]-thiadiazole (Amir et al. 2007). Importantly, very low ulcerogenic index was observed for these derivatives. N-(1,3,4-Thiadiazol-2-yl)-carboxamides treats metabolic and inflammatory or immune-associated diseases or disorders (Yang and Doweyko 2008). Benzotrifluorides bearing the 2-(1,2,4-oxadiazol-3-yl) and 3-(1,3,4-thiadiazol-2-yl) substituents or 1,3,4-thiadiazole-aminoalchol derivatives are patented for the treatment of organ transplant rejection (Deng et al. 2008a,b).

Various 1,3,4-thiadiazole derivatives are active as antituberculosis medications. Thus, 1,3,4-thiadiazoles containing 2-(5-nitrothienyl) and alkyl 5-S-thioglycolate exhibit antituberculosis activity against *Mycobacterium tuberculosis* strains (Foroumadi et al. 2003). *Mycobacterium tuberculosis* can be excellently exterminated upon the action of 3,6-bis(2,4-dichloro-5-fluorophenyl)-1,2,4-triazolo[3,4-b]-1,3,4-thiadiazole (Karthikeyan et al. 2007). Sulfamides of the 1,3,4-thiadizaole series are also active against *Mycobacterium tuberculosis* (Minakuchi et al. 2009). These sulfamides differ in the action mechanism from the existent drugs, to which *Mycobacteria* acquire resistance. 2-(4-Methoxyphenyl) and 2-methylsylfonyl derivatives of 5-N-(sulfamidoamine)-1,3,4-thiadiazole are good at preventing obesity (Smaine et al. 2008). Nitrate esters of 1,3,4-thiadiazole sulfonamide benzoic acid were patented as agents for treating eye disorders and cancer (Supuran et al. 2008). Amino derivatives of 1,3,4-thiadiazoles are claimed for the treatment of cancer, inflammation, and metabolic diseases (Allen et al. 2009).

Azacyclohexyl 1,3,4-thiadiazoles were patented as compounds useful for the prevention and treatment of conditions related to abnormal lipid synthesis and metabolism, including cardiovascular disease, atherosclerosis, obesity, diabetes, metabolic syndrome, insulin resistance (Li et al. 2006), and neurological illness (Dahl et al. 2008). 2-Etherial-5-acylamido- or 2,5-diarylimidazolyl-1,3,4-thiadiazoles were described as potential drugs for the treatment of obesity (Bouillot et al. 2008, Kim et al. 2009).

2.4.4.2 Agricultural Protectors

Some compounds of the 1,3,4-thiadiazole series were recommended as plant-growth regulators (Cebelo and Walde 1969, Wang et al. 2001, Gong et al. 2006, Song et al. 2007). A number of 1,3,4-thiadiazole aryloxyphenylamidines were tested and claimed as insectofungicides. They showed high efficacy against wheat brown rust and powdery mildew on barley (Kunz et al. 2007) as well as against aphids, the plant louses (Wan et al. 2009). 2,5-Dimercapto-1,3,4-thiadiazole is stronger against apple moth than usually used insecticides (Zhivotova 2007). Wan et al. (2008) claimed 2-aryl-5-acylamide-1,3,4-thiadiazole derivatives as insecticides killing mosquitoes or aphids. The compounds also exhibit good bactericidal effects.

2-Mercapto-5-aryl-1,3,4-thiadiazoles keep the soil from unnecessary nitrification. Namely, 3-methylphenyl as 5-aryl substituent confers maximal activity to the inhibitor (Saha et al. 2010).

2.4.5 TECHNICAL APPLICATIONS

2.4.5.1 Reprographic Promoters

Hydrazine compounds containing 2,5-bis(thioalkyl)-1,3,4-thiadiazol moiety assists to acceleration of photographic development (Chen et al. 1993a,b, see also Section 2.4.3.7). 2-Substituted derivatives of 5-[(thiomethylene)-4-vinylphenylene]-1,3,4-thiadiazole is claimed as component of polymerizable photosensitive layers for lithographic printing plates (Hagiwara and Furukawa 2007). Copolymer of acrylamide with 5-acrylamido-2-mercapto-1,3,4-thiadiazole was patented as stabilizer for photosensitive materials. It prevents fogging during storage of the materials (Sakamoto et al. 1985).

2.4.5.2 Light Emitters, Liquid Crystals, and Solar Cells

Yamamoto et al. (2006) patented a copolymer between 2,5-dibromo-1,3,4-thiadiazole and 2,5-bis(trimethylstannyl)thiophene as bipolar charge-transfer material for such devices as light-emitting diodes, transistors, and solar cells.

Aromatic derivatives of 1,3,4-thiadiazole included into polymeric networks form liquid-crystalline phases that possess electroluminescent properties (Sato et al. 2002, Fang He et al. 2007, Yokoyama et al. 2007). Alkyl-5-(4-octyloxyphenyl)-1,3,4-thiadiazole-2-carboxylates form liquid crystals of high thermostability. Being exposed to bright sunlight for over 6 months, they have not shown any changes in color or mesophase transitions. Candidates of this group indicate ability of self-organization and very good photochemical and thermal stability (Sybo et al. 2007). The thiadiazole units are responsible for the luminescent emission. However, the most linear structure is needed to obtain enantiotropic nematic liquid-crystalline phase and to organize emission of polarized fluorescent light at room temperature (Sato and Ohta 2007). This is significant for applications in electroluminescent devices. Some 1,3,4-thiadiazole derivatives are used in optical compensation films for liquid crystal displays. The films provide good control of wavelength dispersion (Nagata and Nishikawa 2007).

Fusco et al. (2008) synthesized polymers based on chromophores containing the 1,2,4-triazolo-[3,4-*b*]-[1,3,4]-thiadiazole segments. These segments confer NLO activity. NLO signals are kept 75%–85% of their intensity after monthly storage.

2.4.5.3 Vulcanizers

The use of 2,5-dimercapto-1,3,4-thiadiazole was patented for vulcanizing and crosslinking of polymers containing acyl chloride or sulfonyl chloride groups (Ren 2008). To achieve satisfactory speed of vulcanization for tire treads, the Japanese patent recommends including 1:1 salt of 2,5-dimercapto-1,3,4-thiadiazole with cyclohexylamine into the mixture of butadiene rubbers (Fukutani et al. 2007).

2.4.5.4 Applications to Analysis and Treatment of Industrial Wastes

2,5-Bis(2-hydroxyphenyl)-1,3,4-thiadiazole forms a stable 3:1 complex with silver. The complex is colored. A color reaction was used to spectrophotometric determination of silver trace amounts in wastes (Lu and Zhang 2007). Another highly selective fluorescent and colorimetric sensor to probe Ag^+ in water was synthesized by Qu et al. (2008). This sensor is 5-{3-[2-(naphthalene-3-yloxy)acetyl] thioureido}-1,3,4-thiadiazole-2-carboxylic acid. Upon sonication and at the expense of van der Waals interactions, the thiadiazole-containing molecules form nanoparticles that are stable at room temperature and adsorb the silver cations. Under light irradiation, the $Ag^+ \rightarrow Ag^\circ$ reduction takes place. The reduced silver constitutes very small nanoparticles that coat the organic nanoaggregates and luminescence initially emitted by nanoaggregates is quenched. The method allows the determination of Ag^+ concentration as low as 4.7×10^{-8} M.

Kanemite clay from soil and sedimentary deposits is valuable material for disposal of toxic thorium ions from natural waters and wastes. To prolong the kanemite activity, it must be immobilized. Such an immobilization is performed with 2-amino-5-mercapto-1,3,4-thiadiazole (see Guerra et al. 2009).

2.4.5.5   Corrosion Inhibitors and Lubricating Additives

The anticorrosive properties of 1,3,4-thiadiazole derivatives are well documented. The compounds are nontoxic at treatment and effective for surfaces of diverse metals. They can effectively control silver corrosion caused by jet fuel (Qian et al. 2006, 2008, Zhang et al. 2007). A base oils with 1,3,4-thiadiazole compounds protect steel sheets galvanized with zinc. The fluids reduce zinc cover-plate separation, wear powder generation, and decrease friction coefficient during the metal-working process (Nagase 2008). 2,5-Bis(C_{1-2}-alkyldithio)-1,3,4-thiadiazoles were patented as corrosion inhibitors and metal passivators for lubricating oils, greases, and fuels (Kumar et al. 2004).

2-Mercapto-5-acylamino-1,3,4-thiadiazole is effective inhibitor of bronze corrosion in electrolytes, especially in aggressive aqueous sodium chloride solution (Ignat et al. 2006). 2-Amino-5-alkyl-1,3,4-thiadiazoles prevents copper corrosion in 3% aqueous solutions of sodium hydrogen carbonate (Zhu et al. 2006) or sodium chloride (Sherif and Park 2006). As assumed, the thiadiazole compounds form complexes with copper metal, cuprous or cupric cations. These complexes are polymeric in nature and form adherent protective films on the copper surface. The complexes protect the surface from corrosion by aggressive bicarbonate or chloride ion. 2-Dodecyldithio-4-phenyl-1,3,4-thiadiazole-5-thione is also an effective inhibitor of copper corrosion (Qu et al. 2006).

2-Amino-5-mercapto-1,3,4-thiadiazole performs excellent inhibition effect for the corrosion of mild steel in 0.5 M aqueous solution of hydrochloric acid. Mild-steel is low-carbon (<0.15%) steel or soft-cast steel and is very corrosive in this aggressive medium. 2-Amino-5-mercapto-1,3,4-thiadiazole in concentration of 1.0×10^{-2} M shows inhibition efficiency higher than 99% after 120 h contact (Solmaz et al. 2008). Using a series of techniques, the authors studied kinetics and thermodynamics of the inhibitor adsorption on the metal surface. According to the obtained results, the adsorption proceeds in two phases. At first, physical adsorption of chloride anion on the metal surface takes place (cf., Yurt et al. 2004). Because the thiadiazole derivative is an organic base, it is protonated at nitrogen in the hydrochloric acid solution (in equilibrium with the parental base). Organic cations are adsorbed on the metal surface via chloride ions, which form interconnecting bridges between the metal surface and protonated organic cations. At the second phase, phisisorption is changed by chemisorption: Donor–acceptor interactions take place between, on one hand, the inhibitor π-electrons, the sulfur and/or nitrogen atoms of the 2-amino-5-mercapto-1,3,4-thiadiazole as well as the sulfur atoms of the corresponding thiadiazolium salt and, on the other hand, the vacant d-orbitals of iron metal. In the adsorption process, phisisorption is the base of chemisorption. Without phisisorption, it would be difficult for the thiadiazole to transfer electrons from polar atoms to d-orbital of iron. Coordinative bonds are formed by partial transfer of electrons from electron-donating fragments of the inhibitor to the metal surface. Thermogravimetric analysis showed that the formed protective film has good thermal stability (Solmaz et al. 2008).

2-Amino-5-mercapto-1,3,4-thiadiazole also inhibits corrosion of copper in acidic solutions (Wan et al. 2000). The corrosion mitigation is achieved by the film formation on the copper surface. Copper forms the covalent bond with the amino group and the coordinative bond with the thiadiazole sulfur (Song et al. 2004).

Thiadiazole isomers bearing two thienyl substituents in positions 2 and 5 protect mild steel against corrosion in 0.5 M sulfuric acid (Lebrini et al. 2007). The isomers mentioned had two identical substituents—bis(thien-2-yl) or bis(thien-3-yl). Both isomers inhibit corrosion of the steel, but 2,5-bis(thien-3-yl)-1,3,4-thiadiazole showed better performance. Adsorption of the sulfur-containing molecules takes place on the steel surface. Chemically, an electron is transferred from the thiadiazole to vacant d-orbitals of iron. The lower the energy of one-electron removal from the HOMO of an inhibitor, the better is its anticorrosion efficiency (Lukovits et al. 1997, Khalil 2003). Bis(thien-3-yl)-1,3,4-thiadiazole is better inhibitor than its bis(thien-2-yl) counterpart. As it follows from calculations (Lebrini et al. 2005), the HOMO density is only supported on the sulfur atoms of the thienyl rings just in the case of the bis(thien-3-yl) isomer. On the contrary, the HOMO density is distributed along the whole molecule of the bis(thien-2-yl) isomer, from one thienyl fragment to another through the 1,3,4-thiadiazole center with no localization on the thienyl sulfur. Accordingly, the 3-thienyl isomer is bound with the steel surface stronger than the 2-thienyl counterpart.

For the sake of verity, it should be mentioned that some bis-substituted 1,3,4-thiadiazoles stimulate the steel corrosion process. In 1 M aqueous hydrochloric acid solution, bis-(4-nitrophenyl)- or bis(4-chlorophenyl)-1,3,4-thiadiazole accelerates mild steel corrosion especially at low concentration of the additive (Bentiss et al. 2007). The main cause of this effect is ability of such additives to operate as electron-transfer mediators. Electron transfer converts the adsorbed iron–thiadiazole complex, $[Fe(OH)·TD]_{ads}$, into the soluble form according to the equilibrium: $[Fe(OH)·TD]_{ads} = [Fe(OH)·TD]^+_{sol} + e^-$. The soluble form is readily removed from the metal surface. This removal stimulates continuing corrosion of steel.

2,5-Dimercapto-1,3,4-thiadiazole was used in formulation of corrosion-inhibition and metal-passivation mixtures resulting from reactions with C_4–C_{12} alkyldisulfide-rich waste from petroleum oil refinery. The reactions proceed spontaneously or in the presence of hydrogen peroxide (Bhatnagar et al. 2000). Schaefer et al. (2006) claimed lubricating additives based on epoxidized fatty acid ester intramolecularly cross-linked with 2,5-dimercapto-1,3,4-thiadiazole and (2-ethylhexyl)dithiophosphate. 2,2′-Dimercapto-bis(1,3,4-thiadiazol-5-yl)disulfide acts as lubricating grease additive and exhibits good anticorrosion properties (Hu et al. 2005). The same disulfide is claimed as a component of lubricating mixture for drill bits (Lockstedt and Denton 2007). S,S′-bis(alkylethyleneglycol) derivatives of 2,2′-dimercapto-bis(1,3,4-thiadiazol-5-yl)disulfide was proposed to be mixed with copper nanopowder and used as grease. The nanocomposite grease exhibits enhanced lubricating effect and also enforces wear resistance (Kang et al. 2008). 5-Dodecylthio-3-phenyl-1,3,4-thiadiazole-2-thione possesses good thermal stability. Being added to the rapeseed oil, it is able to decrease the wear and friction coefficients (Lu et al. 2007). 2-Mercapto-5-amino-1,3,4-thiadiazole was claimed as a passivation agent for copper and cobalt (Rath et al. 2006).

2.4.5.6 Oil Improvers

Derivatives of 2,5-dimercapto-1,3,4-thiadiazole were patented by Habeeb et al. (2006) as additives to improve release of air from lubricating oils, including hydraulic oil and crankcase oils. These are often used in environments, in which the oil is subject to mechanical agitation in the presence of air. As a consequence, the air becomes entrained in the oil and also forms foam. Air entrainment reduces viscosity of a lubricating composition. Both air entrainment and foaming deteriorate the fluid due to enhanced oil oxidation. The patent cited proposes a good solution for the problem.

2.4.5.7 Dyestuffs

Applications of 1,3,4-thiadiazole to dying compositions are also documented. 2-Amino-5-mercapto-1,3,4-thiadiazole derivatives were widely used in preparation of dispersed dyes for hydrophobic fibers (Maradiya and Patel 2002). Dyes based on 1,3,4-thiadiazoles are good at toning keratin fibers, including human hair. Thus, an azodye prepared by coupling of diazotized 2-amino-5-ethylthio-1,3,4-thiadiazole with *N*-phenylethanolamine was patented as a component of such a coloring composition (Benicke and Hoeffkes 2007).

2.4.6 Conclusion

The distinctive features of 1,3,4-thiadiazole are the diversities in the methods of their synthesis and relative scarcity of their reactivity. The scarcity is predefined by their structure that excludes aromatic annulation. But on the other hand, building of the thiadiazoles having a common nitrogen atom with other heterocycles or spirocompounds having a common carbon atom is possible. A significant aspect of 1,3,4-thiadiazoles is their partial aromaticity that determines chemical reactivity and selectivity of coordination metals.

Molecules of the 1,3,4-thiadiazole family are seemed to be especially suited for metabolism. They manifest themselves as bioactive compounds. Many of the newly synthesized and tested 1,3,4-thiadiazole derivatives have shown biomedical applicability and may have very good future uses. There is a wealth of patents and journal publications revealing the biomedical importance of 1,3,4-thiadiazoles. (This book's volume allows citing not all of them but the typical ones only.)

Extensive studies were devoted to the technical applicability of 1,3,4-thiadiazoles. They found applications to lithography, photoprinting, solar conversion, vulcanization, corrosion and wear prevention, toxic metal determination and disposal from water and wastes.

The 1,3,4-thiadiazole field is developing and attractive for diversified professionals.

REFERENCES TO SECTION 2.4

Abdelhamid, A.O.; Sallam, M.M.M.; Amer, S.A. (2001) *Heteroat. Chem.* **12**, 468.

Abdel-Rahman, T.M. (2006) *Phosphorus Sulfur* **181**, 1737.

Alemagna, A.; Bacchetti, T. (1972) *Gazz. Chim. Ital.* **102**, 1977.

Alemagna, A.; Bacchetti, T.; Rizzi, C. (1972) *Gazz Chim Ital.* **102**, 311.

Allen, J.G.; Bourbeau, M.P.; Domingues, C.; Fotsch, Ch.H.; Han, N.; Hong, F.-Ts.; Huang, X.; Lee, M.R.; Liu, Q.; Reichelt, A.; Tadesse, S.; Wang, X.; Yao, G.; Yuan, Ch.Ch.; Zeng, Q. (2009) WO Patent 011,871.

Al-Soud, Y.A.; Al-Masoudi, N.A.; Loddo, R.; la Colla, P. (2008) *Ach. Pharm.* **341**, 365.

Amir, M.; Kumar, H.; Javed, S.A. (2007) *Bioorg. Med. Chem. Lett.* **17**, 4504.

Anis'kov, A.A.; Sazonov, A.A.; Klochkova, I.N. (2009) *Mendeleev Commun.* **19**, 52.

Augustine, J.K.; Vairaperumal, V.; Narasimhan, Sh.; Alagarsamy, P.; Radhakrishnan, A. (2009) *Tetrahedron* **65**, 9989.

Bastidas, J.M.; Otero, E. (1966) *Mater. Corros.* **47**, 333.

Behrouzi-Fardmoghadam, M.; Poorrajab, J.; Kaboudanian, A.; Emami, S.; Shafiee, A.; Foroumadi, A. (2008) *Bioorg. Med. Chem.* **16**, 4509.

Benicke, W.; Hoeffkes, H. (2007) Germ. Patent 102,006,036,898.

Bentiss, F.; Lebrini, M.; Lagrenee, M.; Traisnel, M.; Elfarouk, A.; Vezin, H. (2007) *Electrochim. Acta* **52**, 6865.

Bhatnagar, A.K.; Tuli, D.K.; Sarin, R.; Arora, A.; Mondal, P.K. (2000) Ind. Patent MU 00,171.

Bird, C.W. (1985) *Tetrahedron* **41**, 1409.

Bouillot, A.M.J.; Boyer, Th.; Daugan, A. C.-M.; Dean, A.W.; Filmor, M.C.; Lamotte, Y. (2008) WO Patent 104,524.

Budarina, E.V.; Dolgushina, T.S.; Petrov, M.L.; Labeish, N.N.; Kol'tsov, A.A.; Bel'skii, V.K. (2007) *Zh. Org. Khim.* **43**, 1521.

Bursavich, M.G.; Gilbert, A.M.; Lombardi, S.; Georgiadis, K.E.; Reifenberg, E.; Flannery, C.R.; Morris, E.A. (2007) *Bioorg. Med. Chem. Lett.* **17**, 5630.

Carvalho, S.A.; Ferreira da Silva, E.; de Castro, S.L.; Fraga, C.A.M. (2004) *Bioorg. Med. Chem. Lett.* **14**, 5967.

Carvalho, S.A.; Lopes, F.A.S.; Salomao, K.; Romeiro, N.C.; Wardell, S.M.S.V.; de Castro, S.L.; Ferreira da Silva, E.; Fraga, C.A.M.(2008) *Bioorg. Med. Chem.* **16**, 413.

Cebelo, T.; Walde, R. (1969) Germ. Patent 1,912,543.

Chauviere, G.; Bouteille, B.; Enanga, B.; de Albuquerque, C.; Croft, S.L.; Dumas, M.; Perie, J. (2003) *J. Med. Chem.* **46**, 427.

Chen, C.-J.; Song, B.-A.; Yang, S.; Xu, G.-F.; Bhaduri, P.S.; Jin, L.-H.; Hu, D.-Y.; Li, Q.-Zh.; Liu, F.; Xue, W.; Lu, P.; Chen, Zh. (2007) *Bioorg. Med. Chem.* **15**, 3981.

Chen, S.-L.; Ji, S.-X.; Zhu, Zh.-H.; Yao, Z.-G. (1993a) *Dyes Pigments* **23**, 275.

Chen, Sh.-L.; Zhu, Zh.-H.; Yao, Z.-G.; Xue, J.S. (1993b) *Dyes Pigments* **23**, 197.

Cowdrey, W.A.; Davies, D.S. (1952) *Chem. Soc. Quart. Rev.* **6**, 358.

Curphey, T.J.; Prasad, K.S. (1972) *J. Org. Chem.* **37**, 2259.

Dahl, B.H.; Peters, D.; Olsen, G.M.; Timmermann, D.B.; Joergensen, S. (2008) WO Patent 049,864.

Danilova, E.A.; Melenchuk, T.V.; Islyaykin, M.K.; Kolesnikov, N.A. (2007) Russ. Patent 2,313,523.

De, S.K.; Stebbins, J.L.; Chen, L.-H.; Riel-Mehan, M.; Machleidt, Th.; Dahl, R.; Yuan, H.; Emadadi, A.; Barile, E.; Chen, V.; Murphy, R.; Pellecchia, M. (2009) *J. Med. Chem.* **52**, 1943.

Deng, H.-f.; Evindar, G.; Bernier, S.; Yao, G.; Coffin, A.; Yang, H.-f.; Acharya, R.A. (2008a) WO Patent 016,674.

Deng, H.-f.; Evindar, G; Yao, G. (2008b) WO Patent 091,967.

Doss, S.H.; Mohareb, R.M.; Elmegeed, G.A.; Louca, N.A. (2003) *Pharmazie* **58**, 607.

Eldavy, M.A.; Shams El-Dine, S.A.; El-Brembaly, K.M. (1979) *Pharmazie* **34**, 248.

Elhazazi, S.; Baouid, A.; Hasnaoui, A.; Compain, Ph. (2003) *J. Marocaine Chim. Heterocycl.* **2**, 61.

Elmegeed, G.A.; Khalil, W.K.B.; Raouf, A.A.; Abdelhalim, M.M. (2008) *Eur. J. Med. Chem.* **43**, 763.

El-Shekeil, A.G.; Al-Shuja'a, O.M. (2007) *J. Macromol. Sci. A* **44**, 931.

El-Shiekh, S.M.; Abd-Elzaher, M.M.; Eweis, M. (2006) *Appl. Organometal. Chem.* **20**, 505.

Ettore, R.; Marton, D.; Nodari, L.; Russo, U. (2006) *J. Organometal. Chem.* **691**, 805.

Ettorre, R.; Longato, B.; Zangrando, E. (2008) *Inorg. Chim. Acta.* **361**, 2985.

Fabretti, A.C.; Franchini, G.; Malvasi, W.; Peyronel, G. (1985) *Polyhedron* **4**, 989.

Fabretti, A.C.; Franchini, G.; Peyronel, G. (1980) *Spectrochim. Acta* **36A**, 689.

Fang He, C.; Richards, G.F.; Kelly, S.M.; Contoret, A.E.A.; O'Neill, M. (2007) *Liq. Cryst.* **34**, 1249.

Ferreira da Silva, E.; Carvalho, S.A.; de Castro, S.L.; Fraga, C.A.M. (2006) Braz. Patent 005,518.

Ferreira da Silva, W.; Freire-de-Lima, L.; Saraiva, V.B.; Alisson-Silva, F.; Mendoca-Previato, L.; Previato, J.O.; Echevarria, A.; Freire de Lima, M.E. (2008) *Bioorg. Med. Chem.* **16**, 2984.

Fleischhauer, J.; Beckert, R.; Hornig, D.; Guenter, W.; Goerls, H.; Klimesova, V. (2008) *Z. Naturforsch. B* **63**, 415.

Foroumadi, A.; Rineh, A.; Emami, S.; Siavoshi, F.; Masarrat, S.; Safari, F.; Radjabalian, S.; Falahati, M.; Lotfali, E.; Shafiee, A. (2008) *Bioorg. Med. Chem. Lett.* **18**, 3315.

Foroumadi, A.; Kiani, Z.; Soltani, F. (2003) *Formaco* **58**, 1073.

Fukutani, Sh.; Asukai, T.; Onuki, T. (2007) Jpn. Patent 254,496.

Funatsukuri, G.; Veda, M. (1966) Jpn. Patent 020,944.

Fusco, S.; Centore, R.; Riccio, P.; Quatela, A.; Stracci, G.; Archetti, G.; Kuball, H.-G. (2008) *Polymer* **49**, 186.

Glotova, T.E.; Dvorko, M.Yu.; Samoilov, V.G.; Ushakov, I.A. (2008) *Zh. Org. Khim.* **44**, 875.

Gong, Y.-x.; Zhang, Y.; Xiong, L.; Wang, Y.-g. (2006) *Huxue Shiji* **28**, 496.

Guerra, D.L.; Carvalho, M.A.; Leidends, V.L.; Pinto, A.A.; Viana, R.R.; Airoldi, C. (2009) *J. Colloid Interface Sci.* **338**, 30.

Habeeb, J.J.; Baillargeon, D.J.; Deckman, D.E. (2006) US Patent 281,643.

Hadizadeh, F.; Vosooghi, R. (2008) *J. Heterocycl. Chem.* **45**, 1477.

Hafez, H.N.; Hegab, M.I.; Ahmed-Farag, I.S.; El-Gazzar, A.B.A. (2008) *Bioorg. Med. Chem. Lett.* **18**, 4538.

Hagen, H.; Kohler, R.-D.; Fleig, H. (1980) *Liebigs Ann.* **1980**, 1216.

Hagiwara, T.; Furukawa, A. (2007) Jpn. Patent 264,497.

Hangan, A.; Borras, J.; Lui-Gonzalez, M.; Oprean, L. (2007) *Z. Anorg. Allg. Chem.* **633**, 1837.

Hegelund, F.; Larsen, R.W.; Aitken, R.A.; Palmer, M.H. (2008) *J. Mol. Spectrosc.* **248**, 161.

Heindl, J.; Schroeder, E.; Kelm, H.W. (1975) *Eur. J. Med. Chem.* **10**, 171.

Hemmateenejad, B.; Miri, R.; Niroomand, U.; Foroumadi, A.; Shafiee, A. (2007) *Chem. Biol. Drug Des.* **69**, 435.

Heravi, M.M.; Ramezanian, N.; Sadeghi, M.M.; Ghassemzadeh, M. (2004) *Phosphorus Sulfur* **179**, 1469.

Hipler, F.; Fisher, R.A.; Mueller, J. (2002) *J. Chem. Soc. Perkin Trans.* 2, 1620.

Honda, T.; Fujisawa, K.; Aono, H.; Ban, M. (2008) WO Patent 093,674.

Hough, T.L.; Jones, G.P. (1984) *J. Heterocycl. Chem.* **21**, 1377.

Hu, J.; Wang, L.; Ouyang, X.; Ji, F.; Wei, X. (2005) *Fushi Yu Fanghu* **26**, 63.

Ignat, I.; Varvara, S.; Muresan, L.M. (2006) *Studia Universitatis Babes-Bolyai Chemia* (Roumania) **51**, 127.

Jatav, V.; Jain, S.K.; Kashaw, S.K.; Mishra, P. (2006) *Indian J. Pharm. Sci.* **68**, 360.

Jatav, V.; Mishra, P.; Kashaw, S.; Stables, J.P. (2008a) *Eur. J. Med. Chem.* **43**, 135.

Jatav, V.; Mishra, P.; Kashaw, S.; Stables, J.P. (2008b) *Eur. J. Med. Chem.* **43**, 1945.

Jung, K.-Y.; Kim, S.K.; Gao, Zh.-G.; Gross, A.S.; Melman, N.; Jacobson, K.A.; Kim, Y.-Ch. (2004) *Bioorg. Med. Chem.* **12**, 613.

Juszczak, M.; Matysiak, J.; Brzana, W.; Niewiadomy, A.; Rzeski, W. (2008) *Arzneimittel-Forshung (Drug Res.)* **58**, 353.

Kadirova, Sh.A.; Ishankhodzhaeva, M.M.; Parpiev, N.A.; Karimov, Z.; Tozhiboev, A.; Tashkhodzhaev, B. (2007) *Zh. Obshch. Khim.* **77**, 1734.

Kang, Y.; Sun, H.; Long, J. (2008) Chin. Patent 101,205,217.

Kappel, J.C.; Yokum, T.M.; Barany, G. (2004) *J. Combinator. Chem.* **6**, 746.

Karthikeyan, M.S.; Holla, B. Sh.; Kalluraya, B.; Kumari, N.S. (2007) *Monatshefte Chem.* **138**, 1309.

Keil, S.; Schoenafinger, K.; Matter, H.; Urmann, M.; Glien, M.; Wendler, W.; Scaefer, H.-L.; Falk, E. (2007) WO Patent 039,171.

Khalil, N. (2003) *Electrochim. Acta* **48**, 2635.

Kilburn, J.P.; Lau, J.; Jones, R.C.F. (2003) *Tetrahedron Lett.* **44**, 7825.

Kim, J.Y.; Seo, H.J.; Lee, S.-H.; Jung, M.E.; Ahn, K.; Kim, J.; Lee, J. (2009) *Bioorg. Med. Chem. Lett.* **19**, 142.

Kiryanov, A.A.; Sampson, P.; Seed, A.J. (2001) *J. Org. Chem.* **66**, 7925.

Kovalenko, S.N.; Vlasov, S.V.; Fedosov, A.I.; Chernykh, V.P. (2007) *Zh. Org. Farm. Khim.* **5**(3), 34.

Kuecuekguezel, S.G.; Kuecuekguezel, I.; Tatar, E.; Rollas, S.; Sahin, F.; Guelluece, M.; de Clercq, E.; Kabasakal, L. (2007) *Eur. J. Med. Chem.* **42**, 893.

Kumar, A.; Bhati, S.K.; Rajput, Ch.S.; Singh, J. (2007) *Orient. J. Chem.* **23**, 183.

Kumar, M.P.; Chand, D.Kh.; Rakesh, S.; Kumar, T.D.; Prakash, V.R.; Kumar, B.A. (2004) Ind. Patent MU 660.

Kundu, T.K.; Rangappa, K.S.; Sailaja, B.S.; Varier, R.A.; Shivananju, N.; Basappa (no initials) (2007) WO Patent 034,510.

Kunz, K.; Gruel, J.; Guth, O.; Hartmann, B.; Ilg, K.; Moradi, W.; Seitz, Th.; Dahmen, P.; Voerste, A.; Wachendorf-Neumann, U.; Dunkel, R.; Ebbert, R.; Franken, E.-M.; Malsam, O. (2007) WO Patent 031,513.

Lachance, N.; Li, Ch.S.; Leclerc, J.-Ph.; Ramtohul, Y.K. (2008) WO Patent **128**, 335.

le Meo, R.; Noto, M.; Gruttadauria, M.; Weber, G. (1999) *J. Heterocycl. Chem.* **36**, 667.

Lebrini, M.; Lagrenee, M.; Traisnel, M.; Gengembre, L.; Vezin, H.; Bentiss, F. (2007) *Appl. Surface Sci.* **253**, 9267.

Lebrini, M.; Lagrenee, M.; Vezin, H.; Gengembre, L.; Bentiss, F. (2005) *Corros. Sci.* **47**, 485.

Li, Ch.S.; Ramtohul, Y.K.; Huang, Zh.; Lachance, N. (2006) WO Patent 6,130,986.

Li, D.-L.; Fu, H.-Q. (2007b) *Heterocycl. Commun.* **13**, 347.

Li, Y.J.; Sun, Y.Zh.; Xu, Y.T.; Jin, K.; Wen, L.P.; Hou, N.B.; Sun, X.X. (2007a) *Chinese Chem. Lett.* **18**, 1047.

Li, Zh.; Feng, X.; Zhao, Y. (2008a) *J. Heterocycl. Chem.* **45**, 1489.

Li, Zh.-H.; Lin, P.; Du, Sh.-W. (2008b) *Polyhedron* **27**, 232.

Linganna, L.; Lokanatha Rai, K.M. (1998) *Synth. Commun.* **28**, 4611.

Liu, J.; Que, W.; Jin, J.; Wu, Y. (2005a) Chin. Patent 1,948,295.

Liu, J.; Que, W.; Jin, J.: Wu, Y. (2005b) Chin. Patent 1,948,307.

Lockstedt, A.W.; Denton, R. (2007) US Patent 7,107,940.

Lopez-Torres, E.; Cowley, A.R.; Diworth, J.R. (2007) *Dalton Trans.* 1194.

Lu, P.-Zh.; Zhang, F.-T.; Qu, L.; Duan, T.-B.; Ye, W.-Y. (2007) *Huaxue Yanjiu* **18**, 65.

Lu, Y.; Zhang, Y. (2007) *Yejin Fenxi* **27**, 45.

Lu, Y.-Q.; Yan, Zh.-N.; Ye, B.-X.; Guo, Y.-P. (2008) *Henan Shifan Daxue Xuebao Ziran Kexueban* **36**, 73.

Lukovits, I.; Palif, K.; Bako, I.; Kolman, E. (1997) *Corrosion* **53**, 915.

Ma, Ch.; Li, J.; Zhang, R.; Wang, D. (2006) *J. Inorg. Organometal. Polym. Mater.* **16**, 147.

Ma, Ch.; Sun, J.; Zhang, R.; Wang, D. (2007) *J. Organometal. Chem.* **692**, 4029.

Macdonald, G.J.; Bartolome-Nebreda, J.M.; van Gool, M.L.M. (2008) WO Patent 128,996.

Maradiya, H.R.; Patel, V.S. (2002) *Polym. Plast. Technol. Eng.* **41**, 735.

Maren, T.H. (1967) *Physiol. Rev.* **47**, 595

Mashelkar, U.C.; Audi, A.A. (2006) *Indian J. Chem.* **45B**, 1463.

Mathew, M.; Palenik, G.J. (1974) *J. Chem. Soc. Perkin Trans.* 2, 532.

Matiychuk, V.S.; Pokhodilo, N.T.; Krupa, I.I.; Obushak, M.D. (2007) *Ukrain. Bioorg. Acta* **5**, 3.

Matysiak, J. (2007) *Eur. J. Med. Chem.* **42**, 940.

Matysiak, J. (2008) *QSAR Combinator. Sci.* **27**, 607.

Mayer, H.; Lauerer, D. (1970) *Liebigs Ann.* **731**, 142.

Mazzone, G.; Puglisi, G.; Bonina, F.; Corsaro, A. (1983) *J. Heterocycl. Chem.* **20**, 1399.

Mekuskiene, G.; Dodonova, J.; Burbuliene, M.M.; Vainilavicius, P. (2007) *Heterocycl. Commun.* **13**, 267.

Melenchuk, T.V.; Danilova, E.A.; Stryapan, M.G.; Islyaykin, M.K. (2008) *Zh. Obshch. Khim.* **78**, 495.

Minakuchi, T.; Nishimori, I.; Vullo, D.; Scozzafava, A.; Supuran, C.T. (2009) *J. Med. Chem.* **52**, 2226.

Mirzaei, J.; Siavoshi, F.; Emami, S.; Safari, F.; Khoshayand, M.R.; Shafiee, A.; Foroumadi, A. (2008) *Eur. J. Med. Chem.* **43**,1575.

Mohammadhosseini, N.; Lefavat, B.; Siavoshi, F.; Emami, S.; Safari, F.; Shafiee, A.; Foroumadi, A. (2008) *Med. Chem. Res.* **17**, 578.

Moriarty, R.M.; Chin, A. (1972) *J. Chem. Soc. Chem. Commun.* 1300.

Moss, S.F.; Taylor, D.R. (1982) *J. Chem. Soc. Perkin Trans.* 1, 1987.

Nagase, N. (2008) Jpn. Patent 138,092.

Nagata, I.; Nishikawa, H. (2007) Jpn. Patent 246,672.

Ohta, X.; Hagiwara, Y.; Nizushima, Y. (1953) *Yakugaku Zasshi (J. Pharm. Soc. Jpn)* **73**, 701.

Ortega, P.A.; Vera, L.R.; Campos-Vallette, M.; Diaz Fleming, G. (1996) *Specrosc. Lett.* **29**, 477.

Ortega, P.A.; Vera, L.R.; Guzman, M.E. (1997) *Macromol. Chem. Phys.* **198**, 2949.

Ortega-Luoni, P.; Vera, L.R.; Astudillo, C.; Guzman, M.E.; Ortega-Lopez, P. (2007) *J. Chilean Chem. Soc.* **52**, 1120.

Pandey, V.K.; Negi, H.S.; Joshi, M.N.; Bajpai, S.K. (2003) *Indian J. Chem.* **42B**, 206.

Pintilie, O.; Profire, L.; Sunel, V.; Popa, M.; Pui, A. (2007) *Molecules* **12**, 103.

Polshettiwar, V.; Varma, R.S. (2008) *Tetrahedron Lett.* **49**, 879.

Prasad, D.J.; Ashok, M.; Karegoudar, P.; Poojary, B.; Holla, B.Sh.; Kumari, N.S. (2009) *Eur. J. Med. Chem.* **44**, 551.

Prasad, M.R.; Raziya, S.; Kishore, D.P. (2007) *J. Chem. Res.*, 133.

Qian, J.-h.; Liu, L.; Wang, D.-l.; Xing, J.-j. (2006) *Shiyou Huagong Gaodeng Xuexiao Xuebao* **19**(4), 5.

Qian, J.-H.; Liu, L.; Xing, J.-J.; Zhao, Sh.; Wang, D.-L. (2008) *Youji Haxue* **28**, 160.

Qu, F.; Liu, J.; Yan, H.; Peng, L.; Li, H. (2008) *Tetrahedron Lett.* **49**, 7438.

Qu, L.; Duan, T.; Zhang, F.; Lu, P.; Ye, W. (2006) *Huaxue Yanjiu* **17**(2), 65.

Rao, R.; Srinivasan V.R. (1970) *Indian J. Chem.* **8**, 509.

Raslan, M.A.; Khalil, M.A. (2003) *Heteroat. Chem.* **14**, 114.

Rath, M.; Bernhard, D.D.; Minsek, D.W.; Baum, Th.H. (2006) WO Patent 133,253.

Ren, P. (2008) Chin. Patent 101,100,461.

Rodrigues, R.F.; da Silva, E.F.; Echevarria, A.; Fajardo-Bonin, R.; Amaral, V.F.; Leon, L.L.; Canto-Cavalheiro, M.M. (2007) *Eur. J. Med. Chem.* **42**, 1039.

Rostamizadeh, Sh.; Aryan, R.; Ghaieni, H.R.; Amani, A.M. (2008) *Heteroat. Chem.* **19**, 320.

Rostamizadeh, Sh.; Aryan, R.; Ghaieni, H.R.; Amani, A.M. (2010) *J. Heterocycl. Chem.* **47**, 616.

Rostkowska, H.; Lapinski, L.; Nowak, M.J. (2010) *J. Phys. Org. Chem.* **23**, 56.

Rostom, S.A.F.; Shalaby, M.A.; El-Demellawry, M.A.E. (2003) *Eur. J. Med. Chem.* **38**, 959.

Rzeski, W.; Matysiak, J.; Kandefer-Szerszen, M. (2007) *Bioorg. Med. Chem.* **15**, 3201.

Saha, A.; Kumar, R.; Kumar, R.; Devakumar, C. (2010) *J. Heterocycl. Chem.* **47**, 838.

Sakamoto, H.; Yoshimoto, Sh.; Mizukura, N. (1985) Jpn. Patent 141,118.

Sancak, K.; Unver, Y.; Er, M. (2007) *Turkish J. Chem.* **31**, 125.

Sato, M.; Ohta, R. (2007) *Liq. Cryst.* **34**, 295.

Sato, M.; Ohta, R.; Handa, M.; Kasuga, K. (2002) *Liq. Cryst.* **29**, 1441.

Schaefer, V.; Lindstaedt, B.; Jarre, W. (2006) WO Patent 131,363.

Schoenafinger, K.; Keil, S.; Urmann, M.; Matter, H.; Glien, M.; Wendler, W. (2007a) WO Patent 039,173.

Schoenafinger, K.; Keil, S.; Urmann, M.; Matter, H.; Glien, M.; Wendler, W. (2007b) WO Patent 039,174.

Seed, A. (2007) *Chem. Soc. Rev.* **36**, 2046.

Seferoglu, Z.; Ertan, N.; Hokelek, T.; Sahin, E. (2008) *Dyes Pigments* **77**, 614.

Shafiee, A.; Naimi, E.; Mansobi, P.; Foroumadi, A.; Shekari, M. (1995) *J. Heterocycl. Chem.* **32**, 1235.

Sharba, A.H.K.; Al-Bayati, R.H.; Aouad, M.; Rezki, N. (2005b) *Molecules* **10**, 1161.

Sharba, A.H.K.; Al-Bayati, R.H.; Rezki, N.; Aouad, M. (2005a) *Molecules* **10**, 1153.

Sharma, K.; Fernandes, P.S. (2007) *Indian J. Heterocycl. Chem.* **16**, 361.

Sharma, R.; Sainy, J.; Chaturvedi, S.Ch. (2008b) *Acta Pharm.* **58**, 317.

Sharma, Sh.; Joshi, A.; Talesara, G.L. (2008a) *Indian J. Heterocycl Chem.* **17**, 315.

Shawali, A.S.; Sayed, A.R. (2007) *J. Sulfur Chem.* **28**, 23.

Shen, X.-Q.; Yao, H.-Ch.; Yang, R.; Li, Zh.-J.; Zhang, H.-Y.; Wu, B.-L.; Hou, H.-W. (2008) *Polyhedron* **27**, 203.

Sherif, E.M.; Park, S.-M. (2006) *Corros. Sci.* **48**, 4065.

Singh, D.V.; Mishra, A.M.; Mishra, R.M. (2004) *Indian J. Heterocycl. Chem.* **14**, 43.

Singh, P.; Kumar, P.; Rani, P.; Singh, J.; Yeshowardhan (no initials); Rajput, Ch.S.; Bhati, S.R.; Kumar, A. (2007a) *Acta Ciencia Indica* **33C**, 287.

Singh, P.; Kumar, P.; Yeshowardhan (no initials); Singh, J.; Kumar, A. (2007b) *Acta Cien. Ind.* **33C**, 331.

Smaine, F.-Z.; Pacchiano, F.; Rami, M.; Barragan-Montero, V.; Vullo, D.; Scozzafava, A.; Winum, J.-Y.; Supuran, C.T. (2008) *Bioorg. Med. Chem. Lett.* **18**, 6332.

Soares-Bezzera, R.J.; Ferreira da Silva, E.; Echevarria, A.; Gomes-da-Silva, L.; Cysne-Finkelstein, L.; Pereira Monteiro, F.; Leon, L.L.; Genestra, M. (2008) *J. Enzyme Inhib. Med. Chem.* **23**, 328.

Solmaz, R.; Kardas, G.; Yazici, B.; Erbil, M. (2008) *Colloids Surf. A* **312**, 7.

Somogyi, L.; Batta, G.; Gunda, T.; Benyei, A.C. (2008) *J. Heterocycl. Chem.* **45**, 489.

Song, X.-J.; Tan, X.-H. (2008) *Phosphorus Sulfur* **183**, 1755.

Song, X.-J.; Wang, Sh.; Tan, X.-H.; Wang, Z.-Y.; Wang, Y.-G. (2007) *Youji Huaxue* **27**, 72.

Song, X.-L.; Qiu, G.-Zh.; Wang, H.-B; Wu, X.-L., Qu, X-H. (2004) *Zhongguo Youse Jinshu Xuebao* **14**, 291.

Spalinska, K.; Foks, H.; Kedzia, A.; Wierzbowska, M.; Kwapisz, E.; Gebska, A.; Ziolkowska-Klinkosz, M. (2006) *Phosphorus Sulfur* **181**, 609.

Stolle, R.; Johannissien, A. (1904) *J. Prakt. Chem.* **69**, 474.

Stolle, R.; Stevens, M.P. (1904) *J. Prakt. Chem.* **69**, 366.

Supuran, C.; Benedini, F.; Biondi, S.; Ongini, E. (2008) WO Patent 071,421.

Supuran, C.T. (2007) *Curr. Top. Med. Chem.* **7**, 825.

Sybo, B.; Bradley, P.; Grubb, A.; Miller, S.; Proctor, K.J.W.; Clowes, L.; Lawrie, M.R.; Sampson, P.; Seed, A.J. (2007) *J. Mater. Chem.* **17**, 3406.

Takamizawa, A.; Matsushita, Y.; Harada, H. (1980) *Chem. Pharm. Bull.* **28**, 447.

Tan, Ch.X.; Feng, R.F.; Peng, X.X. (2007) *Chinese Chem. Lett.* **18**, 505.

Thomas, H.; Toledo-Sherman, L.M. (2007) US Patent 213,378.

Thomasco, L.M.; Gadwood, R.C.; Weaver, E.A.; Ochoada, J.M.; Ford, C.W.; Zurenko, G.E.; Hamel, J.C.; Stapert, D.; Moerman, J.K.; Schaadt, R.D.; Yagi, B.H. (2003) *Bioorg. Med. Chem. Lett.* **13**, 4193.

Tseng, B.-Ch.; Wu, Y.-L.; Lee, G.-H.; Peng, Sh.-M. (2007) *New J. Chem.* **31**, 199.

Wan, R.; Wang, J.; Han, F.; Yin, L.; Wang, B. (2009) Chin. Patent 101,357,909.

Wan, R.; Wang, J.; Wu, F.; Han, F.; Yin, L.; Wang, B. (2008) Chin. Patent 101,148,440.

Wan, X.-sh.; Zhu, Y.-f.; Shi, B.-b. (2000) *Mater. Protect.* **33**, 37.

Wang, X.; Li, Zh.; Da, Y.; Chen, J. (2001) *Indian J. Chem.* **40B**, 422.

Wang, Y.; Zhang, L.; Fan, Y.; Hou, H.; Shen, X. (2007) *Inorg. Chim. Acta* **360**, 2958.

Warrener, R.N.; Margetic, D.; Tiekinik, E.R.T.; Russell, R.A. (1997) *SYNLETT*, 196.

Werber, G.; Buccheri, F.; Gentile, M. (1977) *J. Heterocycl. Chem.* **14**, 823.

Wheeler, W.J.; Blanchard, W.B. (1992) *J. Labelled Compd. Radopharm.* **31**, 495.

Yamamoto, R.; Yasuda, T.; Sakai, Y.; Aramaki, Sh. (2006) Jpn. Patent 131,799.

Yang, B.; Doweyko, L.M. (2008) WO Patent 057,859.

Yang, H.; Song, W.; Ji, J.; Zhu, X.; Sun, Y.; Yang, R.; Zhang, Z. (2008) *Appl. Surf. Sci.* **255**, 2994.

Yarovenko, V.N.; Shirokov, A.V.; Zavarzin, I.V.; Krupinova, O.N.; Ignatenko, A.V.; Krayushkin, M.M. (2004) *Synthesis* 17.

Yokoyama, N.; Akisada, T.; Kusano, Sh. (2007) WO Patent 119,461.

Yoshida, Y.; Matsuda, K.; Sasaki, H.; Matsumoto, Y.; Matsumoto, S.; Tawara, S.; Takasugi, H. (2000) *Bioorg. Med. Chem.* **8**, 2317.

Yurt, A.; Balaban, A.; Kandemir, S.U.; Bereket, G.; Erk, B. (2004) *Mater. Chem. Phys.* **85**, 420.

Yusuf, M.; Khan, R.A.; Ahmed, B. (2008) *Bioorg. Med. Chem.* **16**, 8029.

Yuye, Y. (2007) *Asian J. Chem.* **19**, 3141.

Zhang, Q.-Ch.; Xing, J.-J.; Liu, L.; Quan, J.-J. (2007) *Bonhai Daxue Xuebao Ziran Kexueban* **28**, 123.

Zhivotova, T.S. (2007) *Izv. Nats. AN Resp. Kazakhstan Ser. Khim.*, 66.

Zhu, L.; Liu, R.; Wang, J. (2006) *Zhongguo Fushi Yu Fanghu Xuebao* **26**, 125.

LOOKING OUT OVER CHAPTER 2: COMPARISON OF ISOMERIC THIADIAZOLES

There are four combinations of heteroatoms in thiadiazoles. Scheme 2.129 represents 1,2,3-, 1,2,4-, 1,2,5-, and 1,3,4-isomers, respectively.

According to the alternation of heteroatoms, electronegativity of isomeric thiadiazoles changes in order 1,2,5-thiadiazole > 1,2,4-thiadiazole > 1,2,3-thiadiazole > 1,3,4-thiadiazole. According to calculations (Bakulev and Dehaen 2004, Glossman-Mitnik 2006), 1,2,3-isomer is the least stable. Note that the 1,2,3-isomer is the only one that is capable of eliminating nitrogen upon heating or photoirradiation, with the formation of thioketene.

Bird (1985) calculated aromaticity indexes (I_A) for isomeric thiadiazoles. The calculation was based on bond length in the corresponding molecules. Benzene was

SCHEME 2.129

the understandable reference compound with $I_A = 100$. This author obtained the following sequence: 1,2,5-thiadiazole ($I_A = 84$) > 1,2,4-thiadiazole ($I_A = 72$) > 1,3,4-thiadiazole ($I_A = 63$) > 1,2,3-thiadiazole ($I_A = 54$). Later, Friedman and Ferris (1997) performed ab initio calculations of aromaticity for 1,1-dioxides of isomeric thiadiazoles. According to the aromaticity degree, the thiadiazole dioxides considered were arranged in the following manner: 1,2,5-thiadiazole > 1,2,3-thiadiazole > 1,3,4-thiadiazole. Among sulfur-containing five-membered heterocycles, thiophene is the most aromatic compounds, but its 1,1-dioxide has the lowest degree of aromaticity (Friedman and Ferris 1997). Obviously, the relative aromaticity of thiadiazole dioxides depends on the ability of the nitrogen atoms to work against electron withdrawal from the ring by the sulfur dioxide moiety. By and large, the data obtained for 1,1-dioxides of isomeric thiadiazoles are most probably relevant to relative aromaticity of the thiadiazoles themselves.

From all of the correlations mentioned above, 1,2,5-thiadiazole is the most perfect in the sense of aromaticity and it prevails above 1,3,4-thiadiazole. This is an intrinsic peculiarity of 1,2,5-*thia*diazole: $I_A = 43$ of 1,2,5-*oxa*diazole is lower than $I_A = 50$ of 1,3,4-*oxa*diazole and $I_A = 47$ of 1,2,5-*selena*diazole is lower than $I_A = 53$ of 1,3,4-*selena*diazole (Bird 1985).

Kunz et al. (1997) compared bioactivity of S-methyl benzo-1,2,3-thiadiazole-7-carbothioate with its isomeric counterpart containing the 1,2,5-thiadiazole moiety. Whereas the 1,2,3-isomer manifested high and diverse bioactivity in plants, the 1,2,5-isomer was inactive. The majority of published data on drug design is devoted to compounds of the 1,3,4-thiadiazole family although biomedical activity was also observed for representatives of other thiadiazoles. Partial aromaticity of the 1,3,4-ring assists in keeping its integrity during transportation within organisms but does not prevent the ultimate participation in metabolic reactions.

REFERENCES TO CHAPTER 2: OVERLOOK

Bakulev, V.A.; Dehaen, W. (2004) *The Chemistry of 1,2,3-Thiadiazoles*. Wiley, Hoboken, NJ.
Bird, C.W. (1985) *Tetrahedron* **41**, 1409.
Friedman, P.; Ferris, K.F. (1997) *TEOCHEM* **418**, 119.
Glossman-Mitnik, D. (2006) *J. Mol. Graph. Model.* **25**, 455.
Kunz, W.; Schurter, R.; Maetzke, T. (1997) *Pesticide Sci.* **50**, 275.

3 Selenadiazoles

3.1 1,2,3-SELENADIAZOLES

Selenadiazoles of this series are well-known sources of alkynes and cycloalkynes, especially strained ones. The alkynes are products of the 1,2,3-seleneadiazole thermolysis resulting in elimination of nitrogen and selenium. Ring-opening reactions are also known to give linear seleno derivatives. This section will consider principal cases of preparation, structural features, and reactivity of 1,2,3-selenadiazoles along with their importance in the sense of biomedical and technical applications.

3.1.1 FORMATION

Since the selenadiazoles under consideration can extrude selenium and nitrogen, mild conditions are needed during their preparation.

3.1.1.1 From vic-Arylselenol Diazonium Salts

Scheme 3.1 illustrates this method; derivatives with substituents at position 4 or 5 in the benzene ring are also available (Keimatsu and Satoda 1935).

Kreis et al. (2005) proposed to perform the reaction depicted in Scheme 3.1 using the diazo component on solid support (Merrifield resin with piperazine as a linker). In the published example (X=H), the yield was moderate, 63%, and the reaction condition was complicated (usage of butyl lithium in argon in water-free tetrahydrofuran, temperature −40°C, duration more than 12 h). Nevertheless, this method should be mentioned because of its possible future development.

3.1.1.2 From Arylisoseleno Cyanates and Diazocompounds

The method consists of diazomethane cycloaddition to the Se=C bond of arylisoselenocyanates. As part of the reaction development, selenadiazoline derivatives are initially formed. These derivatives undergo 1,3 hydrogen transfer and give final products, Scheme 3.2 (Suchar and Kristian 1975). Diazomethane is commonly used reactant, but care should be taken having in mind its explosive ability when initiated by shock, friction, or heat. Inhalation with diazomethane is also dangerous.

For this reason, a modification was proposed that consists of a reaction between aroyl chloride and potassium isoselenacyanate and ethyl diazoacetate to form 5-(aroylimino)-2,5-dihydro-1,2,3-selenadiazole-4-carboxylate (Zhou and Heimgartner 2000). Scheme 3.3 depicts a mechanism consisting of the initial formation of the aroyl isoselenocyanate followed by 1,3-dipolar cycloaddition of diazo compound to the C=Se bond.

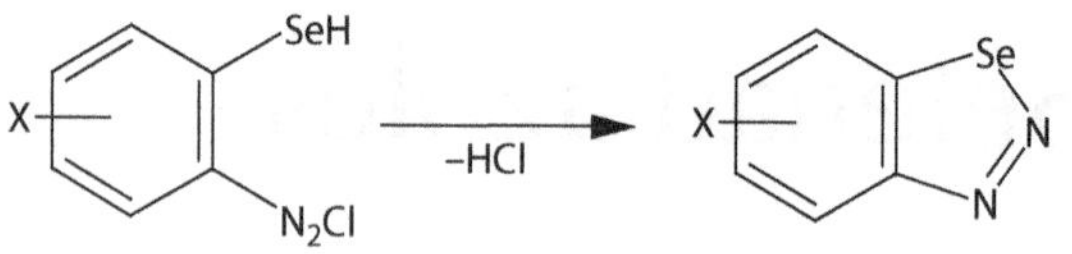

SCHEME 3.1

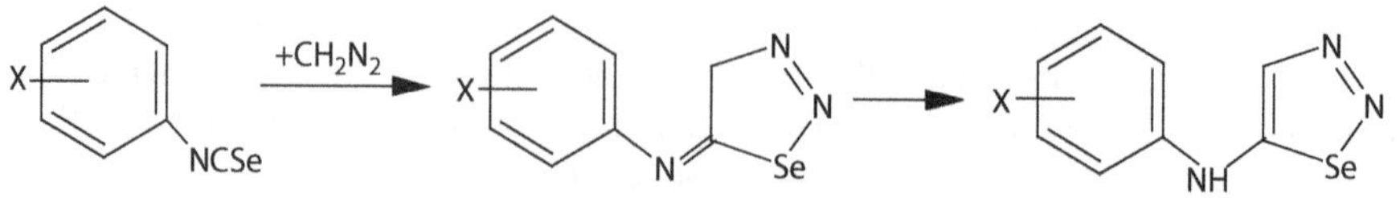

SCHEME 3.2

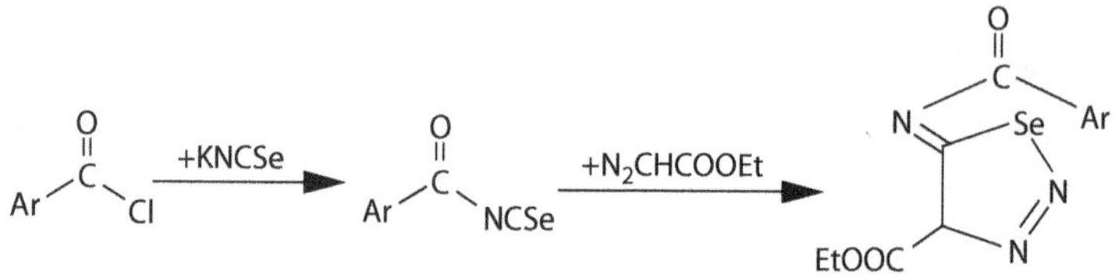

SCHEME 3.3

3.1.1.3　From Hydrazones and Diselenium Dichloride or Selenium Oxychloride

This reaction is distinguished by perfect regioselectivity (see Scheme 3.4) (Grandi and Vivarelli 1989).

　　Because hydrazones are the predecessors of 1,2,3-selenadiazoles, the availability of the hydrazones is a key problem in the selenadiazole preparation. Attanasi et al. (2003) proposed expeditious synthesis of 1,2,3-selenadiazoles using 1,2-diaza-1, 3-butadienes as hydrazone precursors and further reactions with selenium dioxide (Lalezari method, see Lalezari et al. 1973a) or selenium oxychloride (Hurd and Mori 1955). The overall reaction is seen from Scheme 3.5. Namely, α-substituted hydrazones (obtained from 1,2-diaza-1,3-butadiene) and substrates (containing activated methylene or methyne moiety) react with selenium dioxide to yield 4-substituted 1,2,3-selenadiazoles. Reaction of such hydrazones with selenium oxychloride leads to 4-substituted 2,3-dihydro-1,2,3-selenadiazoles. Yields are high enough. The reaction conditions marked in Scheme 3.5 define the product structures and the possibility of aromatization (Attanasi et al. 2003).

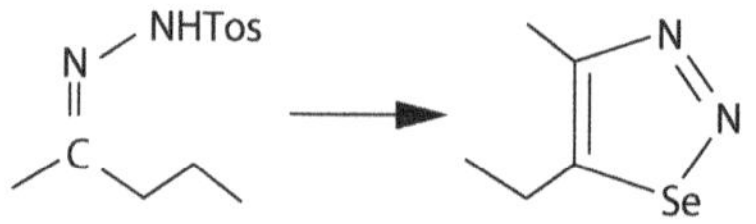

SCHEME 3.4

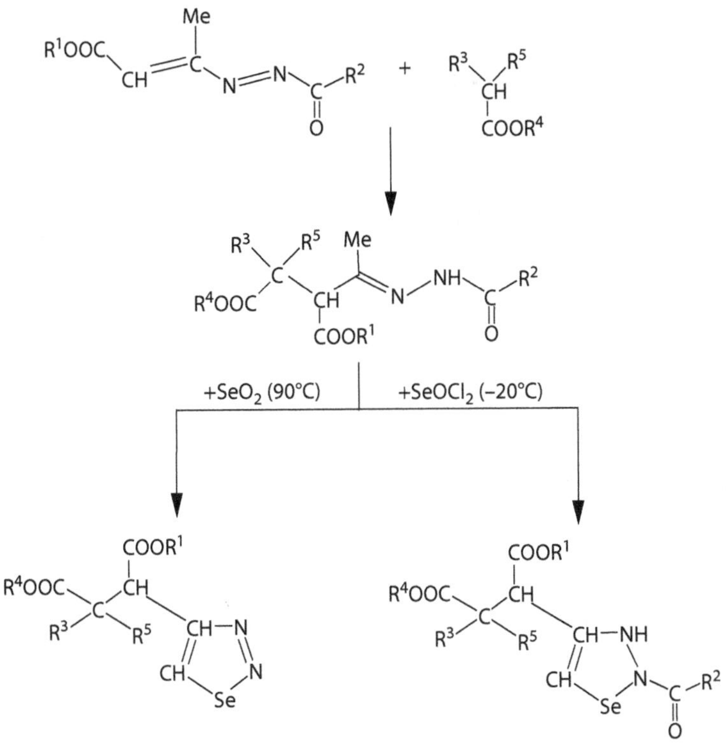

SCHEME 3.5

3.1.1.4 From Semicarbazones and Selenium Dioxide

The semicarbazones of aldehydes and ketones react with selenium dioxide to give 1,2,3-selenadiazoles. Beginning from Lalezari et al. (1969), most authors use either acetic acid as the solvent with suspended selenium dioxide or the solution of selenium dioxide in aqueous dioxane (see also Meier and Voigt 1972). Scheme 3.6 presents one typical reaction.

Scheme 3.7 shows that the ring closure does not depend on the Z/E-ratio of the semicarbazone (Meier and Voigt 1972, Schuhmacher and Meier 1992). The E conformer is involved in cyclization predominantly, and methylene hydrogens of a semicarbazone are more reactive than the methyl hydrogens (Zimmer and Meier 1981).

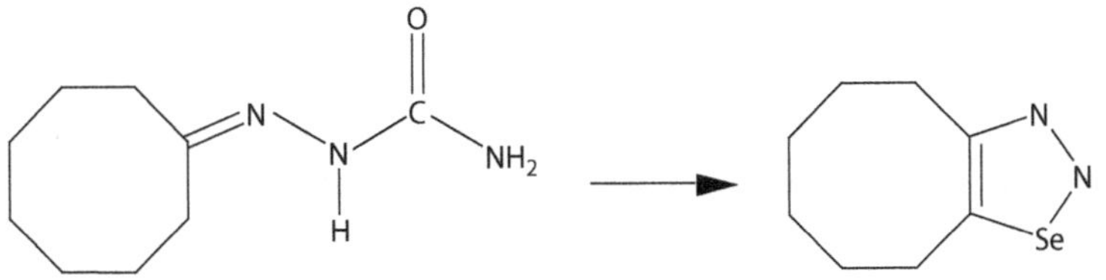

SCHEME 3.6

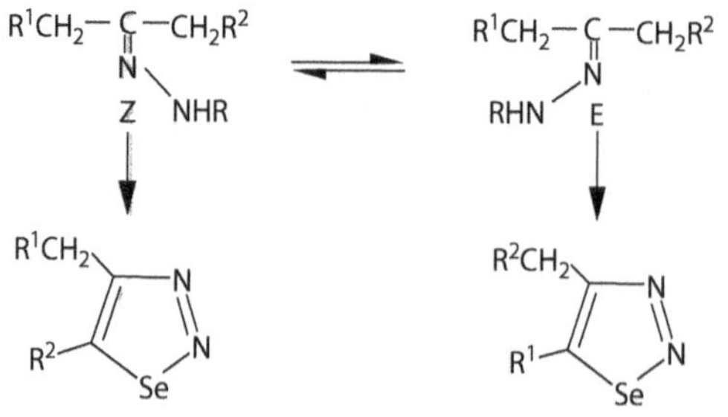

SCHEME 3.7

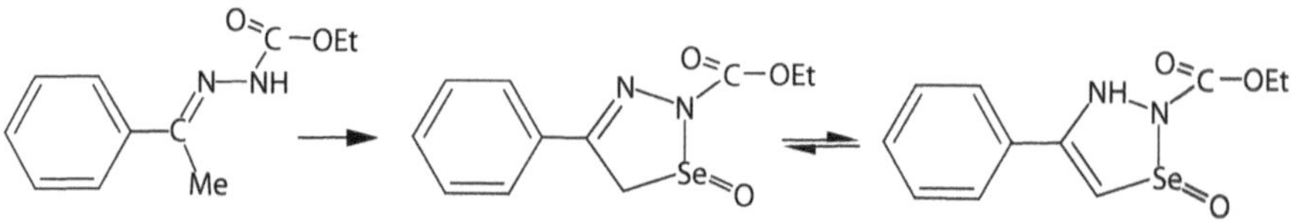

SCHEME 3.8

One recent example of the semicarbazide reaction deserves to be mentioned here (see Scheme 3.8). In this case, the final product is not the corresponding selenadiazole, but Se-oxide (Al-Smadi and Ratrout 2004).

Importantly, oxidation of semicarbazones from (+)3-carene or α-pinene with selenium dioxide leads to chiral 1,2,3-selenadiazole derivatives (Morzherin et al. 2001). It means that reactions of this type are not accompanied with epimerization.

At present time, reactions in dry media under microwave irradiation become very popular. As applied to the preparation of 1,2,3-selenadiazole derivatives, such reactions require precautions because the selenadiazole are prone to thermal destruction. Thus, heterocyclization of diarylpiperidone semicarbazone was proposed to perform under microwave dry heating. Sodium bisulfate and silica were also added to the reaction mixture (Gopalakrishnan et al. 2008a).

3.1.2 FINE STRUCTURE

Data on electronic structure of 1,2,3-selenadiazoles are scarce. Based on ^1H and ^{13}C NMR spectra, Calpin (1973, 1974), Sharma et al. (1994) concluded that there is direct polar conjugation between 4-aryl substituents and $C_3=C_4$ double bond of heterocyclic ring. Duddeck et al. (1990) studied ^{15}N and ^{77}Se spectra of cycloalkeno-1,2,3-selenadiazoles and observed an extremely large ^{77}Se chemical shift and significant $^{15}N–^{77}Se$ coupling constants. (Duddeck and Hotopp (1995) proposed ^{77}Se chemical shifts of 1,2,3-selenadiazole derivatives as suitable tools for chiral discrimination in diastereomeric compounds.) The NMR results mentioned were ascribed to electron interaction between nitrogen and selenium within the selenadiazole parts. There are neither measurements of ring current within the heterocycle nor theoretical calculations of aromaticity indexes for 1,2,3-selenadiazole. Therefore, it is difficult to judge on the degree of aromaticity of the 1,2,3-selenadiazole ring.

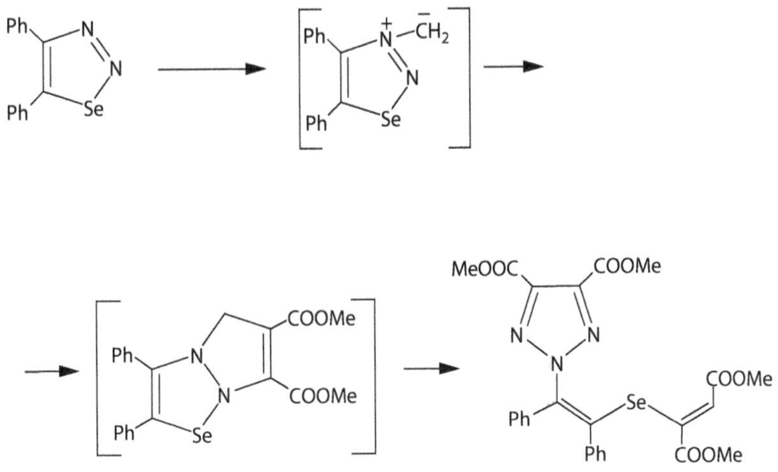

SCHEME 3.9

Crystallographic studies (Gunasekaran et al. 2007, Marx et al. 2007, 2008) note the weak intramolecular interaction between C–H and Se, but such an interaction is usual for compounds containing so voluminous an atom as selenium.

3.1.3 Reactivity

Distinctive feature of 1,2,3-selenadiazoles is their sensitivity to heat and light and high chemical reactivity toward strong bases. This makes them starting materials for various synthetic procedures leading to numerous new organic compounds. Selenadiazoles with 1,2,3 alternation of the hetero atoms are readily decomposed by extrusion of nitrogen, or selenium, or both of these elements. The resulting products are alkynes, 1,3-diselenafulvenes, selenophenes, selenoesters, selenoamides, or alkylseleno carboxylic acids. In the recent review, Dehaen et al. (2004) combined examples of the corresponding transformations. The most representative reactions will be listed here.

3.1.3.1 Salt Formation

4,5-Diphenyl-1,2,3-selenadiazole was alkylated with $Me_3SiCH_2OSO_2CF_3$ at N_3 and N_2, respectively. Scheme 3.9 traces the following transformations of the selenadiazolium obtained using the formation of N_3 isomer as a representative example. The following desilylation generates a transient selenadiazolium methanide (ylide) intermediate. The latter can be trapped with dimethyl acetylene dicarboxylic acid to give the final product depicted by Scheme 3.9 (Butler and Fox 2001).

3.1.3.2 Ring Cleavage

Malek-Yazdi and Yalpani (1977), then Zachinyaev and Orlov (1980) prepared potassium arylethynylselenolates, $Ar–C\equiv C–SeK$, from aryl-1,2,3-selenadiazoles. The reaction consisted of treating substrate with potassium hydroxide in dioxane.

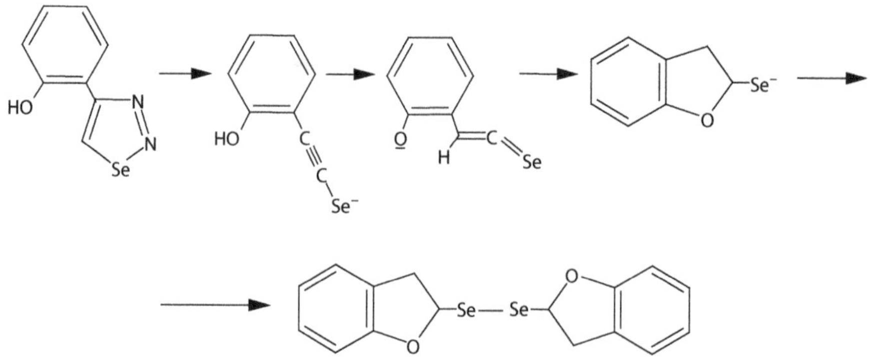

SCHEME 3.10

The selenolates then easily transform into seleno and thio amides (Malek-Yazdi and Yalpani 1977). The latter transformation is an example of the 1,2,3-selenadiazole usage as a synthon for variable organic substances. Laishev et al. (1981) claimed an approach to aryl(alkyl) ethynyl selenides, RC≡CSeR′, by treatment of 4-R-1,2,3-selenadiazoles with BuLi in dioxane or in tetrahydrofuran and then with alkyl halides, R′Hal. Here, R′ stands for alkyl and phenyl. R′ can also be halo, nitro, alkoxy, or dialkylamino group. Petrov et al. have published a series of papers and described the transformation of 4-(2-hydroxyphenyl)-1,2,3-selenadiazole into sodium coumarone 2-selenide under treatment with sodium carbonate in acetonitrile (Petrov et al. 1999, 2000a,b, 2001, 2002, 2006). The sodium selenide was alkylated to give the corresponding alkylselenide or oxidized with iodine to form diselenide by the usual way (see Scheme 3.10; coumarone is benzofuran).

Base-initiation of 1,2,3-selenadiazoles decomposition was used for preparation of E/Z diphenyltetraselenafulvalene according to Scheme 3.11 (Jackson et al. 1987, Rajagopal et al. 2002). This opens new ways to prepare components of organic metals.

3.1.3.3 Photolysis

Light irradiation of 1,2,3-selenadiazole results in the formation of selenoketene (Scheme 3.12) (Meier and Menzel 1972, Harrit et al. 1985).

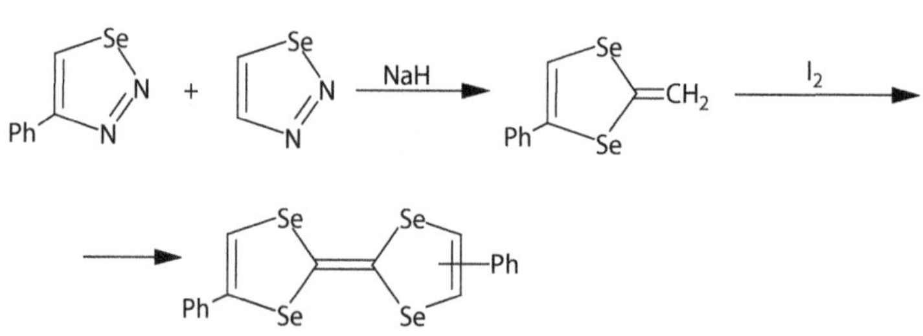

SCHEME 3.11

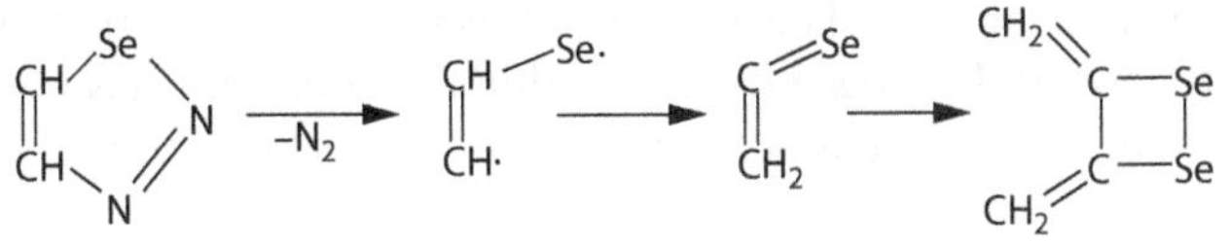

SCHEME 3.12

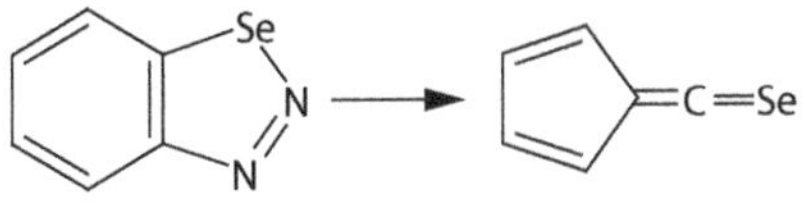

SCHEME 3.13

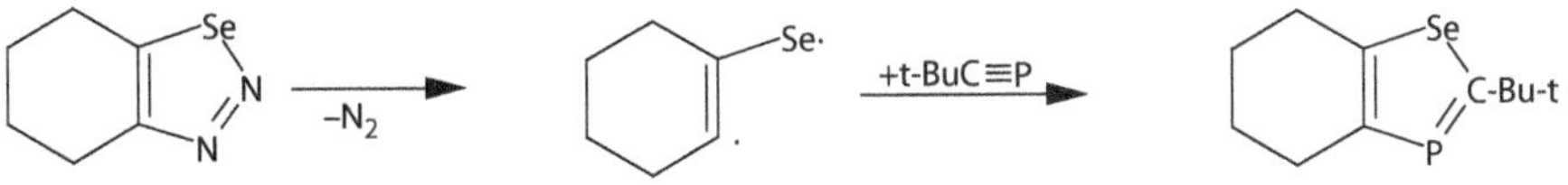

SCHEME 3.14

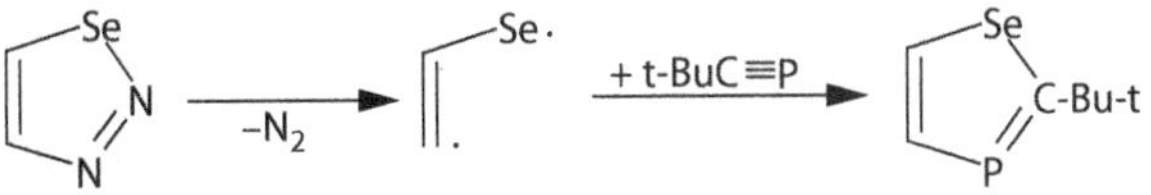

SCHEME 3.15

In the case of benzo-condensed selenadiazole, photolysis also leads to the seleno-ketene-like product, namely, 6-fulvenoselenone (Scheme 3.13) (Schulz and Schweig 1984a,b).

Photolysis of tetrahydrobenzo-1,2,3-selenadiazole in the presence of the kinetically stabilized phosphaalkyne t-BuC≡P gives rise to the corresponding 1,3-selenaphosphole (Burkhart et al. 1991). Biradical of CR=CR–Se is presumed to be responsible for this cycloaddition (Scheme 3.14) (cf. thermolytic reaction, Scheme 3.15).

In contrast to Scheme 3.14, photolytic reaction of ethyl 4-(4-methylphenyl)-1,2,3-selenadiazole-5-carboxylate with N-bromosuccinimide in the presence of benzoyl peroxide leads to elimination of selenium, nitrogen, and ethyl (4-methylphenyl)pro-pynyl carboxylate. No methyl-bromination of the starting selenadiazole takes place (Foroumadi et al. 2005).

3.1.3.4 Thermolysis

Gas-phase pyrolysis of 1,2,3-selenadiazoles results in the elimination of the two ring-nitrogen atoms and in the formation of selenoketenes which were isolated at −196°C (Holm et al. 1979, Bock et al. 1980, Schulz and Schweig 1980). The seleno-ketenes were identified by various spectral methods (see also Bak et al. 1978, 1981, Sander and Chapman 1985).

These selenoketenes were not found in liquid-phase thermolysis experiments (Schulz and Schweig 1984a). Scheme 3.12 also describes the gas-phase thermolysis of

the parent 1,2,3-selenadiazole. Namely, after nitrogen is expelled, ·CH=CH–Se· biradical is initially formed. Hydrogen transfer within the biradical leads to CH_2=C=Se
selenoketene. The latter is unstable and thermally cyclodimerizes according to
Scheme 3.12.

The biradical formed in reaction (3.12) can be intercepted with phosphaalkynes.
The resulting products are 1,3-selenaphospholes (Scheme 3.15) (Regitz and Krill
1996).

In condensed phase, thermal decomposition of 1,2,3-selenadiazoles proceeds in
quite another way. Lalezari et al. (1973b) established that 4-aryl-1,2,3-selenadiazole
on heating afforded 2,5-diarylselenophenes and small quantities of 2,4-diarylsel-
enophenes. Heating of monoarylacetylenes with elemental selenium led to the same
products. The same products were also obtained during heating of a mixture of
4-aryl-1,2,3-selenadiazole with monoarylacetylene. A large excess of arylacetylene
permits to obtain 2,5-diarylselenophenes as the sole products with yields up to 80%
(Arsenyan et al. 2002). All such thermal reactions proceed through liberation of all
of the heteroatoms and generate alkynes. Another product, elemental selenium forms
in a very active form. It is the active form of selenium that attacks monoarylacetylene. The latter two combine and form diarylselenophene. The primary products of
the 1,2,3-selenadiazole thermolysis, alkynes and active selenium, are both important
for further aimed synthesis. The following examples illustrate this statement.

Thermolysis of cycloalkeno-1,2,3-selenadiazoles allows obtaining cycloalkynes
even if they are sterically hindered. Of course, the thermolysis should be performed
with precautions. The precautions (usual for thermal decompositions) demand usage
of small amounts of the starting materials. It is better to perform thermolysis using
"solid diluents" such as sand, glass, and copper-powder. For instance, cyclodecyne
was obtained with 90% yield after heat of the cyclodecene-1,2,3-selenadiazole dispersed in sand at 180°C (Scheme 3.16) (Meier and Menzel 1972).

The pyrolytic decomposition depicted in Scheme 3.16 is typical but not solely possible. Scheme 3.13 is also applicable to describe the results of flash vacuum pyrolysis
of benzo-1,2,3-selenadiazole: Instead of benzyne, 6-fulvenoselenone was fixed as
the product (Schulz and Schweig 1980b).

The ketone-semicarbazide-1,2,3-selenadiazole-(thermolysis)-alkyne sequences
are used specially to introduce a triple bond during the syntheses of very complicated organic and organoelemental compounds. Coordination of cyclo-fused
selenadiazole compounds with cobalt cyclopentadienyl does not prevent the cyclo-
alkyne generation (Roers et al. 2000). Thus, such a sequence was a key approach to
belt-like macrocycles with four and eight (cyclopentadienyl)(cyclobutadienyl)cobalt
building blocks, formed by stepwise oligomerization of 1,6-cyclodecadiyne units
(Schaller et al. 2002). The reaction of 4-phenyl-1,2,3-selenadiazole with triphenyl phosphine in boiling benzene leads to triphenyl selenophosphorane (Ph_3P=Se)

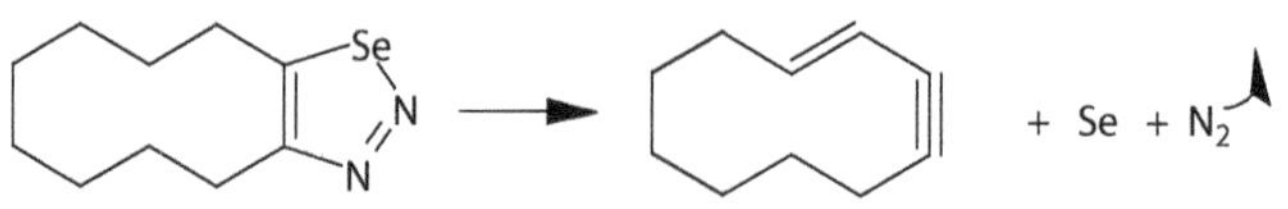

SCHEME 3.16

and phenylacetylene. When tributyl phosphine was used instead of triphenyl phosphine, the reaction could be performed in the same solvent (benzene) but at room temperature. Tributyl selenophosphorane and phenylacetylene are formed in this case (Arsenyan et al. 2004). As seen, activity of a reactant defines the selenadiazole decomposition.

Nishiyama et al. (2000) examined the reaction of 1,2,3-selenadiazoles with excess amount of olefins at 130°C in autoclave. With the 1,2,3-selenadiazole derived from cyclic ketones, the reaction with olefins results in the formation of dihydroselenophenes in moderate to good yields. On the other hand, in reaction of the 1,2,3-selenadiazoles derived from aromatic and linear ketones, the corresponding alkynes are formed as the sole product. Obviously, reactants of variable activity differ in their ability to intercept intermediates that form during the 1,2,3-selenadiazol thermal decomposition. The intermediates also differ with respect to their relative activity.

Selenium that is formed in its highly reactive form (see Scheme 3.16) is now finding applications in the synthesis of metal selenide nanopowders. Thus, thermolysis of cyclohepteno-1,2,3-selenadiazole in the presence of silver nitrate results in the formation of pure silver selenide as particles of the size less than 500 nm (Khanna and Das 2004). The cadmium selenide nanopowder was analogously prepared from cyclohepteno- (Khanna et al. 2004a) or cycloocteno-1,2,3-selenadiazole (Khanna et al. 2003, 2004b) and cadmium chloride or acetate. Metal selenides are important materials in optoelectronics. Thermolysis of 1,2,3-selenadiazoles opens a greener way in their preparations as compared to the usually used oxidation of toxic hydrogen selenide.

3.1.3.5 Coordination

Although the key heteroatom, selenium, has an unshared electron pair, boron trifluoride etherate and phenyldichloroborate are complexing to the adjacent nitrogen atom. The complexes formed are stable, the boron atom is tetracoordinated (Scheme 3.17) (Arsenyan et al. 2007a).

The same manner of coordination was observed for the (4-methyl-1,2,3-selenadiazole)chromium pentacarbonyl complex (Baetzel and Boese 1981, Pannell et al. 1983). Coordination to transition metals proceeds at the expense of the selenadiazole-N_3 in 4-(2-pyridyl)-1,2,3-selenadiazole. Involving this N_3 into coordination becomes possible due to the metal chelation with the pyridinyl nitrogen and the closest selenadiazole-N_3 (Richardson and Steel 2002). The authors postulated the selenium coordination to silver as the first step of interaction between 4-(2-pyridyl)-1,2,3-selenadiazole and silver nitrate in acetonitrile. Formylbenzotriazole is the final product of the sequence represented by Scheme 3.18 (Richardson and Steel 2002).

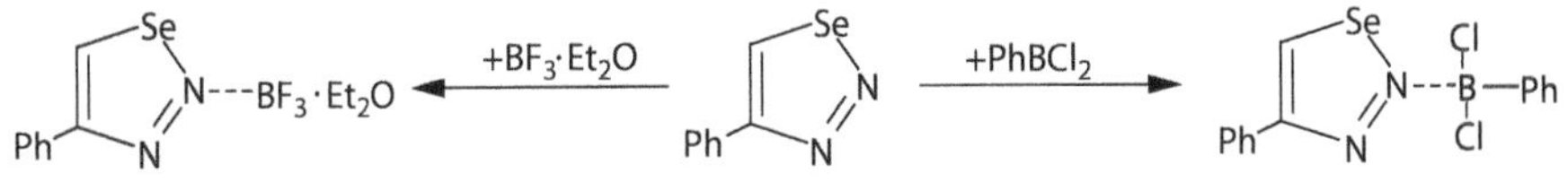

SCHEME 3.17

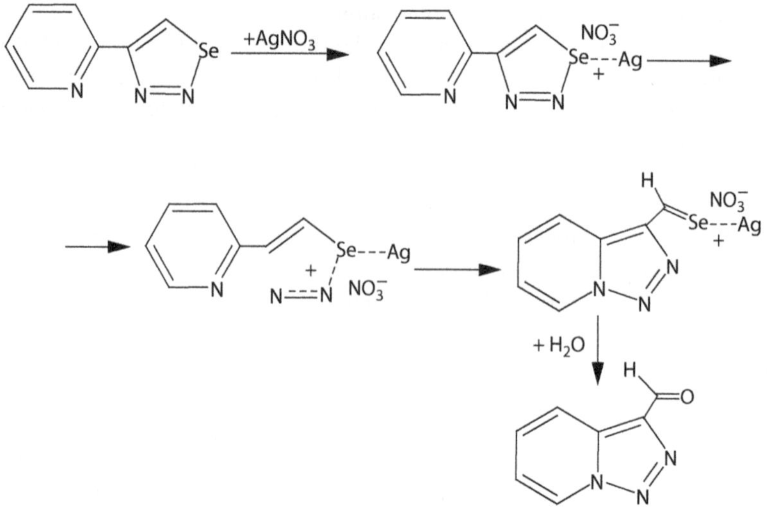

SCHEME 3.18

3.1.3.6 Ring Transformation

Scheme 3.19 describes melting of a condensed 1,2,3-selenadiazole with elemental sulfur. The reaction leads to a mixture of 1,3,5-trithiole and 1,2,3,4-pentathiepin (Ando et al. 1987). Pentathiepin is produced from the trithiole but not from the starting selenadiazole.

Interestingly, the selenadiazole that is the substrate in the reaction of Scheme 3.19, reacts with elemental selenium in hot (130°C) tributylamine giving 4,4,9,9-tetram ethyl-4,9-dihydro-1,2,3-triselenacyclopenta[*b*]naphthalene (Scheme 3.20) (Tokitoh et al. 1988). Seemingly, tributylamine acts as a solvent and as donor component facilitating the reaction here.

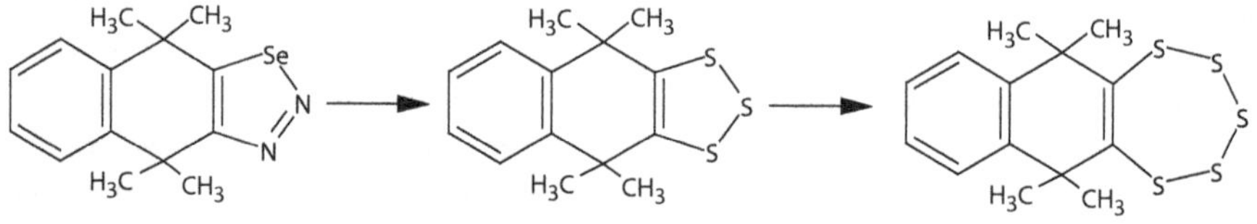

SCHEME 3.19

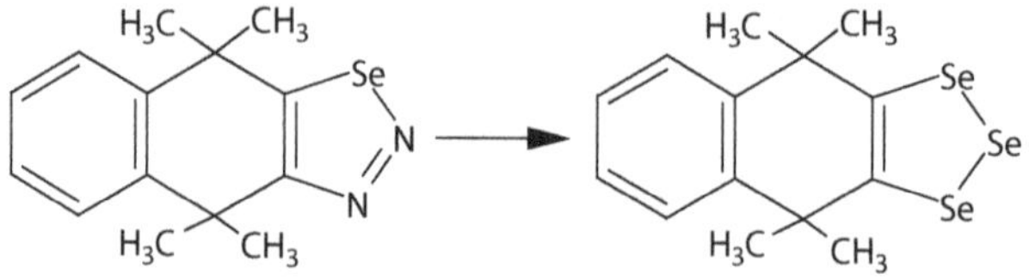

SCHEME 3.20

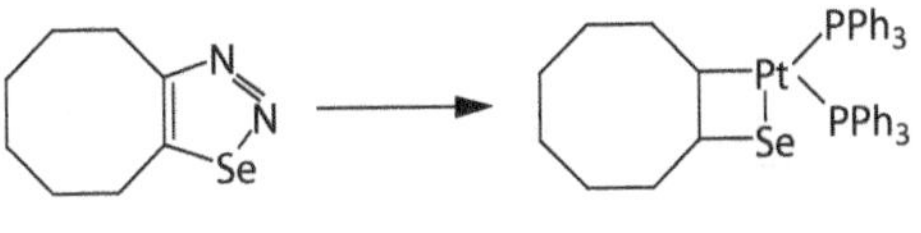

SCHEME 3.21

Reacting with tetrakis(triphenylphosphine) platinum, the 1,2,3-selenadiazole ring transforms into the selena-platinum four-membered ring (Scheme 3.21) (Khanna and Morley 1995).

Using the same reactant, namely tetrakis(triphenylphosphine) platinum, and ethyl 4-methyl-1,2,3-selenadiazole-2-carboxylate, Arsenyan et al. (2005) obtained diethyl 3,5-dimethyl-1,7-diselena-3a,4-diaza-7a-platinaindene-2,6-dicarboxylate. The structure of this complex was established by x-ray analysis. The complex effectively catalyzes hydrosilylation acetylenes with dimethylphenylsilane. Scheme 3.22 represents this platinum complex and the hydrosilylation mentioned (Arsenyan et al. 2005).

In contrast to the reaction of Scheme 3.21, tetrakis(triphenylphosphine) *palladium* forms the cyclo(diselenapalladium) dimer according to Scheme 3.23 (Ford et al. 1999).

The difference is obviously defined by the reaction mechanism. The intimate mechanism of these transformations is unknown. It is only possible to note that the nitrogen removal should lead to selenoketene, probably in its biradical form. What proceeds further is now unclear. It can be added that the nature of the product formed depends on the phosphine basicity used as a reagent. Thus, the same cyclooctenoselenadiazole from Scheme 3.23 reacts with the mixture of tibutylphosphine and palladium dibenzylidene acetone [Pd$_2$(dba)$_3$] forming the product analogous to that of Scheme 3.22. Having been established by x-ray method, the structure of this product is shown in Scheme 3.24 (Ford et al. 1998, Morley et al. 2004).

Reaction of cyclohepteno-1,2,3-selenadiazole and (η^5-C$_5$H$_5$)$_2$Mo$_2$(CO)$_4$ leads to the molybdenum inclusion in the hetero ring instead of the nitrogen atom at position 2. The product is isolatable and depicted by Scheme 3.25 (Cervantes-Lee et al. 1998).

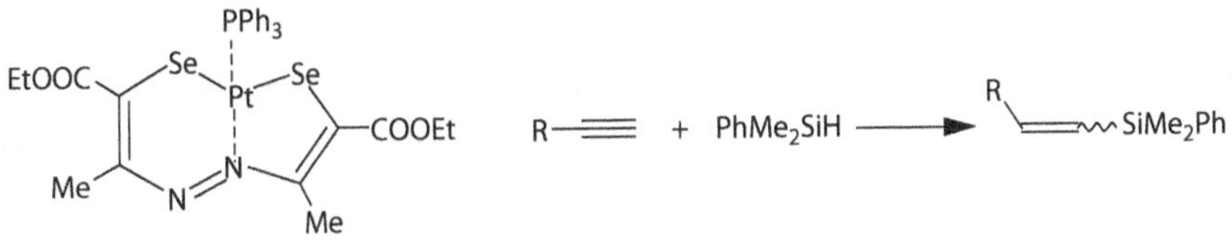

SCHEME 3.22

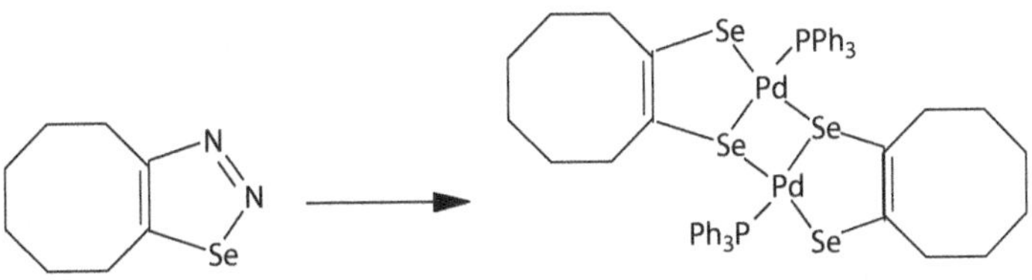

SCHEME 3.23

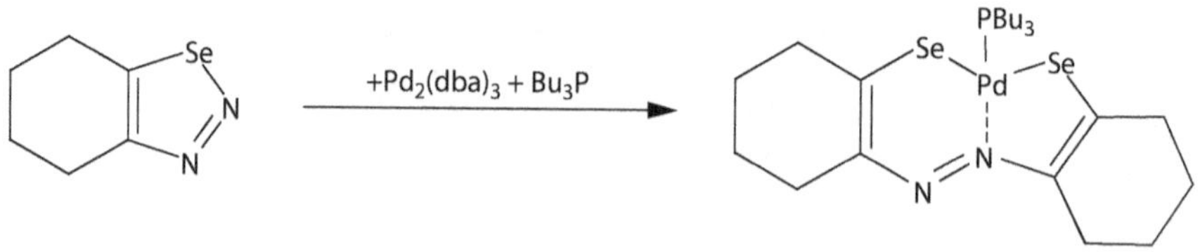

SCHEME 3.24

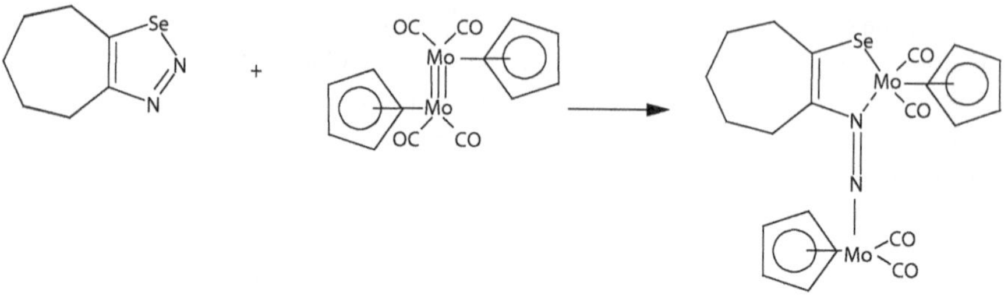

SCHEME 3.25

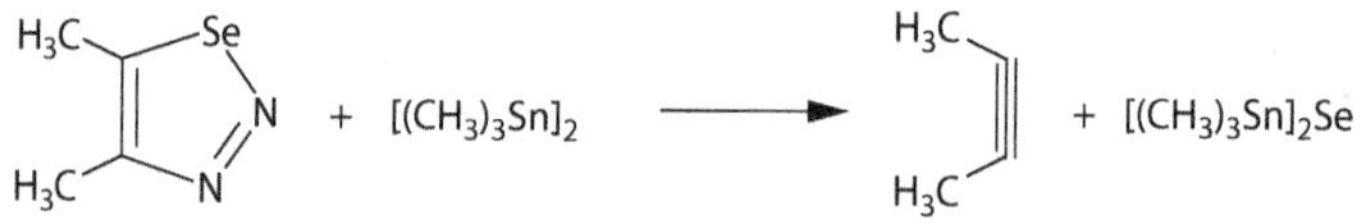

SCHEME 3.26

The reaction of 4,5-dimethyl-1,2,3-selenadiazole with hexamethyl ditin affords dimethylacetylene and bis(trimethyltin) selenide (Arad-Yellin and Wudl 1985) (Scheme 3.26).

The reaction with tributylstannane and vinyl derivatives is initiated with azoisobi-sisobutyronitrile. The tributylstannyl radical formed attacks the initial substrate that results in disclosure of the 1,2,3-selenadiazole ring with the diazonyl intermediate formation. The diazonyl intermediate loses nitrogen and transforms into a carboradi-cal. Subsequent reaction with the vinyl reactant leads to removal of tributylstannyl and to the final dihydroselenophene (Scheme 3.27) (Nishiyama et al. 1999, 2002).

3.1.4　Biomedical Importance

3.1.4.1　Medical Significance

In oncology, current interest lies in the prevention of certain cancers by supplementation organisms with selenium. Selenium is essential for cell metabolism as a component of glutathione peroxidase and other enzyme systems. According to one of the proposed mechanisms, the cytotoxic effect of selenium is caused by its interaction with tumor cells (Mugesh et al. 2001). The ability of 1,2,3-selenadiazoles to liberate selenium (mentioned in Section 3.1.3) can also show up in metabolism. Therefore, selenadia-zoles are objects of significant research in oncomedicine. A Japanese patent claimed

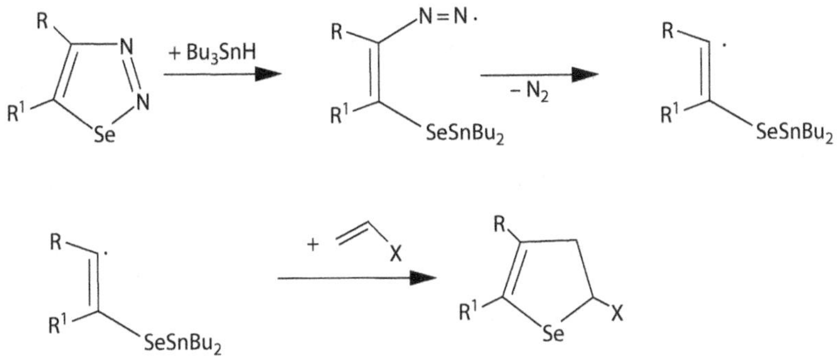

SCHEME 3.27

selenadiazoles active against Ehrlich ascite tumor cells (Hidaka and Sakakahara 1984). Cytotoxic activity of 4-methyl-1,2,3-selenadiazole-5-carboxamides were also described for human fibrosarcoma HT-1080, mouse hepatoma MG-22A, and mouse fibroblasts 3T3 cell lines. In vivo evaluation of these amides on mouse sarcoma S-180 confirms high (58%–85%) antitumor activity (Aresenyan et al. 2007b).

There is an extensive body of works directed to find 1,2,3-selenadiazoles that can be used as bactericides, disinfectants, and antiseptics. This is a task of medicine in general. The substantial increase in the incidence of fungal infection during the past two decades is the major case of morbidity and mortality among immunocompromised patients. Aggressive immunosuppression, HIV-1 infection, cancer, myelotoxic therapeutic regimens, and organ transplantation—all these contemporary features open the door to pathogenic fungal organisms. Despite remarkable efficacy of presently used drugs in the treatment of invasive mycoses, severe toxic reactions and development of fungal resistance to these drugs caused the widening of search for new curative agents. Some recently published relevant examples need to be cited here. Thus, 5,7-bis(4-chlorophenyl)-4,4-dimethyl-4,5,6,7-tetrahydropyridino[3,4-*d*]-1,2,3-selenadiazole is more effective against *Aspergillus flavus* and *Mucor* (Gopalakrishnan et al. 2008b). Activity of 5-(4-biphenylyl)-7-(4-fluorophenyl)-[3,4-*d*]-1,2,3-benzo-selenadiazole against *Staphylococcus aureus* and *Bacillus subtilis* is higher than that of standard medicines (Balasankar et al. 2007). High activity against *Cryptococcus neoformans* of 7-nitro-4,5-dihydronaphtho[1,2-*d*]-1,2,3-selenadiazole is accompanied by low toxicity with respect to brine shrimps (Jalilian et al. 2003). 9-Methyl-11-bromo-[1,2,3]-selenadiazolo-[4,5-*b*]-quinoline may also be listed among prospective bactericides. The compound shows high activity against Eurostrains of the bacteria *S. aureus* and *Escherichia coli* (Bhojya et al. 2006).

3.1.4.2 Imaging Agents

1,2,3-Selenadiazole derivatives inhibit actions of the adrenocorticosteroid hydrolase enzymes. For the reason, search of their biodistribution is constantly kept as an actual task. Thus, ^{75}Se-labeled 4-phenyl-1,2,3-selenadiazole was prepared in more than 98% of radiochemical purity and its biodistribution was studied. Namely, the concentration of the radiolabeled compound in the adrenal glands of rats over a

0.25–24 h period was detected and compared with that in the blood, liver, and kidneys. The concentration in the adrenal glands and the target to nontarget ratios were much lower than those reported for other adrenocorticosteroid inhibitors (Hanson and Davis 1979, 1981). Finding new ^{75}Se-1,2,3-selenadiazole adrenocorticosteroid imaging agents remains an actual task. In this sense, the work of Jalilian et al. (2004) deserves to be mentioned. The authors developed the synthesis of ethyl 4-methyl-1,2,3-selenadiazole-^{75}Se-5-carboxylate. The radiochemical and chemical purity of the product was greater than 98% without application of high-performance liquid chromatography. As the authors mentioned, the reaction can be easily automated to facilitate large-scale synthesis of other ^{75}Se- or ^{73}Se-1,2,3-selenadiazoles.

3.1.5 TECHNICAL APPLICATIONS

3.1.5.1 Optoelectronics

Preparation of metal selenide nanopowders during thermolysis of cycloalkeno-1,2,3-selenadiazoles is an effective way to synthesize materials for optoelectronics. The process has already been considered in Section 3.1.3. The 1,2,3-selenadiazole was mentioned as a "greener source" of selenium as compared to the more toxic hydrogen selenide (Khanna et al. 2003). Selenium deposition is also important in the manufacture of selenium films. 4-Methyl-1,2,3-selenadiazole was used as a precursor for the low-temperature plasma-assisted chemical vapor deposition of selenium. The process is characterized by the following advantages: Growth rate is 50 nm/min at 40°C and plasma power density is 1 W/cm^2 (Jackson et al. 1993).

3.1.5.2 Organic Metals

1,2,3-Selenadiazoles are useful sources to organic metals. Thus, thermolysis of cyclopentano-1,2,3-selenadiazole in carbon disulfide leads to the corresponding annulated sym-diselenadithiafulvalene. The latter with tetracyanoquinodimethane forms a charge-transfer complex of high electric conductivity (Spencer et al. 1975).

3.1.5.3 Catalysts

Ionkin (2002) claimed selective hydrogenation of adipodinitrile for the manufacture of aminocapronitrile. The catalyst for the process was 1,2,3-selenadiazole. Ethyl 4-methyl-1,2,3-selenadiazole-2-carboxylate is the precursor of diethyl 3,5-dimethyl-1,7-diselena-3a,4-diaza-7a-platinaindene-2,6-dicarboxylate, an effective catalyst of hydrosilylation acetylenes with dimethylphenylsilane. Scheme 3.22 has already been introduced to represent this platinum complex and the hydrosilylation mentioned (Arsenyan et al. 2005).

3.1.6 CONCLUSION

Compounds containing the 1,2,3-selenadiazole ring easily eliminate the nitrogen molecule and the selenium atom. This reaction is an obvious route to prepare long-chain ethynic monomers (for dendrimers, in particular) from 1,2,3-selenadiazoles connected by alkyl chains. Synthesis of the latter was described by Al-Smadi (2007).

Various derivatives of 1,2,3-selenadiazoles are widely used to prepare other acetylenic and heterocyclic compounds. This reaction provides possibilities to study reaction mechanisms and leads to practically important substances. In particular, the reaction allows preparing the isotopomer of acetylenic compounds. Thus, phenylacetylene 1-(^{14}C) and 1,4-diphenylbutadiyne 1,4-(^{14}C$_2$) were prepared by Dhawan and Kagan (1982) as follows: Chlorination of Me^{14}COOH followed by condensation with PhH gave Ph^{14}COOH. The latter was converted to its semicarbazone and was cyclocondensed with SeO$_2$ to give 4-phenyl-1,2,3-selenadiazole-(4-^{14}C). Pyrolysis of the selenadiazole at 180°C and 67 Pa gave 1-phenylacetylene 1-(^{14}C). Oxidative dimerization of this acetylene leads to 1,4-diphenylbutadiyne 1,4-(^{14}C$_2$).

Elimination of selenium in active form is seemingly the cause for the biomedical importance of 1,2,3-seleneadiazoles. The same possibility is central to economical and green methods of chemical vapor deposition or preparation of metal selenide nanoparticles for microelectronics.

REFERENCES TO SECTION 3.1

Al-Smadi, M. (2007) *J. Heterocycl. Chem.* **44**, 915.

Al-Smadi, M.; Ratrout, S. (2004) *Molecules* **9**, 957.

Ando, W.; Kumamoto, Y.; Tokitoh, N. (1987) *Tetrahedron Lett.* **28**, 4833.

Arad-Yellin, R.; Wudl, F. (1985) *J. Organometal. Chem.* **280**, 197.

Arsenyan, P.; Oberte, K.; Belyakov, S. (2007a) *Khim. Gheterotsikl. Soed.* 289.

Arsenyan, P.; Oberte, K.; Rubina, K.; Belyakov, S. (2005) *Tetrahedron Lett.* **46**, 1001.

Arsenyan, P.; Oberte, K.; Rubina, K.; Lukevics, E. (2004) *Khim. Gheterotsikl. Soed.* 599.

Arsenyan, P.; Pudova, O.; Lukevics, E. (2002) *Tetrahedron Lett.* **43**, 4817.

Arsenyan, P.; Rubina, K.; Shestakova, I.; Domracheva, I. (2007b) *Eur. J. Med. Chem.* **42**, 635.

Attanasi, O.A.; De Crementini, L.; Favi, G.; Fillipone, P.; Giorgi, G.; Mantellini, F.; Santeusanio, S. (2003) *J. Org. Chem.* **68**, 1947.

Baetzel, V.; Boese, R. (1981) *Zeitschr. Naturforsch.* **36B**, 172.

Bak, B.; Kristiansen, N.A.; Svanholt, H.; Holm, A.; Rosenkilde, S. (1981) *Chem. Phys. Lett.* **78**, 301.

Bak, B.; Nielsen, O.J.; Svanholt, H.; Holm, A. (1978) *Chem. Phys. Lett.* **53**, 374.

Balasankar, T.; Gopalakrishnan, M.; Nagarajan, S. (2007) *J. Enzyme Inhibit. Med. Chem.* **22**, 171.

Bhojya, N.; Halehatty, R.; Machenhalli, S.; Swetha, B.; Roopa, Th. (2006) *Phosphorus Sulfur* **181**, 533.

Bock, H.; Aygen, S.; Rosmus, P.; Solouki, B. (1980) *Chem. Ber.* **113**, 3187.

Burkhart, B.; Krill, S.; Okano, Y.; Ando, W.; Regitz, M. (1991) *SYNLETT*, 536.

Butler, R.N.; Fox, A. (2001) *J. Chem. Soc., Perkin Trans. 1*, 394.

Calpin, A. (1974) *J. Chem. Soc., Perkin Trans. 1*, 30.

Calpin, G.A. (1973) *Org. Magn. Reson.* **5**, 169.

Cervantes-Lee, F.; Parkanyi, L.; Kapoor, R.N.; Mayr, A.J.; Pannell, K.H.; Pang, Y.; Barton, Th.J. (1998) *J. Organometal. Chem.* **562**, 29.

Dehaen, W.; Bakulev, V.A.; Taylor, E.C.; Wipf, P., Eds. (2004) *The Chemistry of Heterocyclic Compounds*, Vol. 62, John Wiley & Sons, Sussex, U.K.

Dhawan, S.N.; Kagan, J. (1982) *J. Label. Compds. Radiopharm.* **19**, 331.

Duddeck, H.; Hotopp, T. (1995) *Magn. Reson. Chem.* **33**, 490.

Duddeck, H.; Wagner, P.; Mueller, D.; Jaszberenyi, J.C. (1990) *Magn. Reson. Chem.* **28**, 549.

Ford, S.; Khanna, P.K.; Morley, C.P.; di Vaira, M. (1999) *J. Chem. Soc., Dalton Trans.* 791.

Ford, S.; Morley, Ch. P.; di Vaira, M. (1998) *Chem. Commun.* 1305.

Foroumadi, A.; Amini, M.; Vosooghi, M.; Vahdatizde, H.; Shafiee, A. (2005) *Chemistry (Rajkot, India)* **1**, 707.

Gopalakrishnan, M.; Thanusu, J.; Kanagarajan, V. (2008a) *J. Korean Chem. Soc.* **52**, 47.

Gopalakrishnan, M.; Sureshkumar, P.; Thanusu, J.; Kanagarajan, V. (2008b) *J. Enzyme Inhibit. Med. Chem.* **23**, 347.

Grandi, R.; Vivarelli, P. (1989) *J. Chem. Res.* 186.

Gunasekaran, B.; Saravanan, S.; Manivannan, V.; Muthusubramanian, S.; Nethaji, M. (2007) *Acta Crystallogr.* **E63**, So 4167.

Hanson, R.N.; Davis, M.A. (1979) *J. Label. Compds. Radiopharm.* **16**, 31.

Hanson, R.N.; Davis, M.A. (1981) *J. Pharm. Sci.* **70**, 91.

Harrit, N.; Rosenkilde, S.; Larsen, B.D.; Holm, A. (1985) *J. Chem. Soc., Perkin Trans. 1*, 907.

Hidaka, H.; Sakikahara, N. (1984) Jpn. Patent 59,020,274.

Holm, A.; Berg, C.; Bjerre, C.; Bak, B.; Svanholt, H. (1979) *J. Chem. Soc., Chem. Commun.* 99.

Hurd, C.D.; Mori, R.I. (1955) *J. Am. Chem. Soc.* **77**, 5359.

Ionkin, A.S. (2002) US Patent 645,523.

Jackson, A.D.; Jones, Ph.A.; Lickiss, P.D.; Pilkington, R.D. (1993) *J. Mater. Chem.* **3**, 429.

Jackson, Y.A.; White, Ch.L.; Lakshmikantham, M.V.; Cava, M.P. (1987) *Tetrahedron Lett.* **28**, 5635.

Jalilian, A.R.; Rowshanfarzad, P.; Afarideh, H.; Shafiee, A.; Sabet, M.; Kyomarsi, M.; Raisali, Ch.R. (2004) *Appl. Radiat. Isotopes* **60**, 659.

Jalilian, A.R.; Sattari, S.; Bineshmarvasti, M.; Daneshtalab, M.; Shafiee, A. (2003) *Farmaco* **58**, 63.

Keimatsu, S.; Satoda, I. (1935) *Yakugaku Zhasshi* **55**, 233.

Khanna, P.K.; Das, B.K. (2004) *Mater. Lett.* **58**, 1030.

Khanna, P.K.; Gorte, R.M.; Gokhale, R. (2004a) *Mater. Lett.* **58**, 966.

Khanna, P.K.; Gorte, R.M.; Morley, C.R. (2003) *Mater. Lett.* **57**, 1464.

Khanna, P.K.; Morley, Ch.P. (1995) *J. Chem. Res.* 64.

Khanna, P.K.; Morley, Ch.P.; Gorte, R.M.; Gokhale, R.; Subbarao, V.V.V.S.; Satyanarayana, C.V.V. (2004b) *Mater. Chem. Phys.* **83**, 323.

Kreis, M.; Nising, C.F.; Schroen, M.; Knepper, K.; Braese, S. (2005) *Org. Biomol. Chem.* **3**, 1835.

Laishev, V.Z.; Petrov, M.L., Petrov, A.A. (1981) Russ. Patent 859, 361.

Lalezari, I., Shafiee, A.; Yalpani, M. (1969) *Tetrahedron Lett.* **10**, 5105.

Lalezari, I.; Shafiee, A.; Yalpani, M. (1973a) *J. Org. Chem.* **38**, 338.

Lalezari, I.; Shafiee, A.; Rabet, F.; Yalpani, M. (1973b) *J. Heterocycl. Chem.* **10**, 953.

Malek-Yazdi, F.; Yalpani, M. (1977) *Synthesis*, 328.

Marx, A.; Manivannan, V.; Saravanan, S.; Muthusubramanian, S.; Sridhar, B. (2007) *Acta Crystallogr.* **E63**, o4676.

Marx, A.; Saravanan, S.; Muthasubramanian, S.; Manivannan, V.; Rath, N.P. (2008) *Acta Crystallogr.* **E64**, o349.

Meier, H.; Menzel, I. (1972) *Tetrahedron Lett.* **36**, 445.

Meier, H.; Voigt, E. (1972) *Tetrahedron* **28**, 187.

Morley, Ch.P; Ford, S.; di Vairo, M. (2004) *Polyhedron* **23**, 2967.

Morzherin, Yu.; Glukhareva, T.V.; Mokrushin, V.S.; Tkachev, A.V.; Bakulev, V.A. (2001) *Heterocycl. Commun.* **7**, 173.

Mugesh, G.; du Mont, W.-W.; Sies, H. (2001) *Chem. Rev.* **101**, 2125.

Nishiyama, Y.; Hada, Y.; Anjiki, M.; Hanita, S.; Sonoda, N. (1999) *Tetrahedron Lett.* **40**, 6293.

Nishiyama, Y.; Hada, Y.; Anjiki, M.; Miyake, K; Hanita, S.; Sonoda, N. (2002) *J. Org. Chem.* **67**, 1520.

Nishiyama, Y.; Hada, Y.; Iwase, K.; Sonoda, N. (2000) *J. Organometal. Chem.* **611**, 488.

Pannell, K.H.; Mayr, A.J.; Hoggard, R.; McKennis, J.S.; Dawson, J.C. (1983) *Chem. Ber.* **116**, 230.

Petrov, M.L.; Abramov, M.A.; Abramova, I.P.; Dehaen, W.; Lyakhovetskiy, Yu, I. (2001) *Zhurn. Org. Khim.* **37**, 1713.

Petrov, M.L.; Abramov, M.A.; Androsov, D.A.; Dehaen, W. (2000a) *Zhurn. Obshch. Khim.* **70**, 1755.

Petrov, M.L.; Abramov, M.A.; Androsov, D.A.; Dehaen, W.; Lyakhovetsky, Yu, I. (2002) *Zhurn. Obshch. Khim.* **72**, 1365.

Petrov, M.L.; Abramov, M.A.; Dehaen, W. (2000b) *Zhurn. Org. Khim.* **36**, 629.

Petrov, M.L.; Abramov, M.A.; Dehaen, W.; Topper, S. (1999) *Tetrahedron Lett.* **40**, 3903.

Petrov, M.L.; Androsov, D.A.; Abramov, M.A.; Abramova, I.P.; Dehaen, W.; Lyakhovetsky, Yu.I. (2006) *Zhurn. Org. Khim.* **42**, 1521.

Rajagopal, D.; Lakshmikantham, M.X.; Cava, M.P. (2002) *Sulfur Lett.* **25**, 129.

Regitz, M.; Krill, S. (1996) *Phosphorus Sulfur* **115**, 99.

Richardson, Ch.; Steel, P.J. (2002) *Austral. J. Chem.* **55**, 783.

Roers, R.; Rominger, F.; Nuber, B.; Gleiter, R. (2000) *Organometallics* **19**, 1578.

Sander, W.W.; Chapman, O.L. (1985) *J. Org. Chem.* **50**, 543.

Schaller, R.J.; Gleiter, R.; Hofmann, J.; Rominger, F. (2002) *Angew. Chem. Intl. Ed.* **41**, 1181.

Schuhmacher, H.; Meier, H. (1992) *Zeitschr. Naturforsch.* **47B**, 563.

Schulz, R.; Schweig, A. (1980) *Angew. Chem.* **92**, 52.

Schulz, R.; Schweig, A. (1984a) *Zeitschr. Naturforsch.* **39B**, 1536.

Schulz, R.; Schweig, A. (1984b) *Tetrahedron Lett.* **25**, 2337.

Sharma, K.S.; Kumari, S.; Kumari, Sh. (1994) *Indian J. Heterocycl. Chem.* **4**, 137.

Spencer, H.K.; Lakshmikantham, M.V.; Cava, M.P.; Carito, A.F. (1975) *J. Chem. Soc. Chem. Commun.* 867.

Suchar, G.; Kristian, P. (1975) *Chem. Zvesti* **29**, 244.

Tokitoh, N.; Ishizuka, H.; Ando, W. (1988) *Chem. Lett.* 657.

Zachinyaev, Ya.,V.; Orlov, D.S. (1980) *Khim. Promyshl.* No. 6, 51.

Zhou, Y.; Heimgartner, H. (2000) *Helv. Chim. Acta* **83**, 539.

Zimmer, O.; Meier, H. (1981) *Chem. Ber.* **114**, 2938.

3.2   1,2,4-SELENADIAZOLES

Materials on the compounds of this class remain scarce. Nevertheless, they will be reviewed in the framework of a separate section for the sake of unity in consideration.

3.2.1   FORMATION

3.2.1.1   From Selenocyanates and *N*-Haloamidines

The typical example of the cyclization is represented in Scheme 3.28 (Goerdeler et al. 1963). The yields of the cyclized product are moderately high.

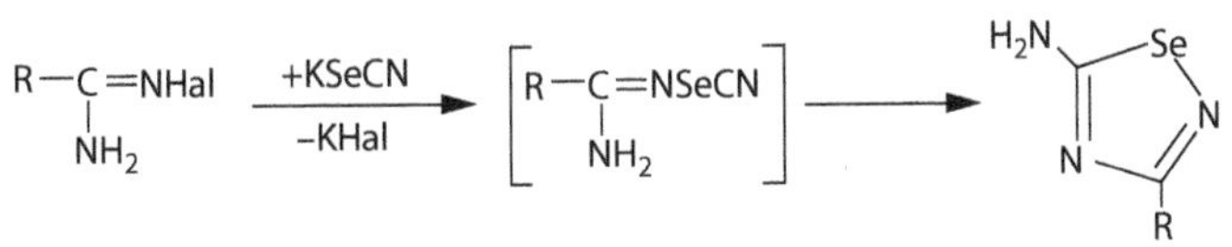

SCHEME 3.28

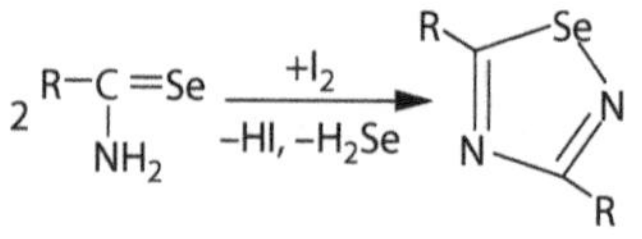

SCHEME 3.29

3.2.1.2 From Selenocarbamides

Cyclization with the formation of the 1,2,4-selenadiazole ring can also be achieved by selenocarboxamide oxidation by reagents which are not incorporated into the hetero ring. The typical example is given in Scheme 3.29 (Becker and Meyer 1904, Cohen 1978). As seen, dimerization of selenocarboxamide is accompanied by removal of hydrogen and selenium. These elements remove as hydrogen iodide and hydrogen selenide.

Besides iodine, other oxidants were used, such as hydrogen peroxide (Shimada et al. 1991), α-arylsulfonyl-α-bromoacetophenons (Shafiee et al. 1999), arylsulfonyl chlorides (Zhao and Yu 2002), 3-choroperoxybenzoic acid (Shimada et al. 1991), diacetoxyiodobenzene (Zhdankin and Stang 2002), and N-bromosuccinimide (Shimada et al. 1991). With these oxidants, yields were often moderate and, in the majority of cases, the desired 1,2,4-selenadiazoles were contaminated by the corresponding by-products. This complicates the preparation because a chromatographic technique should be used to purify the desired product. Poly[styrene(iododiacetate)] was recently proposed as a new oxidant. Its usage offers a simple and convenient workup. Poly[styrene(iododiacetate)] can be regenerated and reused with no loss of activity. No chromatographic separation is needed. Ease of manipulation, short reaction times, and high yields are additional advantages of this method (Huang and Chen 2003).

There is one more, very simple and effective, method of oxidation of arylselenocarbamides by air oxygen upon catalytic action of disodium tetrachloropalladate, Na_2PdCl_4 (Al-Rubaie et al. 2002).

3.2.2 Fine Structure

Considering 1,2,4-selenadiazole as a system containing 2,4-diazadiene, Takikawa et al. (1991) treated 3,5-disubstituted 1,2,4-selenadiazoles with dienophiles. Maleic anhydride, tetracyanoethylene, and phenylacetylene did not react at all even at temperatures of about 200°C. Prolonged reaction time at the elevated temperature caused decomposition of the starting material. With dimethyl acetylene dicarboxylate, however, 1,2,4-selenadiazoles form dimethyl 2,4-disubstituted 1,3-diazine-5,6-dicarboxylate, elemental selenium, and tetramethyl selenophene tetracarboxylate. Scheme 3.30 illustrates the reaction. In the case of 3,5-diphenyl-1,2,4-selenadiazole, the reaction proceeded in melt at 150°C 12 h and the yield of the corresponding diazine was 54%. In the case of 3,5-bis(N,N-dimethylamino)-1,2,4-selenadiazole, the reaction proceeded in benzene solution at room temperature in 2 h and the yield of the corresponding diazine was 99%. As an intermediate for these reactions, the authors postulated the product of cycloaddition. This intermediate is included in Scheme 3.30, although no proof of its existence was given. In any event, inclusion of the intermediate makes selenium extrusion understandable.

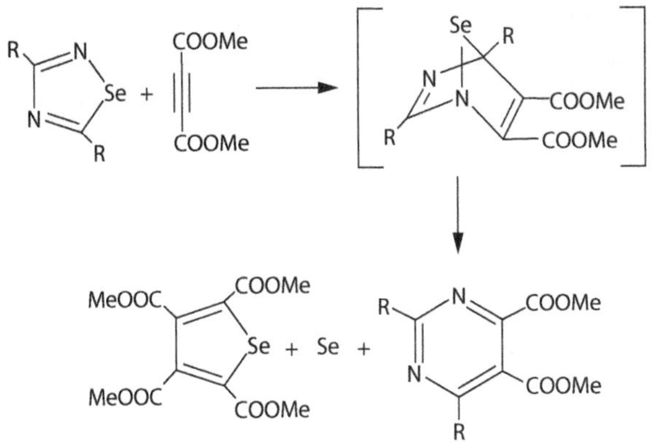

SCHEME 3.30

3.2.3 REACTIVITY

3.2.3.1 Salt Formation

Quaternization of 3,5-diphenyl-1,2-4-selenadiazole with Me_3SiCH_2 OSO_2CF_3 eventually leads to ring expansion with the formation of 1,3,5-selenadiazine. The reaction occurs at N_2 and initially gives methylene ylide that undergoes ring opening to a conjugated selenotriene. Ring closure of the latter leads to the final selenadiazine (Scheme 3.31) (Butler and Fox 2001). It is reasonable to compare Schemes 3.9 and 3.31. It can be seen, that alternation of heteroatoms in the selenadiazole ring changes the reaction directions due to changes in relative stability of the intermediary compounds.

3.2.3.2 Photolysis

Photolysis of 3,5-diphenyl-1,2,4-selenadiazole confined in a poly(vinylchloride) film at 85 K leads to phenylcyanoselenide, PhCNSe. The latter transforms into benzonitrile, PhCN, and expels selenium (Pedersen and Hacker 1977).

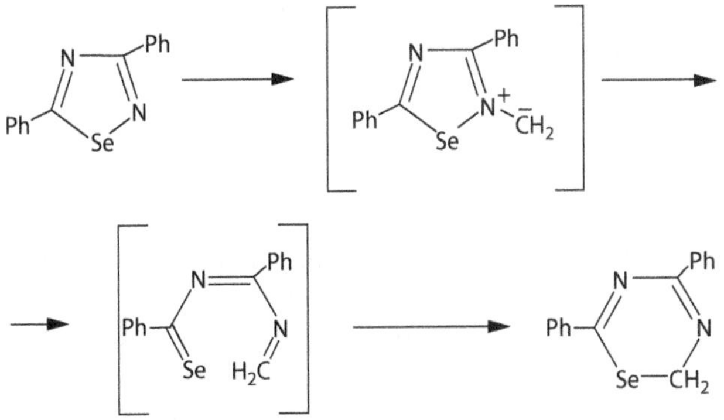

SCHEME 3.31

3.2.3.3   Coordination

3,5-Diaryl-1,2,4-selenadiazoles (L) coordinate to palladium dichloride forming $[PdCl_2 \cdot L]_2$ dimers (Al-Rubaie et al. 2002). It remains unclear whether π-coordination takes place or the metal is bound with some of the nitrogen atoms directly.

3.2.4   CONCLUSION

Despite scarcity of the literature data on 1,2,4-selenadiazoles, it is obvious that they are relatively stable compounds and the methods of their preparation are diverse enough. Reactivity of 1,2,4-selenadiazoles differs from that of their isomers having other alternation of the atoms in the heterocyclic fragments. All these features give hope for widening of studies on their fine structures and reactivity, in the near future.

REFERENCES TO SECTION 3.2

Al-Rubaie, A.Z.; Yousif, L.Z.; Al-Hamad, A.J.H. (2002) *J. Organometal. Chem.* **656**, 274.
Becker, W.; Meyer, J. (1904) *Chem. Ber.* **37**, 2553.
Butler, R.N.; Fox, A. (2001) *J. Chem. Soc., Perkin Trans. 1*, 394.
Cohen, V.I. (1978) *Synthesis*, 768.
Goerdeler, J.; Gross, D.; Klinke, H. (1963) *Chem. Ber.* **96**, 1289.
Huang, X.; Chen, J. (2003) *Synth. Commun.* **33**, 2823.
Pedersen, Ch.L.; Hacker, N. (1977) *Tetrahedron Lett.* **41**, 3982.
Shafiee, A.; Ebrahimzadeh, M.A.; Maleki, A. (1999) *J. Heterocycl. Chem.* **36**, 901.
Shimada, K.; Matsuda, Y.; Hikage, Sh.; Takeishi, Y.; Takikawa, Y. (1991) *Bull. Chem. Soc. Jpn.* **64**, 1037.
Takikawa, Yu.; Hikage, Sh.; Matsuda, Y.; Higashiyama, K.; Takeishi, Y.; Shimada, K. (1991) *Chem. Lett.*, 2043.
Zhao, H.R.; Yu, Q.S. (2002) *Chinese Chem. Lett.* **13**, 729.
Zhdankin, V.V.; Stang P.J. (2002) *Chem. Rev.* **102**, 2523.

3.3   1,2,5-SELENADIAZOLES

Due to their diverse applicability and easiness of preparation, 1,2,5-selenadiazoles, especially their benzo-[1,2-*c*]-fused derivatives named piazoselenols, are the most scrutinized compounds. They are widely used in various aspects of academia, biology, medicine, and technique: Subsections 3.3.4 and 3.3.5 introduce relevant examples.

3.3.1   FORMATION

3.3.1.1   From vic-Diamines and Diselenium Dichloride or Selenium Dioxide

The synthesis of 1,2,5-selenadiazoles from 1,2-diamines with diselenium dichloride proceeds at ambient temperature but the yield is low (7%, Weinstock et al. 1967). This synthesis was significantly improved by using selenium dioxide instead of diselenium dichloride and carrying out the reaction in boiling dimethylformamide so that the yield reached 43% (Scheme 3.32, Bertini 1967a,b).

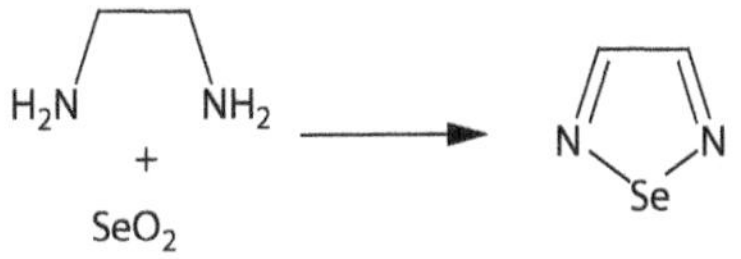

SCHEME 3.32

3.3.1.2 From vic-Diamines and Dithosyl Selenium Bis(Imide)

One of the reactants, dithosyl selenium bis(imide), is in situ formed from selenium and chloramine-T (sodium *N*-chloro-4-toluenesulfonamide). This kind of cyclization does not require heating, but the product yield does not exceed 20% (Bertini et al. 1979).

3.3.1.3 From vic-Bis(Imines) or vic-Dioximes with Diselenium Dichloride, or Selenium Dioxide, or Selenium Oxychloride

The synthesis from 1,2-diimines and diselenium dichloride gives rise to 1,2,5-selenadiazoles with yields up to 60%. Being treated with selenium dichloride, 1,2-diaryl-1,2-bis(trimethylsilyl)imines give the corresponding 1,2,5-selenadiazoles with yields of 30%–50% (Buchwald and Ruehlman 1979). The yield of 70% was reported for replacing diselenium dichloride by selenium oxychloride (Roesky et al. 1984, 1986). A higher yield of 80% was obtained using the pair of selenium oxychloride with 1,2-bis(trimethylsilyl)imine (Neidlein and Knecht 1987).

When 1,2-dioximes are the starting materials, the yields of 1,2,5-selenadiazoles come to 6% with diselenium dichloride and 50% with selenium dioxide (Bertini 1967a). The reaction between 1,2-benzoquinone dioxime and diselenium dichloride produces benzo-[1,2-*c*]-[1,2,5]-selenadiazole N-oxide with 68% yield (Scheme 3.33) (Pedersen 1974).

In 2008, one unusual case of "diimide" transformation into 1,2,5-selenadiazole was described by Bacon et al. (2008). Treatment of the *N*-lithiopyridylketimine derivative, Li[Ph(C=N)Py], with selenium dichloride affords the fused selenadiazolium chloride. The fused structure and chloride coordination to the selenium center was confirmed by x-ray crystallographic analysis. Scheme 3.34 presents the synthesis sequence and the two possible resonance forms of the condensed product. A natural bond orbital analysis of the cation reveals that the resonance form with the coordinative bond between selenium and the pyridine nitrogen contributes most significantly to the total structure. (The selenium center bears a formal positive charge of +0.95.)

As mentioned, selenium dioxide is a typical reactant for 1,2-diaminoarenes and their derivatives with other functional groups. Important exception is devoted to triaminoarenes. Selenium dioxide is a rather strong oxidant. Interaction of selenium

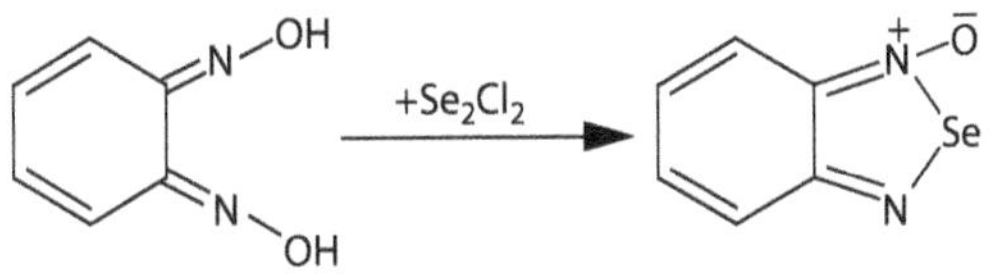

SCHEME 3.33

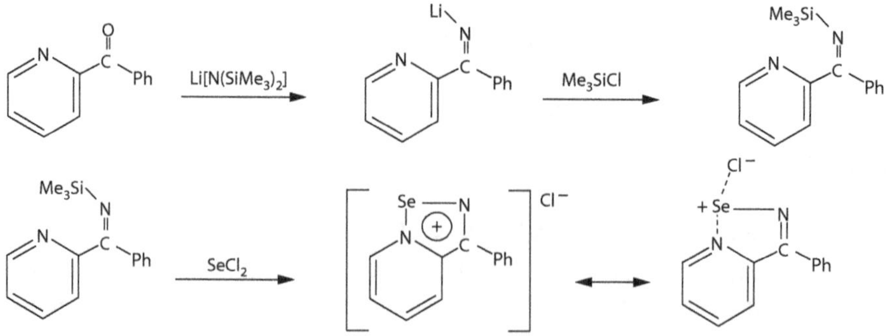

SCHEME 3.34

dioxide with 1,2,3- or 1,2,4-triaminobenzene leads to tarring of the reaction mixture. To avoid such an undesirable result, one may use triaminobenzene hydrochloride as a substrate and sodium selenite as a cyclizing agent (Efros and Todres 1957a,b). Reducing the basicity of amino groups allows using of selenium dioxide for the selenadiazole formation. Thus, 4,5,6-triaminopyrimidine reacts with selenium dioxide in water solution and gives 7-amino-[1,2,5]-selenadiazolo-[3,4-*d*]-pyrimidine (Zima and Mohr 1959). Further, weakly basic diamines do not react with selenium dioxide at all. In these cases, selenium oxychloride (Komin and Carmac 1976) or selenium tetrachloride (Zibarev and Miller 1990) is purportedly applicable. Representative examples are reactions of 4,5-dimethyl-1,2-diaminobenzene with selenium oxychloride (100% yield, Cillo and Lash 2004) and of 3,4,5,6-tetrafluoro-1,2-diaminobenzene with selenium tetrachloride (90% yield, Zibarev et al. 1990). Scheme 3.35 visualizes both examples mentioned.

It needs to note that activity of selenium oxychloride is not always sufficient for the formation of the 1,2,5-selenadiazole cycle. Thus, selenium oxychloride with hexaminobenzene forms tris(selenadiazole)benzene in traces only, although thionyl chloride with hexaaminobenzene forms tris(thiadiazole)benzene in a one-pot synthesis with 70% yield (Praefke et al. 1990).

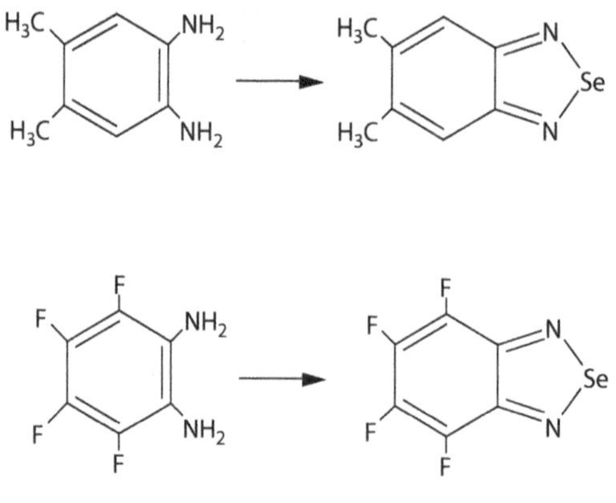

SCHEME 3.35

The reaction of selenium dioxide with deactivated diamines can be accelerated by sonication or grinding of the solid reactants. Thus, microwave irradiation (Zhang et al. 2004a,b) or adequately grinding of powder mixtures containing selenium dioxide and 4,5-diamino-6-hydroxypyrimidine leads to 7-hydroxy-1,2,5-selenadiazolo-[3,4-*d*]-pyrimidine with 60% yield (Zou et al. 2004, Zheng et al. 2007). Mechanisms of activation by grinding or microwave irradiation are discussed in Todres' monographs of 2006 and 2009, respectively. These solid-state reactions proceed at room temperature and they are benign to the environment, completed with higher yields and more convenient work-up.

3.3.1.4   From Diazabutadienes and Selenium Tetrachloride with Tin Dichloride

The reaction between a stoichiometric mixture of 1,4-[2,6-bis(isopropyl)phenyl]-1,4-diaza-1,3-bitadiene and tin dichloride with one equivalent of selenium tetrachloride leads to the dicationic salt of 2,5-[2,6-bis(isopropyl)phenyl]-1,2,5-selenadiazolium with the tin hexachloride anion (Dutton et al. 2006a, 2009). The structure of the reaction product was established by x-ray crystallographic analysis. The authors consider this reaction product as two-coordinative, dicationic, N-heterocyclic selenium carbenoid.

3.3.1.5   From Spirocyclic Aminals with Selenium Dioxide

This unusual method involves the reaction depicted in Scheme 3.36 (Hazelton et al. 1992). The corresponding 1,2,5-selenadiazole is obtained with almost quantitative yield.

3.3.1.6   From 1,2,5-Thiadiazoles and Organyl Magnesium Halides with Selenium Monochloride

These reactions result in substitution of selenium for sulfur as the key heteroatom. The process was described by Bertini et al. (1987).

3.3.2   FINE STRUCTURE

Because the selenium key heteroatom can come out in the bivalent or tetravalent state, selenadiazoles are considered in the two possible forms (see Scheme 3.37) (Sergeev et al. 1989).

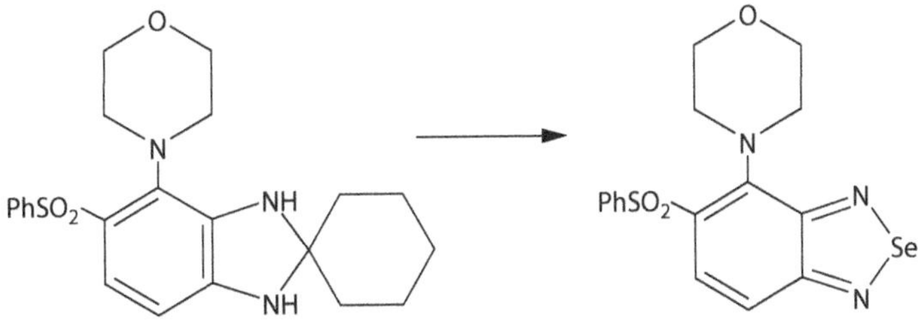

SCHEME 3.36

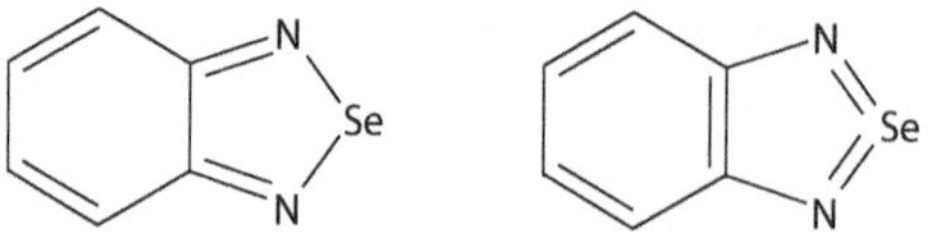

SCHEME 3.37

In Subsection 3.3.2, materials on fine structure will be separately considered for monocyclic 1,2,5-selenadiazoles and for 1,2,5-selenadiazoles fused with aromatic fragments.

3.3.2.1 Monocyclic 1,2,5-Selenadiazoles

Blackman et al. (1967) recorded the microwave spectra of the ^{80}Se and ^{78}Se isotopomers and inferred that the 1,2,5-selenadiazole molecule is planar and has C_{2v} symmetry. In 1968, Benedetti and Bertini confirmed the planarity and C_{2v} symmetry type based on infrared and Raman spectra of 1,2,5-selenadiazole. The corresponding vibrational frequencies were theoretically reproduced using just the C_{2v} symmetry constraint (El-Azhary and Al Kahtany 2001, Hegelund et al. 2006).

Bond lengths in 1,2,5-selenadiazole were used to calculate its aromaticity index, I_A (Bird 1985). Another aromaticity index, I_U, was calculated on the basis of the 1,2,5-selenadiazole heat of formation (Bird 1992). In the both calculations, benzene was used as a reference compound with its $I_A = I_U = 100$. 1,2,5-Selenadiazole was characterized by $I_A = 47$ and $I_U = 58$. It is as a "half-aromatic" compound.

Structurally, 1,2,5-selenadiazole consists of 1,4-diazabutadiene, the diaza ends of which meet at the key selenium atom. Takikawa et al. (1991) tried to introduce this closed diazabutadiene fragment into cycloaddition using dimethyl acetylenedicarboxylate as a dienophile. Starting with 3,4-diphenyl-1,2,5-selenadiazole and the reaction proceeding for 20 h at 150°C in benzene (a sealed tube), dimethyl 2,3-diphenyl-1,4-diazine-5,6-dicarboxylate was obtained with 16% yield. Scheme 3.38 shows the transformation and includes a cycloaddition product as an assumed intermediate. No data on accompanying products were mentioned by the authors.

To estimate 1,2,5-selenadiazole as a substituent, the interaction was considered between phenyl and selenadiazole moieties. Studies on crystal and molecular structure of 3,4-diphenyl-1,2,5-selenadiazole showed that only very weak, if any, conjugation exists between the heterocyclic moiety and the phenyl substituents (Mellini and Merlino 1976). This can be caused by a strong unfolding of the neighboring and bulky phenyl group. This can result in loss of conjugation. In 3-phenyl-1,2,5-selenadiazole, such a mutual steric effect is absent. Being partially oriented in the nematic

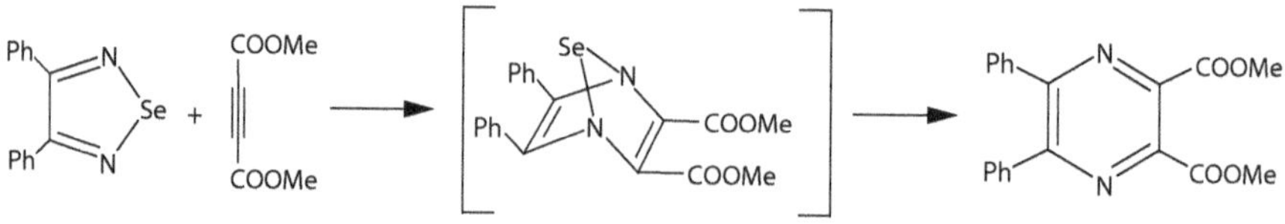

SCHEME 3.38

phase, the monophenyl compound accepts twisted conformation with respect to the bond between the two rings, the selenadiazole and the phenyl. As followed from high-resolved NMR, the twisting angle is 18°–21° (Veracini et al. 1977). As known, conjugation is kept possible up to the twisting angle of 40°. This means that free rotation of the selenadiazole and phenyl counterparts in a non-oriented solvent should not prevent conjugation. At a moment, the mutual orientation of these groups, needed for conjugation, can be achieved. From this point of view, it is principal to consider the selenadiazole ring as an orienting substituent with respect to the phenyl nucleus. Due to its electrophilic nature, the 1,2,5-selenadiazole fragment should be a meta-orientant during electrophilic substitution. As it turned out, the selenadiazole-substituted phenyl nucleus is brominated in the para position, is mononitrated in the para or ortho position, and is dinitrated in the both ortho and para positions. The electrophilic substitution never involves the heterocyclic ring in its 4 position (de Munno et al. 1978). The heterocyclic ring is an ortho–para directing group (Scheme 3.39). In this sense, the selenadiazole resembles chlorine as a substituent. Despite strong negative inductive effect of the chlorine substituent, chlorobenzene is mono-nitrated in ortho or para position, although slower than benzene itself.

1,2,5-Selenadiazole annulated with 1,2,5-thiadiazole is of special interest. In this molecule, only one chalcogen atom can be bivalent, the other one must be tetrava-lent. Recently prepared potassium salt of the [1,2,5]-selenadiazolo-[3,4-*c*]-[1,2,5]-thiadiazole anion-radical is a completely π-delocalized species (Bagryanskaya et al. 2007). Such a kind of delocalization can be specific for the anion-radical due to the possibility for an unpaired electron to pass the selenium and sulfur key heteroatoms using its vacant d or p orbitals. For the parent molecule, the question of its aromatic-ity remains open.

3.3.2.2   Fused 1,2,5-Selenadiazoles

Benzo-[1,2-*c*]-[1,2,5]-selenadiazole (piazoselenol) is a typical compound of this kind.

1,2-diaminobenzene is the predecessor of piazoselenol. The formula with the tetravalent selenium supposes preservation of the initial aromatic character of the benzene ring. The formula with the bivalent selenium presumes quinoidization of the benzene ring (see Scheme 3.37).

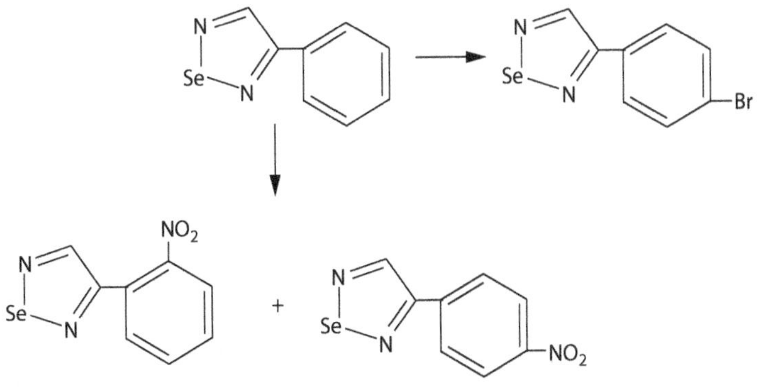

SCHEME 3.39

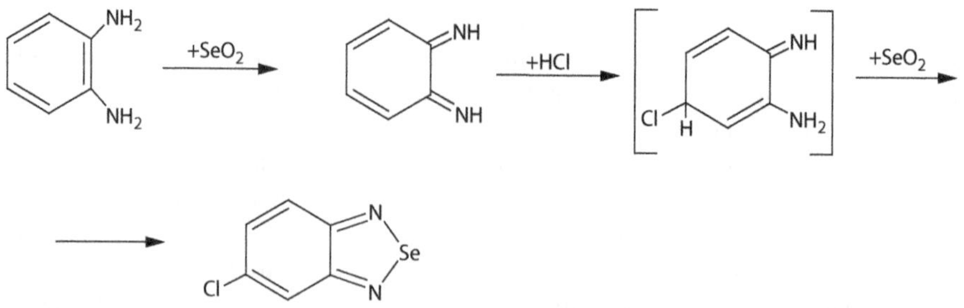

SCHEME 3.40

The selenadiazole ring formation proceeds during mixing of water solutions of 1,2-diaminobenzene and selenium dioxide at room temperature. Piazoselenol immediately settles out in the quantitative yield. If the wet reaction between 1,2-diaminobenzene and selenium dioxide is performed in concentrated hydrochloric acid, 5-chloropiazoselenol is formed, instead of the unsubstituted piazoselenol as it was observed in the water medium. At the same time, attempts to add hydrogen chloride to piazoselenol performed under various conditions invariably led to the formation of its hydrochloride salt (Efros and Todres 1957a). These results testify for the formation of the intermediary 1,2-benzoquinone bis(imine) in the way from the initial to final product, according to Scheme 3.40.

Interestingly, the reverse transformation, that is, reduction of piazoselenol into 1,2-diaminobenzene also passes through a stage of the 1,2-quinoinediimine formation (Zhdanov et al. 1967, Tsveniashvili et al. 1968).

The quinonoid structure of piazoselenol finds its confirmation in x-ray crystallography (Luzzati 1951). This was confirmed for several piazoselenol derivatives (see Gieren et al. 1980 and Huebner et al. 1984 for a compilation of structural data). In particular, Se–N bond lengths occurred to be intermediate between those of typical single and double bonds.

Calculations of π-electron structure of piazoselenol showed that its six-membered ring contains only two carbon–carbon bonds having the double-bond order. 1,2-Benzoquinone bis(imines) has the same bond system. In the five-member ring of piazoselenol, nitrogen-selenium bonds have the lowest order whereas carbon-nitrogen bonds have an order close to that for the corresponding double bonds (Gul'maliev et al. 1973).

Meanwhile, piazoselenol reacted neither with maleic anhydride (dienophile) nor with 1,4-butadiene (diene). Such inactivity in Diels–Adler reactions was tried to be explained with the electron withdrawing effect of the 1,2,5-selenadiazole ring (Efros and Todres 1957a). For instance, butadiene bearing the nitro group is also incapable of cycloaddition to dienophiles (Kataev 1955). Addition of maleic anhydride to anthracene positions 9 and 10 is a classical example of the Diels–Alder reaction. However, the presence of the fused 1,2,5-selenadiazole keeps the anthracene 9,10 unsubstituted positions from this reaction. It was shown for anthra-[1,2-c]-[1,2,5]-selenadiazole (Gorelik and Bogdanov 1960). Note, anthracene itself readily adds maleic anhydride just at 9,10 positions (Scheme 3.41). The latter reaction is

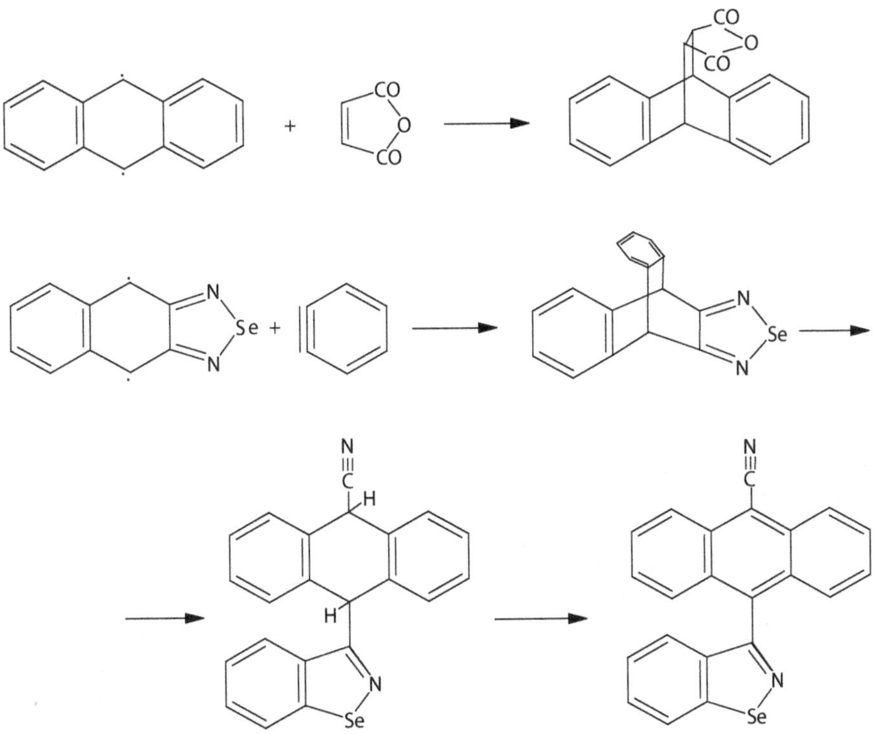

SCHEME 3.41

interesting to compare with a reaction of naphtho-[2,3-c]-[1,2,5]-selenadiazole with benzyne. The naphthoselenadiazole and benzenediazonium 2-carboxylate were kept in boiling tetrahydrofuran. Benzyne (generated in situ) added to the meso (unsaturated) positions of naphtho-[2,3-c]-[1,2,5]-selenadiazole according to Scheme 3.41. Rearrangement of the primary addition product eventually led to 3-(10-cyano-9-anthryl)-1,2-benzisoselenazole (Vernon et al. 1983).

All the data on the fine structure of piazoselenol confirm its likeness to 1,2-benzoquinone bis(imine). However, there is one important difference: In piazoselenol, the nitrogen atoms are locked in a cycle by the key heteroatom. Therefore, a question arises whether or not this heteroatom participates in electron conjugation within the molecular contour. To answer this question, electrochemical and physicochemical studies of piazoselenol and its derivatives were undertaken.

Ring current in these substances were studied by the method of proton magnetic resonance (Fedin and Todres 1968, Breier et al. 1969). From the ring-current point of consideration, piazoselenol resembles naphthalene. As distinct from naphthalene, the ring-current intensity in piazoselenol is 0.4 from that in naphthalene. The ring-current is less intense, but does exist. Along with α-nitroso-β-hydroxynaphthalene, α-nitroso-β-hydroxypiazoselenole can izomerize to the quinone monoxime form according to Scheme 3.42 (Kukushkin et al. 1973, Belen'kaya et al. 1992a). This compound forms a palladium chelate in the usual way (Kukushkin et al. 1973).

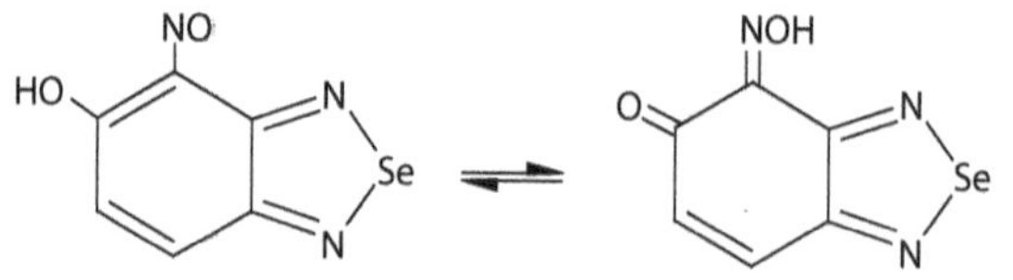

SCHEME 3.42

Analysis of x-ray structural data for naphtho-[2,3-*c*]-[1,2,5]-selenadiazole-4,9-dione led to the conclusion that significant interaction exists between the "chalcogen-bis(imide)" moiety and the π-system of the condensed naphthoquinone system (Gieren and Lamm 1982).

Spectra in fields of ultraviolet (Fajer 1965, Todres and Zaitsev 1971) and infrared absorption (Pozdyshev et al. 1960, Bird and Cheeseman 1964, Pesin 1969, Todres and Zaitsev 1971), electrochemical studies (Zhdanov et al. 1967, Todres et al. 1968a,b, 1969, Tsveniashvili et al. 1968), and registration of ^{14}N quadrupole resonance (Krause and Whitehead 1973) do not refute the presence of some conjugation within benzoselenadiazoles.

For anion-radicals, electron spin resonance methods were employed (Strom and Russell 1965, Kursanov and Todres 1967, Solodovnikov and Todres 1967, 1968, Todres et al. 1968a, 1969). As turned out, benzo-[1,2-*c*]-[1,2,5]-selenadiazole and naphtho-[2,3-*c*]- [1,2,5]-selenadiazole are reduced to anion-radicals. Neither consecutive reactions nor dimerization takes place. Similarly to other aromatic systems, these compounds form anion-radicals, which are stable and can be reversibly oxidized into initial neutral forms. These properties are typical for anion-radicals of aromatic compounds. Analysis of electron spin resonance spectra shows that the selenium key atom is not a barrier to delocalization of the unpaired electron within the molecular skeleton of the anion-radical. In particular, π-electron density in the anion-radicals of 1,2,5-selenadiazole and benzo-[1,2-*c*]-[1,2,5]-selenadiazole are close to 1.9 on one selenium atom and 2.8 on the two nitrogen atoms (Gul'maliev et al. 1975). Whether or not an unpaired electron can fall on the empty d orbital of selenium in anion-radicals of 1,2,5-selenadiazoles—this question had already been put forward in 1968 by Fajer et al. However, no answer has been received till now. Models of π and σ orbitals (Gul'maliev et al. 1975) or p orbitals all satisfactorily describe the pertinent experimental data. By hook or by crook, the selenium key atom participates in the formation of the closed π-electron contour.

To compare the 1,2,5-selenadiazole ring's electrophilicity with that of the nitro group, stepwise reduction of 4- and 5-nitropiazoselenols were performed (Todres et al. 1968b, 1972). As turned out, the nitro group captures one electron before single-electron reduction of the electrophilic hetero ring. Captured one electron, the group NO_2^- transforms in a donating substituent like the group NH_2. The phenomenon is comprehensively considered in Todres' monograph of 2009 (see pages 2–6). Despite high inclination of areneselenadiazoles to be reduced to 1,2-arenediamines, the corresponding nitro derivatives are reduced to areneselenadiazole amines by tin with ammonium bromide (Elmaaty 2008) or iron with acetic acid (Efros and Todres 1957b, Edin and Grivas 2000).

3.3.3 REACTIVITY

3.3.3.1 Monocyclic 1,2,5-Selenadiazoles

3.3.3.1.1 Salt Formation

Reacting with alkyl halides, 1,2,5-selenadiazoles undergo quaternization. One recent and unusual example has been provided by Dutton et al. (2006b). The selenium tetrachloride reaction with 1,4-bis(*tert*-butyl)-1,4-diazabuta-2,3-diene lead to the salt containing two 2-(*tert*-butyl)-1,2,5-selenadiazolium cations and one selenium hexachloride dianion, according to Scheme 3.43 (Dutton et al. 2006b). This is a novel redox route to selenium–nitrogen heterocycles. The reaction proceeds with loss of *tert*-butyl chloride and molecular chlorine. Selenium tetrachloride concomitantly abstracts Cl_2 forming the selenium hexachloride dianion. The remaining selenium fastens the two azabutadiene nitrogens, one of which keeps the *tert*-butadiene moiety. The monobutyl selenadiazolium is formed. Two of these cations are bound with the selenium hexachloride dianion. The salt was separated with 92% yield. Scheme 3.43 shows the over-all reaction.

3.3.3.1.2 Ring Cleavage

As claimed by Begland (1974), dimethyl 1,2,5-selenadiazol-3,4-dicarboxylate is readily cleaved with hydrogen sulfide to form dimethyl ester of diaminomaleic acid. The product is used as a cross-linking agent for epoxy resins.

Lithium alkyls attack 3-methyl- or 3-phenyl-1,2,5-selenadiazole at the selenium atom and also cleave the heterocycle. The reaction yields 1,2-diimine salts. The salts were hydrolyzed to 1,2-dicarboxyl compounds, RCOCOR (Bertini et al. 1974).

An alkyl magnesium bromide or (ethylene) platinum bis(triphenylphosphine) opens the heterocycle as well. With alkyl magnesium bromide, dialkyl selenide is formed, after hydrolysis, along with other products of the selenadiazole cleavage (Bertini et al. 1974, Rykowski and Sczesna 1989). As seen from Scheme 3.44, selenadiazole acts as a source of selenium in converting Grignard reagents into the corresponding dialkyl selenides.

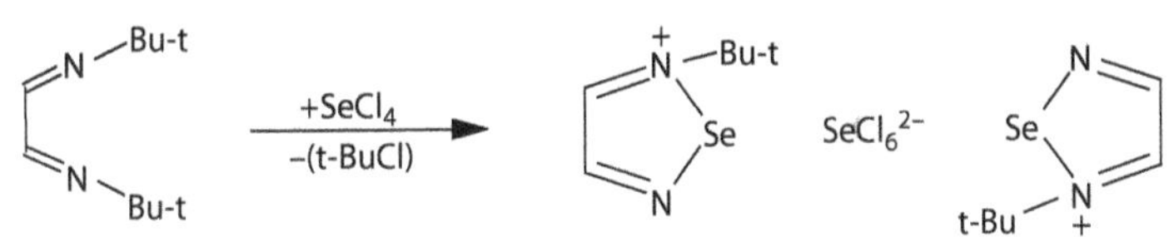

SCHEME 3.43

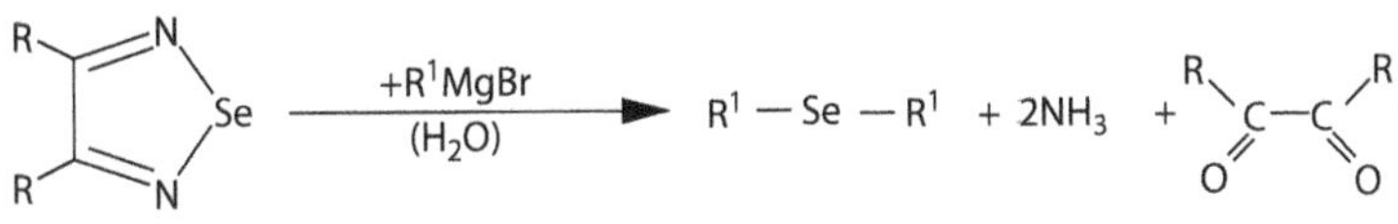

SCHEME 3.44

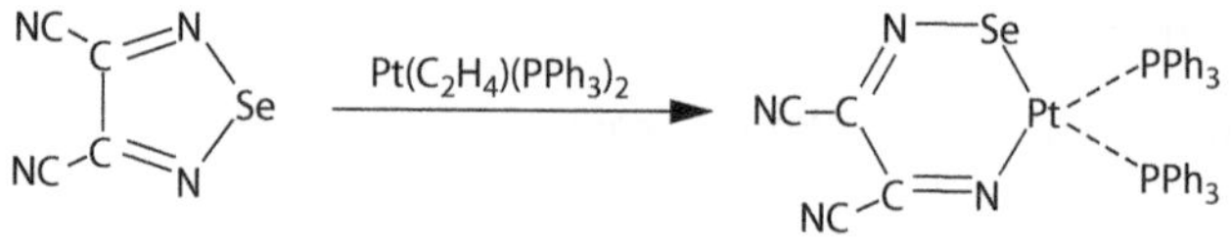

SCHEME 3.45

With the platinum complex of Scheme 3.45, the selenadiazole ring is cleaved and the platinum six-membered ring is formed (Roesky et al. 1986).

3.3.3.1.3 Photolysis

Upon photolysis at 85 K, 3,4-diphenyl-1,2,5-selenadiazole confined in poly(vinyl-chloride) film, transforms into benzonitrile selenide, PhCNSe, which loses selenium and gives rise to benzonitrile, PhCN (Pedersen and Hacker 1977, Pedersen et al. 1977).

3.3.3.1.4 Coordination

Re-examination of published structural data and, then, computational modeling led to the conclusion on self-coordination of 1,2,5-selenadiazoles in crystalline state (Cozzolino et al. 2005). According to this type of association, four-membered cycles are formed as a result of nitrogen–selenium coordination between two antiparallel selenadiazole rings. This coordination is repeated to link another molecule, and a ribbon polymer grows (see Scheme 3.46).

The selenium possibility to participate in coordination was excellently demonstrated with interaction of N-(tert-butyl)-1,2,5-selenadiazolium with triphenylphosphine (Scheme 3.47) (Dutton and Ragogna 2009). Based on ^{31}P NMR results, the authors substantiate that the coordination of the donor (PPh$_3$) constituent takes place at the selenium atom rather than at the positively charged nitrogen atom. Accordingly, the addition of one stoichiometric equivalent of PAlk$_3$ to a dichloromethane solution

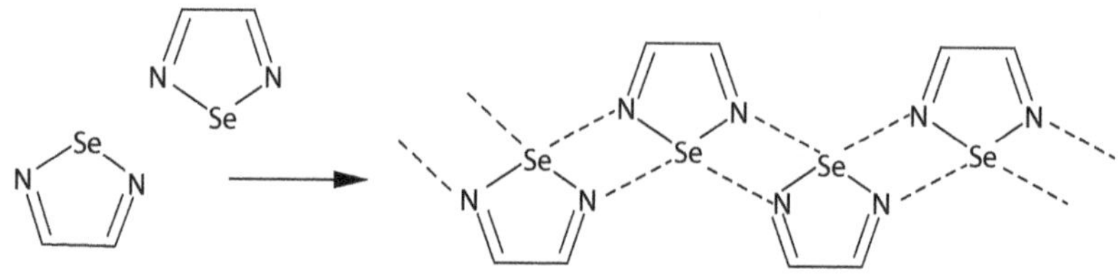

SCHEME 3.46

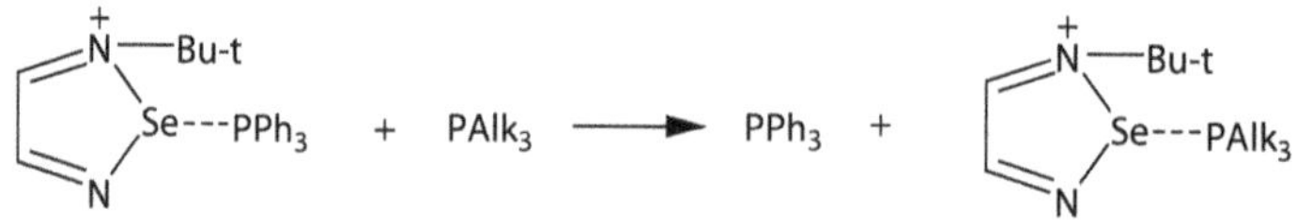

SCHEME 3.47

SCHEME 3.48

of the initial complex leads to ligand exchange at selenium for the formation of a new Se–P bond.

3.3.3.1.5 Ring Transformation

When 3,4-diphenyl-1,2,5-selenadiazole is treated with ethyl magnesium bromide followed by tellurium tetrachloride and then triethylamine, 3,4-diphenyl-1,2,5-telluradiazole is formed with a yield up to 80% (Bertini et al. 1982) (Scheme 3.48).

Saturation with hydrogen sulfide of a pyridine–chloroform solution containing the 1,2,5-selenadiazole fused to one of the pyrazole rings of hexaphenylthiadiazoloporphyrazine complex with iron leads to direct replacement of selenium by sulfur. The resulting 1,2,5-thiadiazole derivative was comprehensively characterized (Ul-Haq et al. 2007).

3.3.3.2 Fused 1,2,5-Selenadiazoles

3.3.3.2.1 Salt Formation

Saturation with hydrogen chloride of an ethanolic solution containing benzo-[1,2-c]-1,2,5-selenadiazole leads to the formation of the hydrochloride salt (Efros and Todres 1957a). Trifluoroacetic acid carries out diprotonation of this substrate (Sawicki and Carr 1957, Brown and Bladon 1968). Reaction of benzo-[1,2-c]-1,2,5-selenadiazole with dimethyl sulfate according to the method of Schotten–Baumann leads to the quantitative formation of the 2-methyl derivative (Nunn and Ralph 1965). Only one nitrogen atom acquires the onium state despite diprotonation found for piazoselenol in trifluoroacetic acid. The presence of the nitro group in α or β positions does not prevent the quaternization. In alkaline mediums, the nitro N-methylselenadiazoliums undergo ring opening that affords 1-(N-methylamino)-2-amino-3- or 4-nitrobenzene according to Scheme 3.49 (Bella and Milata 2008).

3.3.3.2.2 Reduction

Reductive ring opening is the most practically important reaction of these compounds. They are readily prepared from the corresponding aromatic 1,2-diamines

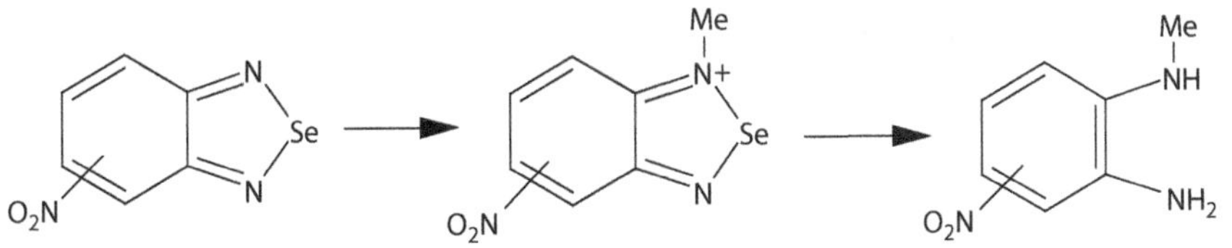

SCHEME 3.49

and easily cleaved back by treatment with hydroiodic acid (Sawicki and Carrr 1957, Wright and McClure 2004). In the presence of sodium hydroxide, phenanthro[1,2-*c*]- and anthra[1,2-*c*]-[1,2,5]-selenadiazol-6,11-diones undergo cleavage of the selenadiazole rings simply under heating of their solutions in aqueous organic solvents (Gorelik 1968, 1971). The easiness of this cleavage is connected with enhanced reactivity of the selenadiazole ring in these diones. X-ray diffraction study of the selenadiazole-dione by Klimasenko et al. (1973) established that the heterocycle has a distorted structure (the anthraquinone remaining rings do not exhibit any distortion).

Reduction of piazoselenol with phenyl lithium results in the formation of *N*-phenyl-1,2-diaminobenzene and diphenylselenide (Lane and Williams 1955). The generalized version of the method was claimed by the authors in 1957.

Importantly, the selenadiazole formation is a way to protect 1,2-diamine against undesirable oxidation. Such a danger is encountered during nitration, so that it is possible to prepare otherwise inaccessible aromatic nitro-1,2-diamines (Grivas 2000).

Interestingly, tetrafluoropiazoselenol reacts with ethyl magnesium bromide with cleavage of the selena-containing ring, but the product is the bis(magnesium bromide) derivative of tetrafluoro-1,2-benzoquione bis(imine) (Kovtonyuk et al. 1996). At this point, it is reasonable to compare Schemes 3.40, 3.44, and 3.50.

The reductive opening of a 1,2,5-selenadiazole ring leads to vicinal diamino species and hydrogen sulfide has been reported as convenient reducing agent (Bertini 1967a, Begland 1974). Cleavage of the 1,2,5-selenadiazole ring condensed to metal porphirazines can also be easily performed by hydrogen sulfide in the presence of pyridine (Bauer et al. 1999, Kudrik et al. 2001, Goslinski et al. 2006). Preparation of such selenadiazoles (Moerkved et al. 1995, Ercolani et al. 1999) and their further reduction is a route to the corresponding 1,2-diamines. The latter, in their turn, are the starting materials in the syntheses of bis(azomethynes) (Zhao et al. 2003) or metal porphirazines with annulated diazamacrorings of the diazepine type (Baum et al. 2003). Concerning the reduction of 1,2,5-selenadiazole ring with hydrogen sulfide, the following obstacle should be noted: As Ul-Haq et al. (2007) noted, this reduction process can sometimes be stopped at the stage of the formation of the 1,2,5-thiadiazole analog. Such unexpected ring transformation was observed in the case of hexaphenyl-(1,2,5-selenadiazolo)-porphyrazine iron(II) complex. The thiadiazole analog formed can however be reduced to the vicinal diamine during further reduction with excessive hydrogen sulfide.

In piazoselenol, the heteroring cleavage was performed upon action of active thiols such as glutathione (γ-glutamylcysteinylglycine [GSH]). This reaction was

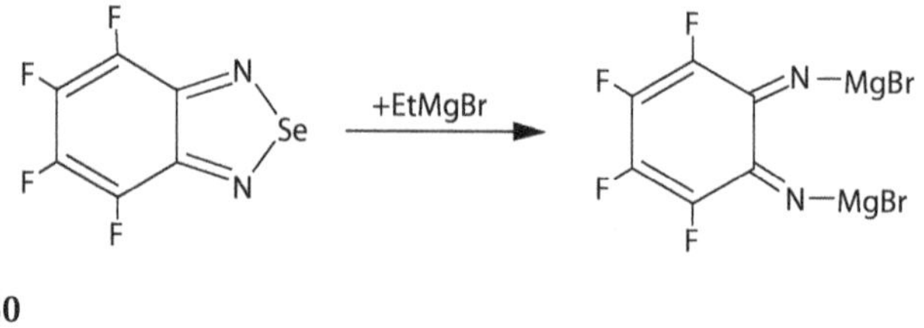

SCHEME 3.50

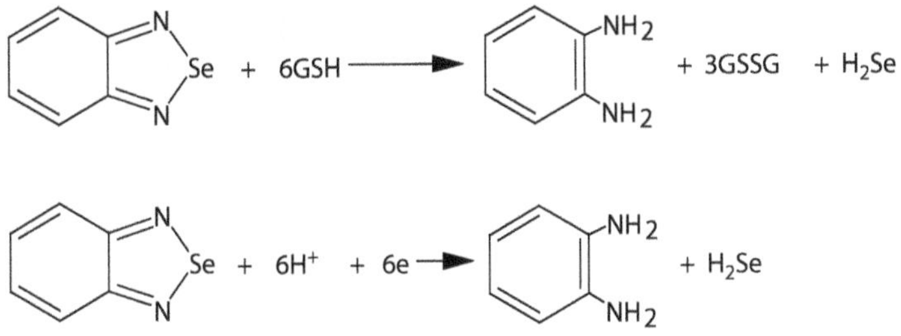

SCHEME 3.51

proposed for quantitative determination of glutathione which is the most abundant cellular thiol. Scheme 3.51 introduces the cleavage reactions with GSH (Wang et al. 2008) and in conditions of electrochemical reduction in proton-donor mediums (Tsveniashvili et al. 1967). Namely, GSH is added to the buffered solution of piazoselenol that results in the heteroring cleavage according to Scheme 3.51. After that, the amount of the lasting piazoselenol is determined by means of electrochemistry (the peak current of piazoselenol decreases in linear dependence of the GSH amount). This specific electrochemical probe of piazoselenol is applicable to detect GSH and related cellular thiols, including those extracted from rat breast cancer cells 4T-1 (Wang et al. 2008).

3.3.3.2.3 Oxidation

Piazoselenol is stable in oxidative conditions: It is not oxidized by 30% hydrogen peroxide in acetic acid at 100°C (Sawicki and Carr 1957). Nevertheless, under these conditions [1,2,5]-selenadiazolo-2-oxide-[3,4-*d*]-pyrimidin-5,7-(4H,6H)-dione gives rise to a fused furazan according to Scheme 3.52 (Yavolovsky et al. 2000).

3.3.3.2.4 Electrophilic Substitution

At the first glance, electrophilic effect of the fused 1,2,5-selenadiazole cycle should weaken the benzene ring capability of electrophilic ring substitution. Nevertheless, nitration and sulfonation of piazoselenol proceeds under convenient conditions. Calculations of piazoselenol by the Pariser–Parr–Popple method or within the frontier orbital approach (Gul'maliev et al. 1973) were established that α-positions are the most favorable for electrophilic attack. Indeed, mononitration (Efros and Todres 1957a) and monosulfonation (Todres et al. 1968b) lead to the corresponding α-substituted derivatives. Scheme 3.53 shows the direction of nitration of

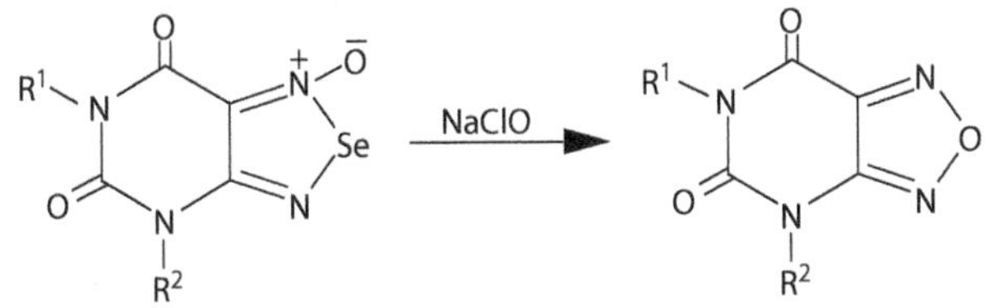

SCHEME 3.52

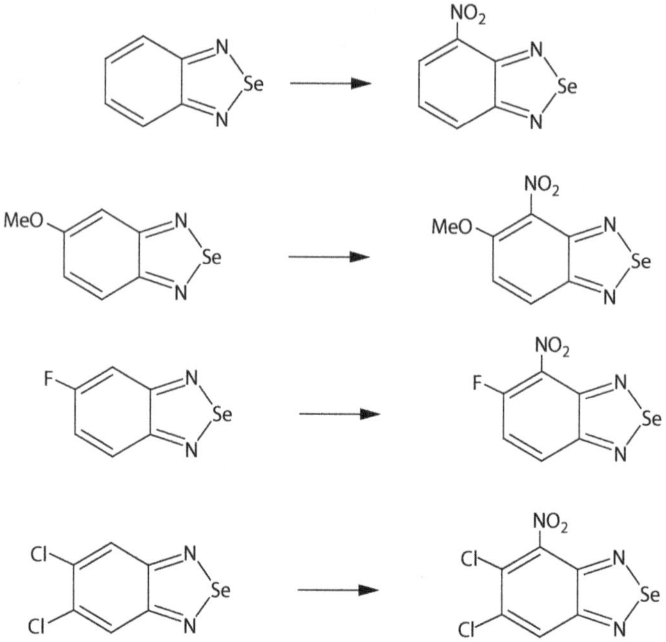

SCHEME 3.53

piazoselenol (Efros and Todres 1957a), its β-methoxy (Grivas and Tian 1992), β-fluoro (Wright and McClure 2004), or β,β′-dichloro (Sergeev et al. 1989) derivatives. These four examples of nitration were chosen in order to follow the condition of the reaction. For the ring nitration of piazoselenol and β-methoxypiazoselenol, the usual nitrating mixture was sufficient (namely, nitric and sulfuric acids). Nitration of β-fluoropiazoselenol by the usual nitrating mixture leads to substitution of the hydroxyl for fluorine. To circumvent this obstacle, nitric acid or sodium nitrate in Eaton's reagent as medium were used. Eaton's reagent is a 10% solution of phosphorus pentoxide in methanesulfonic acid. Room temperature is sufficient for this kind of nitration. The method is applied to the multi-gram preparation of α-nitro-β-fluoropiazoselenol. (The nitro product was used without purification to reduce it to 4-fluoro-3-nitrobenzene-1,2-diamine with 94% yield.) To obtain α-nitro-β,β′-dichloropiazoselenol, a mixture of sodium nitrate and concentrated sulfuric acid is needed and heating at 100°C is required.

The examples of Scheme 3.53 describe electrophilic substitution in α position of the substrates. If α position is blocked, the other, α′ position, is attacked. Electron-withdrawing effect of the heterocycle manifests itself on the ability of α-sulfo group to be substituted by chlorine upon action of sodium chlorate in hydrochloric acid (Todres et al. 1968b). Such ability can be compared to the ability of anthraquinone sulfonic acids to transfer into chloroanthraquinones under the same conditions. For anthraquinone sulfonic acid, this transformation proceeds quantitatively and is a convenient analytical method. For piazoselenol-α-sulfonic acid, the substitution of chlorine for the sulfo group does proceed, but to a lower extent (not more 5%), and oxidative destruction is the main direction.

Azo coupling reactions of aminopiazoselenols proceed easily and regioselectively. α-Aminopiazoselenol gives α-amino-α′-(azobenzene)piazoselenol (Efros and Todres 1957b). β-Aminopiazoselenol yields α-(azobenzene)-β-aminopiazoselenol (Efros and El'tsov 1958). If positions α and α′ are occupied, no such coupling is found (Pesin et al. 1962).

3.3.3.2.5 Nucleophilic Substitution

Substitution of the ring bromine with cyanide (from copper monocyanide) is illustrated by Scheme 3.54 (yield 23%, Kumari et al. 1990).

Reaction of α,β-bis(iodo)piazoselenol with the malodinitrile in the presence of sodium hydride and palladium leads to the bis(dicyanomethane) derivative. The latter gives rise to the corresponding quinodimethane after oxidation with lead dioxide (Scheme 3.55) (Suzuki et al. 1997).

Suzuki coupling reaction was also used to obtain various N-phenylaminopiazoselenols from α,α′-dibromopiazoselenol (Velusamy et al. 2005a, Yasuda et al. 2005). These compounds contain piazoselenol electron-accepting units as the central bridge and display distinct optical, electrochemical, and thermal properties. The power conversion efficiency can reach 3.8% (Velusamy et al. 2005b). The result achieved puts these dyes forward as candidates for dye-sensitized solar cells.

The kinetics of the α- or β-chloropiazoselenol reaction with methoxide anion was studied by Casoni and Sandri (1963). As turned out, the β-chloro compound is more reactive than the α-chloro isomer. This again underlines the high bond order between α- and β-carbon atoms of the six-membered ring and the higher electron density at α position compared to that at β position. Note, the dipole moment of piazoselenol is 0.94 D. For β-chloropiazoselenol, this magnitude drops to 0.05 D

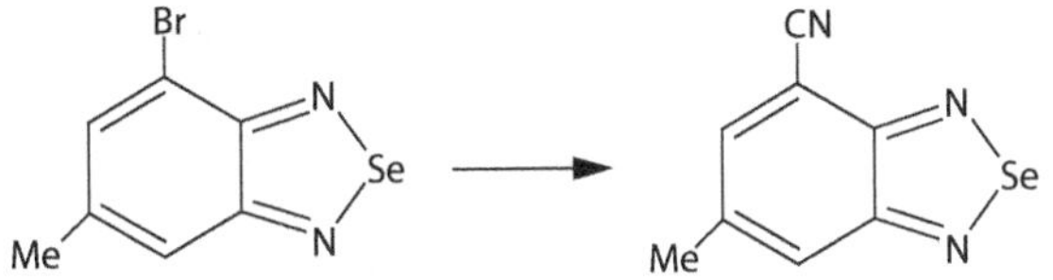

SCHEME 3.54

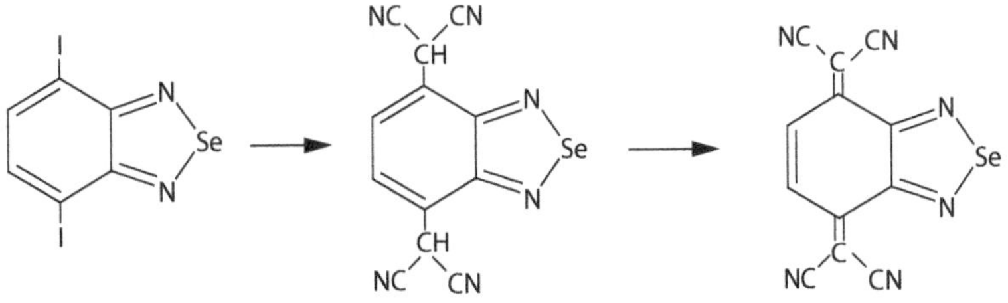

SCHEME 3.55

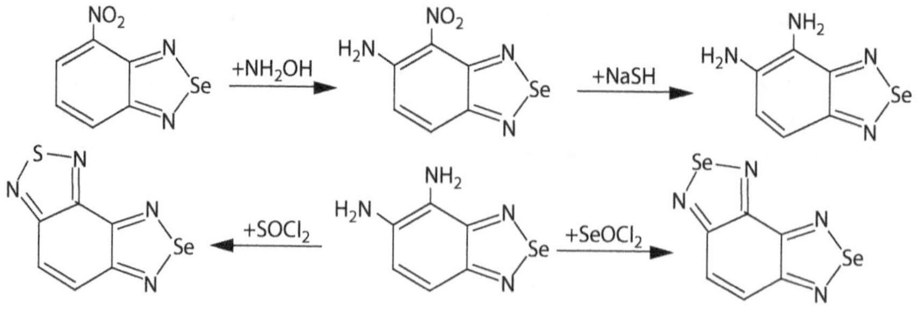

SCHEME 3.56

(Hill and Sutton 1949). In other words, the heterocyclic ring carries the negative end of the piazoselenol dipole.

The piazoselenol amination with hydroxylamine sulfate under the divanadium pentoxide catalysis leads to α-aminopiazoselenol with 25% yield (Sergeev et al. 1972). Direct amination by hydroxylamine was described for α- or β-nitropiazoselenol (Brizzi et al. 1964). In both the cases, the entering amino group occupies only neighboring position with respect to the nitro group, see also Cillo and Lash (2004). It is the amino-nitro mutual disposition that occurs to be preferential thermodynamically due to combination of the orientation and field effects. The obvious possibility of hydrogen bonding between the adjacent nitro and amino groups may also govern the orientation during this amination. Amination of α-nitropiazoselenol proceeds in β-position, although α′-position is free. The amino nitro compound is obtained with 80% yield and is readily reduced to α,β-piazoselenol diamine with sodium hydrosulfite (the selenadiazole ring is not touched during this reduction!). The final diamine can be cyclized by action of SeOCl$_2$ or SOCl$_2$ (Cillo and Lash 2004). Scheme 3.56 presents the whole sequence of transformations. In Scheme 3.56, the final tricyclic compounds containing two diazole rings fused with the benzenoid one attract attention as chromophores of the donor–acceptor–donor charge-transfer type (see Section 3.3.5).

The reaction of α-nitropiazothiol with methoxide in methanolic dimethylsulfoxide leads to the isomeric Meisenheimer complexes depicted in Scheme 3.57. This nucleophilic reaction is reversible, but the rate of decomposition is different for the isomers: The complex with the methoxy group in ortho position with respect to the nitro group decomposes 1700 times faster than the complex containing the methoxy group in the para position (Deicha and Terrier 1981). The higher stability of the para adduct relative to the ortho isomer may reasonably be attributed to the fact that the nitro group can better accommodate the negative charge when it occupies a para position, rather than ortho, to the point of attachment of the reagent. Furthermore, the para adduct may take advantage of a larger extent of electron delocalization due to more remoteness of the methoxy group from the nitro group as it is obvious from comparison of structures for the isomeric Meisenheimer complexes in Scheme 3.57.

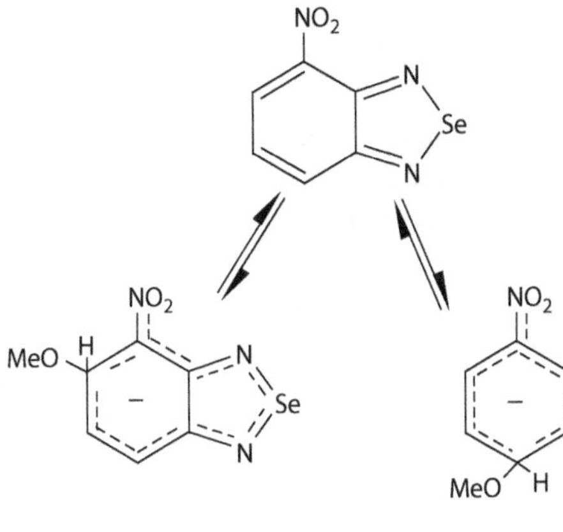

SCHEME 3.57

3.3.3.2.6 Photolysis

According to Pedersen et al. (1977), ultraviolet irradiation of piazoselenol at room temperature produces nitriles and elemental selenium. The latter originate from the intermediary formed benzonitrile selenide, PhCNSe.

3.3.3.2.7 Coordination

Crystallographic x-ray analysis had shown that 1,2,5-selenadiazole fused with naphthoquinodimethane is characterized by intermolecular coordinative interaction between the selenium key atoms and cyano groups to form a two-dimensional network (see Scheme 3.58) (Suzuki et al. 1987a,b). When this finely powdered

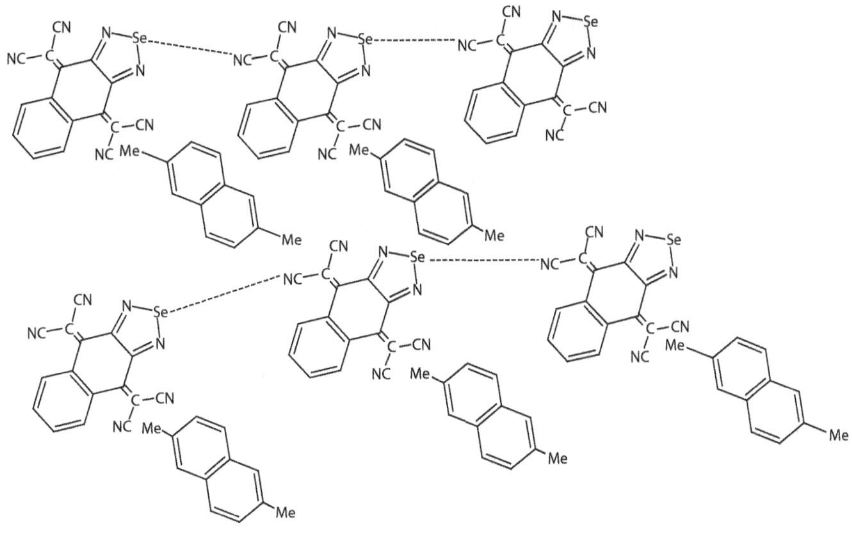

SCHEME 3.58

species is suspended in a mixture of isomeric dimethyl naphthalenes, charge-transfer coordination takes place so that dimethylnaphthalene occupies hollows in the two-dimensional network. Each of these hollows has a size that is compatible with the size of 2,6-dimethyl naphthalene only. After separation the newly formed charge-transfer complex and then its thermal decomposition, both constituents are obtained separately. As to separation selectivity, it reaches 98% in respect of 2,6-dimethyl naphthalene (Suzuki et al. 1992). Scheme 3.58 shows the space structure of the charge-transfer complex responsible for the separation.

1,2,5-Selenadiazole fused with the benzene ring crystallizes in dimers containing two antiparallel selenium–nitrogen bonds and these dimers are not destroyed during coordination to silver nitrate. Silver is bound with the nitrogen atoms that remain free in the dimer (Zhou et al. 2005).

In spite of electronegativity of the 1,2,5-selenadiazole ring, lone pairs of its nitrogen atoms keep they ability of coordination to metals. One of the two nitrogens can be involved in the coordination with cupric chloride (Bezzubets et al. 1984) or cupric salts of fatty acids (D'yachenko et al. 1991). Copper bis(trimethylacetate) forms polymer-chain complex with piazoselenol. In this complex, piazoselenol coordinates to copper via its both heterocyclic nitrogen atoms whereas trimethylacetate, Me_3CCOO^-, acts as a bridge between the metallic centers (Bel'skii et al. 1984).

The presence of the nitro group in α- or β-position of the benzene fused ring does not prevent coordination to metals. In the case of α-aminopiazoselenol, however, two nitrogen atoms, one of the amino group and other of the heteroring, react with cupric chloride. The presence of the amino group makes the bidentate coordination possible (Bezzubets et al. 1984). The bidentate coordination was also observed for (anthrapiperidino)selenadiazole quinones in their coordination to gallium and indium halides. In this case, both the carbonyl group and the nearest nitrogen atom of the heterocycle are bound with a metal (Rudnitskaya et al. 1981).

As a ligand, piazoselenol is characterized by both donor and acceptor activities. In this regard, it is important to point out that, in its complexes with copper, the metal is stabilized in the highest oxidation degree: The complexes obtained exhibit paramagnetism (Solozhenkin et al. 1976, Bel'skii et al. 1984). Coordinating to the vinyl-thiocarbonyl-triphenylposphine chlororuthenium of Scheme 3.59, piazoselenol does not disturb the geometry of the starting complex and does not force the vinyl or any other ligand to migrate (Cowley et al. 2007).

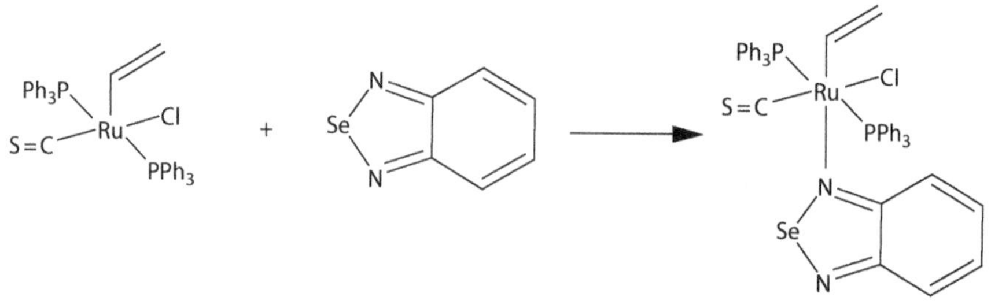

SCHEME 3.59

For the piazoselenol complexes with ruthenium, osmium, and iridium, Herberhold and Hill (1989) observed that the coordinated piazoselenol ligand is kinetically very liable. In other words, very weak coordination takes place.

While the piazoselenol neutral coordinates to a metal through the nitrogen atom according to Scheme 3.59, the more basic piazoselenol anion-radical coordinates through the selenium atom. It was shown for the reactive pair of the anion-radical and tungsten hexacarbonyl. In the complex formed, piazoselenol occupies the place of one carbonyl at the tungsten center. This anion-radical complex is persistent (Kaim and Kasack 1982, Kaim 1984, Kaim et al. 1989). The electron spin resonance spectrum suggests C_{2v} symmetry that can be just at the case of Se–W coordination. It is the presence of the delocalized unpaired electron in the piazoselenol ligand that makes Se–W coordination possible. This is a probable explanation, although Solozhenkin et al. (1976) suggested rapid fluctuation of the $W(CO)_5$ fragment between the two nitrogen sites according to Scheme 3.60. (Attempts to freeze out a static structure were hampered by the typical broadening effect of the resonance line at temperatures below 0°C.)

Rapid fluctuation depicted in Scheme 3.60 does not take place in cases of piazothiol anion-radical complexes with $Cr(CO)_5$ or $Mo(CO)_5$ (Kaim and Kasak 1982, Kaim et al. 1989). In these cases, only nitrogen–metal coordination was observed, whereas, in the tungsten complex, the selenium atom comes to the scene.

3.3.3.2.8 Ring Transformation

Being exemplified by phenanthroselenadiazole, this transformation is depicted in Scheme 3.61. The reaction involves the substrate reaction with ethyl magnesium bromide, followed by treatment with tellurium tetrachloride and triethylamine. The yield of the final product is 35% (Scheme 3.61) (Neidlein et al. 1987).

Being prepared according to Scheme 3.33, selenadiazole N-oxide can transform into oxadiazole derivatives. Upon thermolysis, piazoselenol N-oxide gives rise to

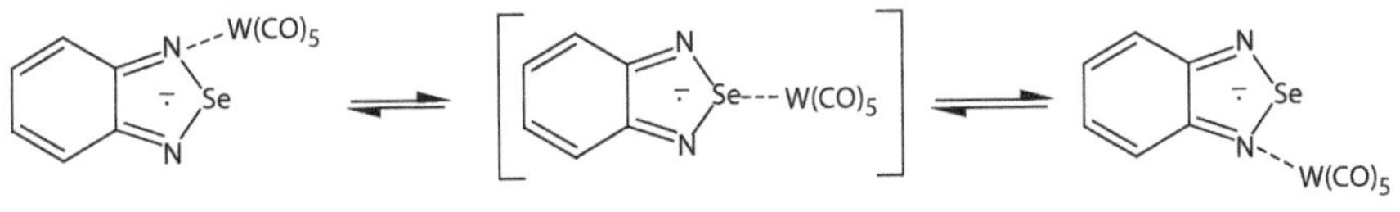

SCHEME 3.60

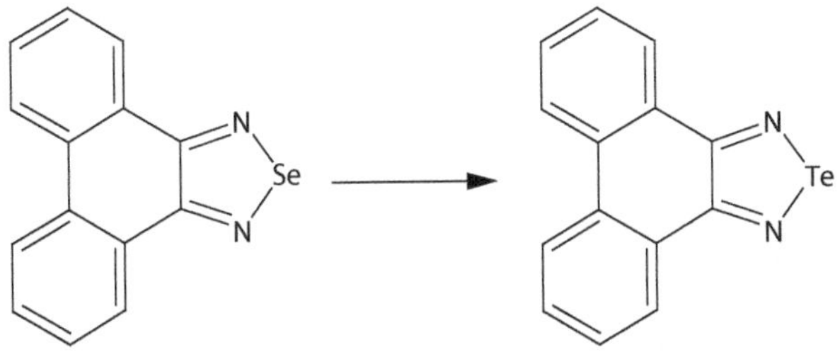

SCHEME 3.61

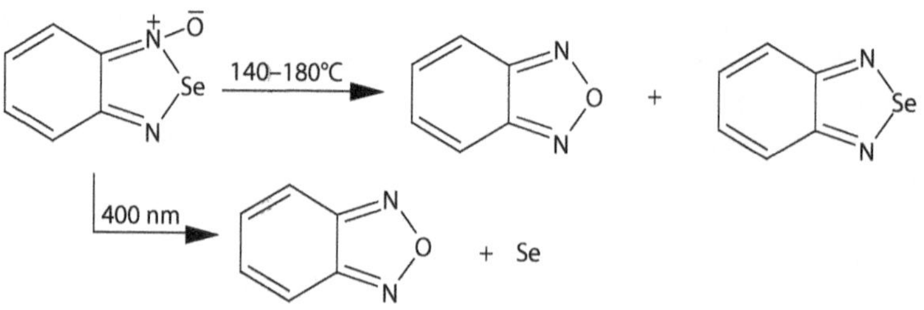

SCHEME 3.62

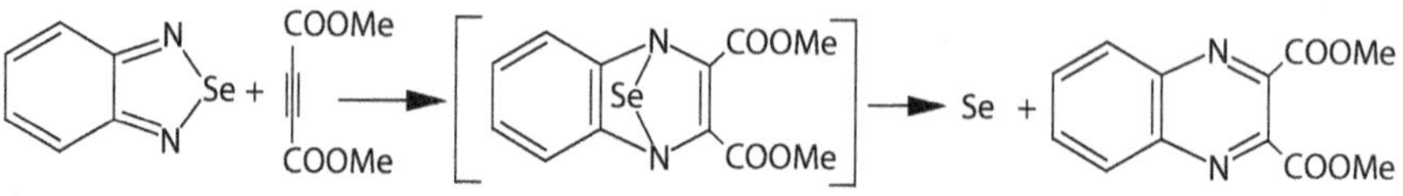

SCHEME 3.63

benzofurazan (32%) and piazoselenol itself (51%). Upon photolysis, this N-oxide is converted into benzofurazan only with 96% yield (Pedersen 1974, 1976, 1979). Scheme 3.62 illustrates these heteroring transformations.

When [1,2,5]-selenadiazolo-2-oxide-[3,4-*d*]-pyrimidine-5,7-(4*H*,6*H*)diones react with sodium hypochlorite, they are converted into the corresponding oxadiazoles with good yields (Scheme 3.52) (Yavolovsky et al. 2000).

Campbell et al. (1978) obtained dimethyl quinoxaline-2,3-dicarboxylate from piazoselenol and dimethylacetylene dicarboxylate. The reaction was performed at 70°C through 10 days. Free selenium was another product of the reaction. There are no data on the mechanism of the reaction, but the authors supposedly included a step of cycloaddition to the N=CH–CH=N diene system without any proofs for the sequence of transformations depicted by Scheme 3.63. They obtained proofs for the final product only, the intermediary ("Diels–Alder") product was included presumably.

3.3.4 BIOMEDICAL IMPORTANCE

3.3.4.1 Medical Significance

Inhibitory activity of 4-amino-1,2,5-selenadiazole-3-carboxamide, its *N*-methyl-amino and *N*-butylamino derivatives were established against bacteria, yeasts, filamentous fungi, and algae. Comparative studies with inorganic selenium salts showed that the inhibitory activity of selenadiazoles is not due to the selenium toxicity. Single therapeutic doses exerted significant in vivo activity in mice infected with a *S. aureus* (Hunt and Pittillo 1966, cf. Shealy et al. 1966 as well as Shealy and Clayton 1967). Piazoselenol bearing piperidino and *N,N*-diethylamino groups in α and β positions, respectively, are very active against *B. subtilis* and moderately active against *Candida albicans* (Sharma et al. 1984).

Zheng et al. (2007) claimed the microwave-assisted solid-state preparation of pyrimidine[3,4-*d*]-[1,2,5]-selenadiazole-5,7-(4*H*,6*H*)-dione (more than 98% of purity). This patent characterizes the compound as an anticancer drug. The comprehensive study (Chen et al. 2008) identified this selenadiazole dione as an antiproliferative agent against the human breast-adrenocarcinoma, hepatoma, and melanoma cells. The authors proved that the drug reacts with caspase-8 and caspase-9 enzymes and thereby induces mitochondria-mediated apoptosis in the cells. (Apoptosis means the suicide-active mode of cell death.)

3.3.4.2 Agricultural Protectors

Belen'kaya et al. (1992b, 1995) found biological activity of 1,2,5-selenadiazole fused to benzene-bearing amino, ethereal, esterial groups. As noted, these compounds showed nematocidal activity, controlling gall nematode on cucumber roots. Methoxy derivatives were the most effective insecticides against housefly and rice weevil imagoes. Halosubstituted esters showed fungicidal activity, inhibited lettuce germination, and showed antiviral activity against influenza virus A2 Victoria in 10-day old chicken embryos.

3.3.5 Technical Applications

3.3.5.1 Dyestuffs

In the presence of piazoselenol moieties, many dyestuffs shift their light-absorption maxima toward the longer wavelength. Specifically, the effect was described for the case of the selenadiazole-ring attachment to the benzothiazole nucleus in the cyanine dyes (Fridman 1961). Formulations for two-component hair dyes are claimed that contains amines and piazoselenol derivatives (Pasquier et al. 2002, Umbricht et al. 2002). The piazoselenol derivative supposedly participates in the amine oxidation (leading to the formation of a hair dyestuff) and transforms itself to species useful for hair strengthening.

3.3.5.2 Light Emitters

Copolymers containing areneselenadiazole and fluorene fragments were described as perspective organic electroluminescent materials for display applications (Tian et al. 2003, Yang et al. 2003, 2004, 2005, Huang et al. 2006, Cao et al. 2007). Some of these polymers are depicted in Scheme 3.64. A mechanistic study (Chien et al. 2007) revealed that energy transfer from polyfluorene backbone to the benzoselenadiazole segments, rather than charge trapping, is mainly operative in the electroluminescence process. In comparison to the piazothiol units, the piazoselenol ones provide red shift of the luminescence emission and decreases emission in the infrared region of the spectrum. (The naphthalene analog furnishes stronger red shift than its benzene analog and strongly enhances the luminous efficiency.) The same red emission was also described for non-polymeric electroluminescent compounds, for example, 4,8-bis{[4-N,N-bis(4-octyloxyphenyl)-2-thiophene]benzo-[1,2-c:4,5-c']-bis(1,2,5-selenadiazole)} compared to its thia counterpart (Quan et al. 2008). The already mentioned isosteric effect plays a role in optical properties too. In this case, the effect is mainly caused by the sulfur–selenium differences in atomic size. The selenium analogs are capable of closer contacts between different copolymer chains in the material films.

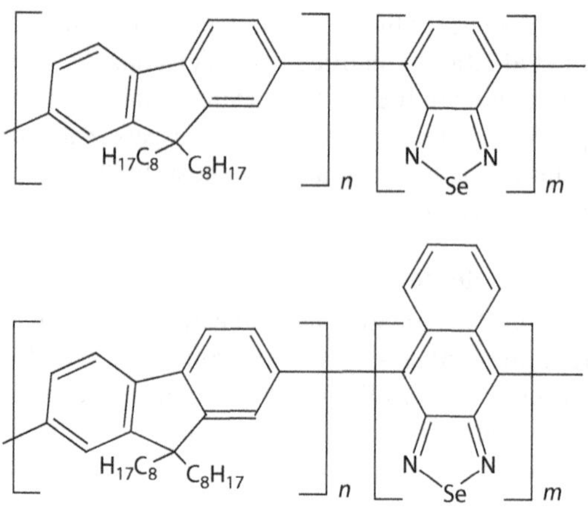

SCHEME 3.64

3.3.5.3 Organic Metals

Scheme 3.55 illustrates the synthesis of tetracyanoquinodimethane fused with 1,2,5-selenadiazole. The same reactions were performed to prepare the 1,2,5-thiadiazole and 1,2,5-oxadiazole analogs. Comparison of x-ray structures indicated that the packing arrangements of the diverse chalcogenadiazole derivatives are quite different from each other in spite of their similar molecular geometries. In the crystal of a selenadiazole derivative, are observed very short intermolecular contacts between selenium and nitrogen to form infinite sheet-like network, whereas coplanar diads are formed by C–H⋯N=C hydrogen bonds in an oxadiazole and a thiadiazole derivatives. All of the tetracyanoquinodimethane–chalcogenadiazole derivatives afforded stable anion-radicals salts with tetrathiafulvalenes. In the crystal state, these salts manifest high electric conductivity, the selenadiazole species are superior (Suzuki et al. 1997).

Yamashita et al. (1992) measured electric conductivities of the two analogs depicted in Scheme 3.65. The conductivity was compared for single components taken as compressed pellets. The selenium analog manifests better conductivity by one order than its sulfur isosteric analog (2.3×10^{-5} and 5.3×10^{-6} S/cm, respectively). The difference originates from the stronger intermolecular coordination caused by the selenium atom (note, isosterism in action!).

Generally speaking, annulation of 1,2,5-selenadiazole ring to aromatic derivatives is prospective to construct acceptor components of ion-radical salts with high electroconductivity. Thus, extremely high electron affinity was observed for [1,2,5]-selenadiazolo-[3,4-b]-quinoxalinium with the N-methylated nitrogen in the quinoxaline ring (Yamashita et al. 1989).

3.3.5.4 Lubricating Additives

As found, β-(N,N-dimethylamino)piazoselenol is an effective antioxidant and can be used as a component of lubricating mixtures working at high temperatures (Dueltgen et al. 1962).

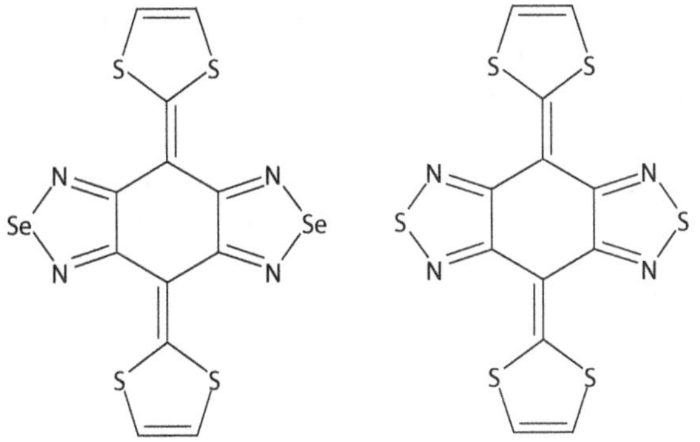

SCHEME 3.65

3.3.6 ANALYTICAL APPLICATIONS

Selenadiazoles condensed with the benzene of naphthalene ring found specific uses in analysis of selenium and environmental monitoring. It is based on the selenadiazole spectral properties that permit analytical determinations at very low concentrations of selenium in samples or tissues. Spectrofluorimetric determination of 1,2,5-selenadiazole derivatives is the most popular procedure. The enhancement of the fluorescence intensity is possible by addition of a surfactant and a hall-containing complexone to aqueous solutions under analysis. Thus, the intensity of the naphtho-[2,3-c]-[1,2,5]-selenadiazole occurs to be ca. 30 times lower in water than that in the aqueous solution containing sodium dodecyl sulfate and β-cyclodextrin (Zheng and Lu 1992). (When dodecyl sulfate is a micellizating component, cyclodextrin provides its cavity to accept the analyte. The fluorescent enhancement is due to the altered microenvironment experienced by naphthoselenadiazole upon its transfer from water to the hydrophobic cavity of cyclodextrin and the formation of the host–guest complex.)

When the selenadiazole derivatives are formed within analysis, selenium involves in its 4+ oxidation form, that is, as selenous acid or as a metal selenite. In these cases, arene-1,2-diamines are used to form the corresponding selenadiazoles. All the diamines readily react with Se^{4+}. At the same time, they differ in limits of spectrophotometric or fluorimetric determination of the selenadiazole formed. The highest sensitivity was observed for selenadiazoles formed from 3,4-dichloro-1,2-phenylenediamine (Goto and Toei 1965) or 2,3-diaminonaphthalene (see, e.g., Grant 1981, Koike et al. 1993). The latter analytical reagent works excellently even in strongly acidic solutions (Parker and Harvey 1962). For determination of selenium in soil composts, 1,2-diamino-4-nitrobenzene was recommended. The β-nitropiazoselenol formed was extracted and analyzed by graphite-furnace atomic-absorption spectrometry (Shan et al. 1982).

The usage of 2,3-diaminonaphthalene was successful for the selenium determination in blood and in urine, in tissues of kidney, liver, in wool, in pasture-grass, in milk, in food supplements as well as in soil samples. For examples, see Grant (1981), Lebedev and Lebedev (1996), Pyatkova et al. (2003). Importantly, this diamine was used for spectrophotometric analysis of selenium in blood plasma and blood serum at lowered levels such as in patients with selenium deficiency cardiomyopathy or ischemic heart disease. The same method was applied for monitoring the selenium level during supplementation of the patients with 300 mg of selenium a day for 30 day to normalize this level (Lebedev and Lebedev 1996). Koike et al. (1993) used this diamine to study Se^{4+} contents in marine planktons as biological material and in open-sea layers. Contents of Se^{4+} in the sea layers and in the corresponding samples of planktons coincide. (Decomposition of planktons is a source of Se^{4+} in sea waters.) Because naphtho-[2,3-c]-[1,2,5]-selenadiazole has strong fluorescent characteristics, it is applicable to determination of selenium microgram quantities in foods. For instance, it was established that selenium content in rice and tea comes to 6.9×10^{-3} and 2.1×10^{-3} μg/g, respectively (Wang and Chen 1998). High-performance liquid chromatography is also good at quantitative determination of the naphthoselenadiazole formed (Zhang and Chen 1996).

Indirect atomic-absorption method is used for the determination of selenium, too. The method involves the formation of naphtho-[2,3-c]-[1,2,5]-selenadiazole and its coordination to palladium dichloride. Both reactions proceed quantitatively and the palladium absorption is recalculated to the selenium content (Lau and Lott 1971).

N-alkyl derivatives of arene-1,2-diamines are also used as reagents for the spectrophotometric determination of Se^{4+}. In these cases, the following observation is important: In 1.5–6 M HCl, Se^{4+} reacts as $SeOCl_2$, 1-N-(β-hydroxypropyl)-1,2-diaminobenzene is doubly protonated, and the reaction becomes to be reversible (Kasterka 1979).

Sometimes, the selenadiazole derivatives are used as themselves or formed as a part of an analytical procedure.

As has already been mentioned, naphtho-[2,3-c]-[1,2,5]-selenadiazole quantitatively coordinates to palladium dichloride. Accordingly, this ready-made selenadiazole is used to determine palladium. Atomic absorption, gravimetry, spectrophotometry, fluorimetry, and radiometry all are applicable for the analysis (Lau and Lott 1970). Piazoselenol bearing the (3,4-diaminophenyl) at α-position, was recommended for analysis of the nitrite and nitrate content in natural waters (Canney et al. 1974). This analytical reagent forms the corresponding triazole under action of nitrite. The triazole amounts are determined by methods of fluorimetry and spectrophotometry. Interferences from high concentration of manganese dioxide and ferric cations are eliminated by filtration and ion-exchange treatments. The method is compatible with nitrate determination after its reduction to nitrite within a cadmium–mercury column.

Cukor et al. (1964) proposed the method to determine contents of ^{75}Se in oats, cornflakes, rice, and grass seeds. The method consists of the sample combustion, extraction of the selenous acid isotopomer at pH 2.0 and treatment of the solution obtained with 2,3-diaminonaphthalene dissolved in aqueous ethanol. The naphtho-[2,3-c]-[1,2,5]-selenadiazole formed is extracted with cyclohexane and fluorescence of the extract is measured at 540 nm.

The fluorimetric determination of selenium via naphtho-[2,3-c]-[1,2,5]-selenadiazole was also proposed for the analysis of cast iron (Clarke 1970). There are several patents claimed by Tsveniashvili et al. (1982, 1983, 1984) and their peer-reviewed generalization (Tsveniashvili 1985) on determination of the water traces in organic solvents, especially in aprotic ones. Namely, piazoselenol complex with cobalt dichloride decomposes in the presence of water and formed the cobalt dichloride hexahydrate. The reaction proceeds stoichiometrically and the water amounts are calculated according to the amount of a selenadiazole liberated. The selenadiazole determination can be performed by spectrophotometry or polarography. In total, the approach has advantages over commonly used water determinations including the Fischer method. Because the liberated cobalt cation binds water, this decomplexation process is applicable for drying of organic solvents. The complexes are used as they are if the water content is less than 0.01%. When the water content in a solvent is in the range from 2% to 10%, this selenadiazole complex is recommended for use as an admixture to other drying agents (e.g., to anhydrous cupric sulfate).

The electrochemical reduction of piazoselenol is the basic method of the quantitative determination. The method introduced by Scheme 3.51 makes understandable the approach to determination of glutathione and other cellular thiols in biochemical studies. In this approach, the thiol content is calculated from the selenadiazole depreciation (Wang et al. 2008).

A Japanese patent claimed preparation of fluorescent dyes based on 1,2,5-selenadiazoles fused with substituted benzene, pyridine, or pyridinium. These dyes are prospective for high-sensitivity detection of biomolecules (Isobe and Mataga 2008).

3.3.7 CONCLUSION

Due to their specific chemistry, 1,2,5-selenadiazoles bring remarkable opportunities. In this sense, aromatic-annulated 1,2,5-selenadiazoles stand out specifically. They are similar to the related annulated aromatics, but have reduced ring current characteristics. In their anion-radicals, delocalization of an unpaired electron embraces the whole molecular contour. Accordingly, the neutral aromatic-annulated 1,2,5-selenadiazoles coordinate to metals via the heterocyclic nitrogen atoms, while the corresponding anion-radicals acquire a possibility to coordinate through the key selenium atom.

Arene-selanadiazoles are easily formed from aromatic 1,2-diamines. These selenadiazoles give diverse derivatives with the substituents in aromatic parts of the molecules. The selenadiazole ring in these chemically changed derivatives is readily reduced. The possibility appears to prepare modified 1,2-diamines inaccessible by other methods. Analytical use of arene-selenadiazoles is miscellaneous and the corresponding methods of quantitative determination are very easy in their performance.

Compounds containing the 1,2,5-selenadiazole moiety attract increased attention as oil additives, light emitters, organic metals, species for medical tests, and, especially, as drugs for preventive and therapeutic medical treatment. Ability to transform into cycles with key heteroatoms other than selenium also opens interesting synthetic routes.

REFERENCES TO SECTION 3.3

Bacon, C.E.; Eisler, D.J.; Melen, R.L.; Rawson, J.M. (2008) *Chem. Commun.* 4924.

Bagryanskaya, I.Yu.; Gatilov, Yu.V.; Gritsan, N.P.; Ikorskii, V.N.; Irtegova, I.G.; Lonchakov, A.V.; Lork, E.; Mews, R.; Ovcharenko, V.I.; Vasilieva, N.V.; Zibarev, A.V. (2007) *Eur. J. Inorg. Chem.* 4751.

Bauer, E.M.; Ercolani, C.; Galli, P.; Popkova, I.A.; Stuzhin, P.A. (1999) *J. Porphyrins Phthalocyanines* **3**, 371.

Baum, S.M.; Trabanko, A.A.; Montblan, A.G.; Micallef, A.S.; Zhong, Ch.; Meunier, H.G.; Suhling, K.; Phillips, D.; White, A.J.P.; Williams, D.F.; Barrett, A.J.M.; Hoffman, B.M. (2003) *J. Org. Chem.* **68**, 1665.

Begland, R.W. (1974) US Patent 3,849,479.

Belen'kaya, I.A.; Sirik, S.A.; Shapiro, Yu.E.; D'yacheko, E.K. (1992a) *Khim. Geterotsikl. Soed.* 1135.

Belen'kaya, I.A.; Sirik, S.A.; Yasinskaya, O.G.; Ivanova, V.V.; Krokhina, G.P.; Shashenkova, D.Kh.; Prokhorchuk, E.A.; Uskova, L.A.; Grib, O.K. (1992b) *Fisiolog.-Aktiv. Veshchestva*, No. 24, 39.

Belen'kaya, I.A.; Yasinskaya, O.G.; Ivanova, V.V.; Sirik, S.A. (1995) *Khim.-Farm. Zh.* **29**, 54.

Bella, M.; Milata, V. (2008) *J. Heterocycl. Chem.* **45**, 425.

Bel'skii, V.K.; Ellert, D.G.; Seifulina, Z.M.; Novotortsev, V.M.; Tsveniashvili, V.Sh.; Garnovskii, A.D. (1984) *Izv. AN SSSR, Ser. Khim.* 1914.

Benedetti, E.; Bertini, V. (1968) *Spectrochim. Acta, Part A* **24**, 57.

Bertini, V. (1967a) *Gazz. Chim. Ital.* **97**, 1870.

Bertini, V. (1967b) *Angew. Chem. Intl. Ed.* **6**, 563.

Bertini, V.; de Munno, A.; Menconi, A.; Fissi, A. (1974) *J. Org. Chem.* **39**, 2294.

Bertini, V.; de Munno, A.; Piccci, N.; Lucchesini, F.; Pocci, M. (1987) *Heterocycles* **26**, 2153.

Bertini, V.; Lucchesini, F.; de Munno, A. (1979) *Synthesis* 979.

Bertini, V.; Lucchesini, F.; de Munno, A. (1982) *Synthesis* 681.

Bezzubets, E.A.; D'yachenko, E.K.; Fadeeva, I.I.; Ostapkevich, N.A. (1984) *Zh. Obshch. Khim.* **54**, 910.

Bird, C.W. (1985) *Tetrahedron* **41**, 1409.

Bird, C.W. (1992) *Tetrahedron* **48**, 335.

Bird, C.W.; Cheeseman, G.W.H. (1964) *Tetrahedron* **20**, 1701.

Blackman, G.l.; Brown, R.D.; Burden, F.R.; Kent, J.E. (1967) *Chem. Phys. Lett.* **1**, 379.

Breier, L.; Petrovskii, P.V.; Todres, Z.V.; Fedin, E.I. (1969) *Khim. Geterotsikl. Soed.* 62.

Brizzi, C.; dal Monte, D.; Sandri, E. (1964) *Ann. Chim.* **54**, 476.

Brown, N.M.D.; Bladon, P. (1968) *Spectrochim. Acta, Part A* **24**, 1869.

Buchwald, H.; Ruehlman, K. (1979) *J. Organometal. Chem.* **166**, 25.

Campbell, C.D.; Rees, Ch.W.; Bryce, M.R.; Cooke, M.D.; Hanson, P.; Vernon, J.M. (1978) *J. Chem. Soc. Perkin Trans. 1*, 1006.

Canney, P.J.; Armstrong, D.E.; Wiersma, J.H. (1974) Govt. Report Announce (U.S.) **47**, 49.

Cao, W.; Dong, H.; Huang, F.; Shen, H.; Cao, Y. (2007) *Gaofenzi Xuebao* 566.

Casoni, D.D.M.; Sandri, E. (1963) *Ann. Chim. Rome* **83**, 1697.

Chen, T.; Zheng, W.; Wong, Y.-Sh.; Yang, F. (2008) *Biomed. Pharmacol.* 77.

Chien, Ch.-H.; Shih, P.-I.; Shu, Ch.-F. (2007) *J. Polymer Sci., Part A* **45**, 2938.

Cillo, C.M.; Lash, T.D. (2004) *J. Heterocycl. Chem.* **41**, 955.

Clarke, W.E. (1970) *Analyst* **95**, 65.

Cowley, A.R.; Hector, A.L.; Hill, A.F.; White, A.J.P.; Williams, D.J.; Wilton-Ely, D.E.T. (2007) *Organometallics* **26**, 6114.

Cozzolino, A.F.; Vargas-Baca, I.; Mansour, S.; Mahmoudkhani, A.H. (2005) *J. Am. Chem. Soc.* **127**, 3184.

Cukor, P.; Walzcyk, J.; Lott, P.F. (1964) *Anal. Chim. Acta* **30**, 473.

D'yachenko, E.K.; Obozova, L.A.; Lyubomirova, K.N.; Razukrantova, N.V. (1991) *Khim.-Farm. Zh.* **25**, 37.

de Munno, A.; Bertini, V.; Rasero, P.; Picci, N.; Bonfanti, L. (1978) *Atti Accad. Nation. dei Lincei, Classe di Sci. Fis., Math., Natur., Rendiconti* **64**, 385.

Deicha, C.; Terrier, F. (1981) *J. Chem. Res.* 312.

Dueltgen, R.L.; Lugash, M.N.; Cosgrove, S.L. (1962) *Lubr. Eng.* **18**, 218.

Dutton, J.L.; Martin, C.D.; Sgro, M.J.; Jones, N.D.; Ragogna, P.J. (2009) *Inorg. Chem.* **48**, 3239.

Dutton, J.L.; Ragogna, P.J. (2009) *Inorg. Chem.* **48**, 1722.

Dutton, J.L.; Tuononen, H.M.; Jennings, M.C.; Ragogna, P.J. (2006a) *J. Am. Chem. Soc.* **128**, 12624.

Dutton, J.L.; Tindale, J.L.; Jennings, M.C.; Ragogna, P.J. (2006b) *Chem. Commun.* 2474.

Edin, M.; Grivas, S. (2000) *ARKIVOC* (i), 9.

Efros, L.S.; El'tsov, A.V. (1958) *Zh. Obshch. Khim.* **28**, 2172.

Efros, L.S.; Todres, Z.V. (1957a) *Zh. Obshch. Khim.* **27**, 983.

Efros, L.S.; Todres, Z.V. (1957b) *Zh. Obshch. Khim.* **27**, 3127.

El-Azhary, A.A.; Al-Kahtany, A.A. (2001) *THEOCHEM* **572**, 81.

Elmaaty, T.M.A. (2008) *J. Heterocycl. Chem.* **45**, 1179.

Ercolani, C.; Stuzhin, P.; Donzello, M.P.; Bauer, E. (1999) WO Patent 9,915,533.

Fajer, J. (1965) *J. Phys. Chem.* **69**, 1773.

Fajer, J.; Bielski, B.H.J.; Felton, R.H. (1968) *J. Phys. Chem.* **72**, 1281.

Fedin, E.I.; Todres, Z.V. (1968) *Khim. Geterotsikl. Soed.* 416.

Fridman, S.G. (1961) *Zh. Obshch. Khim.* **31**, 1096.

Gieren, A.; Lamm, V. (1982) *Acta Crystallogr.* **B38**, 2605.

Gieren, A.; Lamm, V.; Haddon, R.C.; Kaplan, M.L. (1980) *J. Am. Chem. Soc.* **102**, 5070.

Gorelik, M.V. (1968) *Zh. VKhO* **13**, 467.

Gorelik, M.V. (1971) *Khim. Geterotsikl. Soed.* 212.

Gorelik, M.V.; Bogdanov, S.V. (1960) *Zh. Obshch. Khim.* **30**, 2949.

Goslinski, T.; Zhong, Ch.; Fuchter, M.J.; Stern, Ch.L.; White, A.J.P.; Barrett, A.G.M; Hoffman, B.M. (2006) *Inorg. Chem.* **45**, 3686.

Goto, M.; Toei, K. (1965) *Talanta* **12**, 125.

Grant, A.B. (1981) *N.Z. J. Sci.* **24**, 65.

Grivas, S. (2000) *Curr. Org. Chem.* **4**, 707.

Grivas, S.; Tian, W. (1992) *Acta Chem. Scand.* **46**, 1109.

Gul'maliev, A.M.; Stankevich, I.V.; Todres, Z.V. (1973) *Khim. Geterotsikl. Soed.* 1473.

Gul'maliev, A.M.; Stankevich, I.V.; Todres, Z.V. (1975) *Khim. Geterotsikl. Soed.* 1055.

Hazelton, J.C.; Iddon, B.; Suschitsky, H.; Wooley, L.H. (1992) *J. Chem. Soc., Perkin Trans. 1*, 685.

Hegelund, F.; Larsen, R.W.; Aitken, R.A.; Palmer, M.H. (2006) *J. Mol. Spectrosc.* **236**, 189.

Herberhold, M.; Hill, A.F. (1989) *J. Organometal. Chem.* **377**, 151.

Hill, R.W.; Sutton, L.E. (1949) *J. Chim. Phys.* **46**, 244.

Huang, F.; Hou, L.; Shen, H.; Yang, R.; Hou, Q.; Cao, Y. (2006) *J. Polym. Sci. A* **44**, 2521.

Huebner, T.; Lamm, V.; Neidlein, R.; Droste, D. (1984) *Z. Naturforsch. B* **39**, 485.

Hunt, D.E.; Pittillo, R.F. (1966) *Antimicrob. Agents Chemother.* 551.

Isobe, Sh.; Mataga, Sh. (2008) Jpn. Patent 156,556.

Kaim, W. (1984) *J. Organometal. Chem.* **264**, 317.

Kaim, W.; Kasack, V. (1982) *Angew. Chem.* **94**, 712.

Kaim, W.; Kohlmann, S.; Lees, A.J.; Zulu, M. (1989) *Zeitschr. Anorg. Allgem. Chem.* **575**, 97.

Kasterka, B. (1979) *Chem. Analityczna* **24**, 329.

Kataev, E.G. (1955) *Mendeleev Chem. Soc. Commun.* 49.

Klimasenko, N.L.; Chetkina, L.A.; Gol'der, G.A. (1973) *Zh. Struct. Khim.* **14**, 515.

Koike, Yu.; Nakaguchi, Yu.; Hiraki, K. (1993) *Bunseki Kagaku* **42**, 285.

Komin, A.P.; Carmac, M. (1976) *J. Heterocycl. Chem.* **13**, 13.

Kovtonyuk, V.N.; Makarov, A.Yu.; Shakirov, M.M.; Zibarev, A.V. (1996) *Chem. Commun.* 1991.

Krause, L.; Whitehead, M.A. (1973) *Mol. Phys.* **25**, 99.

Kudrik, E.V.; Bauer, E.M.; Ercolani, C.; Chiesi-Villa, A.; Rizzoli, C.; Gabercorn, A.; Stuzhin, P.A. (2001) *Medeleev Commun.* 45.

Kukushkin, Yu.N.; D'yachenko, S.A.; Vlasova, R.A.; Glazyuk, N.P. (1973) *Zh. Obshch. Khim.* **43**, 1179.

Kumari, S.; Sharma, K.S.; Singh, S.P. (1990) *Indian J. Chem., Sect. B* **29**, 781.

Kursanov, D.N.; Todres, Z.V. (1967) *Dokl AN SSSR*, **172**, 1086.

Lane, E.S.; Williams, C. (1955) *J. Chem. Soc.* 1468.

Lane, E.S.; Williams, C. (1957) GB Patent 781,790.

Lau, H.K.Y.; Lott, P.F. (1970) *Talanta* **17**, 717.

Lau, H.K.Y.; Lott, P.F. (1971) *Talanta* **18**, 303.

Lebedev, P.A.; Lebedev, A.A. (1996) *Khim.-Farm. Zh.* **30**, 54.

Luzzati, V. (1951) *Acta Crystallogr.* **4**, 193.

Mellini, M.; Merlino, S. (1976) *Acta Crystallogr.* **B32**, 1074.

Moerkved, E.H.; Neset, S.M.; Bjoerlo, O.; Kjoesen, H.; Hvistendahl, G.; Mo, F. (1995) *Acta Chem. Scand.* **49**, 658.

Neidlein, R.; Knecht, D. (1987) *Helv. Chim. Acta* **70**, 1076.

Neidlein, R.; Knecht, D.; Gieren, A.; Ruiz-Perez, C. (1987) *Zeitschr. Naturforsch.* **42B**, 84.

Nunn, A.J.; Ralph, J.T. (1965) *J. Chem. Soc.* 6769.

Parker, C.A.; Harvey, L.G. (1962) *Analyst* **87**, 558.

Pasquier, C.; Charriere, V.; Braun, H.-Ju. (2002) WO Patent 022,093.

Pedersen, Ch.L. (1974) *J. Chem. Soc. Chem. Commun.* 704.

Pedersen, Ch.L. (1976) *Acta Chem. Scand.* **B30**, 675.

Pedersen, Ch.L. (1979) *Tetrahedron Lett.* **43**, 745.

Pedersen, Ch.L.; Hacker, N. (1977) *Tetrahedron Lett.* **41**, 3982.

Pedersen, Ch.L.; Harrit, N.; Poliakoff, M.; Dunkin, I. (1977) *Acta Chem. Scand.* **B31**, 848.

Pesin, V.G. (1969) *Khim. Geterotsikl. Soed.* 235.

Pesin, V.G.; Khaletskii, A.M.; Sergeev, V.A. (1962) *Zh. Obshch. Khim.* **32**, 181.

Pozdyshev, V.A.; Todres, Z.V.; Efros, L.S. (1960) *Zh. Obshch. Khim.* **30**, 2551.

Praefke, K.; Kohne, B.; Kornith, F. (1990) *Lieb. Ann.* **1990**, 203.

Pyatkova, L.N.; Dmitrienko, S.G.; Ul'yanova, E.V.; Bashilov, A.V.; Zolotov, Yu.A. (2003) *Zavodsk. Lab.* **69**, 13.

Quan, G.; Dai, B.; Luo, M.; Yu, D.; Zhan, J.; Zhang, Zh.; Ma, D.; Wang, Zh. Y. (2008) *Chem. Mater.* **20**, 6208.

Roesky, H.W.; Gries, T.; Hofmann, H.; Schimkowiak, J.; Jones, P.G.; Meyer-Baese, K.; Sheldrik, G.M. (1986) *Chem. Ber.* **119**, 366.

Roesky, H.W.; Hofmann, H. (1984) *Zeitschr. Naturforsch.* **39B**, 1315.

Rudnitskaya, O.V.; Zaitsev, B.E.; Gorelik, M.V.; Molodkin, A.K. (1981) *Zh. Neorg. Khim.* **26**, 1261.

Rykowski, Z.; Sczesna, E. (1989) *Pol. J. Chem.* **63**, 307.

Sawicki, E.; Carr, A. (1957) *J. Org. Chem.* **22**, 503.

Sergeev, V.A.; Pesin, V.G.; Kotikova, N.M. (1972) *Khim. Geterotsikl. Soed.* 328.

Sergeev, V.A.; Pesin, V.G.; Papirnik, M.P. (1989) *Zh. Org. Khim.* **25**, 1802.

Shan, X.Q.; Jin, L.Zh.; Ni, Zh. M. (1982) *Atom. Spectrosc.* **3**, 41.

Sharma, K.S.; Lata, S.; Goel, Sh. (1984) *Indian J. Chem.* **23B**, 180.

Shealy, Y.F.; Clayton, J.D. (1967) *J. Heterocycl. Chem.* **4**, 96.

Shealy, Y.F.; Clayton, J.D.; Dixon, G.J.; Dulmudge, E.A.; Pittillo, R.F.; Hunt, D.E. (1966) *Biochem. Pharmacol.* **15**, 1610.

Solodovnikov, S.P.; Todres, Z.V. (1967) *Khim. Geterotsikl. Soed.* 811.

Solodovnikov, S.P.; Todres, Z.V. (1968) *Khim. Geterotsikl. Soed.* 360.

Solozhenkin, P.M.; Tsveniashvili, V. Sh.; Semyonov, E.V.; Khavtasi, N.S. (1976) *Dokl. AN Tadzh. SSR* **19**, 38.

Strom, E.T.; Russell, G.A. (1965) *J. Am. Chem. Soc.* **87**, 3326.

Suzuki, T.; Fuji, H.; Yamashita, Y.; Kabuto, Ch.; Tanaka, Sh.; Harasawa, M.; Mukai, T.; Miyashi, Ts. (1992) *J. Am. Chem. Soc.* **114**, 3034.

Suzuki, T.; Kabuto, Ch.; Yamashita, Y.; Mukai, T. (1987a) *Chem. Lett.* 1129.

Suzuki, T.; Kabuto, Ch.; Yamashita, Y.; Saito, G.; Mukai, T.; Myashi, Ts. (1987b) *Chem. Lett.* 2285.

Suzuki, T.; Yamashita, Y.; Fukushima, T.; Muyashi, Ts. (1997) *Mol. Crystl. Liq. Crystl. Sci. Technol. A* **296**, 165.

Takikawa, Yu.; Hikage, Sh.; Matsuda, Y.; Higasiyama, K.; Takeishi, Y.; Shimada, K. (1991) *Chem. Lett.* 2043.

Tian, R.; Yang, R.; Peng, J.; Cao, Y. (2003) *Synth. Met.* **135–136**, 177.

Todres, Z.V. (2006) *Organic Mechanochemistry and Its Practical Applications.* Taylor & Francis/CRC Press, Boca Raton, FL.

Todres, Z.V. (2009) *Ion-Radical Organic Chemistry.* Taylor & Francis/CRC Press, Boca Raton, FL.

Todres, Z.V.; Lyakhovetsky, Yu.I.; Kursanov, D.N. (1969) *Izv. AN SSSR, Ser. Khim.* 1455.

Todres, Z.V.; Pozdeeva, A.A.; Chernova, V.A.; Zhdanov, S. I. (1972) *Tetrahedron Lett.* **13**, 3835.

Todres, Z.V.; Tsveniashvili, V.Sh.; Zhdanov, S.I.; Kursanov, D.N. (1968a) *Dokl. AN SSSR* **181**, 906.

Todres, Z.V.; Zaitsev, B.E. (1971) *Khim. Geterotsikl. Soed.* 1036.

Todres, Z.V.; Zhdanov, S.I.; Tsveniashvili, V.Sh. (1968b) *Izv. AN SSSR, Ser. Khim.* 975.

Tsveniashvili, V.Sh. (1985) *Elektrokhimiya* **21**, 1142.

Tsveniashvili, V.Sh.; Gaprindashvili, V.N.; Khavtasi, N.S.; Malashkhiya, M.V. (1982) Russ. Patent 979,995.

Tsveniashvili, V.Sh.; Gaprindashvili, V.N.; Malashkhiya, M.V.; Khavtasi, N.S.; Belen'kaya, I.A. (1984) Russ. Patent 1,117,295.

Tsveniashvili, V.Sh.; Khavtasi, N.S.; Malashkhiya, M.V.; Gaprindashvili, V.N. (1983) Russ. Patent 1,058,966.

Tsveniashvili, V.Sh.; Todres, Z.V.; Zhdanov, S.I. (1968) *Zh. Obshch. Khim.* **38**, 1894.

Tsveniashvili, V.Sh.; Zhdanov, S.I.; Todres, Z.V. (1967) *Zeitschr. Analyt. Chem.* **224**, 389.

Ul-Haq, A.; Donzello, M.P.; Stuzhin, P.A. (2007) *Mendeleev Commun.* **17**, 337.

Umbricht, G.; Braun, H.-J.; Oberson, S.; Mueller, C. (2002) Germ. Patent 10,114,426.

Velusamy, M.; Thomas, K.R.J.; Lin, J.T.; Wen, Y.S. (2005a) *Tetrahedron Lett.* **46**, 7647.

Velusamy, M.; Thomas, K.R.J.; Lin, J.T.; Hsu, Y.-Ch.; Ho, K.-Ch. (2005b) *Org. Lett.* **7**, 1899.

Veracini, C.A.; de Munno, A.; Bertini, V.; Longeri, M.; Chidichimo, G. (1977) *J. Chem. Soc. Perkin Trans. 2*, 561.

Vernon, J.M.; Bryce, M.R.; Dransfield, T.A. (1983) *Tetrahedron* **39**, 835.

Wang, Sh.; Chen, Y. (1998) *Guangpu Shiyanshi* **15**, 34.

Wang, W.; Li, L.; Liu, Sh.; Ma, C.; Zhang, Sh. (2008) *J. Am. Chem. Soc.* **130**, 10846.

Weinstock, L.M.; Davis, P.; Mulvey, D.M.; Schaeffer, J.C. (1967) *Angew. Chem. Intl. Ed.* **6**, 364.

Wright, S.W.; McClure, L.D. (2004) *J. Heterocycl. Chem.* **41**, 1023.

Yamashita, Y.; Saito, K.; Mukai, T.; Miyashi, Ts. (1989) *Tetrahedron Lett.* **30**, 7071.

Yamashita, Y.; Tanaka, Sh.; Imaeda, K.; Inokuchi, H.; Sano, M. (1992) *J. Org. Chem.* **57**, 5517.

Yang, J.; Jiang, Ch.; Zhang, Y.; Yang, R.; Yang, W.; Hou, Q.; Cao, Y. (2004) *Macromolecules* **37**, 1211.

Yang, R.; Tian, R.; Hou, Q.; Yang, W.; Cao, Y. (2003) *Macromolecules* **36**, 7453.

Yang, R.; Tian, R.; Yan, J.; Zhang, Y.; Yang, J.; Hou, Q.; Yang, W.; Zhang, Ch.; Cao, Y. (2005) *Macromolecules* **38**, 244.

Yasuda, T.; Imase, T.; Yamamoto, T. (2005) *Macromolecules* **38**, 7378.

Yavolovsky, A.A.; Ivanov, E.I.; Ivanova, R.Yu. (2000) *Khim. Geterotsikl. Soed.* 1571.

Zhang, J.; Zheng, W.; Zou, J.; Fang, Y.; Yan, B.; Li, Y. (2004a) *Chem. Abstracts* **143**, abstract No 153354.

Zhang, J.; Zou, J.; Zheng, W.; Yang, F.; Bai, Y.; Li, Y. (2004b) *Huaxue Yanjiu Yu Yingyong* **16**, 561.

Zhang, Sh.; Chen, G. (1996) *Zhongguo Yiyao Gongue Zazhi* **27**, 355.

Zhao, M.; Stern, Ch.; Barrett, A.G.M.; Hoffman, B.M. (2003) *Angew. Chem. Intl. Ed.* **42**, 462.

Zhdanov, S.I.; Tsveniashvili, V.Sh.; Todres, Z.V. (1967) *J. Polarogr. Soc.* **13**, 100.

Zheng, W.; Chen, T.; Zhang, J.; Yang, F. (2007) PR Chin. Patent 100,999,527.

Zheng, Y.; Lu, D. (1992) *Microchim. Acta* **106**, 3.

Zhou, A.-J.; Zheng, Sh.-L.; Fang, Y.; Tong, M.-L. (2005) *Inorg. Chem.* **44**, 4457.

Zibarev, A.V.; Fugaeva, O.M.; Miller, A.O.; Konchenko, S.N.; Korobeinicheva, I.K.; Furin, G.G. (1990) *Khim. Geterotsikl. Soed.* 1124.

Zibarev, A.V.; Miller, A.O. (1990) *J. Fluorine Chem.* **50**, 359.

Zima, O.; Mohr, G. (1959) Germ. Patent 1,057,127.

Zou, J.; Yang, F.; Zheng, W.; Bai, Y.; Li, Y. (2004) *Huaxue Shiji* **26**, 289.

3.4   1,3,4-SELENADIAZOLES

In selenadiazoles of this type, the two nitrogen atoms are in immediate proximity and equally separated from the selenium key heteroatom. The literature contains a wide body of synthetic approaches to heterocyclic compounds of this type, but data on their applications are unexpectedly scant.

3.4.1   FORMATION

3.4.1.1   From Azines and Hydrogen Selenide

Hydrogen selenide in the presence of pyridine traces reacts with N,N-dimethylformamidoazine giving rise to 1,3,4-selenadiazole with no substituents (Scheme 3.66) (Kendall and Olofson 1970).

3.4.1.2   From Selenosemicarbazides and Carboxylic Acids

This reaction is performed in the presence of phosphoryl chloride and results in the formation of 2-acetylamino-1,3,4-selenadiazole. The latter gives 2-amino-1,3,4-selenadiazole upon hydrolysis (Scheme 3.67) (Lalezari and Shafiee 1971).

3.4.1.3   From Isoselenocyanates and Acylhydrazines or Selenosemicarbazides

These reactions present another route to N-substituted 2-amino-1,3,4-selenadiazoles, Schemes 3.68 and 3.69 present the formation of 1,3,4-selenadiazole amino derivatives (Bulka and Ehlers 1973a,b, Bulka et al. 1973).

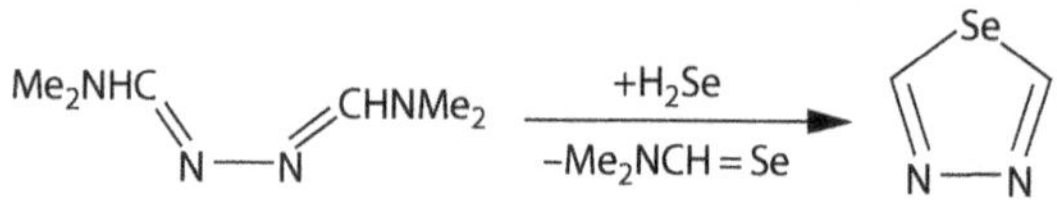

SCHEME 3.66

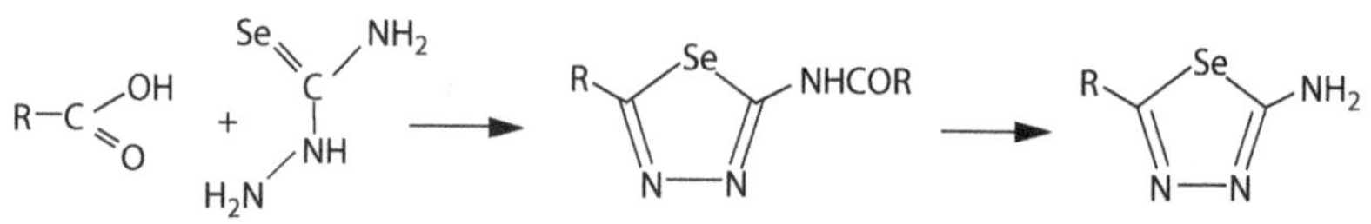

SCHEME 3.67

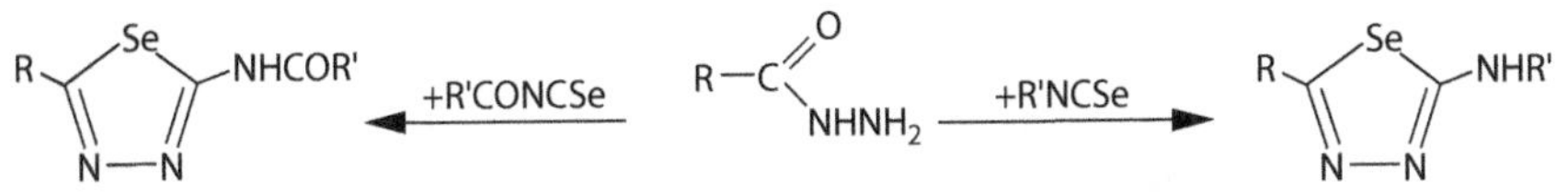

SCHEME 3.68

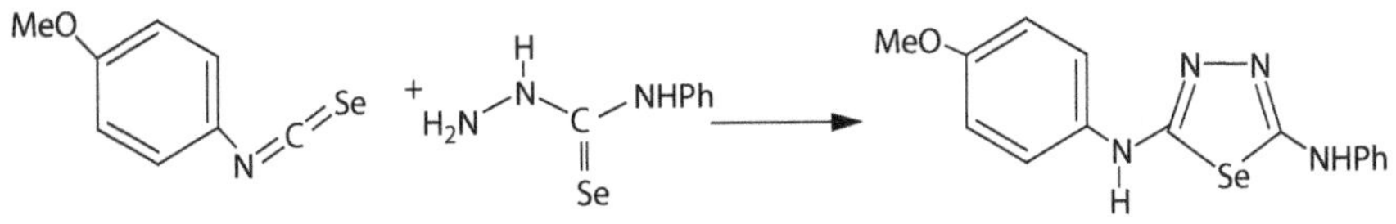

SCHEME 3.69

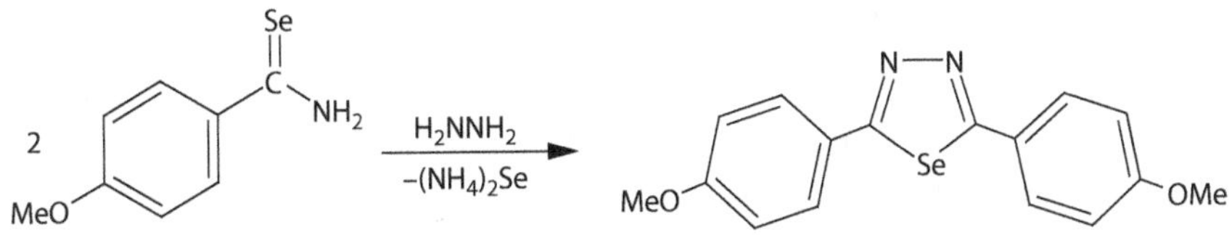

SCHEME 3.70

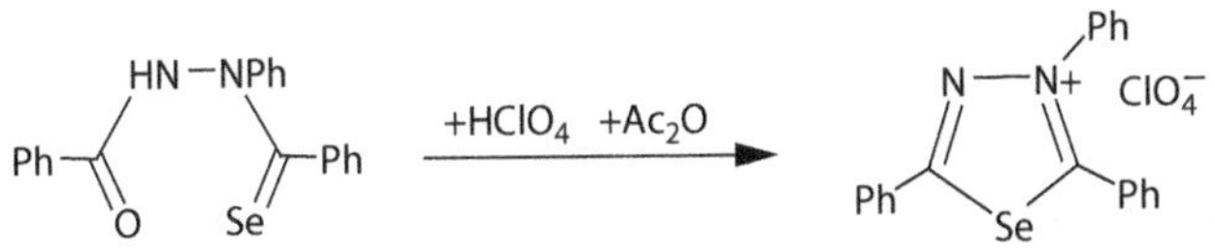

SCHEME 3.71

3.4.1.4   From Selenoamides and Hydrazine

This method involves treatment of aryl and hetaryl selenoamides with hydrazine hydrate to give the corresponding 2,5-diaryl-1,3,4-selenadiazoles (Scheme 3.70) (Cohen 1979).

3.4.1.5   From N-Acyl-N'-Selenoacylhydrazines

Reaction of 2-acyl-1-selenoacyl hydrazines with perchloric acid in acetic anhydride results in dehydrative cyclization to give the 1,3,4-selenadiazolium salt (Scheme 3.71) (Shvaika and Lipitskii 1985).

3.4.1.6   From Hydrazones and Diselenium Dibromide

This reaction is interesting in the sense of preparation of sterically congested 1,3,4-selenadiazoline derivatives. Scheme 3.72 gives an example (Okuma et al. 2003). The example also shows how temperature affects the reaction direction. Namely, 1,1,3,3-tetramethylindan-2-one hydrazone gives the corresponding selenadiazoline as a major product at −20°C while at +25°C the major product turned out to be the corresponding selenone (both products are depicted in Scheme 3.72).

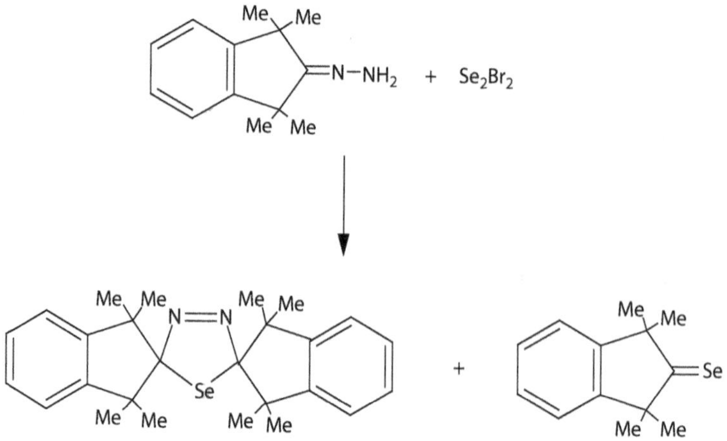

SCHEME 3.72

3.4.1.7 From Selenocyanates or Selenoaldehydes and Aryldiazonium Salts or Hydrazonoyl Halides

Reaction of acyl selenocyanates with aryldiazonium salts is illustrated by Scheme 3.73 (Takahashi and Kurosawa 1980).

When the diazonium chloride from anthranilic acid was used, the imine of Scheme 3.74 was obtained. This imine was eventually transformed into the 1,3,4-selenadiazolo-[2,3-*b*]-quinazolone derivative indicated by Scheme 3.74 (Abdelhamid et al. 1983).

Reaction of hydrazonoyl halides with potassium selenocyanate is illustrated in Scheme 3.75 (Fusco and Musante 1938, Zohdi et al. 1993, Abdelhamid et al. 2001, 2004, Abdel-Riheem et al. 2003, Abdelhamid and Alkhodshi 2005).

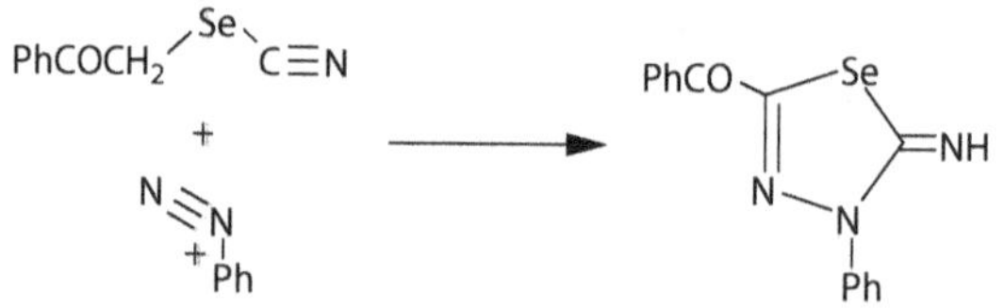

SCHEME 3.73

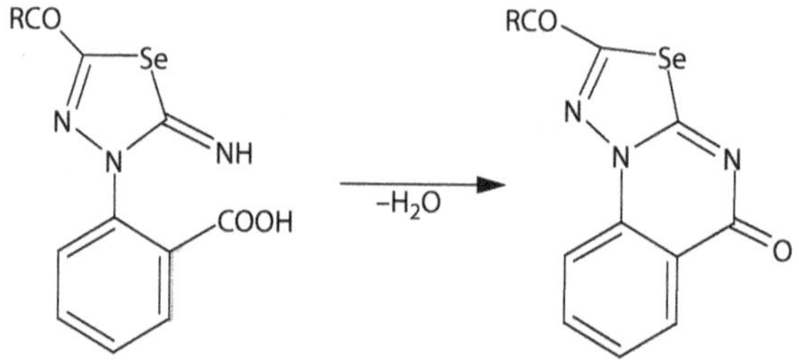

SCHEME 3.74

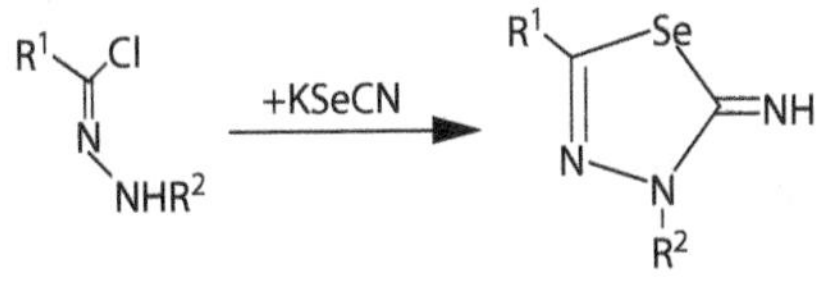

SCHEME 3.75

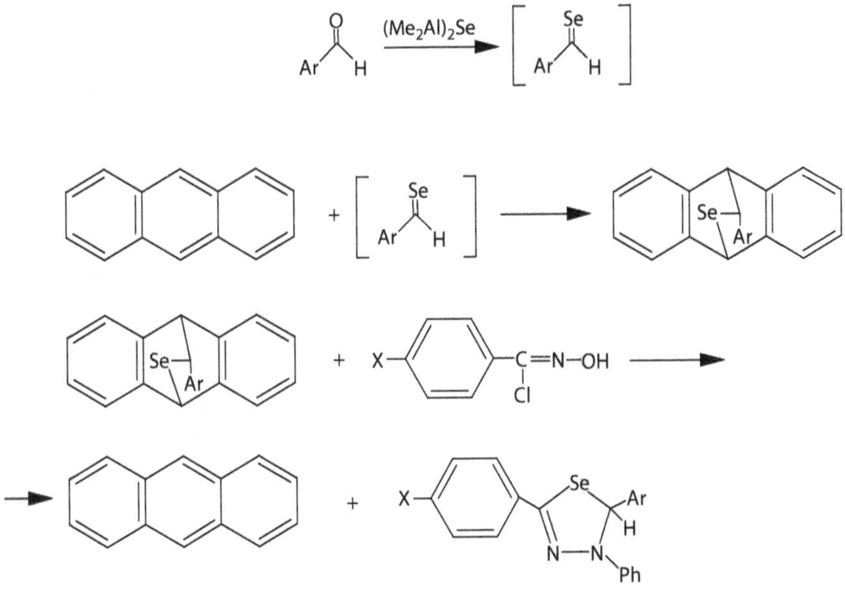

SCHEME 3.76

Reaction of hydrazonoyl halides with selenoaldehydes is realizable if selenoalde-hydes are used in the stable form. To stabilize selenoaldehydes, adducts with anthracene are preliminary prepared. Hydrazonoyl halides "extract" the selenoaldehydes and form substituted 1,3,4-selenadiazolinse with high yields. Scheme 3.76 describes the cycliza-tion reaction that proceeds in boiling toluene (30 min) in the presence of triethylamine (Segi et al. 2007). The efficacy and wide applicability of the reaction makes it a good approach to 1,3,4-selenadiazoles despite apparent "complexity" of the reactants used.

3.4.1.8 From 1,3,4-Selenadiazines

Scheme 3.77 shows the reaction that takes place upon heating in acetic acid. The product is selenadiazoline. Some part of the product undergoes Dimroth rearrange-ment and gives triazinoselenone. The latter is obtained as an admixture to the sele-nadiazoline (Fleischhauer et al. 2008).

3.4.2 Fine Structure

The unsubstituted 1,3,4-selenadiazole ring is planar with C_{2v} geometry as inferred from the molecule microwave spectrum (Levine et al. 1969). This ring can be

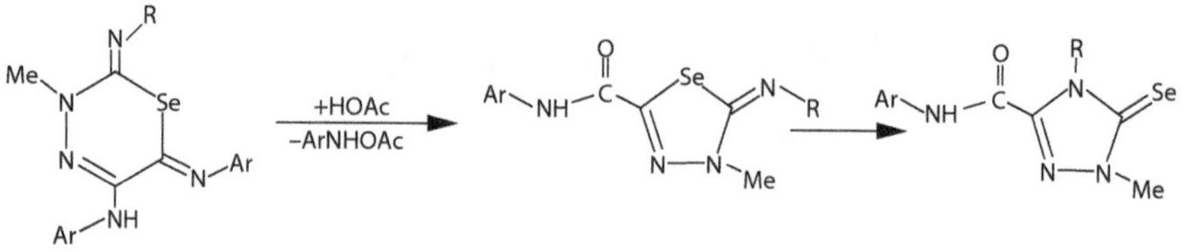

SCHEME 3.77

characterized by an aromaticity index of 53 or 65. The indexes were calculated by Bird on the basis of bond length (1985) and heat of formation (1992), respectively. For benzene, such an index was found to be 100.

It might be well to point out that the 1,3,4-selenadiazole ring does not always keep its planar and more or less aromatic character. X-ray structural analysis of 2-acetyl-4-phenyl-5-benzyl-5-piperidino-5H-1,3,4-selenadiazole established that the selenium-containing ring has an envelope form (Abramov et al. 1992). Spectra of ^{13}C NMR testify that the selenium key heteroatom transfers only inductive effects across the ring (Bartels-Keith et al. 1977).

According to ^{1}H NMR spectra of 2-amino-1,3,4-selenadiazole in benzene, a π-type solvent complex is formed, in which the amino group is involved in intermolecular hydrogen bonding with this solvent (Svanholm 1972). This testifies that there is significant overflow of electron density from the amino-group to the selenadiazole ring.

Diazotization and azocoupling of 2-amino-1,3,4-selenadiazole was described by Bulka et al. (1973). Although these reactions concern reactivity of the amino group, they testify for the definite aromatic character of the hetero ring: Only aromatic amines give stable diazo cations that can couple with those aromatic compounds that bear donor substituents. As established, 2-amino-1,3,4-selenadiazole undergoes diazotization and subsequent coupling with N,N-dialkylanilines forming the corresponding azo dyes. This hint is, however, not very strong: 1,3,4-Selenadiazole thiols exist in the thione forms despite loss of aromaticity (Bartels-Keith et al. 1977).

3.4.3 REACTIVITY

3.4.3.1 Oxidation

Oxidation of one 1,3,4-thiadiazoline with m-chloroperbenzoic acid is depicted in Scheme 3.78. The initially formed cyclic selenoxide loses selenium monoxide and gives rise to the carbon–carbon biradical. Valence isomerization of the latter results in the formation the corresponding azine (Okuma et al. 2003).

3.4.3.2 Photolysis

Photolysis of substituted 1,3,4-selenadiazolines was studied by Guziec et al. (1985). The reaction consists of ring opening with elimination of selenium and the formation of azines.

3.4.3.3 Thermolysis

Just as oxidation of Scheme 3.78, thermolysis of the analogous starting material leads to the formation of the carbon–carbon biradical. In contrast to Scheme 3.78,

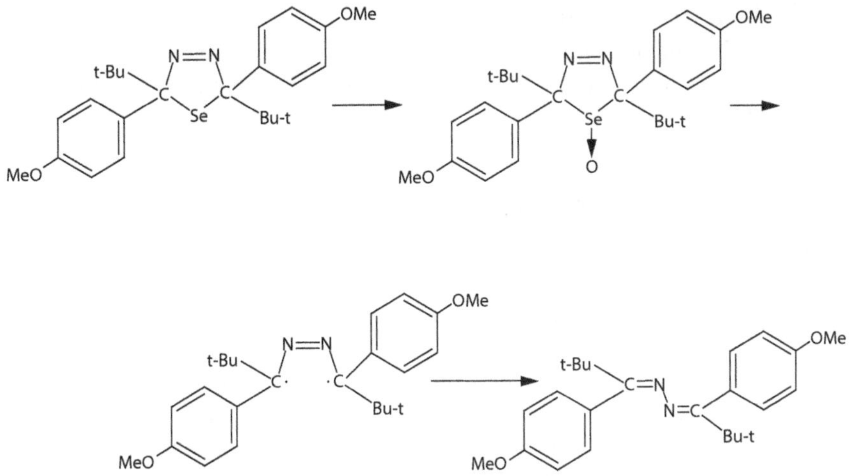

SCHEME 3.78

the biradical eliminates nitrogen and keeps selenium. The ring closure then takes place with the formation of episelenide in the *trans* and *cis* forms. Both of them are too unstable to be isolated and the final product is the corresponding stilbene, also in *trans* and *cis* forms. Scheme 3.79 illustrates the transformation as exemplified by the *trans* isomer for the sake of simplicity (Okuma et al. 2003).

There are other examples of the alkene formation in a result of pyrolysis of 1,3,4-selenadiazoline derivatives (Back et al. 1978, Cullen et al. 1982, Guziec et al. 1985).

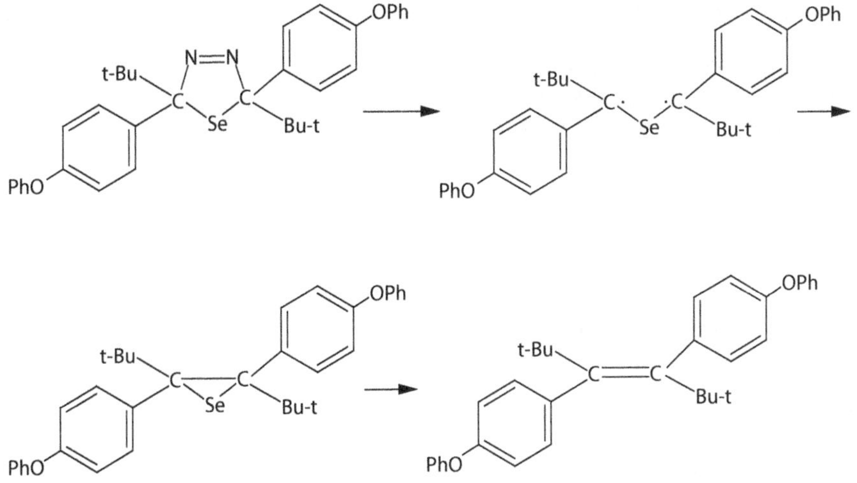

SCHEME 3.79

3.4.3.4   Ring Transformation

Scheme 3.77 has already introduced the transformation of the selenadiazoline to triazinoselenone derivatives (Fleischhauer et al. 2008). This is an example of Dimroth
rearrangement.

3.4.4   BIOMEDICAL IMPORTANCE

Compounds of the 1,3,4-selenadiazole series can be synthesized by wide range of
methods. Such an extensive search of the methods was grounded by hopes to find
new drugs. However, the literature data on their biomedical activity occurred to be
scanty. The compounds have been reported to be of interest as antibacterial agents
(Shafiee et al. 1973). Azomethine derivatives of this selenadiazole manifest activity
against poliovirus, but not against adenovirus. In poliovirus, they, however, inhibit
the cytopathic effect at the concentration of the maximal nontoxic dose (Akihama
et al. 1980). Activity against tuberculosis of selenadiazoline imines was mentioned
in the preliminary report by Fleischhauer et al. (2008), but this activity is similar to
that of the presently used drugs.

3.4.5   CONCLUSION

Compounds of the 1,3,4-selenadiazole series are prone to rearrangements, to react
with ring contraction, to extrude nitrogen and/or selenium. No doubt, some new
transformation of these compounds will be found in the near future. Biomedical and
technical applications are somewhat documented, but their widening is possible in
the near future.

REFERENCES TO SECTION 3.4

Abdelhamid, A.O.; Alkhodshi, M.A.M. (2005) *Phosphorus Sulfur* **180**, 149.
Abdelhamid, A.O.; El-Ghandour, A.H.; Hussein, A.M.; Zaki, Y.H. (2004) *J. Sulfur Chem.* **25**, 329.
Abdelhamid, A.O.; Hassaneen, H.M.; Shawali, A.S. (1983) *J. Heterocycl. Chem.* **20**, 719.
Abdelhamid, A.O.; Sallam, M.M.M.; Amer, S.A. (2001) *Heteroatom. Chem.* **12**, 468.
Abdel-Riheem, N.A.; Rateb, N.M.; Al-Atoom, A.A.; Abdelhamid, A.O. (2003) *Heteroatom. Chem.* **14**, 421.
Abramov, M.A.; Petrov, M.L.; Potekhin, K.A.; Struchkov, Yu.T.; Galishev, V.A.; Petrov, A.A. (1992) *Zhurn. Structur. Khim.* **33**, 180.
Akihama, S.; Fukutomi, K.; Sakamoto, M. (1980) *Meiji Yakka Daigaku Kiyo*, No. 10, 19.
Back, T.G.; Basrton, D.H.R.; Britten-Kelly, M.R.; Guziec, F.C. (1978) *J. Chem. Soc. Perkin Trans. 1*, 2079.
Bartels-Keith, J.R.; Burgess, M.T.; Stevenson, J.M. (1977) *J. Org. Chem.* **42**, 3725.
Bird, C.W. (1985) *Tetrahedron* **41**, 1409.
Bird, C.W. (1992) *Tetrahedron* **48**, 335.
Bulka, E.; Ehlers, D. (1973a) *J. Prakt. Chem.* **315**, 510.
Bulka, E.; Ehlers, D. (1973b) *J. Prakt. Chem.* **315**, 155.
Bulka, E.; Ehlers, D.; Storm, H. (1973) *J. Prakt. Chem.* **315**, 164.
Cohen, V.I. (1979) *J. Heterocycl. Chem.* **16**, 365.

Cullen, E.R.; Guziec, F.C.; Murphy, Ch.J. (1982) *J. Org. Chem.* **47**, 3563.

Fleischhauer, J.; Beckert, R.; Hornig, D.; Guenther, W.; Goerls, H.; Klimesova, V. (2008) *Zeitschr. Naturforsch.* **63B**, 415.

Fusco, R.; Musante, C. (1938) *Gazz. Chim. Ital.* **68**, 665.

Guziec, F.C.; Murphy, Ch.J.; Cullen, E.R. (1985) *J. Chem. Soc. Perkin Trans. 1*, 107.

Kendall, R.V.; Olofson, R.A. (1970) *J. Org. Chem.* **35**, 806.

Lalezari, I.; Shafiee, A. (1971) *J. Heterocycl. Chem.* **8**, 835.

Levine, D.M.; Krugh, W.D.; Gold, L.P. (1969) *J. Mol. Spectrosc.* **30**, 459.

Okuma, K.; Kubo, K.; Yokomori, Y. (2003) *Heterocycles* **60**, 299.

Segi, M.; Tano, K.; Kojima, M.; Honda, M.; Nakajima, T. (2007) *Tetrahedron Lett.* **48**, 2303.

Shafiee, A.; Lalezari, I.; Yazdany, S.; Pornorous, A. (1973) *J. Pharm. Sci.* **62**, 839.

Shvaika, O.P.; Lipitskii, V.F. (1985) *Zhurn. Obshch. Khim.* **55**, 2608.

Svanholm, U. (1972) *Acta Chem. Scand.* **26**, 459.

Takahashi, M.; Kurosawa, M. (1980) *Bull. Chem. Soc. Jpn.* **53**, 1185.

Zohdi, H.F.; Afeefy, H.Y.; Abdelhamid, A.O. (1993) *J. Chem. Res.* 76.

LOOKING OUT OVER CHAPTER 3: COMPARISON OF ISOMERIC SELENADIAZOLES

This chapter presented materials on all possible isomers of selenadiazole. This permits underlining the most important peculiarities that correspond to a specific type of heteroatom alternation. It would be principally needed to bring these most important properties together with comparative data on electronic structure of selenadiazoles with different alternation of heteroatoms. Regrettably, theoretical calculations for all of the enumerated selenadiazoles was up to now not conducted.

The most notable feature of 1,2,3-selenadiazole consists of their ability to easily eliminate the nitrogen molecule and the selenium atom. This provides organic chemists with a simple route to acetylenic compounds. Elimination of selenium in active form is seemingly the reason for the biomedical importance of 1,2,3-seleneadiazoles. The same possibility is central to economical and green methods of chemical vapor deposition or preparation of metal selenide nanoparticles for microelectronics.

Despite the scarcity of the literature data on 1,2,4-selenadiazoles, it is clear that they are relatively stable compounds and the methods of their preparation are diverse enough. Reactivity of 1,2,4-selenadiazoles is different from that of their isomers having other disposition of the atoms in the heterocyclic fragments. All these features raise the hope that there will be an increase in the number of studies on their fine structures and reactivity.

Due to their specific chemistry, 1,2,5-selenadiazoles have remarkable opportunities. In this sense, arene-annulated 1,2,5-selenadiazoles stand out sharply. They form stable anion-radicals differ from the parent neutral molecules with delocalization of an unpaired electron within the whole molecular contour. Accordingly, neutral arene-annulated 1,2,5-selenadiazoles coordinate to metals through the heterocyclic nitrogen atoms, while the corresponding anion-radicals acquire the possibility to coordinate through the key selenium atom. Ease of their formation from aromatic 1,2-diamines and reduction to these diamines modified in results of the selenadiazole reactions, are remarkable for synthetic and analytical practice. Biomedical and technical applications of 1,2,5-selenadiazoles are widely developed and have thoroughly been considered in Subsection 3.3.

Compounds of the 1,3,4-selenadiazole series are prone to rearrangements, to react with ring contraction, to eliminate nitrogen and/or selenium. Their biomedical and technical applications are in the initial stage only. Widening of such data is the subject of anticipation for the near future.

4 Telluradiazoles

Several decades have passed after the first publication on telluradiazoles, but their chemistry is still in the cradle phase. Despite scarcity of the data published, the structure of this chapter is built in the same manner as the preceding parts of the book.

4.1 1,2,3-TELLURODIAZOLES

1,2,3-Telluradiazoles are still represented only by the perchlorate of benzo-[1,2-*c*]-[1,2,3]-telluradiazolium salt of Scheme 4.1. This salt has been prepared from azo-bis(4,4'-dimethylbenzene)-2-tellurochloride (Sadekov et al. 1990). This starting material itself differs with significant interaction between the nitrogen and tellurium atoms: The nitrogen–tellurium distance comes to 0.22 nm (Cobbledick et al. 1979). Up to now, there are no data on chemical behavior of 1,2,3-telluradiazoles or 1,2,3-telluradiazolium salts.

4.2 1,2,5-TELLURADIAZOLES

Containing a super-heavy key heteroatom, compounds of this series have low solubility in common solvents and low thermal stability. This could limit their synthetic use and restrict their practical application. However, the solubility and thermal stability can be improved by choosing appropriate substituents.

4.2.1 FORMATION

4.2.1.1 From Thiadiazoles or Selenadiazoles

1,2,5-Thiadiazoles, 1,2,5-selenadiazoles, and their dimethyl and diphenyl derivatives react with ethylmagnesium bromide, then with tellurium tetrachloride. After treatment with triethylamine, 1,2,5-telluradiazoles can be obtained (Bertini et al. 1982). Scheme 4.2 exemplifies this transition: Phenanthro-[9,10-*c*]-[1,2,5]-selenadiazole initially experiences ring cleavage forming the magnesium derivative of 9,10-phenanthrene quinone bis(imine), cf. Scheme 3.50. The bis(imine) reacts with tellurium tetrachloride in the presence of triethylamine.

Phenanthro-[9,10-*c*]-[1,2,5]-telluradiazole is the final product. Its structure was elucidated by the x-ray method (Neidlein et al. 1987).

4.2.1.2 From Quinone Bis(imines)

This reaction differs from that in Schemes 4.2 by the oxidation state of tellurium in the final product (see Scheme 4.3) (Neidlein and Knecht 1987).

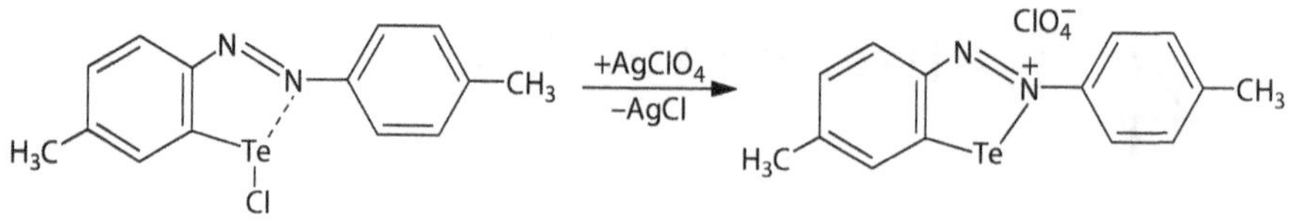

SCHEME 4.1

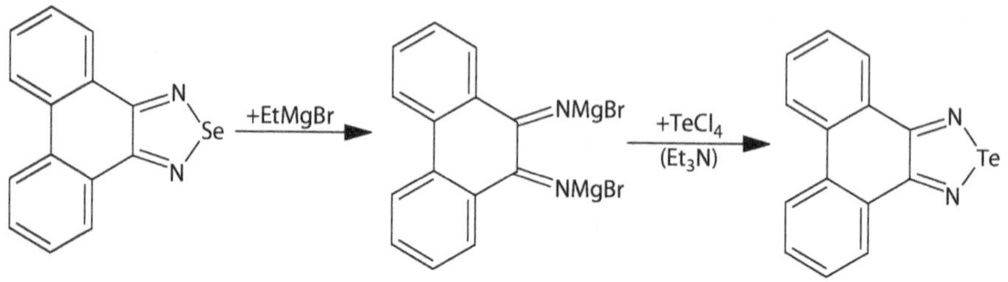

SCHEME 4.2

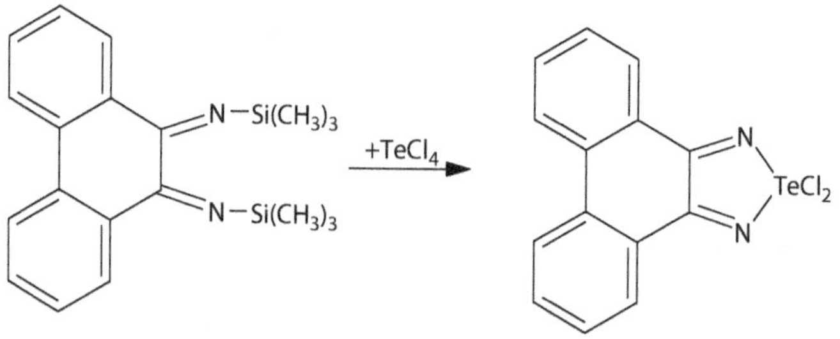

SCHEME 4.3

4.2.1.3 From Diazabutadienes

1,4-N,N'-Bis(*tert*-butyl)-buta-1,3-diene reacts with tellurium tetrabromide and gives the telluradiazolium salt of Scheme 4.4 (Dutton et al. 2008, Dutton and Ragogna 2009). This reaction is accompanied with loss of one *tert*-butyl group.

4.2.1.4 From vic-Diamines

It has already been shown that aromatic 1,2-diamines react with selenium dioxide in aqueous media at room temperature and yield, quantitatively, aromatic-fused 1,2,5-selenadiazoles, see Subsection 3.3.6. Attempts to use tellurium dioxide in the same conditions for preparation of aromatic-fused 1,2,5-telluradiazoles were ineffective, but the use of tellurium tetrachloride as the tellurium source, instead of tellurium dioxide, was successful (Suzuki et al. 1992, Badyal et al. 1998). Scheme 4.5 presents one such reaction. The reaction in Scheme 4.5 was undertaken to obtain an intermediate to telluradiazolotetracyanoquinodimethane as a material conducting electric current. However, the product of Scheme 4.5 exhibited scarce solubility in organic solvents and facile ring opening under the reaction conditions employed (Suzuki et al. 1992).

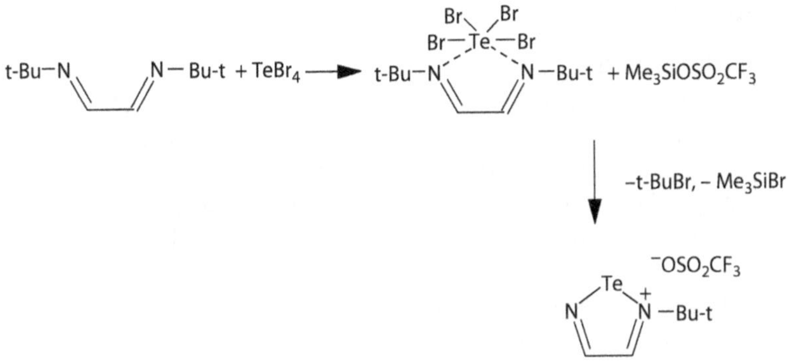

SCHEME 4.4

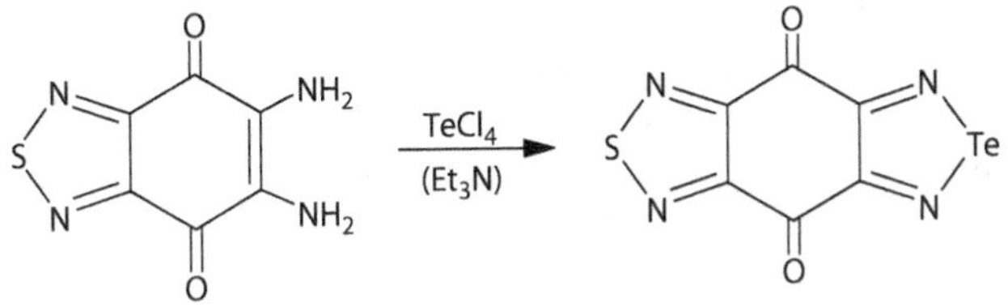

SCHEME 4.5

As to the ability of tellurium dioxide to cyclize vic-diamines, Risto et al. (2008) reported the formation of benzo-[1,2-c]-[1,2,5]-telluradiazole upon the direct reaction of tellurium dioxide with 1,2-phenylenediamine at 180°C.

4.2.1.5 From *N*-Lithium Trialkylanilines and Silylated Tellurium Imides

This reaction (Scheme 4.6) also proceeds with loss of one *tert*-butyl group. Bis(*tert*-butyl)benzotelluradiazole is the final product (Chivers et al. 1996).

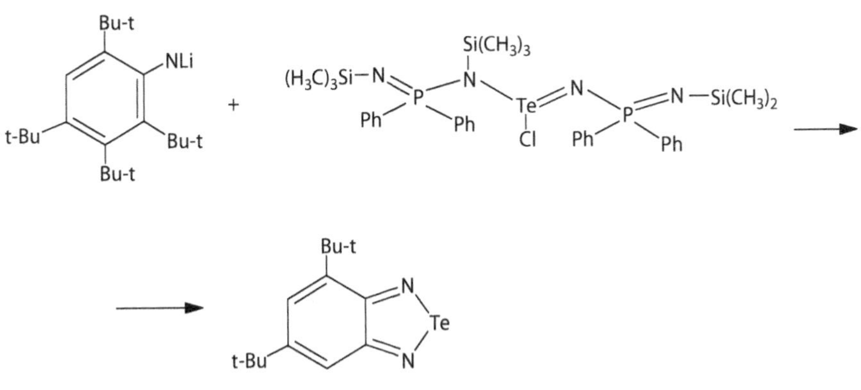

SCHEME 4.6

4.2.2 Reactivity

4.2.2.1 Salt Formation

The telluradiazolium salt of Scheme 4.4 was prepared from *N,N'*-bis(*tert*-butyl)-1,4-diazabutadiene-1,3, using tellurium tetrabromide and (trimethylsilyl)(trifluoro-metane)sulfonate (Dutton et al. 2008, Dutton and Ragogna 2009).

Methylation of benzo-[1,2-*c*]-[1,2,5]-telluradiazole with methyl (trifluoromethane) sulfonate leads to the telluradiazolium salt depicted in Scheme 4.7 (Risto et al. 2008).

4.2.2.2 Hydrolytic Decomposition

The parent 1,2,5-telluradiazole easily cleaves by acidic water. Tetrafluorobenzo-[1,2-*c*]-[1,2,5]-telluradiazole undergoes hydrolysis in aqueous dimethylsulfoxide according to Scheme 4.8 (Kovtonyuk et al. 1996).

4.2.2.3 Coordination

Dutton and Ragogna (2009) described the reaction of 2-*N*-(*tert*-butyl)-1,2,5-telluradiazolium (trifluoromethane)sulfonate with tricyclohexylphosphine. X-ray crystal structure of the product obtained undeniably testifies for coordination to the tellurium key heteroatom (Scheme 4.9).

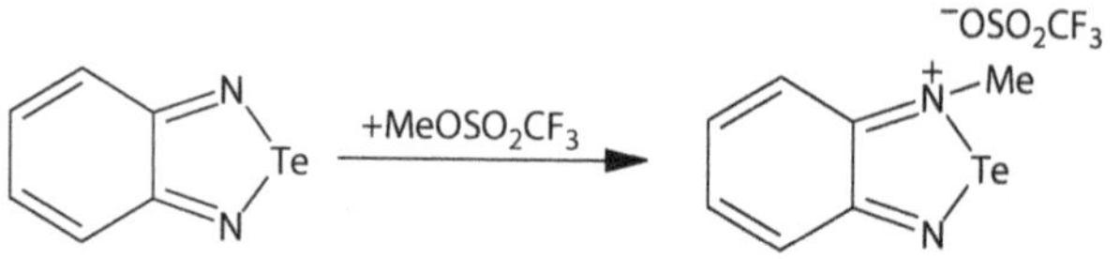

SCHEME 4.7

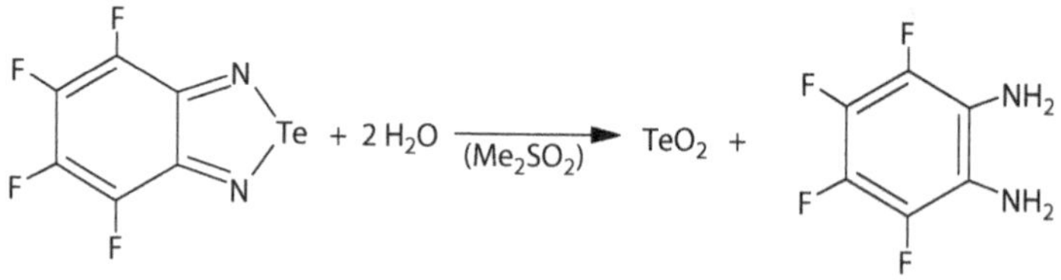

SCHEME 4.8

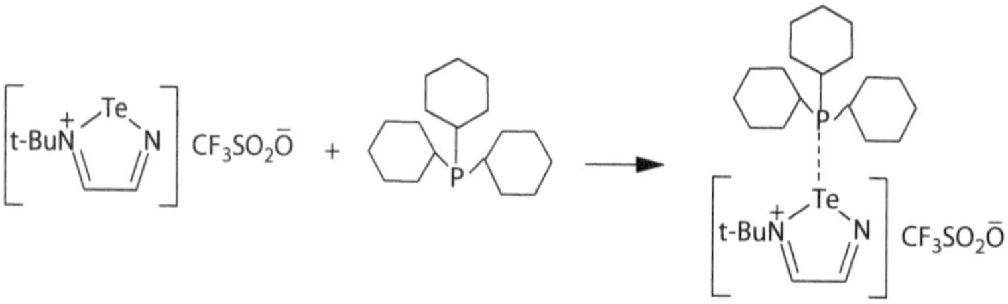

SCHEME 4.9

Reaction with triphenylborane presents quite a different direction of coordination. Benzo-[1,2-c]-[1,2,5]-telluradiazole reacts with triphenylborane forming a coordinative compound, in which boron binds with nitrogen(s). Using the appropriate stoichiometry of the reagents, 1:1 and 1:2 complexes can be prepared (Scheme 4.10). At high temperatures, 1:1 complex undergoes fast shift of the triphenylborane moiety between the two nitrogen atoms (Cozzolino et al. 2009).

One specific type of coordination was found from x-ray study of crystalline 1,2,5-telluradiazole (Bertini et al. 1984). Namely, the two tellurium–tellurium oriented diazoles are bound by the two tellurium–nitrogen coordinative bonds (Scheme 4.11). In this way, a ribbon polymeric complex is produced, which accounts for the exceptionally high melting temperature and low solubility of the compound in common solvents. Theoretical considerations show that the planar conformation of the associate is the most stable. Not only is the telluradiazole intermolecular coordination stronger than that of the selenadiazole analog (see Subsection 3.3.3), but it is more difficult to deform it by elongation of the coordinative bonds. The coordination has a strong donor–acceptor character: The nitrogen atom donates a lone pair into low-lying empty orbitals which are polarized toward a chalcogen (Cozzolino et al. 2005, 2008). The heavier the chalcogen atom, the stronger is this empty orbital polarization. Once formed, the supramolecular telluradiazole complex remains a remarkably rigid structure (Cozzolino and Vargas-Baca 2007). At the same time, practical applicability of telluradiazoles dictates searches of substances whose crystal structures are not so rigid.

Cozzolino et al. (2006) compared crystalline structures of benzo-[1,2-c]-[1,2,5]-telluradiazole and its α,α'-dibromoderivative. The compounds without bromine substituents develop infinite ribbon chains in the solid state. This ribbon is shown in Scheme 4.11. Introducing two bromines into α,α' positions of benzo-[1,2-c]-[1,2,5]-telluradiazole restricts the supramolecular association. The tellurium atom of one molecule occurs to be bound with the nitrogen of the other molecule of dibromo-benzotelluradizole. In the dimer, two such bonds come into being. The dimerization is a primary event. Then, additional tellurium–bromine secondary bonding takes

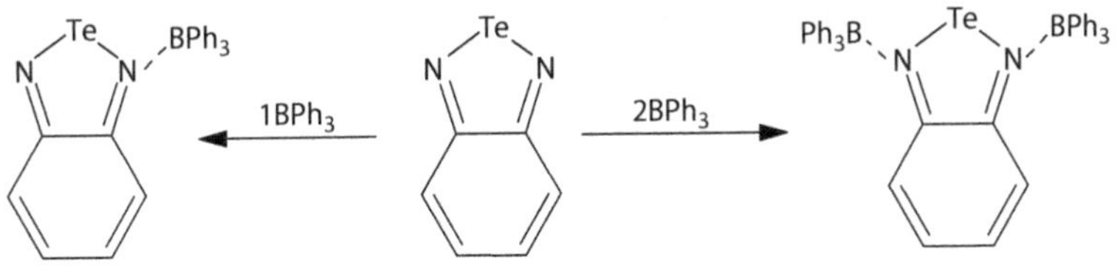

SCHEME 4.10

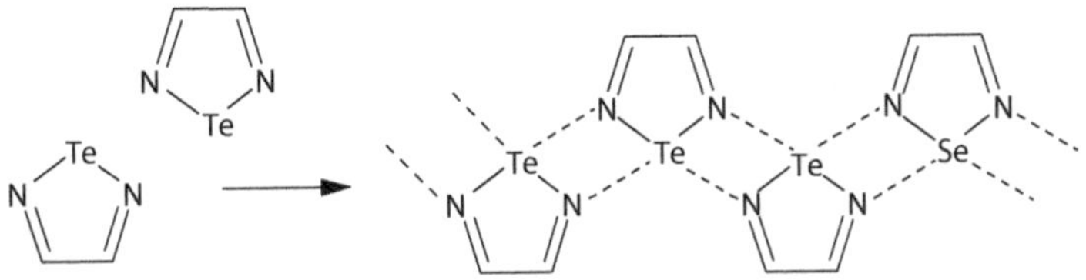

SCHEME 4.11

place and two additional dibromobenzotellurodiazoles are involved. So, a tetramer is formed. The secondary interaction blocks the tellurium's ability to coordinate with nitrogen of other telluradiazole molecules and prevents the formation of the ribbon polymeric structure.

The telluradiazolium of Scheme 4.4 has one of the two nitrogens blocked. Only non-blocked nitrogen can coordinate to tellurium. Accordingly, this salt forms crystals which consist of dimers. No infinite ribbon chains are formed. In these dimers, each constituent is bound with its counterpart by four-center interaction between two different telluriums and two different non-methylated nitrogens (Risto et al. 2008).

4.3 CONCLUSION

Chemistry of telluradiazoles is under initial development. Synthetic methods are run out, but data on telluradiazole reactivity are limited to quaternization, hydrolytic cleavage, complexation, and self-coordination. The self-coordination ability is particularly important for microelectronics. Unfortunately, self-coordination results in low or negligible solubility in common solvents. This presents difficulties for purification of the corresponding species and for their molding. However, some approaches have already been found to tune crystal structures of telluradiazoles. The "modified" objects seem to be promising candidates for use in smart technologies.

REFERENCES

Badyal, K.; Herr, M.; McWhinnie, W.R.; Hamor, Th.A.; Paxton, K. (1998) *Phosphorus Sulfur* **141**, 221.
Bertini, V.; Dapporto, P.; Lucchesini, F.; Sega, A.; de Munno, A. (1984) *Acta Crystallogr.* **C40**, 653.
Bertini, V.; Lucchesini, F.; de Munno, A. (1982) *Synthesis* 681.
Chivers, T.; Gao, X.; Parvez, M. (1996) *Inorg. Chem.* **35**, 9.
Cobbledick, R.E.; Einstein, F.W.B.; McWhinnie, W.R.; Musa, F.H. (1979) *J. Chem. Res.* 145.
Cozzolino, A.F.; Brain, A.D.; Hanhan, S.; Vargas-Baca, I. (2009) *Chem. Commun.* 4043.
Cozzolino, A.F.; Britten, J.F.; Vargas-Baca, I. (2006) *Crystl. Growth Des.* **6**, 181.
Cozzolino, A.F.; Gruhn, N.E.; Lichtenberger, D.L.; Vargas-Baca, I. (2008) *Inorg. Chem.* **47**, 6220.
Cozzolino, A.E.; Vargas-Baca, I. (2007) *J. Organometal. Chem.* **692**, 2654.
Cozzolino, A.F.; Vargas-Baca, I.; Mansour, S.; Mahmoudkhani, A.H. (2005) *J. Am. Chem. Soc.* **127**, 3184.
Dutton, J.L.; Ragogna, P.J. (2009) *Inorg. Chem.* **48**, 1722.
Dutton, J.L.; Sutrisno, A.; Schurko, R.W.; Ragogna, P.J. (2008) *Dalton Trans.* 3470.
Kovtonyuk, V.N.; Makarov, A.Yu.; Shakirov, M.M.; Zibarev, A.V. (1996) *Chem. Commun.*, 1991.
Neidlein, R.; Knecht, D. (1987) *Helv. Chim. Acta* **70**, 1076.
Neidlein, R.; Knecht, D.; Gieren, A.; Ruiz-Perez, C. (1987) *Zeitschr. Naturforsch. B* **42**, 84.
Risto, M.; Reed, R.W.; Robertson, C.M.; Olunkaniemi, R.; Laitinen, R.S.; Oakley, R.T. (2008) *Chem. Commun.*, 3278.
Sadekov, I.D.; Maksimenko, A.A.; Maslakov, A.G.; Minkin, V.I. (1990) *J. Orgamometal. Chem.* **391**, 179.
Suzuki, T.; Fujii, H.; Yamashita, Y.; Kabuto, Ch.; Tanaka, Sh.; Harasawa, M.; Mukai, T.; Miyashi, Ts. (1992) *J. Am. Chem. Soc.* **114**, 3034.

5 Chalcogenadiazoles from the Standpoint of Isosterism

The objective of Chapter 5 is to compare diazoles containing oxygen, sulfur, selenium, or tellurium as a key heteroatom. Materials from Chapters 1 through 4 provide sufficient grounds for the comparison. The reader's attention should also be drawn to some of the problems that are still waiting for their solution. As it happens so often in science, solutions of certain problems put novel others forward.

The best way to compare chalcogenadiazoles consists in their consideration from the standpoint of isosterism. Isosterism is a similarity in properties, resulting from electron arrangements that are identical or similar. As for chalcogenadiazoles, this definition is a subject to refinement. As has already been pointed out, chemical behavior should be more or less similar for the thia, selena, and tellura analogs and differ for the oxa counterparts.

5.1 CHEMICAL ASPECTS

Chacogenadiazoles belong to the group of unsaturated heterocycles. For this group, the concept of aromaticity has long proved useful in rationalizing the corresponding chemistry. Planar, unsaturated heterocycles containing five atoms can be considered as aromatic systems if they have an uninterrupted cycle of p orbitals containing altogether six electrons. In chalcogenadiazoles, this delocalization cannot be very extensive because chalcogen key atoms possess although different, but pronounced electronegativity. If a lone pair on chalcogene is incorporated into a delocalized π-electron sextet, the problem of aromaticity becomes to be actual one. More or less delocalized five-membered heterocycles fused to homoaromatic rings are most conveniently classified as more or less aromatic.

Explanation of relative aromaticity of the heterocycles considered in this book is a challenge. Several calculation approaches point out that **1,2,5-*thia*diazole** is more aromatic than its **1,3,4**-isomer. At the same time, **1,2,5-*oxa*diazole** and **1,2,5-*selena*-diazole** are *less* aromatic than their **1,3,4**-counterparts (Bird 1985, 1992, Friedman and Ferris 1997). Concerning Bird's indexes I_A and I_U, **1,2,5-*thia*diazole** is characterized with $I_A = 84$ ($I_U = 104$), whereas $I_A = 43$ ($I_U = 53$) for **1,2,5-*oxa*diazole** is lower than $I_A = 50$ ($I_U = 62$) for **1,3,4-*oxa*diazole**, as well as $I_A = 47$ ($I_U = 58$) of **1,2,5-*sele*-nadiazole** is lower than $I_A = 53$ ($I_U = 65$) of **1,3,4-*selena*diazole** (Bird 1985, 1992). Obviously, there is subtle difference in orbital interaction of different chalcogen key atoms with other atoms of the corresponding heterocycles. Elucidation of the differences can develop the chemistry of chalcogen compounds in general. Other approaches would also be useful to elucidate aromaticity of benzofused chalcogenadiazoles. In terms of unified aromaticity indexes calculated by Bird (1992), benzo-[1,2-*c*]-[1,2,5]-thiadiazole with $I_U = 115$ is close to the structurally similar bicyclic systems of quinoxaline ($I_U = 132$) and naphthalene ($I_U = 142$). With respect to aromaticity of fused diazoles, the approach described in Subsections 2.3.2 and 3.3.2.2 based on experimental estimation of ring currents from proton magnetic resonance spectra seems to be meaningful.

An assumption of aromaticity does not remove the point concerning the real oxidation state of the sulfur, selenium, or tellurium atoms in the corresponding diazoles. (Oxygen can be bivalent only.) Therewith, three oxidation states of a chalcogen should be taken into account: bivalent, tetravalent states, or trivalent states combined with a positive charge of the chalcogen and a negative one on adjacent nitrogen. The problem originates from the crystallographic data. Thus, the Se–N bond lengths are within the range of 0.178–0.1.81 nm and the Te–N bond lengths are within the range of 0.200–0.205 nm (Bjorvinsson and Roesky 1991). Usual ordinary-bond values are 0.186 and 0.205 nm for Se–N and Te–N, respectively. This suggests that selenium can present in the bivalent state, whereas this state is less characteristic for tellurium. What is the real state is the question.

Comparison of electron affinity within the chalcogenadiazole series has principal importance for development of chemical theory. Dipole moments of 1,2,5-chalcogenadiazoles are symbatic with both Pauling's electronegativity and first ionization potential of the corresponding key chalcogene atom. For 1,3,4-isomers, this dependence is inversely proportional (Kwiatkowski et al. 1997). Dipole moments of piazoselenol, piazothiol, and benzofurazan rise with an increase of electronegativity of a key heteroatom (Hill and Sutton 1949). In the case of a one-electron reduction, this sequence does not observe at all (Zhdanov et al. 1967, Tsveniashvili et al. 1968). Moreover, ring-current intensities do not obey to the sequence of key heteroatom electronegativities (Fedin and Todres 1968). Ionization energies of benzo-[1,2-*c*]-annulated [1,2,5]-thiadiazole, [1,2,5]-selenadiazole, and [1,2,5]-telluradiazole are in a linear, but inverse, antibatic dependence on the first ionization potential of the respective chalcogen (Cozzolino et al. 2008). Analogously, polarization of the C≡N bonds in 3,4-dicyano derivatives of 1,2,5-thia- and 1,2,5-selenadiazoles of 0.13 and 0.14 dimensionless units, respectively (see Stuzhin et al. 2007) is inversely proportional to electron affinity of sulfur (2.07 eV) and selenium (2.02 eV). From these facts, it transpires that real acceptor properties of the key heteroatoms in the chalcogenadiazole family are not the same as those in the series of elemental chalcogenes. Consideration and solving of this fundamental (and in fact, quite general) problem is an important challenge for theoretical (computational) chemists.

Another side of this problem consists in the nature of bonding between chalcogen and nitrogen atoms in chalcogenadiazoles. The type of bonding may well be different. For sulfur, participation of d or p orbitals is possible whereas it is highly unlikely for oxygen. It is a fact that thiadiazoles can in some cases exist in the form with an expanded valence shell of sulfur, whereas such a form is impossible for the oxadiazole oxygen (Akiba and Yamamoto 2007).

The d or p orbitals of selenium and tellurium may not be easily involved due to the large difference in sizes of the nitrogen against the selenium or tellurium counterparts. Estimation of electron delocalization in various chalcogenadiazoles as well as relation between their structures with the two- and four-coordinated sulfur, selenium, and tellurium key heteroatoms remains an actual task for theorists and experimentalists.

It would be also interesting to obtain all-embracing explanation of structure–equilibrium relationship for the whole furoxan family. Ring open/closure equilibrium is the intrinsic property of furoxans, but the state of this equilibrium varies with structural factors. The role of the factors remains to be a challenge.

What is also interesting is the difference between 1,2,3-selenadiazoles and their thia analogs in their reactions of with diiron nonacarbonyl in the presence of ethanol. The selenadiazoles lose both nitrogens and the seleno organyl rests coordinate to iron tricarbonyls. Thiadiazoles open the cycles but keep all of the heteroatoms in the iron tricarbonyl complexes formed (Mayr et al. 1989). The authors pointed out that the selenium and sulfur diazoles, despite their similar constitutions, are dramatically different in the reactivity toward transition-metal complexes. This is not surprising, however, when the donor qualities of the chalcogen atoms are considered. Seemingly, physicochemical and, especially, thermodynamic properties of the species eliminated should be considered too.

Many anion-radical salts of fused thiadiazoles and selenadiazoles are stable (see Todres 2009). They can be interesting as magnetically active materials. Search in this direction seems to be promising. Apparently, selena representatives are more prospective than the thia counterparts. At this point, it is relevant to mention the wonderful difference between behavior of the perfluorobenzoselena- and perfluorobenzothiadiazoles anion-radicals in acetonitrile solution: the former are stable, and the latter lose fluoride ions from α and β positions (Vasilieva et al. 2010). This solvent-assisted hydrodefluorination of perfluorobenzothiadiazole is still to be explained. It should be useful to take into account the available data on aromaticity of benzo-fused selena- and thiadiazoles.

Hydrolytic stability of new materials based on telluradiazole presents another important challenge. It should be carefully checked. Subsection 4.2.2 gives examples of hydrolytic cleavage of 2,1,3-telluradiazole and tetrafluorobenzo-[1,2-c]-[1,2,5]-telluradiazole.

5.2 TECHNICAL ASPECTS

Chalcogenadiazoles find a wide application in technology as components of microelectronic devices or solar cells, as vulcanizers, lubricating additives, or corrosion inhibitors. The compounds of this family are used in dyeing processes and in analytical monitoring of industrial operations.

Strong intermolecular interaction in chalcogenadiazole single-component crystalline stacks is crucial for their applicability in molecular electronics. Relative polarization of chalcogen plays an important role. Another important factor is their atomic size. The larger atomic sizes of selenium comparatively to that of sulfur provide closer contacts between components of the stacks. The enhanced intermolecular interaction was indeed observed when sulfur is substituted by selenium. Selenium exhibits more extensive self-overlap properties (Beer et al. 2001). Correspondingly, electric conductivity of bis[1,2,5]-selenadiazolo-1,4-quinonyl derivatives is higher than that of the analogous thiadiazole derivatives (Yamashita et al. 1992).

Crystalline samples of benzo-[1,2-c]-[1,2,5]-telluradiazole differ with nitrogen–tellurium-shortened bonds and form coplanar ribbon chains. In contrasts, the isomorphic crystals of the corresponding selena and thia analogs only display longer intermolecular nitrogen–chalcogen distances and weaker (Se) or negligible (S) coplanar association (Cozzolino et al. 2005, Cozzolino and Vargas-Baca 2007). Even N-alkylated benzo[1,2-c]-[1,2,5]-telluradiazoliums are similarly prone to form short contacts between heterorings both in crystals and in high dielectric solvents (Risto et al. 2008). This feature of tellurium diazoles and diazolium salts makes possible electron transfer within ensembles to realize electric conductivity. As a related example, tetrakis(methyltelluro)tetrathiafulvalene is a single-component organic semiconductor of high efficiency (Inokuchi et al. 1987). For benzo-[1,2-c]-[1,2,5]-chalcogenadiazoles, quantum-chemical comparative calculations predict strong tellurium–nitrogen intermolecular interaction. Charge transfer is a marked component of this association (Pomogaeva et al. 2010).

Consequently, telluradiazoles as well as their N-quaternized derivatives present an attractive challenge with respect to search new materials for organic microelectronics and nonlinear optics. Self-association in the solids is strong. Their use in polymer films requires preparation of telluradiazole derivatives with enhanced solubility even though in organic solvents.

Synthesis and design of microelectronic systems based on selena- and thiadiazoles sometimes can include preparation of the corresponding diazolium salts. For these cases, a challenge consists in establishing relations between structures of neutral diazoles employed and capability of the corresponding diazolium salts to coordinate at the expense of one or both nitrogens and a key chalcogen (Dutton and Ragogna 2009).

Regulation of the chalcogenadiazole crystal structure is possible through the heteroring annulation. According to x-ray analysis of aromatic diamines fused with thiadiazole units, no short sulfur–nitrogen contacts were observed. In crystalline states, thiadiazoles develop intermolecular charge-transfer complexes. Such a complexation prevents the sulfur–nitrogen interactions (Suzuki 1994). On the other hand, electrostatic interaction through selenium–nitrogen contacts results in a unique catenation. Suzuki (1994) explained such a difference in the molecular packing by stronger chalcogen–nitrogen interaction for selenadiazoles as compared to thiadiazoles. In other words, the electrostatic interaction between the thiadiazole rings is not strong enough to control the molecular arrangement. At the same time, the thiadiazole cation-radicals are endowed with the sulfur–nitrogen communication because weakening of the charge-transfer interaction.

5.3 BIOMEDICAL ASPECTS

The medical basis of isosterism was reviewed by Schatz (1960). Chalcogenadiazoles are chemically flexible and can readily respond to the many biomedical demands:

They contain fragments resembling those in amides, ethers, carboxylic acid, etc. Bioisosterism presupposes that a chemical group in a biologically active molecule can be replaced by another chemical group without loss of activity. In contrast to parent compounds, chalcogenadiazoles include isosteric groups that are more stable against hydrolysis under conditions of metabolism, provide better pharmakinetics, or impart new mechanisms of action as compared to medicines of long-time use. Substitution of selenium for oxygen and sulfur in chemotherapeutically active compounds is impeded by the toxic nature of selenium. In general, when selenium is a part of a functional group, namely, $-SeH$ and $-SeO_3H$, it tends to be more toxic than their sulfur analogs (Klayman and Gunter 1973). But when selenium is part of a ring system, the toxicity of sulfur and selenium does not differ widely. For example, sulfides of bis(4-aryl-1,2,3-selenadiazole) and bis(4-aryl-1,2,3-thiadiazole) manifest practically equal antiseptic activity in respect of Gram-positive and Gram-negative bacteria (Padmavathi et al. 2008). From the view point of histology, selenium derivatives are more adoptable than sulfur analogs.

Widening of search of chalcogenadiazole biomedical applications requires creation of their library. Continuous-flow methods of synthesis show promise in this respect. One example of continuous-flow synthesis (namely, the synthesis of 3,5-diaryl-1,2,4-oxadiazoles from arylcarbonitriles and hydroxylamine) has been considered in detail within Subsection 1.2.1.1. This example cited the paper by Grant et al. (2008). The review, titled "Continuous-Flow Syntheses of Heterocycles," by Glasnov and Kappe (2010) provides convincing evidence of the efficiency of such an approach. "It remains to be seen if continuous-flow reactors will be used more often in the future in both academic and industrial laboratories. As there are numerous continuous-flow devices commercially available today, it is clearly up to the synthetic chemist to decide if a molecule should be synthesized in continuous flow or batch mode."—Glasnov and Kappe (2010).

REFERENCES

Akiba, K.-y.; Yamamoto, Y. (2007) *Heteroat. Chem.* **18**, 161.

Beer, L.; Britten, J.F.; Cordes, A.W.; Clements, O.P.; Oakley, R.T.; Pink, M.; Reed, R.W. (2001) *Inorg. Chem.* **40**, 4705.

Bird, C.W. (1985) *Tetrahedron* **41**, 1409.

Bird, C.W. (1992) *Tetrahedron* **48**, 335.

Bjorvinsson, M.; Roesky, H.M. (1991) *Polyhedron* **10**, 2353.

Cozzolino, A.F.; Vargas-Baca, I.; Mansour S.; Mahmoudkhani, A.H. (2005) *J. Am. Chem. Soc.* **127**, 3184.

Cozzolino, A.F.; Britten, J.F.; Vargas-Baca, I. (2006) *Crystl. Growth Des.* **6**, 181.

Cozzolino, A.F.; Gruhn, N.E.; Lichtenberger, D.L.; Vargas-Baca, I. (2008) *Inorg. Chem.* **47**, 6220.

Cozzolino, A.F.; Vargas-Baca, I. (2007) *J. Organometal. Chem.* **692**, 2654.

Dutton, J.L.; Ragogna, P.J. (2009) *Inorg. Chem.* **48**, 1722.

Fedin, E.I.; Todres, Z.V. (1968) *Khim. Geterotsikl. Soed.* 416.

Friedman, P.; Ferris, K.F. (1997) *TEOCHEM* **418**, 119.

Glasnov, T.; Kappe, C.O. (2011) *J. Heterocycl. Chem.* **48**, 11.

Grant, D.; Dahl, R.; Cosford, N.D.P. (2008) *J. Org. Chem.* **73**, 7219.

Hill, R.W.; Sutton, L.E. (1949) *J. Chim. Phys.* **46**, 244.

Inokuchi, H.; Imaeda, K.; Enoki, T.; Mori, T.; Maruyama, Y.; Saito, G.; Okada, N.; Yamochi, H.; Seki, K.; Higuchi, Y.; Yasuoka, N. (1987) *Nature* **329**, No. 6134, 39.

Klayman, D.L.; Gunter, W.H.H. (1972) *Organo Selenium Compounds: Their Chemistry and Biology*. Wiley, New York.

Kwiatkowski, J.S.; Leszczyn'ski, J.; Teca, I. (1997) *J. Mol. Struct.* **436–437**, 451.

Mayr, A.J.; Pannel, K.H.; Carrasco-Flores, B.; Cervantes-Lee, F. (1989) *Organometallics* **8**, 2961.

Padmavathi, V.; Mahesh, K.; Rangayapalle, C.V.S.; Padmaja, A. (2008) *Heteroatom Chem.* **19**, 261.

Pomogaeva, A.; Gu, F.L.; Imamura, A.; Aoki, Yu. (2010) *Theor. Chem. Acc.* **125**, 453.

Risto, M.; Reed, R.W.; Robertson, C.M.; Olunkaniemi, R.; Laitinien, R.S.; Oakley, R.T. (2008) *Chem. Commun.* 3278.

Schatz, V.B. (1960) *Medicinal Chemistry*. Wiley, New York.

Stuzhin, P.A.; Pimkov, I.V.; Ul'-Khak, A.; Ivanova, S.S.; Popkova, I.A.; Volkovich, D.I.; Kuz'mitskii, V.A.; Donzello, M.-P. (2007) *Zh. Org. Khim.* **43**, 1854.

Suzuki, T. (1994) *Ashi Garasu Zaidan Josei Kenkyu Seika Hokoku* 149.

Todres, Z.V. (2009) *Ion-Radical Organic Chemistry*. CRC Taylor and Francis, Boca Raton, FL.

Tsveniashvili, V.Sh.; Todres, Z.V.; Zhdanov, S.I. (1968) *Zh. Obshch. Khim.* **38**, 1888.

Vasilieva, N.V.; Irtegova, I.G.; Gritsan, N.P.; Lonchakov, A.V.; Makarov, A.Yu.; Shundrin, L.A.; Zibarev, A.V. (2010) *J. Phys. Org. Chem.* **23**, 536.

Yamashita, T.; Tanaka, Sh.; Imaeda, K.; Inokuchi, H.; Sano, M. (1992) *J. Org. Chem.* **57**, 5517.

Zhdanov, S.I.; Tsveniashvili, V.Sh.; Todres, Z.V. (1967) *J. Polarogr. Soc.* **13**, 100.

Index